Principles of Plant Cell Biology

Principles of Plant Cell Biology

Editor: Agatha Wilson

www.callistoreference.com

Callisto Reference,
118-35 Queens Blvd., Suite 400,
Forest Hills, NY 11375, USA

Visit us on the World Wide Web at:
www.callistoreference.com

ISBN: 978-1-63239-801-7 (Hardback)

Printed in China.

Cataloging-in-publication Data

Principles of plant cell biology / edited by Agatha Wilson.
p. cm.
Includes bibliographical references and index.
ISBN 978-1-63239-801-7
1. Plant cells and tissues. 2. Plants--Cytology. 3. Cytology. 4. Plant ultrastructure. I. Wilson, Agatha.
QK725 .P75 2017
571.6--dc23

Table of Contents

Preface ... **VII**

Chapter 1 **Electron Tomography of Cryo-Immobilized Plant Tissue: A Novel Approach to Studying 3D Macromolecular Architecture of Mature Plant Cell Walls In Situ** ... **1**
Purbasha Sarkar, Elena Bosneaga, Edgar G. Yap Jr., Jyotirmoy Das, Wen-Ting Tsai, Angelo Cabal, Erica Neuhaus, Dolonchampa Maji, Shailabh Kumar, Michael Joo, Sergey Yakovlev, Roseann Csencsits, Zeyun Yu, Chandrajit Bajaj, Kenneth H. Downing, Manfred Auer

Chapter 2 **Soil Temperature Determines the Reaction of Olive Cultivars to *Verticillium dahliae* Pathotypes** ... **17**
Rocío Calderón, Carlos Lucena, José L. Trapero-Casas, Pablo J. Zarco-Tejada, Juan A. Navas-Cortés

Chapter 3 **Copper-Deficiency in *Brassica napus* Induces Copper Remobilization, Molybdenum Accumulation and Modification of the Expression of Chloroplastic Proteins** ... **33**
Vincent Billard, Alain Ourry, Anne Maillard, Maria Garnica, Laurent Coquet, Thierry Jouenne, Florence Cruz, José-Maria Garcia-Mina, Jean-Claude Yvin, Philippe Etienne

Chapter 4 **Latitudinal Environmental Niches and Riverine Barriers Shaped the Phylogeography of the Central Chilean Endemic *Dioscorea humilis* (Dioscoreaceae)** ... **46**
Juan Viruel, Pilar Catalán, José Gabriel Segarra-Moragues

Chapter 5 **Cadmium Telluride Quantum Dots (CdTe-QDs) and Enhanced Ultraviolet-B (UV-B) Radiation Trigger Antioxidant Enzyme Metabolism and Programmed Cell Death in Wheat Seedlings** ... **60**
Huize Chen, Yan Gong, Rong Han

Chapter 6 **Exploring the Genes of Yerba Mate (*Ilex paraguariensis* A. St.-Hil.) by NGS and *De Novo* Transcriptome Assembly** ... **73**
Humberto J. Debat, Mauro Grabiele, Patricia M. Aguilera, Rosana E. Bubillo, Mónica B. Otegui, Daniel A. Ducasse, Pedro D. Zapata, Dardo A. Marti

Chapter 7 **Development of a Heat-Shock Inducible Gene Expression System in the Red Alga *Cyanidioschyzon merolae*** ... **89**
Nobuko Sumiya, Takayuki Fujiwara, Yusuke Kobayashi, Osami Misumi, Shin-ya Miyagishima

Chapter 8 **AtRD22 and AtUSPL1, Members of the Plant-Specific BURP Domain Family Involved in *Arabidopsis thaliana* Drought Tolerance** ... **100**
Vokkaliga Thammegowda Harshavardhan, Le Van Son, Christiane Seiler, Astrid Junker, Kathleen Weigelt-Fischer, Christian Klukas, Thomas Altmann, Nese Sreenivasulu, Helmut Bäumlein, Markus Kuhlmann

Chapter 9 **Arabidopsis mTERF15 is Required for Mitochondrial *nad2* Intron 3 Splicing and Functional Complex I Activity**......114
Ya-Wen Hsu, Huei-Jing Wang, Ming-Hsiun Hsieh, Hsu-Liang Hsieh, Guang-Yuh Jauh

Chapter 10 **Variability among the Most Rapidly Evolving Plastid Genomic Regions is Lineage-Specific: Implications of Pairwise Genome Comparisons in *Pyrus* (Rosaceae) and Other Angiosperms for Marker Choice**......125
Nadja Korotkova, Lars Nauheimer, Hasmik Ter-Voskanyan, Martin Allgaier, Thomas Borsch

Chapter 11 **New Insights into Regulation of Proteome and Polysaccharide in Cell Wall of *Elsholtzia splendens* in Response to Copper Stress**......141
Tingting Liu, Chaofeng Shen, Yi Wang, Canke Huang, Jiyan Shi

Chapter 12 **Morphological and Physiological Responses of Cotton (*Gossypium hirsutum* L.) Plants to Salinity**......154
Lei Zhang, Huijuan Ma, Tingting Chen, Jun Pen, Shuxun Yu, Xinhua Zhao

Chapter 13 **Reduction of the Cytosolic Phosphoglucomutase in Arabidopsis Reveals Impact on Plant Growth, Seed and Root Development, and Carbohydrate Partitioning**......168
Irina Malinova, Hans-Henning Kunz, Saleh Alseekh, Karoline Herbst, Alisdair R. Fernie, Markus Gierth, Joerg Fettke

Chapter 14 **Comparative Analysis of Predicted Plastid-Targeted Proteomes of Sequenced Higher Plant Genomes**......179
Scott Schaeffer, Artemus Harper, Rajani Raja, Pankaj Jaiswal, Amit Dhingra

Chapter 15 **Understanding the Relationship between Cotton Fiber Properties and Non-Cellulosic Cell Wall Polysaccharides**......199
Dhivyaa Rajasundaram, Jean-Luc Runavot, Xiaoyuan Guo, William G. T. Willats, Frank Meulewaeter, Joachim Selbig

Chapter 16 **Phylogeny Reconstruction and Hybrid Analysis of *Populus* (Salicaceae) Based on Nucleotide Sequences of Multiple Single-Copy Nuclear Genes and Plastid Fragments**......210
Zhaoshan Wang, Shuhui Du, Selvadurai Dayanandan, Dongsheng Wang, Yanfei Zeng, Jianguo Zhang

Permissions

List of Contributors

Index

Preface

The study of the eukaryotic plant cells is defined as plant cell biology. This book outlines the processes and applications of this discipline in detail. The text will enable its reader to understand the principles of this field in a comprehensive and lucid manner. The book is appropriate for students seeking detailed information in this area as well as for experts. This book traces the progress of this field and highlights some of its key concepts and applications. It unfolds the innovative aspects of plant cell biology which will be crucial for the progress of this field in the future. It consists of contributions made by international experts.

This book is a comprehensive compilation of works of different researchers from varied parts of the world. It includes valuable experiences of the researchers with the sole objective of providing the readers (learners) with a proper knowledge of the concerned field. This book will be beneficial in evoking inspiration and enhancing the knowledge of the interested readers.

In the end, I would like to extend my heartiest thanks to the authors who worked with great determination on their chapters. I also appreciate the publisher's support in the course of the book. I would also like to deeply acknowledge my family who stood by me as a source of inspiration during the project.

Editor

1

Electron Tomography of Cryo-Immobilized Plant Tissue: A Novel Approach to Studying 3D Macromolecular Architecture of Mature Plant Cell Walls In Situ

Purbasha Sarkar[1,2¤], Elena Bosneaga[1,2], Edgar G. Yap Jr.[2], Jyotirmoy Das[1], Wen-Ting Tsai[2], Angelo Cabal[1], Erica Neuhaus[2], Dolonchampa Maji[1], Shailabh Kumar[1], Michael Joo[2], Sergey Yakovlev[3], Roseann Csencsits[2], Zeyun Yu[4], Chandrajit Bajaj[5], Kenneth H. Downing[2], Manfred Auer[1,2]*

1 Energy Biosciences Institute, University of California, Berkeley, California, United States of America, **2** Life Sciences Division, Lawrence Berkeley National Laboratory, Berkeley, California, United States of America, **3** Materials Sciences Division, Lawrence Berkeley National Laboratory, Berkeley, California, United States of America, **4** Department of Computer Science, University of Wisconsin, Milwaukee, Wisconsin, United States of America, **5** Department of Computer Sciences & The Institute of Computational Engineering and Sciences, University of Texas, Austin, Texas, United States of America

Abstract

Cost-effective production of lignocellulosic biofuel requires efficient breakdown of cell walls present in plant biomass to retrieve the wall polysaccharides for fermentation. In-depth knowledge of plant cell wall composition is therefore essential for improving the fuel production process. The precise spatial three-dimensional (3D) organization of cellulose, hemicellulose, pectin and lignin within plant cell walls remains unclear to date since the microscopy techniques used so far have been limited to two-dimensional, topographic or low-resolution imaging, or required isolation or chemical extraction of the cell walls. In this paper we demonstrate that by cryo-immobilizing fresh tissue, then either cryo-sectioning or freeze-substituting and resin embedding, followed by cryo- or room temperature (RT) electron tomography, respectively, we can visualize previously unseen details of plant cell wall architecture in 3D, at macromolecular resolution (~2 nm), and in near-native state. Qualitative and quantitative analyses showed that wall organization of cryo-immobilized samples were preserved remarkably better than conventionally prepared samples that suffer substantial extraction. Lignin-less primary cell walls were well preserved in both self-pressurized rapidly frozen (SPRF), cryo-sectioned samples as well as high-pressure frozen, freeze-substituted and resin embedded (HPF-FS-resin) samples. Lignin-rich secondary cell walls appeared featureless in HPF-FS-resin sections presumably due to poor stain penetration, but their macromolecular features could be visualized in unprecedented details in our cryo-sections. While cryo-tomography of vitreous tissue sections is currently proving to be instrumental in developing 3D models of lignin-rich secondary cell walls, here we confirm that the technically easier method of RT-tomography of HPF-FS-resin sections could be used immediately for routine study of low-lignin cell walls. As a proof of principle, we characterized the primary cell walls of a mutant (*cob-6*) and wild type Arabidopsis hypocotyl parenchyma cells by RT-tomography of HPF-FS-resin sections, and detected a small but significant difference in spatial organization of cellulose microfibrils in the mutant walls.

Editor: Megan D. Lenardon, University of Aberdeen, United Kingdom

Funding: This work was primarily funded by the Energy Biosciences Institute (http://www.energybiosciencesinstitute.org), grant number 007G18 (MA). The LBNL electron microscopes used in this study were supported by National Institutes of Health (http://www.nih.gov); grant number P01-GM051487-15 (KHD, MA). The automated segmentation done at UT Austin was supported by National Institutes of Health (http://www.nih.gov); grant R01-EB004873 (CB). The funders had no role in study design, data collection and analysis, decision to publish, or preparation of the manuscript.

Competing Interests: The authors have declared that no competing interests exist.

* Email: mauer@lbl.gov

¤ Current address: Joint BioEnergy Institute, Physical Biosciences Division, Lawrence Berkeley National Laboratory, Berkeley, California, United States of America

Introduction

Sugars isolated from non-food plant biomass such as grasses or wood chips have the potential for large-scale biofuel production, but require efficient biomass deconstruction and fermentation for such fuel to be economically viable [1–4]. A variety of efforts are being undertaken to make these deconstruction processes efficient [1,5–8] and the biomass less recalcitrant [2–4,6,9–10]. However, a more rational approach for both biomass deconstruction and feedstock development requires knowledge about the precise three-dimensional (3D) organization of plant cell walls, which are the storehouse of sugars in non-food biomass. Plant cell walls are composed of a relatively small number of carbohydrate- and phenolic-based building blocks, but the relative composition has changed throughout evolution, presumably accompanied by organizational changes in the wall architecture [11–12]. Plant cell walls are commonly distinguished as 'primary cell wall' or 'secondary cell wall', although a gradation of wall properties between these two extremes can be found in different cell types and various fundamentally different definitions have been proposed for the two terms to account for such diversity [13–15]. However, a primary cell wall may be defined as the extendable wall layer deposited in most growing cells, while secondary cell wall may be defined as the strong, non-extendable wall layers deposited between primary cell wall and plasma membrane in cell types such as, xylem tracheary element, fibers, and sclereids, after these cells stop growing [14–17]. Protoxylem is the only exception where secondary walls are deposited during cell growth, in arrangements that allow cell elongation. Secondary cell walls are not deposited in cell types such as cortical or vascular parenchyma cells, even at maturity [16]. Primary cell walls are composed of three groups of polysaccharides - cellulose, hemicelluloses and pectins, with few glycoproteins and enzymes also being present. Secondary cell walls are commonly composed of cellulose, hemicelluloses and large quantities of lignin as found in xylem tracheary element, fibers, and sclereids, although lignin-less secondary wall can be found in cell types such as phloem sieve elements, cotton fibers, and collenchyma cells. A thin cementing wall layer called the "middle lamella" separates the primary walls of two adjacent cells. Middle lamellae are mainly composed of pectins with small amount of hemicelluloses in the growing cells, and additional lignin deposited in the mature cell [16].

Cellulose is an unbranched polysaccharide of ß-1,4-glucose, with individual chains held together by hydrogen bonds to form long, unbranched structures called microfibrils. Microfibrils are reported to be ~3–5 nm in diameter, consisting of multiple individual cellulose polysaccharide chains, with crystalline cellulose at the core surrounded by paracrystalline cellulose and hemicelluloses. The number of cellulose chains within a microfibril is debatable and has been reported to be 36 in older literature and 18–24 in recent years [18–25]. An ordered framework of microfibrils is thought to be the primary load-bearing structural feature of the wall, and the orientation of microfibrils plays a vital role in determining orientation of cell expansion [26–27]. Hemicelluloses are branched polysaccharides composed of neutral pentoses and hexoses that strengthen the cell wall by binding to cellulose via hydrogen bonds, and by binding with lignins in some cases [28]. Xyloglucan, a predominant hemicellulose in non-grass angiosperms, is widely believed to bind and surround the paracrystalline cellulose in the outer layer of microfibrils and to function as a tether between two neighboring microfibrils [22–23,29–30]; however, the tether model has been challenged recently [31]. Pectins, like hemicelluloses, are branched polysaccharides, and contain high proportions of D-galacturonic acids. Their precise organization and respective functions are not fully understood [21,29,32]. Furthermore, glycoproteins and enzymes are also embedded in the cell walls, though their precise location is also unknown [21–22].

Most of our current knowledge about cell walls is based on indirect biochemical analyses of cell wall extracts, and the direct microscopy-based cell wall studies done to date give little insight into the 3D nature of the cell wall organization at macromolecular resolutions. Bulk orientation of cellulose microfibrils within the cell walls has been studied by polarized light, infrared microscopy [33–34], and confocal microscopy [35–36], but these methods have resolution of few hundred nanometers and cannot determine the architecture at individual microfibril level. Transmission electron microscopy (TEM) of surface replicas, prepared by rotary heavy metal shadowing of freeze-fractured, deep etched isolated cell walls samples, was the first method for visualization of cell wall architecture at high-resolution [37–40]. This method could however, focus only on a surface layer of the cell wall sample and was therefore restricted to studying only thin primary cell walls. Furthermore, this approach relies on biochemical isolation of cell wall material and involves substantial chemical treatment, and thus the cell walls may no longer be considered in close to their native state. Another remarkable TEM study done in the 1970s focused on the architecture of unisolated cell walls using two complementary sample preparation methods [41]. Parenchyma as well as collenchyma cell walls were mildly extracted and stained with polysaccharide specific stain: periodic acid-thiocarbohydrazide-silver proteinate (PATAG) to understand the cytochemistry; and the fine ultrastructure of comparable walls were studied by chemically fixing, liquid nitrogen freezing, cryo-sectioning and negatively staining of sections with sodium silicotungstate. Although the findings of this study is relevant even to current date, the presence of fixatives and stains in the cryo-sectioned samples leave a doubt that these chemicals might have altered the wall organization at macromolecular level. Three-dimensional texture of the walls was also evaluated in this study by tilting the stage from −45° to +45°, which can be considered as precursor of electron tomography. However, this method does not provide true 3D information of inside the cell wall sections. In recent years, field emission-scanning electron microscopy (FE-SEM) and atomic force microscopy (AFM) have been used to study the organization of cell wall components, especially the cellulose microfibrils at high resolutions [23,42–46]. These studies have yielded valuable topographical surface information, but neither of these techniques allows access to 3D architectural information.

Electron tomography is the only method currently available that can provide true 3D visual insight into plant cell wall ultrastructure at macromolecular resolution (~2 nm) without isolating the cell walls. Electron tomography relies on the collection of single- or dual-axis tilt series, each containing ~70–280 TEM images recorded at different tilt angles of the specimen in the electron microscope. After alignment of individual images with the help of fiducial markers, and 3D reconstruction using weighted back-projection techniques [47–48], 3D volumes (tomograms) showing exquisite detail in 3D are obtained. Such tomograms need to be visually inspected and features of interest must be segmented out before geometrical parameters such as distances, volumes, angles, and lengths can be determined and a realistic 3D model can be built. Electron tomography has been used in recent years in plant biology to study the 3D ultrastructure of the secondary cell wall of *Pinus* wood tissue [49–50] and to characterize plant cell wall deconstruction during thermo-chemical pretreatment of corn stover biomass [51–52]. However, for these studies, the plant samples were chemically fixed and dehydrated in organic solvents at room temperature, and included additional harsh chemical treatment to remove lignin, before embedding the samples in resin. Such sample preparation protocols can lead to aggregation and extraction artifacts as well as uneven or preferential staining that can profoundly alter the perception of the organization [53].

In this paper, we present two new approaches of studying the macromolecular 3D ultrastructure of plant cell wall that include electron tomography of cryo-immobilized fresh tissue, and avoid the conventionally used harsh chemical treatments. We show that faithfully preserved cell walls can be obtained by self-pressurized rapid freezing (SPRF) of fresh tissue followed by cryo-sectioning, as well as by high-pressure freezing (HPF), freeze-substitution (FS) and resin embedding. With cryo-electron tomography of the unstained cryo-sections of intact unextracted Arabidopsis tissue, we were able to visualize never seen before details of macromolecular 3D architecture of both lignin-less primary cell walls and the lignin-rich secondary cell walls in situ in their near-native state. We also show that high-quality 3D data of lignin-less primary cell walls can be obtained by using the relatively easier method of room temperature (RT) electron tomography of HPF-FS-resin embedded, stained sections. Even though cryo-immobilization approaches have been used to address various biological questions, electron tomographic study of plant cell wall architecture using either of the two cryo-immobilization approaches has never been reported to our knowledge. Our cryo-tomography approach will be the first reported imaging method to visualize the organization of polysaccharides at macromolecular (~2 nm) resolution in unextracted lignin-rich secondary cell walls. Using a semi-automated threshold-based segmentation method we further analyzed relatively larger cell wall volumes qualitatively as well as quantitatively, which has not been done for any previous electron tomography study of plant cell walls. As an example of potential routine application of electron tomography of cryo-immobilized plant cell walls, we characterized the subtle architectural differences in the primary cell walls of mutant (*cob-6*) Arabidopsis hypocotyl parenchyma cells, compared to those of wild type (Col 0), as *cob* mutants have been reported to cause disorganization of cellulose microfibril orientation and reduction of crystalline cellulose in the cell walls of roots [54–56].

Materials and Methods

Plant material

For comparison of sample preparation methods, wild type *Arabidopsis thaliana* (Arabidopsis) seeds from the Colombia ecotype (Col 0) were sterilized in 30% bleach, 0.02% Triton and vernalized at 4°C in water for 48 hours. They were germinated on 0.7% agar plates containing 0.5x Murashige and Skoog medium for 10 d at 21°C under continuous light in a growth chamber. The seedlings were then transferred to pots containing soil mixture and placed in a growth chamber programmed for a 16 h light/8 h dark cycle at 21°C. Stem tissue from 3–4 weeks old plants that had newly growing inflorescence stems (2–3 cm long) were used for the three different sample preparation methods described below. For comparative analysis, cell wall areas from all sample types were randomly selected for electron tomography, from cells within xylem tissue that appeared to be xylem tracheary elements.

Self-pressurized rapid freezing (SPRF), vitreous sectioning and cryo-electron tomography

Young inflorescence stems (2–3 cm long) were rapidly hand sectioned with a sharp razor blade into longitudinal 'hair-thin' segments (Fig. 1A). Each tissue segment was then inserted into individual 15 mm long clean, deoxidized copper capillary tubes (inner and outer diameter of 300 μm and 600 μm respectively), in a manner very similar to threading a needle. The capillary tube was then fitted to a pipette with the sample side facing outward (Fig. 1B) and a cryoprotectant solution (20% dextran) was sucked into the capillary tube (Fig. 1C) until droplets of dextran came out from the other end of the tube. The dextran solution fills the air pockets around the stem and reduces freezing damage, and at the same time helps in pulling the sample all the way up into the capillary tube. Due to the small diameter of the tubes, the sample insertion process was difficult, with most stems being stuck within the tubes and not coming out of the other end. All such samples were discarded. Only those stem segments that smoothly came out of the other end (Fig. 1D) were used for further steps to ensure minimum handling damage. Unlike resin sections, cryo sections cannot be quickly checked for presence of samples at the time of sectioning. To minimize chances of getting blank sections, we used only those capillary tubes that had one long stem tissue spanning the entire length of the tube. More than one short stem segment can be inserted back to back instead, if ease of tissue insertion is preferred over ease of finding the sample within the sections. The sample-containing capillary tubes were then quickly sealed at both ends, one end at a time, with a pair of pliers to build pressure inside the tubes (Fig. 1E). The sealed tubes were immediately plunged sideways in liquid ethane for self-pressurized rapid freezing [57–60] of the stem tissue. It was crucial to do all these steps very quickly in order to minimize tissue damage. The sealed tubes were then stored in liquid nitrogen until cryo-sectioning.

Prior to cryo-sectioning, both sealed ends of each tube were trimmed off under liquid nitrogen using a cryo-trimmer diamond knife (Diatome, Hatfield, PA, USA), leaving the central uncompressed part of the tubes with frozen tissue (Fig. 1F). Transverse sections of a nominal thickness of 70 nm were obtained by cryo-sectioning the frozen samples at −160°C using a Leica EMUC6 ultramicrotome fitted with a Leica EM FC7 cryo chamber attachment (Leica Microsystems Inc.), and a 25° cryo-diamond knife for least possible compression (Diatome, Hatfield, PA, USA). Vitreous cryo-sections were picked on carbon-coated lacey/formvar copper mesh grids (Ted Pella, Redding, CA, USA; Fig. 1G and Video S1) and attached to the grids with a Leica EM CRION ionizer [61]. The cryo-sections were screened using very low dose techniques for selecting areas of interest for tomography at low magnifications to avoid beam damage to the delicate cryo-sections. The very low contrast of cryo-sections makes it challenging to do proper anatomical distinction during rapid screening. Since our objective for this study was to only standardize the method, we randomly chose areas with thick secondary cell walls that are presumably xylem tracheary elements. Single-axis cryo tilt series were then collected from −60° to +60° with 2° increments at higher magnification under low dose conditions. Tilt series were recorded by using a Tecnai F20 FEG TEM (FEI Company, Hilsboro, OR, USA) equipped with a cold stage (Gatan, Pleasanton, CA, USA), operated at 120 kV. Images were recorded with a 4K×4K Gatan Ultrascan 4000 CCD camera (Gatan, Pleasanton, CA, USA) and the automatic data acquisition system Leginon [62] at a final magnification of 25,000x corresponding to a detector pixel size of 0.43 nm at the specimen level, and a defocus set to −6 micron. Alternatively, tilt series were recorded using a JEOL JEM-3100 FFC FEG TEM (JEOL Ltd, Akishima, Tokyo, Japan) equipped with an in-column Omega energy filter and a cryo- transfer stage, operated at 300 kV. Zero-loss images were recorded with a 30 eV energy selecting slit, using a Gatan 795 2K×2K CCD camera (Gatan, Pleasanton, CA, USA) and SerialEM software [63], at a final magnification of 25,000x corresponding to a pixel size of 1.1 nm at the specimen level and a defocus set to −12 micron.

High-pressure freezing (HPF), freeze substitution (FS)

Transverse free-hand sections (<200 μm thick) taken from approximately the middle part of the 2–3 cm long young inflorescence stem were placed in 200 μm deep freezer hats (Ted Pella, Redding, CA, USA) with 1-hexadecene as cryoprotectant, high pressure frozen in a Leica EMPACT2 high-pressure freezer (Leica Microsystems Inc.), and then transferred to liquid nitrogen. The samples were then processed following a modified version of previously published freeze-substitution, resin-embedding protocols for plant samples [64–67]. The frozen samples were freeze substituted in 2% OsO_4 and 0.5 mg/ml ruthenium red in anhydrous acetone at −90°C for 5 d, followed by slow warming to room temperature (RT) over a period of 2 d, in the Leica AFS2 (Leica Microsystems Inc.). After rinsing in several acetone washes, the samples were removed from the holders, and infiltrated with increasing concentrations of Epon-Araldite resin (Ted Pella, Redding, CA, USA) in acetone according to the following

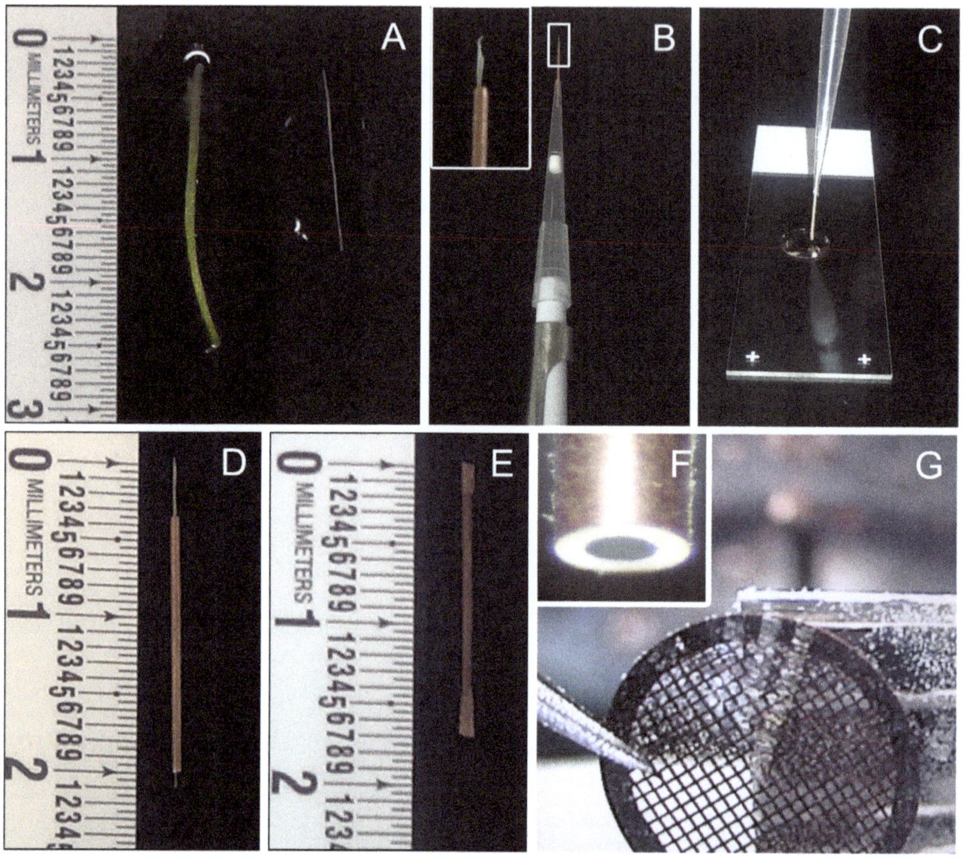

Figure 1. Steps of self-pressurized rapid freezing (SPRF) and cryosectioning of plant tissue. A. Intact stem segment (left) and a dissected longitudinal 'hair-thin' segment of stem tissue (right). B. Tissue inserted on one end of a copper capillary tube, fitted to a pipette on the other end (inset showing close-up). C. Insertion of tissue into the capillary tube by pipetting in cryo-protectant solution. D. Stem segments inserted completely into a capillary tube. E. Capillary tube with sample and cryo-protectant sealed from both end. F. Trimmed capillary tube with frozen sample inside. G. A ribbon of cryo-sections being collected on TEM grids.

schedule: 4 h in 5% resin, 4 h in 10% resin, 12 h in 25% resin, and 24 h in 50%, 75%, and 100% resin, respectively. Polymerization was performed at 60°C for 2–3 days.

Conventional chemical preparation

Stem segments (~2 mm long) from approximately the middle of the 2–3 cm long young inflorescence stem were fixed with 4% paraformaldehyde, 2% glutaraldehyde in 0.03 M phosphate buffer (pH 7.4) containing 0.5 mg/ml ruthenium red overnight at 4°C. After washing with the same buffer, secondary fixation was done with 0.1% osmium tetroxide and 0.5 mg/ml ruthenium red for 1 h at RT. The samples were then dehydrated and infiltrated at RT using a Leica EM AMW automatic microwave tissue processor (Leica Microsystems Inc.). Dehydration was done in ethanol series (25%, 50%, 75%, 95%, 100%, twice in each solution for 45s each), followed by dehydration in 100% acetone (twice for 45s each). The samples were then infiltrated in Epon-Araldite resin-acetone series (5%, 10%, 25%, 50%, 75%, thrice in each solution for 3 min each), infiltrated in 100% resin on a rotor at RT overnight, followed by 3 h infiltration in 100% resin with accelerator, and finally polymerized at 60°C for 2–3 days.

Ultrathin sectioning and room temperature electron tomography of resin embedded samples

Thick transverse sections (~150 nm) were cut from both chemically prepared and HPF-FS samples using the Leica UC6 ultramicrotome (Leica Microsystems Inc.). Sections were picked on formvar-coated copper slot grids (Ted Pella, Redding, CA, USA). All sections were exposed to 5 nm gold fiducials for 4 mins each on both sides followed by several washes in distilled water. The sections were then post stained with 2% uranyl acetate in methanol for 5 mins, followed by Reynold's lead citrate solution for 2 mins. Cell walls of xylem tracheary elements comparable to those selected for cryo-tomography were selected for RT tomography based on cell shape, size, location and wall thickness. Dual axis tilt series were collected from +65° to −65° with 1° increment on a Tecnai F12 FEG TEM (FEI Company, Hilsboro, OR, USA) operated at 120 kV accelerating voltage and equipped with Fischione dual-axis tomography holder (Fischione Instruments, Pittsburg, PA, USA). Images were collected with a 2K×2K Gatan Ultrascan 1000 CCD camera (Gatan, Pleasanton, CA, USA) and SerialEM software [63,68], at a final magnification of 11,000x or 13,000x corresponding to a pixel size of 0.79 nm or 0.92 nm, respectively at the specimen level, and a defocus set to −1 micron.

Image alignment, 3D reconstruction and filtering

Images were aligned and reconstructed with IMOD (The Boulder Laboratory for 3D Electron Microscopy of Cells, University of Colorado Boulder, CO, http://bio3d.colorado.edu/imod/) using the back-projection method [68–69]. All fiducial-less cryo-tomograms were aligned by the 'patch-tracking' method while the resin sections with gold fiducial markers were aligned by the 'fiducial-tracking' method, both available in the IMOD package. The cryo-tomograms collected with a 4K×4K camera were binned by 2 to obtain a pixel size of 0.87 nm, for easy comparison with all other tomograms. All cryo-tomograms were subjected to image filtering using the nonlinear anisotropic diffusion (NAD) filter [70] within the IMOD package, to reduce noise and improve contrast for ease in segmentation.

Selection of segmentation approach

In order to minimize/eliminate any bias that might be introduced from using any one specific feature extraction approach, we first tested three different protocols, namely, manual segmentation, semi-automated threshold segmentation, and an automated custom-built algorithm-based segmentation approach, to extract the cell wall features from a small 3D volume of a tomogram. The manual approach involved precise but labor-intensive tracing of wall features by visual inspection using the commercially available imaging program Amira (Visualization Sciences Group, Burlington, MA, USA). For the semi-automated method, the 'threshold segmentation' tool available in Amira was used, where a threshold value has to be selected by visual inspection so that only the cell wall features denser than the general background are segmented out. For both of the above-mentioned approaches, triangular mesh surfaces were generated from the segmented images with the 'unconstrained smoothing' option in Amira for better visualization of segmented wall features in 3D. For the custom-built automated approach, each voxel in the NAD filtered volume was associated with a pre-computed local structure tensor [71–73] that defines the local feature orientation. A Gaussian-like filter was anisotropically applied to the neighborhood of each voxel in such a way that smoothing was favored in the direction of the local feature orientation given as the eigenvector corresponding to the smallest eigenvalue of the tensor. The filtered volumes were then segmented by topological isosurface selection via the contour spectrum [74] and skeletons were extracted with a topology-preserving thinning algorithm [75] on the segmented volumes. We found no significant difference in the segmented volumes for these very different approaches (Fig. S1A–G), but the semi-automated threshold approach in Amira was the easiest to use. We further tested reliability of the threshold approach by segmenting the same volume with small variations in threshold values, the same volume by 2 different individuals independently, and three different tomograms of three different cell wall samples prepared by the same sample preparation method. After finding consistently reliable threshold ranges for unstained cryo and stained resin samples (Fig. S2A–D, details in Results section), we used the semi-automated threshold segmentation approach for rest of our analyses.

Segmentation and quantitative image analysis of 3D volumes from primary cell walls

For comparative segmentation and quantitative image analysis, we selected nine 3D volumes (500 pixels×500 pixels×Z, where Z = 50, 75, or 100, depending on the number of image slices with clearly distinguishable biological material in each tomogram) from cell wall areas visible in three different tomograms (three volumes from each tomogram) for each of the three sample preparation methods. Cell wall features were then segmented and mesh surfaces were generated for all tomogram volumes by the threshold segmentation approach described above. For quantitative geometric analysis of the cell wall components, we randomly picked two planes within each 3D volume and measured the cross-sectional diameter of randomly picked long filamentous structures, their separation (center-to-center distance) as well as the shortest gap between the filaments (edge-to-edge distance), and the cross-sectional length of short bridge-like cross-links joining the long and typically parallel-running filaments in only the primary cell wall area of each tomogram volume. Secondary cell walls were not included in quantitative analysis, as they appeared featureless in resin-embedded samples.

Chemical treatment - for identification of cell wall features

Wild type Arabidopsis (Col 0) seeds were sterilized and germinated on media, seedlings were transferred and grown in soil for ~3 weeks, and young inflorescence stem segments were fixed, dehydrated, and resin embedded by conventional chemical method, as described above. An additional chemical treatment was done in between the two fixation steps, where stem segments were treated with 0.5% ammonium oxalate at 60°C for 48 hours followed by 4% NaOH at RT for 96 h to remove pectin, hemicelluloses and any non-cellulosic polysaccharides from the cell walls. The extracted samples were sectioned; and cell walls areas were imaged, reconstructed, segmented and analyzed as done for all resin embedded samples above, and compared with non-extracted controls.

Mutant characterization – a proof of concept for possible routine analyses

Wild type Arabidopsis (Col 0) and mutant (*cob-6*) seeds were sterilized and germinated on media as described above. Transverse free-hand sections (<200 μm thick) from the hypocotyls of ~1 week old plants were prepared by the HPF-FS method described above. The samples were sectioned, imaged, reconstructed, segmented and analyzed as done for the HPF-FS samples above, except primary cell walls of cortical parenchyma cells were selected from both WT and mutant samples for this comparative analysis. For each sample, three cell wall 3D volumes (500 pixels×500 pixels×Z, where Z = 50, 75, or 100, depending on the number of image slices with clearly distinguishable biological material in each tomogram) were used for the quantitative analysis.

Results

We present the results of a comparative study of Arabidopsis cell wall 3D ultrastructure obtained by three different methods: (1) Cryo-tomography of unstained vitrified cryo-sections cut from self-pressurized rapidly frozen (SPRF) samples, (2) Room temperature (RT) tomography of stained sections of high pressure frozen (HPF), freeze substituted (FS), resin-embedded samples, and (3) RT tomography of stained sections of microwave-assisted chemically fixed, dehydrated and resin-embedded samples (equivalent to a conventional bench-top protocol).

Comparison of sample preservation quality by 2D transmission electron microscopy (TEM)

Using a 3-day long conventional chemical preparation that included microwave-assisted chemical fixation, dehydration and resin embedding, we obtained fair preservation of Arabidopsis stem tissue (Fig. 2A–D). Cell membranes and most organelles such as nuclei, chloroplasts, mitochondria, Golgi bodies, vesicles, ribosomes, endoplasmic reticulum, and microtubules were preserved (Fig. 2A–C), although the quality of preservation was inferior to the samples prepared by the two cryo-immobilization methods as evidenced by wavy membranes (Fig. 2B, C). We found overall morphology of the cell walls in these samples to be intact without signs of breakage or deformation. The different layers of the cell wall showed different levels of staining, with the middle lamella being the darkest and the secondary cell wall being the lightest (Fig. 2D). When imaged at magnifications over 10,000x, patterns of alternating dark and light filamentous structures were clearly visible in the primary cell walls (Fig. 2D). The secondary walls, however, appeared to be homogeneously light without any detectable texture (Fig. 2D), suggesting that the contrast-generating heavy atom staining solutions did not reach or react with the secondary cell wall material.

High-pressure freezing, freeze substitution (HPF-FS) was carried out with a modified version of a widely reported ~2 week long general protocol for plant samples [64–67]; this provided a much higher quality of preservation of Arabidopsis stem tissue (Fig. 2E–H) compared to the chemical preparation protocol (Fig. 2A–D). Cell membranes and organelles including nuclei, chloroplast, mitochondria, Golgi bodies, ribosomes, endoplasmic reticulum, vesicles, and microtubules appeared to be well preserved (Fig. 2E–G). As was the case for the chemically prepared samples, the cell walls of the HPF-FS samples did not show damage and the different layers of the wall appeared equally differentially stained (Fig. 2H). Upon visual inspection, the primary cell walls of HPF-FS samples appeared to be much more densely packed when imaged at magnifications over 10,000x (Fig. 2H), compared to those in the microwave-assisted chemically prepared samples (Fig. 2D), with a dense pattern of lightly and darkly stained filamentous structures. Secondary walls of HPF-FS samples, on the other hand, appeared light and rather featureless (Fig. 2H), similar to the chemically prepared samples (Fig. 2D).

Upon adaptation of the novel approach of using the self-pressurized rapid freezing (SPRF) method [57–60] to freeze plant samples, vitreous sectioning of the frozen samples, and imaging by cryo-TEM [61,76–77], we succeeded in obtaining several good areas of vitreous Arabidopsis stem tissue sections (Fig. 2I–L). Freezing in capillary tubes provided better sample height during cryo-sectioning compared to using freezer hats that are commonly used for HPF, which made it relatively easier to obtain vitreous sections from the samples. We found that simple insertion of the plant stem segments into unsealed copper capillary tubes and freezing by a standard high-pressure freezer trapped air bubbles within the frozen sample, often resulting in poor preservation of the samples. We overcame this problem by injecting 20% dextran solution into the capillary tubes and manually applying pressure to seal the capillary tube from both sides; this seemed to remove most of the air bubbles trapped around the stem within the capillary tubes in 1 out of every 3 tubes prepared. As our high-pressure freezer was not equipped to freeze sealed capillary tubes, we used liquid ethane plunge freezing, as is typically performed on ultrathin whole mount samples [78–79]. The tubes that were mostly air-bubble free, froze well, although freezing was not always uniform throughout the entire tissue segment, possibly due to factors such as uneven thickness of sample, uneven distribution of dextran, and air trapped within the plant tissue. Only well-frozen samples that produced continuous long ribbons of intact sections were considered "good grids" and used for imaging. Although we frequently encountered a large variety of artifacts, including a waviness of the sections causing compression of the cell in the direction of cutting, occasional breaks, knife marks and contamination with electron-dense ice particles (Fig. 2I) even within the good grids, we found small (~100–200 μm) patches of flat and uncompressed areas all over each section in such grids. The cell walls in such uncompressed areas appeared to be undamaged (Fig. 2L) and only such apparently good cell wall areas that also showed details of primary and secondary cell walls in the projection view were selected for 3D imaging. We collected 23 tilt-series from 5 good grids, each of which had 2–3 ribbons of ~20 apparently flat, undamaged sections. Success rate of getting good alignment and reconstruction of these fiducial-less tilt-series varied due to section waviness or breaks within the region imaged. The 3 data sets that aligned and reconstructed well (error values within acceptable range recommended in IMOD package) and showed unprecedented detail of both primary and secondary cell walls after filtering and segmentation were used for qualitative analysis. Our results have established that cryo-tomography of unisolated, unextracted plant tissue sections can visualize plant cell walls in their near-native frozen-hydrated state. While the cell organelles could not be distinguished at the low magnifications used to survey the grids (Fig. 2I), some organelles such as Golgi bodies and vesicles became visible in some parenchyma cells when imaged at magnifications above 10,000x (Fig. 2J). Other organelles such as microtubules, ribosomes, and endoplasmic reticulum could not be detected with certainty in the sections surveyed due to the low contrast typical to the cryo-images (Fig. 2K). At high magnification, cell wall components appeared densely packed, albeit with low contrast due to the absence of any heavy metal electron-dense staining. We could clearly distinguish the different layers of the cell wall – middle lamella, primary cell wall, and secondary cell wall, with the middle lamella appearing to be much denser than other layers of the wall. Furthermore, we detected well-organized filamentous structures in both primary cell walls and secondary cell walls (Fig. 2L).

Electron tomography and threshold segmentation of plant cell walls

For each sample type, we selected comparable cell wall areas based on cell shape, size, location and wall thickness, and followed identical steps of electron tomography and image analysis. Figure 3 illustrates these steps in a randomly chosen primary cell wall area of a conventionally prepared sample, as an example. Once we identified any promising area with cell walls suitable for tomographic 3D imaging (Fig. 3A), we collected single- or dual-axis tilt series images and reconstructed them into 3D volumes (Fig. 3B, Video S2). Inspection of such 3D volumes, ~1 nm slice at the time, at full resolution revealed that electron-dense cell wall components were arranged in layers that are nearly parallel to each other (Fig. 3C). Using the threshold segmentation tool in Amira, with a threshold chosen to include most of the electron-dense features for each slice above background noise (Fig. 3D), a segmented map was obtained (Fig. 3E) that could be rendered as a mesh surface for 3D visualization (Fig. 3F, Video S3).

Closer inspection of 3D volume (Fig. 3F) and smaller sub-volumes from different orientations (Fig. 3G–I) revealed a layered arrangement of filamentous primary cell wall components, oriented nearly parallel to the plasma membrane running along the direction of the cell elongation. The long filamentous structures observed in our segmented tomograms match the previously published description of cellulose microfibrils as long, unbranched filaments of ~3–5 nm diameter [21,23], although we noticed that these filamentous structures do not appear to be as smooth as seen by previous microscopy studies. The filamentous

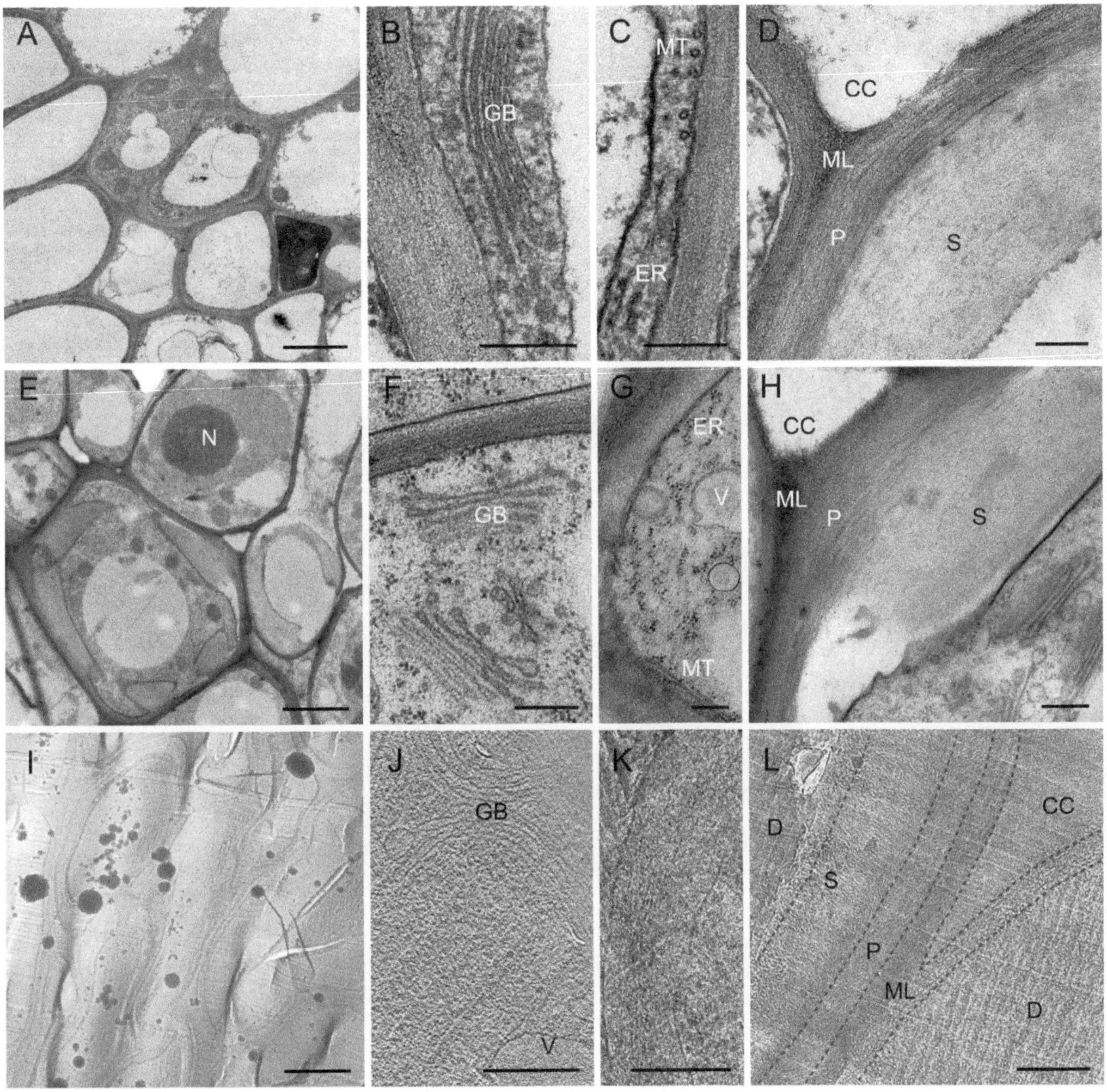

Figure 2. Comparison of sample preparation quality of Arabidopsis stem tissues by transmission electron microscopy. A–D. Chemically prepared samples (conventional method). E–H. High pressure frozen, freeze substituted samples. I–L. Vitreous sections of self-pressurized, rapid frozen samples. A, E, I. Cells and organelles at below 1000x. Bars = 2 µm. B–C, F–G, J–K. Cell organelles at over 10,000x. Bars = 250 nm. D, H, L. Cell wall layers at over 10,000x (Bars = 250 nm). Dotted lines in L. shows the limits of cell wall layers in the low contrast vitreous section. CC = Cell corner, D = Damaged areas, ER = Endoplasmic reticulum, GB = Golgi Bodies, MT = Microtubules, ML = Middle lamella, N = Nucleus, P = Primary cell wall, S = Secondary cell wall, V = Vesicles.

structures detected by our method likely include a crystalline cellulose core surrounded with para-crystalline cellulose (termed elementary fibril by Ding and Himmel, [23]) that possibly has some hemicelluloses and/or pectins attached as well. Following the convention started by Ding and Himmel [23], we will refer to the filamentous structures detected in our tomograms as 'microfibrils' in the remaining text. It must also be noted that the microfibrils in our images appear relatively short compared to previously published length of several microns for cellulose microfibrils because of section geometry, with most microfibrils spanning only the 100–150 nm thickness of each microtome section. In all samples, microfibrils were widely separated from other microfibrils within the same layer (Fig. 2G–H) as well as those in the neighboring layers (Fig. 3I). Bridge-like densities were frequently found between the microfibrils within the same layer as well as between neighboring layers (Fig. 3G–I), which were absent in cell walls of samples extracted with ammonium oxalate and sodium hydroxide that remove hemicellulose and pectin (Fig. S3A–C). We refer to such densities as 'cross-links' in the remaining text as they somewhat match the description of hemicellulose cross-links shown in the well-accepted cell wall models [21,38]. However, these densities could also be pectins as suggested by a recent NMR-based cell wall model [80]. The exact chemical identity, the nature of their chemical interactions and the mechanical roles of these densities cannot be determined from our current data, but can be studied in future using cell wall mutants. The 'empty space'

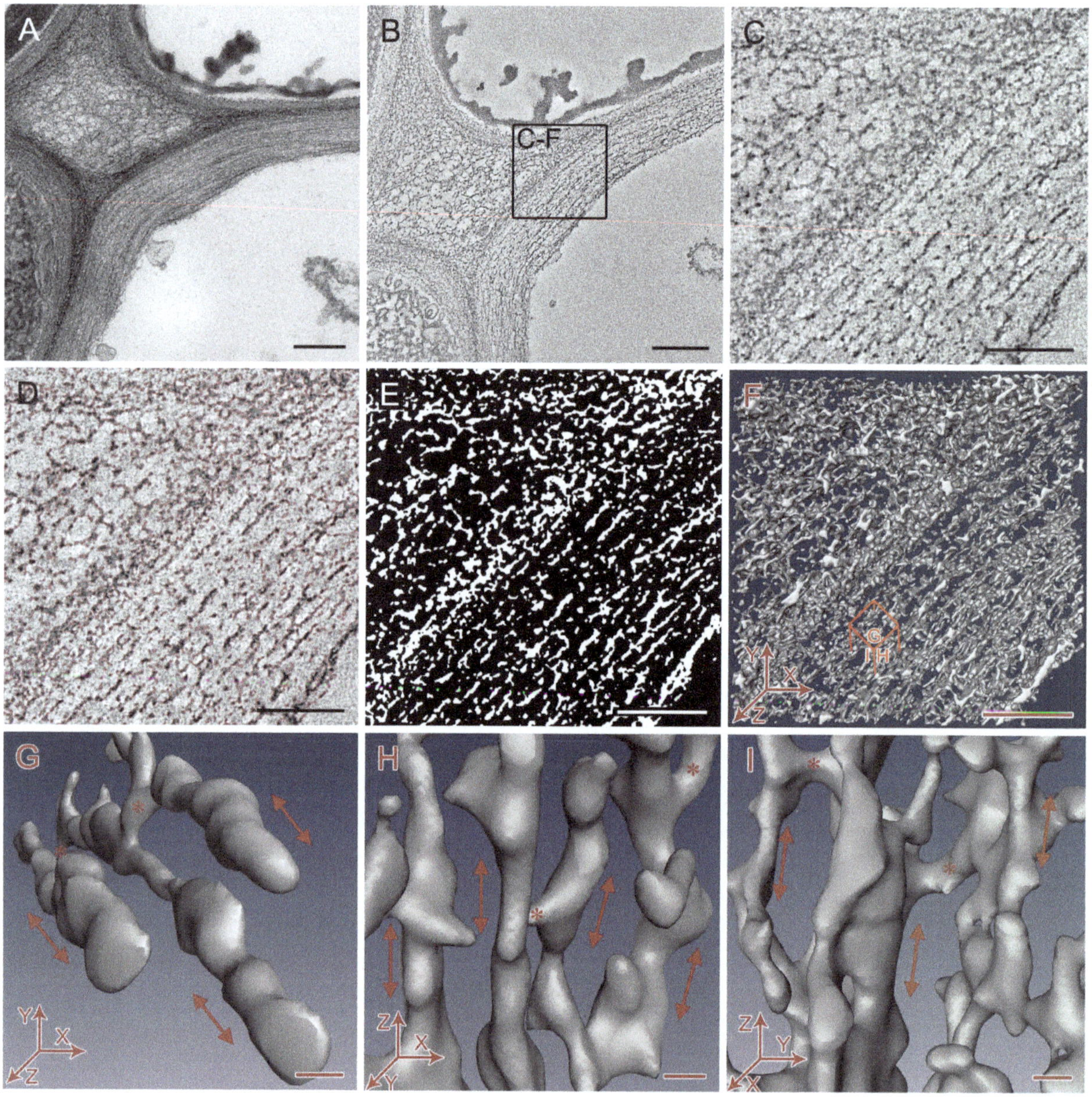

Figure 3. The electron tomography data collection and segmentation process used on Arabidopsis primary cell walls randomly chosen from chemically fixed samples. A. 2D projection image of cell wall. B. Slice of reconstructed tomogram. Bars = 250 nm. C. Sub-area of tomogram. D. Electron dense cell wall components selected by thresholding (selected areas outlined in red). E. Segmentation map of thresholded cell wall components (white). F. Mesh surface rendering of threshold segmentation map. Bars = 100nm. G-I. Small representative 3D volumes of the segmented cell wall showing orientation of filamentous cellulose microfibrils (arrow) and hemicellulose cross-connections (*). G- Top view showing microfibrils running approximately along the axis of cell elongation (Z-axis). H- Side view showing a single layer of microfibrils. I - Side view showing neighboring layers of microfibrils. Bars = 5 nm.

between the electron-dense microfibrils and cross-links could be filled with something that does not get stained (in case of stained sections) or have densities closer to ice (in case of unstained cryo-sections), which could include gases or water in primary walls (that was replaced by resin or amorphous ice in the sections) or lignin in secondary walls.

Qualitative comparison of cell wall preservation quality in electron tomograms

Inspection of individual slices of the raw tomogram (Fig. 4A–C) as well as the threshold segmented density map (Fig. 4D–F) revealed less material in the primary cell wall in samples prepared by conventional chemical protocol (Fig. 4A, D) compared to the HPF-FS-prepared samples (Fig. 4B, E), suggesting that some cell wall components get extracted during the conventional protocol. While in comparison to primary cell walls the pectin-rich middle lamellae appeared denser in both chemically prepared and HPF-FS samples, it was much denser in HPF-FS samples (Fig. 4D–E). At a threshold where the individual structures in the primary cell walls could be segmented well in the HPF-FS samples, the fine features of the middle lamellae could no longer be resolved (Fig. 4E), indicating retention of large amount of pectins and hemicelluloses. Regarding the fine structures of secondary cell walls of both chemically prepared and HPF-FS samples, 2D projection and electron tomography studies did not reveal any fine structural details. However, the filamentous structures in both primary and secondary cell walls were readily visible in the unstained vitreous sections imaged by cryo-electron tomography (Fig. 4C), despite the overall lower contrast of unstained samples. In segmented cryo-tomograms, we were able to detect a highly organized 3D architecture in all layers of the cell wall including the pectin rich-middle lamellae, the primary cell walls, as well as the lignin-rich secondary cell walls (Fig. 4F). To maintain consistency, we restricted the quantitative analysis to the primary cell walls for all sample types.

Apart from differences of the various sample preparation approaches in the overall density of cell wall macromolecules we also noticed that in various parts of the conventionally processed samples, the microfibrils ran in somewhat random directions and occasionally crossed or connected with neighboring microfibrils within the same layer or the neighboring layers (Fig. 4G). In contrast, we found the overall orientation of the microfibrils and layers within the cell wall to be more consistently parallel in the HPF-FS preserved (Fig. 4H) and thus in agreement with cryo-preserved, unstained cell walls (Fig. 4I), suggesting higher order and thus better preservation of the sample in both cryo-immobilization methods compared to conventionally processed

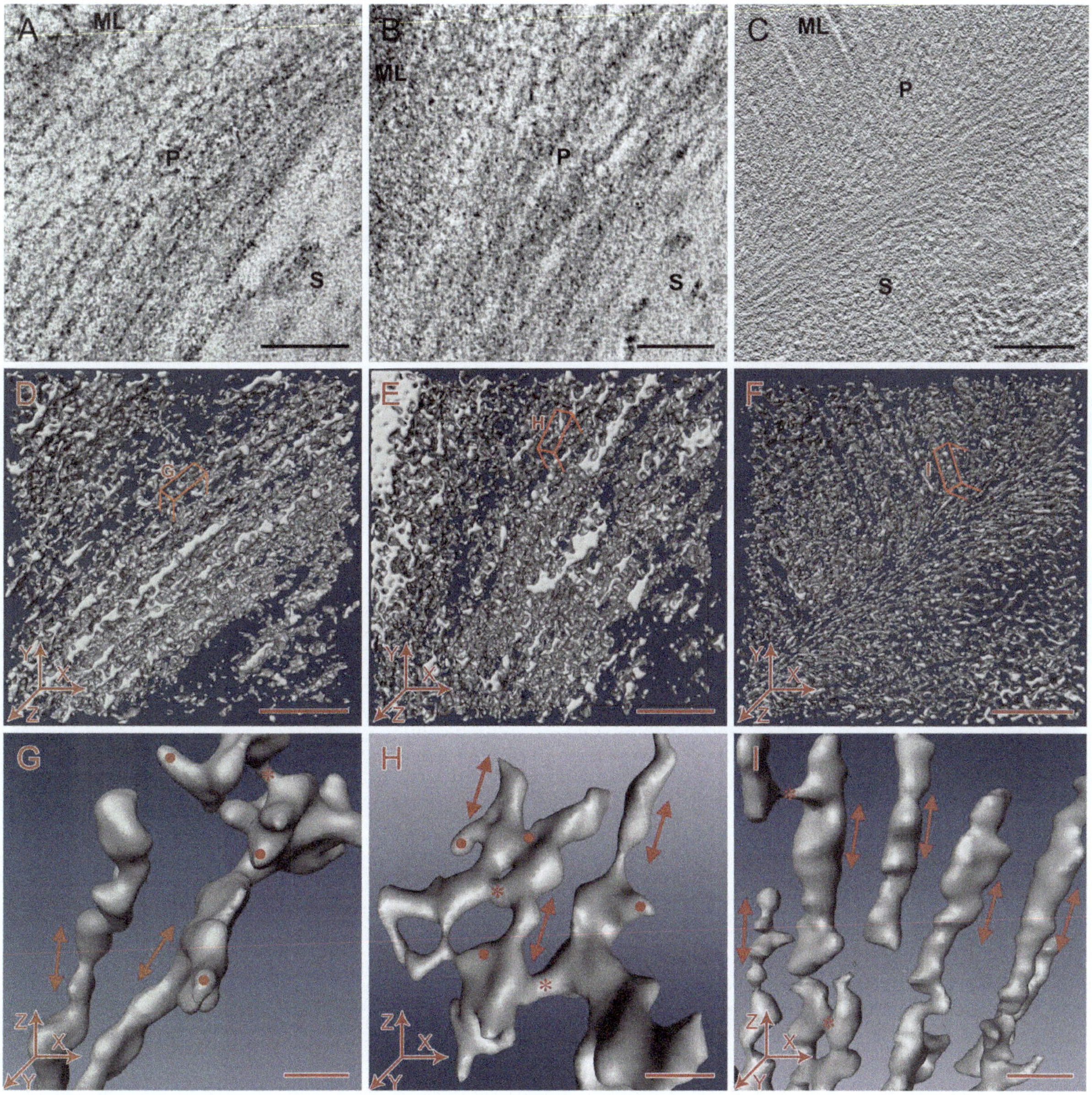

Figure 4. Qualitative comparison of cell wall preservation quality in electron tomograms. A–C. 2D slices taken from the 3D electron tomograms showing overview of cell wall ultrastructure in transverse sections. Bars = 100 nm. D–F. Overview of segmented cell wall volumes. Bars = 100 nm. G–I. Close-up of segmented primary cell wall volumes showing arrangement of cellulose microfibrils (red arrow), cross-links (*), and additional density artifacts (•) from a side view (XZ plane) of the tomograms. Z-axis is approximately the axis of cell elongation. Bars = 10 nm. A, D, G. Chemically prepared samples (conventional method). B, E, H. High pressure frozen, freeze substituted samples. C, F, I. Vitreous sections of self-pressurized, rapid frozen samples. ML = Middle lamella, P = Primary cell wall. S = Secondary cell wall.

samples. The microfibrils in the chemically prepared and HPF-FS samples frequently had additional densities with variable shapes attached to them such as non-filamentous structures, blobs of dense material attached to microfibrils without any obvious pattern, or wide flattened stretch of densities between microfibrils (Fig. 4G, H) that were not present in the cryo-preserved samples (Fig. 4I), which most likely are artifacts from either aggregation of macromolecules or the use of electron-dense stains. The areas with such uncharacteristic additional densities were carefully excluded from the quantitative analyses.

Quantitative geometric analysis of cell wall preservation quality in electron tomograms

Qualitative as well as quantitative analysis of slices of the segmented tomograms showed that the diameter of microfibrils varied significantly along individual microfibrils for all sample preparation schemes (Fig. 4D–F, 5A). We set out to quantify such differences although we realize that (1) there are limitations imposed by the respective sample preparation approaches, e.g. the possibility of preferential staining, (2) the best resolution achievable by electron tomography imaging of ~100–150 nm thin sections is limited to no better than ~2 nm, and (3) it is difficult to establish an objective density threshold, particularly when comparing different data sets. However, for any given data set, all data can be thresholded at the same value allowing a comparison of different microfibrils and cross-links within each data set. We typically applied threshold at a density value where we could visually distinguish the microfibrils and cross-links from background, the microfibrils generally appeared to be long and continuous, and the cross-links clearly bridging two microfibrils. The Segmentation Editor window in Amira software allows simultaneous visual inspection of threshold label image and the original density image, which helps the user to avoid selecting a significantly lower threshold that would include a substantial fraction of noise, or a much higher threshold that would result in disconnected isolated density. We found that selecting a threshold value around '0' in the Amira interface consistently segmented the visible density accurately. A small variation (0±25) yielded essentially the same measurements (Fig. S2A). A larger variation (0±75) resulted in an overall shift of 1–1.5 nm that corresponds to a difference of 1–2 pixels (Fig. S2B), but such differences were consistent between different data sets. Even when 2 different individuals independently chose 'ideal threshold values' and measured independently chosen microfibrils, the overall results were remarkably similar with maximum difference of ~1 nm (Fig. S2C). Such small differences in results were not accounted as significant differences in any of our comparative studies. Even though density maps cannot be completely objectively thresholded for different data sets, a comparison of measurements of microfibril diameter from the same sample preparation protocols but in different data sets yielded remarkably similar numbers (Fig. S2D), suggesting that a comparison between different sample preparation protocols is legitimate, and the same threshold approach can be used to measure cross-link length and edge-to-edge distances between microfibrils as well. Moreover, the above-mentioned limitations do not affect the center-to-center distances of the microfibrils and hence, any significant difference in filament spacing between data sets must therefore represent a real difference.

We found that on average, the microfibrils in chemically prepared samples were slightly thinner than HPF-FS and cryo sample (Fig. 5A), but overall all three approaches match well within the accuracy limitations of electron tomographic imaging. The diameter measurements ranged between ~2.2–8.0 nm (average 4.3±1.3 nm, n = 180) for chemically prepared samples, ~2.2–8.9 nm (average 4.9±1.4 nm, n = 180) for the HPF-FS samples, and ~2.2–8.9 nm (average 5.0±1.5 nm, n = 180) for the cryo samples. Hence, we conclude that filament diameter only differ slightly within the experimental error limit and thus agree well among all of the three sample preparation approaches. When examining the narrowest gap (edge-to-edge distance) between two neighboring microfibrils, we found that for conventionally prepared samples, the gap ranged between 7.2–35.2 nm (average 17.0±4.3 nm, n = 450), whereas in HPF-FS samples and cryo samples, such gaps were significantly narrower, i.e. 2.1–11.8 nm (average 6.9±2.1 nm, n = 450) and 2.7–11.9 nm (average 6.4±1.2 nm, n = 450) respectively (Fig. 5B). The corresponding center-to-center inter-microfibrilar distance ranged from 12.7–40.7 nm (average 22.5±4.3 nm, n = 450), 7.6–17.3 nm (average 12.4±2.1 nm, n = 450) and 8.2–17.4 nm (average 11.9±1.2 nm, n = 450), for chemically prepared, HPF-FS and cryo samples, respectively. The length of cross-links in chemically prepared samples ranged between 2.2–12.6 nm (average 5.6±2.1 nm, n = 90) in chemically prepared samples. Cross-links in HPF-FS and cryo samples were relatively shorter and comparable to each other ranging between 2.2–8.0 nm (average 4.6±1.5 nm, n = 90) in the HPF-FS samples and 2.6–9.1 nm (average 5.0±1.5 nm, n = 90) in cryo-sectioned samples (Fig. 5C). These quantitative data support our qualitative findings from visual inspection that suggested much better preservation of cell wall material in cryo-immobilized samples compared to conventionally processed samples, with remarkably close values for the HPF-FS and cryo-sectioned samples.

Application of electron tomography in cell wall mutant characterization

Encouraged by our finding that HPF-FS approach provided comparable quality of preservation for primary cell wall compared to the technically challenging cryo-sectioning gold standard approach, we applied the HPF-FS method for structural characterization of a cell wall mutant, *cob-6*. Mutations in the *COBRA (COB)* gene of Arabidopsis have been reported to cause disorganization of cellulose microfibril orientation and reduction of crystalline cellulose in the cell walls of roots [54–56]. In our electron tomography study, we observed that in comparable hypocotyl parenchyma cells of wild type (WT) Arabidopsis and *cob-6* mutant, the cell walls, more noticeably the middle lamella, appeared less stained in the WT (Fig. 6A) compared to the mutant (Fig. 6B), indicating possible alteration in pectin composition. The cell membranes were loosened from the cell wall at several places in the mutant (Fig. 6B) suggesting possible disorganization at the wall-membrane interface. Visual inspections of the segmented tomograms (Fig. 6C–D) did not reveal any remarkable difference in the overall organization of cellulose microfibrils and hemicellulose cross-links in the primary cell walls of the mutant (Fig. 6D) compared to that of the WT (Fig. 6C). However, quantitative analysis of tomogram slices revealed that the microfibril diameters were comparable in both samples, ranging from ~2.8–7.7 nm (average 4.5±0.9, n = 60) for the WT and ~2.9–6.2 nm (average 4.6±0.8 nm, n = 60) for the mutant (Fig. 7A). However, the gaps between the microfibrils in the mutant primary walls were more random ranging between ~3.2–20.6 nm (average 8.9±3.8 nm, n = 60) compared to a significantly narrower range of ~3.0–11.0 nm (average 5.2±2.0 nm, n = 60) for the WT walls (Fig. 7B). Cross-links between microfibrils were slightly longer on average in the mutant wall, ranging between ~3.6–6.8 nm (average 4.7±0.8 nm, n = 30) compared to the range of ~2.4–7.1 nm (average 4.2±0.8 nm, n = 30) for the WT wall (Fig. 7C). Our

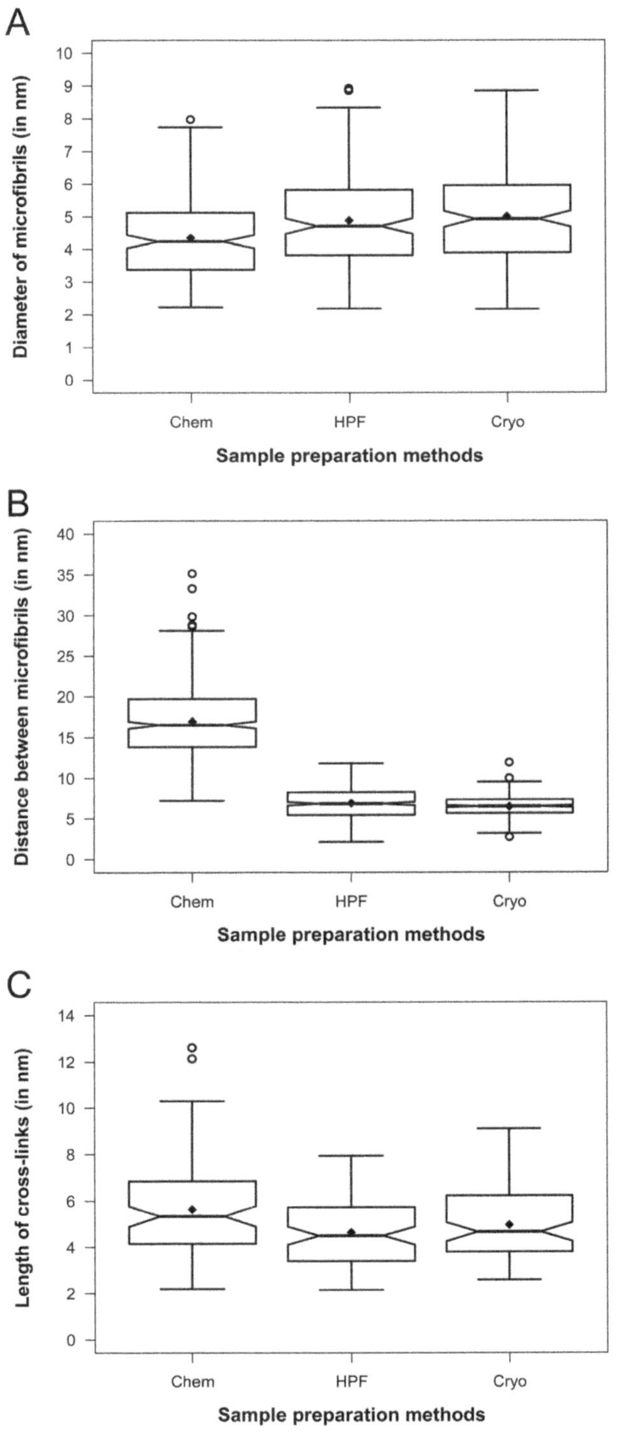

Figure 5. Notched boxplots showing quantitative comparison of cell wall preservation quality in electron tomograms. Cryo = Cryo-tomograms of vitreous sections. HPF = Tomograms of high pressure frozen, freeze-substituted samples. Chem = Tomograms of chemically prepared samples. The thick band inside the box is the median, and the bottom and top of the box are the first quantile (Q1) and the third quantile (Q3), respectively. The ends of the whiskers represent the range of data within 1.5 *IQR (Interquartile range; IQR = Q3−Q1) from the lower quantile (Q1) or the upper quantile (Q3). The notch displays deviation around the median ±1.57×IQR/sqrt of n (where n is the sample size), and approximately shows the 95% confidence interval of median, so that if the notches of two boxes do not overlap, their medians are usually significantly different. The diamond (◆) represents the mean, and the circles (O) represent any outliers. A. Diameter of microfibrils, n = 180. B. Edge-to-edge distance between microfibrils, n = 450. C. Length of cross-links, n = 90.

results show that in comparison to the WT Arabidopsis, the diameter of cellulose microfibrils in primary cell walls essentially remain unchanged in the *cob-6* mutant, indicating no significant change in their individual macromolecular structures. However, the quantitatively detected increase in distances between the microfibrils and cross-link length indicate a subtle but noteworthy loss of microfibril organization in the mutants.

Discussion

Conventional sample preparation protocols for RT electron tomography consist of chemical fixation, RT dehydration, RT resin embedding, heat polymerization, followed by RT ultrathin sectioning and RT electron tomography imaging, and this approach is the most widely used and the least technically challenging protocol currently available for studying molecular resolution 3D ultrastructure of biological samples. Variations of the conventional protocol have been used for all electron tomography studies on plant cell walls reported to date [49–52]. The conventional protocols do not require any expensive equipment for freezing, and tilt series data of samples prepared by this method can be easily acquired on most TEM microscopes that allow stage tilting. Conventional protocols are more commonly done on bench-top and are rather lengthy for plant samples due to slow infiltration of fixatives and resins through cell walls. Using microwave-assisted protocols significantly decreases the time required for each step with preservation being comparable to the bench-top protocols [81]. Using an automated microwave additionally makes the conventional protocol even less labor intensive, and allows a large number of samples to be prepared simultaneously. We consistently obtained fair preservation of Arabidopsis stem tissue samples with our 3-day protocol of microwave-assisted chemical fixation, dehydration and resin embedding. However, compared to cryogenically preserved samples the cell walls visually appeared to be more extracted in these samples, which was confirmed by the quantitative analysis. Such large-scale quantitative analyses of cell wall parameters were not performed in any of the previous electron tomography studies of plant cell wall, although a small-scale quantitative analysis [49] reported the diameter of crystalline cellulose core of a microfibril to be ~2.2 nm and that with a non-crystalline outer layer to be ~3.2 nm. The smallest diameter of microfibril we detected was also ~2.2 nm and a large population of microfibrils we measured in our chemically prepared samples were ~3 to 3.5 nm. We also detected microfibrils with a larger diameter that might include outer layers of non-cellulosic components such as hemicellulose and/or pectins, but overall the microfibrils in chemically prepared samples were slightly smaller in diameter compared to the HPF-FS and cryo samples. Although such differences are well within the resolution limit of the experimental technique and hence may or may not reflect true differences in composition, these differences in diameter greatly resemble the difference between microfibril from normal and pectin extracted cell walls of Arabidopsis imaged in an atomic force microscopy (AFM) study, where the microfibrils in pectin extracted cell wall were reported to be 3.2±0.13 nm while those from normal walls were reported to be 5.8±0.17 nm [44]. A more striking observation was the breakdown in consistency for microfibril orientations in many parts of the sample, as well as the drastically wider space between microfibrils with some of the inter-microfibrilar cross-connections being longer compared to those of

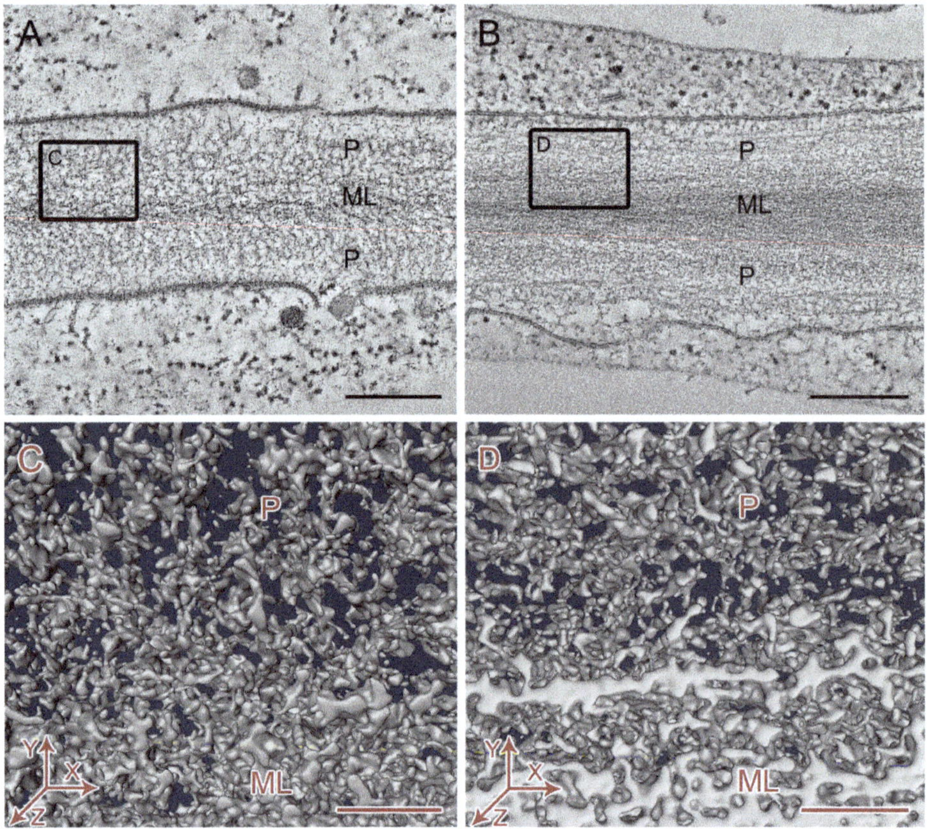

Figure 6. Qualitative comparison of primary cell wall ultrastructure in electron tomograms of high pressure frozen, freeze-substituted wild type hypocotyl parenchyma cells of wild type Arabidopsis (A, C) and cob-6 mutant (B, D). A–B. Slices of electron tomogram showing overview of cell wall ultrastructure in 2D. Bars = 250 nm. C–D. Small volumes of segmented cell wall tomograms showing organization of cell wall components from top view (XY plane). Bars = 50 nm. ML = Middle lamella, P = Primary cell wall.

the HPF-FS and cryo samples. The larger inter-microfibrillar distances in conventionally prepared samples suggest loosening and separation of the microfibril layers, possibly due to extraction of cell wall components that are usually loosely attached to the cellulose microfibrils as well as those that are present in the space between the cellulosic frameworks (such as pectins, hemicelluloses, and glycoproteins) during the lengthy dehydration and resin infiltration steps. The longer cross-links could possibly result from stretching during separation of the microfibril layers. The primary fixative, glutaraldehyde, as well as the secondary fixative/staining reagent, osmium tetroxide, usually target proteins and lipids in biological samples and hence are less effective for carbohydrates [53,82]. The poorly fixed carbohydrates may thus leach out into the organic solvents during the lengthy dehydration steps. Conventional sample preparation protocols that include chemical fixation and dehydration are known to cause aggregation and extraction artifacts [53]. Our results suggest that such protocols are not reliable for studying the high-resolution ultrastructure of plant cell walls by electron tomography.

Cell wall structures were arguably preserved closest to their native state in samples prepared by the cryo method, as the tissue is not subjected to any fixatives, dehydration or resins, and thus is prepared only by self-pressurized rapid freezing (SPRF) of the stem tissue in capillary copper tubes with dextran. However, as the entire cryo-protocol including SPRF, vitreous sectioning of the frozen samples, and imaging by cryo-electron tomography needs to be carried out at liquid nitrogen temperature, the process is technically extremely challenging. Each individual step requires intensive training using specialized equipments, some of which are still in the stage of technology development. Cryo-tomography, while a mature field, is mostly done on whole-mount samples such as small cells (e.g. virus, bacteria) or isolated cell parts (organelles, protein complexes). However, this approach is almost never done on actual tissues, plant or animal, where orientation issues and operator skill development are often required. Though conventional high pressure freezing can solve these problems to some extent, the small sample in the extremely short high pressure freezer hats (100–200 µm deep) made cryo-sectioning very challenging. We found the SPRF in copper capillary tubes to be the most effective way to overcome all of the above mentioned issues, as this method allowed better tracking of tissue orientation and also provided enough sample to be sectioned and screened. By inserting the plant samples in the capillary tubes, quickly introducing 20% dextran solution into the tube, and then closing the tube from each of the ends we were able to remove excess air around the plant sample and create a uniform pressure buildup within the tubes upon rapid plunge-freezing into liquid ethane. Although our cryo-sections still suffered from non-flatness, occasional breaks, knife marks and ice contamination, good areas with intact cell walls could be imaged successfully. However, we could collect only single-axis tilt series from +60° to −60°, with 2° increments for the cryo-sections, as the vitreous sections of Arabidopsis tissue were very unstable even at the low dose used. The cell wall areas started boiling off if subjected to longer data

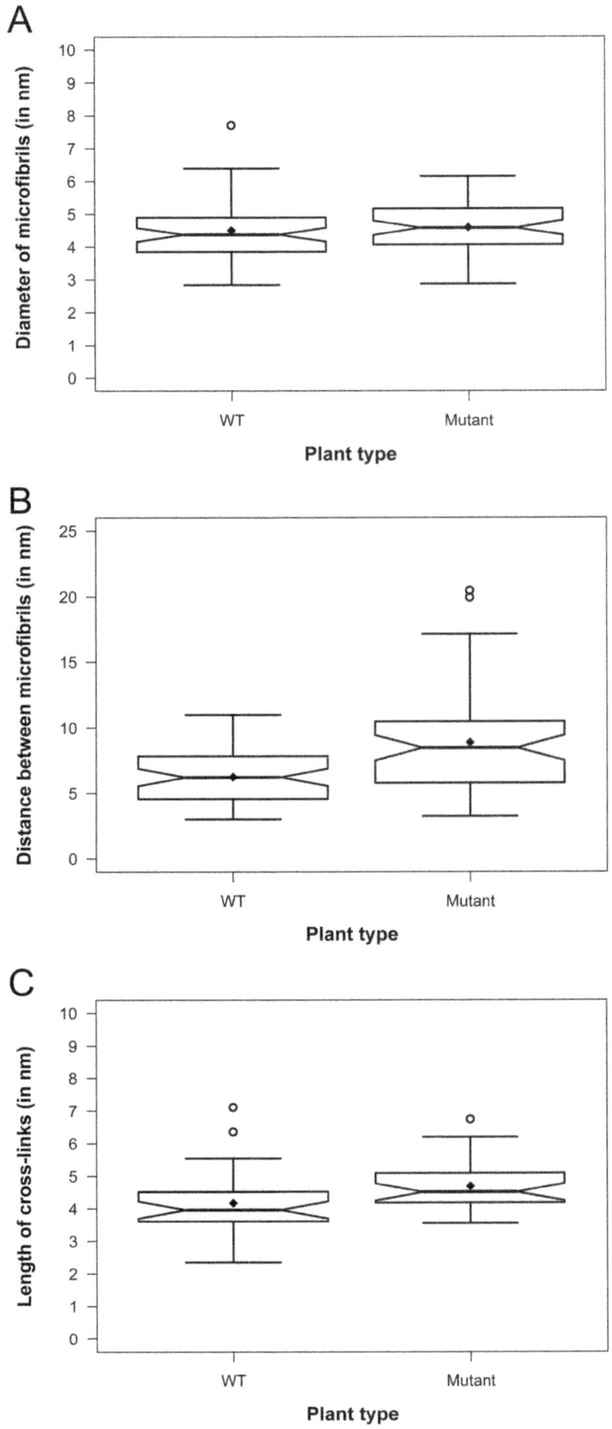

Figure 7. Notched boxplots showing quantitative comparison of primary cell wall ultrastructure in electron tomograms of high pressure frozen, freeze-substituted hypocotyl parenchyma cells of wild type Arabidopsis (WT) and *cob-6* (Mutant) plants. The thick band inside the box is the median, and the bottom and top of the box are the first quantile (Q1) and the third quantile (Q3), respectively. The ends of the whiskers represent the range of data within 1.5 *IQR (Interquartile range; IQR = Q3−Q1) from the lower quantile (Q1) or the upper quantile (Q3). The notch displays deviation around the median ±1.57×IQR/sqrt of n (where n is the sample size), and approximately shows the 95% confidence interval of median, so that if the notches of two boxes do not overlap, their medians are usually significantly different. The diamond (◆) represents the mean, and the circles (O) represent any outliers. A. Diameter of microfibrils, n = 60. B. Edge-to-edge distance between microfibrils, n = 60. C. Length of cross-links, n = 30.

collection times. Collecting dual-axis tilt series with 1° increment comparable to the resin tomograms was not possible as it would have required 1/4th of the dose used, which would further decrease the contrast of the already low-contrast cryo-tomograms. Collecting dual-axis cryo tilt series would also require the region of interest to be away from the grid bars, which was logistically difficult with the mesh grids used for increasing stability of vitreous sections. Reconstruction of our single-axis cryo-tomograms was more difficult than for the dual-axis resin tomograms due to the low contrast, non-flatness and absence of gold fiducials in the tilt-series obtained by the cryo method. Reconstructing tomograms from a dual-axis series would require having flat sections with gold fiducials for accurate alignment combination of the two series. In spite of the low contrast of the tilt series images, segmentation, analysis and measurement of the various cell wall components was surprisingly straightforward in the cryo samples, possibly due to low background noise and lack of any additional electron dense material such as stains, thus providing sufficient contrast of each reconstruction slice. Some caution was needed to detect knife marks in the sample, in order to not confuse knife marks with inter-microfibrillar space, but fortunately, knife marks were found to occur in a direction clearly different from microfibril organization. The final segmented volumes revealed highly organized 3D architecture indicating that the fine structures of cell wall components were well preserved, as close as possible to their native state. The average microfibril diameter recorded in our cryo-sections was 5.0±1.5 nm, which matches well with the 5.8±0.17 nm measurement obtained by AFM of cell walls from Arabidopsis callus tissue [44]. While the cryo-sectioning approach is currently being instrumental to us in developing accurate and comprehensive cell wall models of low-lignin primary as well as lignin-rich secondary cell walls, we think it will be an unnecessarily complicated approach if the objective of a project is to characterize just primary or any lignin-less cell walls.

The HPF-FS method consistently produced high quality preservation of primary cell walls in the Arabidopsis stem tissue samples in our hands. Compared to the chemically prepared samples, the cell walls visually appeared denser, the overall organization of the microfibrils and layers within the cell wall appeared to be more consistent, the spaces between microfibrils were significantly narrower, and cross-connections between microfibrils were shorter and slightly thicker in the HPF-FS samples. Our analyses show that regarding all quantitative volumetric parameters as well as the overall level of order, HPF-FS data compare well with the cryo-sectioning data (typically considered as the gold standard for biological sample preservation), and yet is technically far less challenging than cryo-sectioning. Sectioning and imaging of HPF-FS samples at RT is much easier as resin sections allow cutting well-adherent sections, application of gold fiducial markers for improved alignment of the tilt series images, and dual axis data collection for reducing the data anisotropy. Furthermore, the ability to prescreen the samples for areas of interest as well as storage of the section for subsequent imaging, combined with what appears to be exquisite preservation render the HPF-FS method as an excellent choice for routine cell wall characterization. A major limitation of HPF-FS samples, however, is that unlike cryo-sections, the fine structures of middle lamella and secondary cell walls could not be clearly distinguished in these samples. Presence of high amounts of pectins in the middle lamella of young plants causes deep staining of the middle lamella due to higher affinity of ruthenium red stain for pectins [83]. The fine structure of middle lamella cannot be resolved well in HPF-FS due to the heavy metal staining, particularly if a threshold value is chosen for simultaneous visualization of cell wall components in the primary cell wall. On the other hand, the presence of significant amounts of lignin in secondary cell wall seems to make the walls appear featureless, possibly because lignin may serve as a penetration barrier for the stain and thus prevent contrast generation in the resin sections. Lignin is structurally similar to phenolic compounds present in epoxy resins [84], which could also be the reason of the apparent lack of texture and hence the perceived structural homogeneity of the resin-embedded lignin-rich secondary cell walls. The HPF-FS method can be used as a routine approach for studying cell walls with little or no lignin such as in parenchyma or collenchyma cells, lignin-less pteridophytes, bryophytes, algae, and also pretreated biomass samples. Most laboratories around the world that are familiar with TEM sample preparation can master the HPF-FS method much more easily compared to learning cryo-tomography of vitreous sections. As high-pressure freezing devices are becoming more readily available, and freeze-substitution can be carried out in a homemade low-cost setups [85–86], we submit that HPF-FS may be an appropriate compromise if the objective of a project is to characterize only primary cell walls or any lignin-less cell walls.

To this end, our comparison of WT Arabidopsis and *cob-6* cell wall mutant is a proof-of-concept and demonstrates our ability to detect relatively subtle changes in cell wall architecture. When all the technology and expertise were in place, the comparison of WT to mutant did not take longer than 3 weeks of experimental work, including sample preparation, imaging and image analysis. The whole process could be completed faster (within ~1 week) if recently developed protocols of 'quick freeze-substitution', 'rapid resin embedding and polymerization' [85–87] are used. The *cob* mutants have been shown to display disorganization in cell wall microfibrils in their root cells by FESEM [55]. We detected and quantified a relatively subtle but significant microfibril disorganization in the parenchyma cells of hypocotyl samples by measuring the gaps between individual microfibrils. The powerful combination of localized qualitative and quantitative image analysis possible on the 3D tomography data allowed us to detect and quantify these subtle changes in the mutant wall organization, which cannot be detected by any other imaging method currently used for studying cell wall mutants.

Conclusion

Electron tomography of cryo-immobilized plant tissue can reveal never-seen-before details of the 3D architecture of plant cell walls at macromolecular resolutions, which can be analyzed both qualitatively and quantitatively. Vitreous sectioning of fast-frozen samples provides closest-to-native preservation of both the lignin-less primary and lignin-rich secondary cell walls. Samples prepared by high-pressure freezing, freeze substitution and resin embedding (HPF-FS) also represented faithful preservation of the respective structures and the overall 3D organization of lignin-less primary cell walls. The HPF-FS approach is technically less demanding than the vitreous sectioning approach, yet is clearly superior to the commonly used bench-top or microwave assisted chemical protocols, which suffer from significant extraction artifacts. HPF-FS can hence, be widely adapted by the community as a routine tool for assessing cell walls with little or no lignin, such as in seedlings of cell wall mutants, lower plant groups and algae, and deconstructed biomass. Cryo-tomography of vitreous tissue sections should however be chosen if the goal of a project is to study the architecture of lignin-rich secondary cell walls and developing accurate and comprehensive 3D cell wall models.

Supporting Information

Figure S1 Comparison of segmentation approaches showing reliability of threshold segmentation. A–D. Semi-automated threshold-based segmentation in Amira (A, C) compared with manual tracing of density in Amira (B, D). A–B. Slice of segmented tomogram in Amira. C–D. Segmented microfibrils in Amira. E–G. Microfibrils segmented by semi-automated threshold-based approach in Amira (E) compared with skeletons extracted from microfibrils segmented by algorithm-based automated approach (F, G). Bars (A–E, G) = 5 nm; (F) = 100 nm.

Figure S2 Reliability of threshold segmentation approch for measument of microfibril diatemeter. A. Insignificant difference in measurements with small variability in threshold value (n = 75). B. Shift of ~1 nm only with a larger variability in threshold value (n = 75). C. Comparable results obtained by two independent users from segmenting and analyzing the same 3D volume (n = 60). D. Comparable results obtained from segmenting and analyzing three different tomograms of three different cell wall samples prepared by the same sample preparation method (n = 60). The thick band inside each notched box is the median, and the bottom and top of the box are the first quantile (Q1) and the third quantile (Q3), respectively. The ends of the whiskers represent the range of data within 1.5 *IQR (Interquartile range; IQR = Q3−Q1) from the lower quantile (Q1) or the upper quantile (Q3). The notch displays deviation around the median ±1.57×IQR/sqrt of n (where n is the sample size), and approximately shows the 95% confidence interval of median, so that if the notches of two boxes do not overlap, their medians are usually significantly different. The diamond (◆) represents the mean, and the circles (O) represent any outliers.

Figure S3 Chemically extracted Arabidopsis cell walls to remove hemicelluloses and pectins. A. Slice of electron tomogram showing overview of extracted cell wall. Bar = 250 nm. B. Sub-area of tomogram. Bar = 100 nm. C. Segmented cell wall showing orientation of filamentous cellulose microfibrils. No cross-connections detected. Bar = 10 nm.

Video S1 The process of cryo-sectioning of frozen samples and collection of vitreous sections on a EM grid.

Video S2 A reconstructed 3D volume (tomogram) of an Arabidopsis cell wall showing a z-stack of ~1 nm thin slices.

Video S3 The process of threshold segmentation in Amira, to obtain a segment map that include most of the electron-dense features above background noise for each slice, and rendering as a mesh surface for 3D visualization and further analyses.

Acknowledgments

We thank Dr. Kent McDonald, Reena Zalpuri and Dr. Guangwei Min of UC Berkeley Electron Microscopy Facility; Dr. Bridget Carragher, Joel

Quispe, Erica Jacovetty, Dr. Anchi Cheng of National Resource for Automated Molecular Microscopy (NRAMM), The Scripps Research Institute, La Jolla; and Dr. Birgit Luef and Dr. Luis Comolli of Lawrence Berkeley National Laboratory, for their help in microscopy data collection. Special thanks to Dr. Peter Peters and Jason Pierson of The Netherlands Cancer Institute, Amsterdam for advice on cryo-sectioning, and Dr. Puey Ounjai of Lawrence Berkeley National Laboratory for advice on reconstruction of cryo-tomograms. Thanks to Dr. Andrew Carroll and Dr. Nadav Sorek from Prof. Chris Somerville lab at Energy Biosciences Institute, UC Berkeley for generously providing plant samples.

Author Contributions

Conceived and designed the experiments: PS EB MA. Performed the experiments: PS EB AC EN DM SK RC. Analyzed the data: PS EB EGY JD WT MJ ZY CB. Contributed reagents/materials/analysis tools: KHD ZY CB. Contributed to the writing of the manuscript: PS EB EGY RC CB MA. Standardized the SPRF method for freezing plant tissue: EB SY. Designed the automated segmentation approach used in this study: ZY CB.

References

1. Ragauskas AJ, Nagy M, Kim DH, Eckert CA, Hallett JP, et al. (2006) From wood to fuels: Integrating biofuels and pulp production. Industrial Biotechnol 2: 55–65.
2. McCann MC, Carpita NC (2008) Designing the deconstruction of plant cell walls. Curr Opin Plant Biol 11: 314–320.
3. Pauly M, Keegstra K (2008) Cell-wall carbohydrates and their modification as a resource for biofuels. Plant J 54: 559–568.
4. Sticklen MB (2008) Plant genetic engineering for biofuel production: towards affordable cellulosic ethanol. Genetics 9: 433–443.
5. Mosier N, Wyman C, Dale B, Elander R, Lee YY, et al. (2005) Features of promising technologies for pretreatment of lignocellulosic biomass. Biores Tech 96: 673–686.
6. Himmel ME, Ding SY, Johnson DK, Adney WS, Nimlos MR, et al. (2007) Biomass recalcitrance: engineering plants and enzymes for biofuels production. Science 315: 804–807.
7. Banerjee G, Scott-Craig JS, Walton JD (2010) Improving enzymes for biomass conversion: a basic research perspective. Bioenerg Res 3: 82–92.
8. Elkins JG, Raman B, Keller M (2010) Engineered microbial systems for enhanced conversion of lignocellulosic biomass. Curr Opin Biotech 21: 657–662.
9. Yuan JS, Tiller KH, Al-Ahmad H, Stewart NR, Stewart CN Jr. (2008) Plants to power: bioenergy to fuel the future. Trends Plant Sci 13: 421–429.
10. Xin Z, Watanabe N, Lam E (2011) Improving efficiency of cellulosic fermentation via genetic engineering to create "smart plants" for biofuel production. In: Buckeridge, MS, Goldman, GH, editors. Routes to cellulosic ethanol. New York: Springer. 181–197.
11. Popper ZA (2008) Evolution and diversity of green plant cell walls. Curr Opin Plant Biol 11: 286–292.
12. Sarkar P, Bosneaga E, Auer M (2009) Plant cell walls throughout evolution: towards a molecular understanding of their design principles. J Exp Bot 60: 3615–3635.
13. Keegstra K (2010) Plant cell walls. Plant Physiol 154: 483–486.
14. Lee KJ, Marcus SE, Knox JP (2011) Cell wall biology: perspectives from cell wall imaging. Mol Plant 4: 212–219.
15. Leroux O (2012) Collenchyma: a versatile mechanical tissue with dynamic cell walls. Ann Bot 110: 1083–1098.
16. Albersheim P, Darvill A, Roberts K, Sederoff R, Staehelin A (2010) Plant cell walls. New York: Garland Science. 1–42.
17. Cosgrove DJ, Jarvis MC (2012) Comparative structure and biomechanics of plant primary and secondary cell walls. Front Plant Sci 3: 204.
18. Ohad I, Danon D (1964) On the dimensions of cellulose microfibrils. J Cell Biol 22: 302–305.
19. Frey-Wyssling A (1968) The ultrastructure of wood. Wood Sci Technol 2: 73–83.
20. Heyn AN (1969) The elementary fibril and supermolecular structure of cellulose in soft wood fiber. J Ultrast Res 26: 52–68.
21. Somerville C, Bauer S, Brininstool G, Facette M, Hamann T, et al. (2004) Toward a systems approach to understanding plant cell walls. Science 306: 2206–2211.
22. Cosgrove DJ (2005) Growth of the plant cell wall. Nat Rev Mol Cell Biol 6: 850–861.
23. Ding S-Y, Himmel ME (2006) The maize primary cell wall microfibril: a new model derived from direct visualization. J Agric Food Chem 54: 597–606.
24. Newman RH, Hill SJ, Harris PJ (2013) Wide-angle x-ray scattering and solid-state nuclear magnetic resonance data combined to test models for cellulose microfibrils in mung bean cell walls. Plant Phys 163: 1558–1567.
25. Thomas LH, Forsyth VT, Sturcová A, Kennedy CJ, May RP, et al. (2013) Structure of cellulose microfibrils in primary cell walls from collenchyma. Plant Physiol 161: 465–476.
26. Somerville C (2006) Cellulose synthesis in higher plants. Annu Rev Cell Dev Biol 22: 53–78.
27. Taylor NG (2008) Cellulose biosynthesis and deposition in higher plants. New Phytol 178: 239–252.
28. Scheller HV, Ulvskov P (2010) Hemicelluloses. Annu Rev Plant Biol 61: 263–289.
29. McCann MC, Roberts K (1991) Architecture of the primary cell wall. In CW Lloyd, editor. The cytoskeletal basis of plant growth and form. London: Academica Press. 109–129.
30. Rose JKC, Bennett AB (1999) Cooperative disassembly of the cellulose-xyloglucan network of plant cell walls: parallels between cell expansion and fruit ripening. Trends Plant Sci 4: 176–183.
31. Park YB, Cosgrove DJ (2012) A revised architecture of primary cell walls based on biomechanical changes induced by substrate-specific endoglucanases. Plant Physiol 158: 1933–1943.
32. Cosgrove DJ (2000) Expansive growth of plant cell walls. Plant Physiol Biochem 38: 109–124.
33. Baskin TI, Meekes HTHM, Liang BM, Sharp RE (1999) Regulation of growth anisotropy in well-watered and water-stressed maize roots. II. Role of cortical microtubules and cellulose microfibrils. Plant Physiol 119: 681–692.
34. MacKinnon IM, Sturcova A, Sugimoto-Shirasu K, His I, McCann MC, Jarvis MC (2006) Cell-wall structure and anisotropy in procuste, a cellulose synthase mutant of *Arabidopsis thaliana*. Planta 224: 438–448.
35. Verbelen JP, Kerstens S (2000) Polarization confocal microscopy and congo red fluorescence: a simple and rapid method to determine the mean cellulose fibril orientation in plants. J Microscopy 198: 101–107.
36. Anderson CT, Carroll A, Akhmetova L, Somerville C (2010) Real-time imaging of cellulose reorientation during cell wall expansion in Arabidopsis roots. Plant Physiol 152: 787–796.
37. Preston RD, Nicolai E, et al. (1948) An electron microscope study of cellulose in the wall of Valonia ventricosa. Nature 162: 665–667.
38. McCann MC, Wells B, Roberts K (1990) Direct visualization of cross-links in the primary plant cell wall. J Cell Sci 96: 323–334.
39. Fujino T, Itoh T (1994) Architecture of the cell wall of a green alga, *Oocystis apiculata* Protoplasma 180: 39–48.
40. Fujino T, Sone Y, Mitsuishi Y, Itoh T (2000) Characterization of cross-links between cellulose microfibrils, and their occurrence during elongation growth in pea epicotyl. Plant Cell Physiol 41: 486–494.
41. Roland J C, Vian B, Reis D (1975) Observations with cytochemistry and ultracryotomy on the fine structure of the expanding walls in actively eleongating plant cells. J Cell Sci 19: 239–259.
42. Sugimoto K, Williamson RE, Wasteneys GO (2000) New techniques enable comparative analysis of microtubule orientation, wall texture, and growth rate in intact roots of Arabidopsis. Plant Physiol 124: 1493–1506.
43. Thimm JC, Burritt DJ, Ducker WA, Melton LD (2000) Celery (*Apium graveolens L.*) parenchyma cell walls examined by atomic force microscopy: effect of dehydration on cellulose microfibrils. Planta 212: 25–32.
44. Davies LM, Harris PJ (2003) Atomic force microscopy of microfibrils in primary cell walls. Planta 217: 283–289.
45. Ding SY, Liu YS, Zeng Y, Himmel ME, Baker JO, et al. (2012) How does plant cell wall nanoscale architecture correlate with enzymatic digestibility? Science 338: 1055–1060.
46. Radotic K, Roduit C, Simonovic J, Hornitschek P, Fankhauser C, et al. (2012) Atomic force microscopy stiffness tomography on living Arabidopsis thaliana cells reveals the mechanical properties of surface and deep cell-wall layers during growth. Biophys J 103: 386–394.
47. Lucic V, Forster F, Baumeister W (2005) Structural studies by electron tomography: from cells to molecules. Annu Rev Biochem 74: 833–865.
48. McIntosh R, Nicastro D, Mastronarde D (2005) New views of cells in 3D: an introduction to electron tomography. Trends Cell Biol 15: 43–51.
49. Xu P, Donaldson LA, Gergely ZR, Staehelin LA (2006) Dual-axis electron tomography: a new approach for investigating the spatial organization of wood cellulose microfibrils. Wood Sci Technol 41: 101–116.
50. Xu P, Liu H, Donaldson LA, Zhang Y (2011) Mechanical performance and cellulose microfibrils in wood with high S2 microfibril angles. J Mat Sci 46: 534–540.
51. Chundawat SPS, Donohoe BS, da Costa Sousa L, Elder T, Agarwal UP, et al. (2011) Multi-scale visualization and characterization of lignocellulosic plant cell wall deconstruction during thermochemical pretreatment. Energy Environ Sci 4: 973.
52. Ciesielski PN, Matthews JF, Tucker MP, Beckham GT, Crowley MF, et al. (2013) Electron tomography of pretreated biomass informs atomic modeling of cellulose microfibrils. ACS Nano 7: 8011–8019.
53. McDonald KL, Auer M (2006) High-pressure freezing, cellular tomography, and structural cell biology. Biotechniques 41: 137–143.
54. Schindelman G, Morikami A, Jung J, Baskin TI, Carpita NC, et al. (2001) COBRA encodes a putative GPI-anchored protein, which is polarly localized

and necessary for oriented cell expansion in Arabidopsis. Genes Develop 15: 1115–1127.
55. Roudier F, Fernandez AG, Fujita M, Himmelspach R, Borner GH, et al. (2005) COBRA, an Arabidopsis extracellular glycosyl-phosphatidyl inositol-anchored protein, specifically controls highly anisotropic expansion through its involvement in cellulose microfibril orientation. Plant Cell 17: 1749–1763.
56. Gu Y, Kaplinsky N, Bringmann M, Cobb A, Carroll A, et al. (2010) Identification of a cellulose synthase-associated protein required for cellulose biosynthesis. Proc Natl Acad Sci USA 107: 12866–12871.
57. Leunissen JL, Yi H (2009) Self-pressurized rapid freezing (SPRF): a novel cryofixation method for specimen preparation in electron microscopy. J Microscopy 235: 25–35.
58. Yakovlev S, Downing KH (2011) Freezing in sealed capillaries for preparation of frozen hydratedsections. J Microscopy 244: 235–247.
59. Yakovlev S, Luef B, Comolli LR, Downing KH (2010) Self Pressure Freezing of Caulobacter crescentus for the Preparation of Frozen Hydrated Sections. Microscopy Microanal 16: 860–861.
60. Han H-M, Huebinger J, Grabenbauer M (2012) Self-pressurized rapid freezing (SPRF) as a simple fixation method for cryo-electron microscopy of vitreous sections. J Struct Biol 178: 84–87.
61. Pierson J, Fernandez JJ, Bos E, Amini S, Gnaegi H, et al. (2010) Improving the technique of vitreous cryo-sectioning for cryo-electron tomography: electrostatic charging for section attachment and implementation of an anti-contamination glove box. J Struct Biol 169: 219–225.
62. Suloway C, Pulokas J, Fellmann D, Cheng A, Guerra F, et al. (2005) Automated molecular microscopy: the new Leginon system. J Struct Biol 151: 41–60.
63. Mastronarde DN (2005) Automated electron microscope tomography using robust prediction of specimen movements. J Struct Biol 152: 36–51.
64. Otegui MS, Mastronarde DN, Kang BH, Bednarek SY, Staehelin LA (2001) Three-dimensional analysis of syncytial-type cell plates during endosperm cellularization visualized by high resolution electron tomography. Plant Cell 13: 2033–2051.
65. Segui-Simarro JM, Austin JR, II, White EA, Staehelin LA (2004) Electron tomographic analysis of somatic cell plate formation in meristematic cells of Arabidopsis preserved by high-pressure freezing. Plant Cell 16: 836–856.
66. Donohoe BS, Kang BH, Staehelin LA (2007) Identification and characterization of COPIa- and COPIb-type vesicle classes associated with plant and algal Golgi. Proc Natl Acad Sci USA 104: 163–168.
67. Donohoe BS, Mogelsvang S, Staehelin LA (2006) Electron tomography of ER, Golgi and related membrane systems. Methods 39: 154–162.
68. Mastronarde DN (1997) Dual-axis tomography: an approach with alignment methods that preserve resolution. J Struct Biol 120: 343–352.
69. Kremer JR, Mastronarde DN, McIntosh JR (1996) Computer visualization of three-dimensional image data using IMOD. J Struct Biol 116: 71–76.
70. Frangakis AS, Hegerl R (2001) Noise reduction in electron tomographic reconstructions using nonlinear anisotropic diffusion. J Struct Biol 135: 239–250.
71. Fernandez JJ, Li S (2003) An improved algorithm for anisotropic nonlinear diffusion for denoising cryo-tomograms. J Struct Biol 144: 152–161.
72. Yu Z, Bajaj C (2008) Computational approaches for automatic structural analysis of large biomolecular complexes. IEEE/ACM Trans Comput Biol Bioinform 5: 568–582.
73. Zhang Q, Bettadapura R, Bajaj C (2012) Macromolecular structure modeling from 3D EM using VolRover 2.0. Biopolymers 97: 709–731.
74. Bajaj CL, Pascucci V, Schikore DR (1997) The contour spectrum. In Proceedings of IEEE Visualization 1997, Phoenix, AZ, 167–173.
75. Németh G, Palágyi K (2011) Topology preserving parallel thinning algorithms. Int J Imag Syst Tech 21: 37–44.
76. Al-Amoudi A, Chang JJ, Leforestier A, McDowall A, Salamin LM, et al. (2004a) Cryo-electron microscopy of vitreous sections. EMBO J 23: 3583–3588.
77. Al-Amoudi A, Norlen LP, Dubochet J (2004b) Cryo-electron microscopy of vitreous sections of native biological cells and tissues. J Struct Biol 148: 131–135.
78. Eltsov M, Dubochet J (2005) Fine structure of the *Deinococcus radiodurans* nucleoid revealed by cryoelectron microscopy of vitreous sections. J Bact 187: 8047–8054.
79. Resch GP, Brandstetter M, Pickl-Herk AM, Konigsmaier L, Wonesch VI, et al. (2011) Immersion freezing of biological specimens: rationale, principles, and instrumentation. Cold Spring Harbor protocols 2011: 778–782.
80. Dick-Perez M, Zhang Y, Hayes J, Salazar A, Zabotina OA, et al. (2011) Structure and interactions of plant cell-wall polysaccharides by two- and three-dimensional magic-angle-spinning solid-state NMR. Biochem 50: 989–1000.
81. Humphrey E (2006) Microwave processing in a modern microscopy facilty. Microscopy Microanal 12: 194.
82. Hayat MA (2000) Principles and techniques of electron microscopy: Biological applications, Ed 4th. Cambridge: Cambridge University Press. pp. 4–84.
83. Krishnamurthy KV (1999) Methods in cell wall cytochemistry. Boca Raton: CRC Press. pp. 29–150.
84. El Mansouri NE, Yuan Q, Huang F (2011) Characterization of alkaline lignins for use in phenol-formaldehyde and epoxy resins. BioResources 6: 2647–2662.
85. McDonald KL, Webb RI (2011) Freeze substitution in 3 hours or less. J Microscopy 243: 227–233.
86. McDonald KL (2014) Out with the old and in with the new: rapid specimen preparation procedures for electron microscopy of sectioned biological material. Protoplasma 251: 429–448.
87. McDonald KL (2013) Rapid Embedding Methods into Epoxy and LR White Resins for Morphological and Immunological Analysis of Cryofixed Biological Specimens. Microsc Microanal 19: 1–12.

2

Soil Temperature Determines the Reaction of Olive Cultivars to *Verticillium dahliae* Pathotypes

Rocío Calderón, Carlos Lucena, José L. Trapero-Casas, Pablo J. Zarco-Tejada, Juan A. Navas-Cortés*

Instituto de Agricultura Sostenible (IAS), Consejo Superior de Investigaciones Científicas (CSIC), Apartado 4084, Campus de Excelencia Internacional Agroalimentario, Córdoba, Spain

Abstract

Background: Development of Verticillium wilt in olive, caused by the soil-borne fungus *Verticillium dahliae*, can be influenced by biotic and environmental factors. In this study we modeled i) the combined effects of biotic factors (i.e., pathotype virulence and cultivar susceptibility) and abiotic factors (i.e., soil temperature) on disease development and ii) the relationship between disease severity and several remote sensing parameters and plant stress indicators.

Methodology: Plants of Arbequina and Picual olive cultivars inoculated with isolates of defoliating and non-defoliating *V. dahliae* pathotypes were grown in soil tanks with a range of soil temperatures from 16 to 32°C. Disease progression was correlated with plant stress parameters (i.e., leaf temperature, steady-state chlorophyll fluorescence, photochemical reflectance index, chlorophyll content, and ethylene production) and plant growth-related parameters (i.e., canopy length and dry weight).

Findings: Disease development in plants infected with the defoliating pathotype was faster and more severe in Picual. Models estimated that infection with the defoliating pathotype was promoted by soil temperatures in a range of 16 to 24°C in cv. Picual and of 20 to 24°C in cv. Arbequina. In the non-defoliating pathotype, soil temperatures ranging from 16 to 20°C were estimated to be most favorable for infection. The relationship between stress-related parameters and disease severity determined by multinomial logistic regression and classification trees was able to detect the effects of *V. dahliae* infection and colonization on water flow that eventually cause water stress.

Conclusions: Chlorophyll content, steady-state chlorophyll fluorescence, and leaf temperature were the best indicators for Verticillium wilt detection at early stages of disease development, while ethylene production and photochemical reflectance index were indicators for disease detection at advanced stages. These results provide a better understanding of the differential geographic distribution of *V. dahliae* pathotypes and to assess the potential effect of climate change on Verticillium wilt development.

Editor: Malcolm Bennett, University of Nottingham, United Kingdom

Funding: Financial support for this research was provided by Project P08-AGR-03528 from "Consejería de Economía, Innovación y Ciencia" of Junta de Andalucía and the European Social Fund (JANC), and projects AGL-2012-37521 (JANC) and AGL2012-40053-C03-01 (PJZT) from the Spanish "Ministerio de Economia y Competitividad" and the European Social Fund. R. Calderón is a recipient of research fellowship BES-2010-035511 from the Spanish "Ministerio de Ciencia e Innovación" and C. Lucena was a recipient of a JAE-DOC postdoctoral contract from "Consejo Superior de Investigaciones Científicas" (CSIC) co-funded by the European Social Fund. The funders had no role in study design, data collection and analysis, decision to publish, or preparation of the manuscript.

Competing Interests: The authors have declared that no competing interests exist.

* Email: j.navas@csic.es

Introduction

Verticillium wilt (VW) of olive (*Olea europea* L.), caused by the fungus *Verticillium dahliae* Kleb., is the most important soil-borne disease affecting olive trees worldwide [1,2] and can cause severe yield losses and plant death [3]. The disease was first observed in Italy in 1946 [4] and is now present in many Mediterranean countries and in California, USA [1]. In Spain, the spread of Verticillium wilt in olive trees has been associated with the expansion of olive cultivation and changes in cropping practices aimed at increasing yields [5]. Such changes include the use of self-rooted planting stocks to establish high-density plantings, drip irrigation, reduced or no tillage, and high inputs of fertilizer in newly cultivated soils or fertile soils [6] previously cropped with plants susceptible to *V. dahliae*, such as cotton [5].

Microsclerotia, the long-lasting surviving structures of *V. dahliae*, constitute the main potential infective inoculum of the pathogen in field soils, where it can survive for up to 15 years [7]. These structures germinate multiple times in response to root exudates [8] and favorable soil environmental conditions, forming hyphae that penetrate the plant root, grow across the root cortex, and upon reaching the xylem vessels facilitate the rapid upward spread of the pathogen by conidia transported in the transpiration stream [9]. As a result of xylem colonization by the pathogen, water flow decreases, leading to water stress [10]. Infection with *V. dahliae* in olive trees has resulted in two main disease syndromes

-defoliating (D) and non-defoliating (ND), which are induced by specific D and ND *V. dahliae* pathotypes, respectively [1,11]. The D syndrome is characterized by early drop of asymptomatic green leaves from individual twigs and branches, eventually leading to complete defoliation and necrosis [1]. These symptoms can develop from late fall to late spring [11]. Conversely, the ND syndrome comprises two symptom complexes: (i) apoplexy, a rapid and extensive dieback of twigs and branches of olive trees without loss of leaves occurring in late winter, and (ii) slow decline, mainly characterized by flower mummification and necrosis of inflorescences along with leaf chlorosis and necrosis [1], which occurs during springtime [11]. Infections with the D pathotype can be lethal to the plant, whereas ND-infected olive trees may eventually show remission from symptoms [3,12]. D and ND *V. dahliae* pathotypes also have different modes of dispersal and produce different spatial patterns of disease [11]. Infections with the D pathotype are of greater concern because the pathogen can spread rapidly over short and relatively long distances through windblown infected leaves that fall early and in large numbers from diseased trees [11]. Planting resistant cultivars is the most effective measure for controlling and limiting the spread of Verticillium wilt [1]. However, the most widely used olive cultivars in Spain (i.e., Picual and Arbequina) have been found to be highly susceptible and susceptible to D *V. dahliae*, respectively, and susceptible and moderately resistant to ND *V. dahliae*, respectively, under controlled conditions in artificial inoculation tests [13].

A recent study on Verticillium wilt in olive trees has shown that overall disease incidence is related to initial inoculum density in the soil [14]. However, the development of symptoms in relation to inoculum density is variable and strongly influenced by environmental and soil conditions [15]. Soil temperature is a critical factor for the development of Verticillium wilt and fungal growth [16]. Using soil tanks, McKeen [17] found that infection of potato by *V. albo-atrum* occurred between 12 and 32°C, but symptom expression was greatest between 20 and 28°C. In southern Spain, Bejarano-Alcázar et al. [18] reported that the optimal temperature for in vitro growth of *V. dahliae* isolates from cotton ranged from 24 to 27°C in the D pathotype and from 21 to 24°C in the ND pathotype. In China, Xu et al. [19] determined that the optimal growth temperature for *V. dahliae* isolates of the D and ND pathotype was 25°C, although D pathotype isolates can adapt well to high temperatures and severely infect cotton at temperatures ranging from 25 to 30°C. In olive plants, development of Verticillium wilt is favored by air and soil temperatures close to the optimal growth range of *V. dahliae* [12]. In Mediterranean-type climates, severity of Verticillium wilt attacks is favored by moderate air temperatures during spring, but high summer temperatures suppress further development of the disease [3,11]. Nevertheless, very little is known about the influence of the physical environment on Verticillium wilt in olive trees, which limits our understanding of the disease [1].

In plants, water stress caused either by *V. dahliae* infection or drought induces stomatal closure, which reduces the transpiration rate [10]. As a result, evaporative cooling decreases and leaf temperature increases. In the past, this increase in leaf temperature has been detected early using thermal infrared radiation [20] recorded by spectrometers at ground level. The visible part of the spectrum has also been used for early water stress detection based on indices that use bands at specific wavelengths in which photosynthetic pigments are affected by stress. An example of such indices is the photochemical reflectance index (PRI), a narrow-band spectral index [21] that is sensitive to the epoxidation state of xanthophyll-cycle pigments and to photosynthetic efficiency, serving as a proxy for water stress detection at leaf level [22]. Another indicator of water stress is chlorophyll fluorescence emission, as shown by several laboratory studies that have found it to be strongly correlated with photosynthesis and other physiological processes [23]. Over the last ten years, scientific interest in steady-state chlorophyll fluorescence (Fs) (i.e., fluorescence emitted under constant illumination without saturation flashes) has increased because measurements of Fs do not require high-energy sources and can be conducted remotely using active or passive methods [24]. In particular, leaf-level Fs measurements obtained with instruments known as pulse amplitude modulating (PAM) fluorometers have been used successfully to detect plant water stress [24]. In fact, *Minolta Corporation* developed a portable chlorophyll meter (SPAD) in the 1990s to take rapid measurements of chlorophyll content in leaves. This instrument uses two light-emitting diodes (650 and 940 nm) and a photodiode detector to measure transmission of red and infrared light through plant leaves. Given that there is a close relationship between leaf chlorophyll content and the output of the SPAD meter [25], such measurements have been used to assess stress in crops [25,26], revealing that water stress levels decrease chlorophyll content in leaves and consequently SPAD readings.

Leaf-level remote sensing studies have been conducted to detect and assess diseases in various crops. Most of these studies have focused on foliar pathogens in annual crops. Leaf temperature has been shown to successfully detect water stress caused by soil-borne pathogens, as mentioned above. In fact, Pinter et al. [27] recorded leaf temperatures 3–4°C higher than those of healthy plants in sugar beet and cotton crops affected by *Pythium aphanidermatum* and *Pymatotrichum omnivorum*, respectively. Other examples of leaf temperature measurements for the detection of root diseases include beans infected by *Fusarium solani*, *Pythium ultimum*, and *Rhizoctonia solani* [28], soybeans affected by brown stem rot caused by *Phialophora solani* [29], and wheat with moderate take-all symptoms caused by *Gaeumannomyces graminis* var. *tritici* [30]. As regards *V. dahliae* infection, Nilsson [31] reported that infected oilseed rape plants exhibited leaf temperatures 5–8°C higher than those of non-infected plants. Calderón et al. [32] found increases of 2°C at early stages of Verticillium wilt development in olive trees under field conditions. Chlorophyll fluorescence was also found to be a good indicator for detecting Verticillium wilt at early stages of the disease, while the photochemical reflectance and chlorophyll indices were good indicators for detecting the disease at advanced stages [32].

The objectives of this study were (i) to quantify the combined effects of biotic factors (i.e., pathotype virulence and cultivar susceptibility) and abiotic factors (i.e., soil temperature) on the development of Verticillium wilt and olive tree growth and (ii) to determine the relationship between stress indicators including several remote sensing parameters and VW severity according to the hypothesis that thermal, reflectance, and fluorescence measurements are sensitive to physiological changes induced by infection and colonization by *V. dahliae* pathotypes.

Materials and Methods

1. *Verticillium dahliae* isolates, olive plants, inoculation, and growth conditions

Olive plants were inoculated with *V. dahliae* isolates V138 (D) and V176 (ND), which have been characterized in previous studies [33] and are deposited in the culture collection of the *Departamento de Protección de Cultivos* (Crop Protection Department) of the *Instituto de Agricultura Sostenible* (IAS-CSIC) in Cordoba, Spain. Isolates were stored by covering cultures on plum extract agar with liquid paraffin and keeping them at 4°C in the dark. Active cultures of isolates were obtained on chlortetracycline-amended water agar (1 l of distilled water, 20 g of agar, 30 mg of chlortetracycline) and were further subcultured on Potato Dextrose Agar (PDA; Difco Laboratories, Detroit, USA). Cultures on PDA were grown for 7 days at 24°C in the dark.

Eight-month-old plants of olive cvs. Arbequina and Picual were used. Plants were obtained by micropropagation techniques and provided by Cotevisa (L'Alcudia, Valencia, Spain). Arbequina and Picual are olive cultivars grown extensively throughout Spain [34]. Picual plants were inoculated with both *V. dahliae* pathotypes, while Arbequina plants were inoculated with the D isolate only.

Inocula of *V. dahliae* were produced in an autoclaved cornmeal-sand (CMS) mixture in flasks incubated at 24±1°C in the dark for 6 weeks. Infested CMS substrate was mixed thoroughly with an autoclaved soil mixture (clay loam/sand/peat at 1:1:1 vol/vol/vol) at a rate of 1:10 (wt/wt) to reach an inoculum density of approximately 4×10^4 CFU/g of soil for each of the *V. dahliae* isolates. Non-infested CMS mixed with the autoclaved soil mixture at the same rate explained above served as control. Plants were grown in soil tanks (Frisol S.A., Córdoba, Spain) placed inside a walk-in growth chamber adjusted to 24±1°C, 40 to 70% relative humidity, and a 14-h photoperiod of fluorescent light at 360 µE $m^{-2}\cdot s^{-1}$ for 3 months. Pots with soil and plant roots were set inside the soil tanks at a constant temperature of 16, 20, 24, 28 and 32°C, with a maximum variation of ±1°C for each of temperatures. Plants were watered daily as needed and fertilized weekly with 100 ml of Hoagland's nutrient solution [35].

The experiment consisted of a three-way factorial treatment design with pathotype of *V. dahliae*, olive cultivar, and soil temperature as factors. For each soil temperature there were ten and six replicated pots (one plant per pot) for inoculated and non-inoculated plants, respectively, in a completely randomized design. The full experiment was repeated twice.

2. Disease assessment

Incidence (0 = plant showing no disease symptoms; 1 = plant showing disease symptoms) and severity of symptoms were assessed at 2- to 3-day intervals throughout the duration of the experiment, 3 months after inoculation. Disease severity was assessed by visually observing foliar symptoms in each individual plant and rating them on a 0 to 4 scale according to the percentage of foliage with disease symptoms, where 0 = 0%, 1 = 1 to 33%, 2 = 34 to 66%, 3 = 67 to 100%, and 4 = dead plant [33]. Symptoms cause by the ND pathotype consisted on a dieback of twigs and branches where leaves turned light-brown, rolled back toward the abaxial side, dried up and remained attached to the symptomatic shoots; whereas, those caused by the D pathotype characterized by early drop of green, infected leaves that eventually gave rise to complete defoliation and necrosis of branches. Upon termination of the experiments, colonization of plant tissues by *V. dahliae* was determined in each plant by isolating the fungus on water agar amended with Aureomycin (1 l of distilled water, 20 g of agar, 30 mg of aureomycin). For each plant, six 5-mm-long stem pieces representative of the lower, middle and upper parts were thoroughly washed under running tap water for 30 min. The surface of the samples was disinfested in 0.5% NaClO for 1.5 min; next, the samples were rinsed with sterile water, plated onto the medium, and incubated at 24°C in the dark for 7 days [11,33]. Colonies of *V. dahliae* were identified by microscopic observation of verticillate conidiophores and formation of microsclerotia. Data from the stem isolations of the pathogen were used to calculate the intensity of stem vascular colonization, determined as the percentage of stem pieces from which the pathogen was isolated.

3. Disease, stress, plant growth-related parameters, and data analyses

3.1. Relationship between disease development and parameters associated with disease progress curves. Disease progress curves were obtained from accumulated disease severity scores over time in days from the date of inoculation. The nonlinear form of the Gompertz model was evaluated for goodness of fit to disease severity progress data using nonlinear regression analyses. In the Gompertz equation, $DS(t) = K \exp[-B \exp(-r\,t)]$, where DS = disease severity, K = asymptote parameter, B = constant of integration, r = relative rate of disease increase, and t = time of disease assessment in days after inoculation.

To further assess disease development, four additional variables associated with disease progression were explored. These variables included (i) the incubation period (IP), established as the time in days to first symptoms or its reciprocal ($IP_R = 1/IP$); (ii) the final disease incidence and severity assessed at the end of the experiment; (iii) the standardized area under the disease severity progress curve (SAUDPC), calculated using the trapezoidal integration method standardized by duration of disease development in days [36]; (iv) the intrinsic rate of disease increase (*rho*) parameter estimates of the Gompertz model fitted to the disease severity progress data; and (v) the intensity of stem vascular colonization determined as the percentage of stem pieces from which the pathogen was isolated.

3.2. Leaf-level stress-related parameters. In addition to disease-associated parameters, several stress-related parameters were measured in six leaves of each of ten (inoculated) and six (non inoculated, control) plants for each soil temperature and olive cultivar-*V. dahliae* pathotype combination. All measurements were taken at 2-week intervals, starting 20 days after inoculation until the end of the experiment, and comprised (i) leaf temperature; (ii) leaf chlorophyll fluorescence; (iii) leaf photochemical reflectance index; (iv) leaf chlorophyll content (through SPAD readings); and (v) ethylene production. For each stress-related parameter, average daily increase was determined as the standardized area under the parameter value progress curve over the observation period, calculated with the trapezoidal method. Such increase was used to assess the relationship between stress-related parameters and soil temperature and differences between experimental treatments in such parameters. Leaf temperature and steady-state chlorophyll fluorescence measurements were conducted with the PAM-2100 pulse-amplitude modulated fluorometer (Heinz Walz GMBH, Effeltrich, Germany). Steady-state chlorophyll fluorescence was also assessed separately with a second instrument designed to measure chlorophyll fluorescence (FluorPen, Photon System Instruments, Brno, Czech Republic). In addition, measurements of the leaf photochemical reflectance index [21], calculated as $(R_{570}-R_{531})/(R_{570}+R_{531})$ [37,38,39], were obtained with a custom-designed instrument to measure the R531 and R570 spectral bands with a bandwidth of 10 nm (PlantPen, Photon System Instrument, Brno, Czech Republic). Leaf chlorophyll content was obtained with the SPAD-502 chlorophyll meter (Minolta Corp., Ramsey, NJ, USA). This instrument was used preferentially because of the strong relationship between its digital readings and real leaf chlorophyll content, as demonstrated by several authors (e.g., [40,41]). Ethylene production was determined in the two youngest fully expanded leaves in the upper part of the plant. Leaves were separated from the stem and enclosed in 3-ml test tubes containing 50 µl tap water. Tubes were sealed with rubber caps and incubated in the dark at 24°C for 24 h. Before sampling, the test tubes were stirred to favor the diffusion of ethylene gas into the water. Gas samples were withdrawn from the incubation tubes with a 1 ml gas-tight syringe and assayed with a Hewlett Packard gas chromatograph (Model 5890A), as previously described [42].

3.3. Plant growth-related parameters. At the end of the experiments, plants were removed from the soil and their roots were washed free of soil. We measured the weight of fresh roots and dry and fresh aerial plant parts as well as the length of stems and shoots of individual plants. For this latter parameter, we calculated a daily growth rate relative to the initial values.

3.4. Relationship between disease, stress and plant growth-related parameters, and soil temperature. Three functions were used to describe the effects of soil temperature on disease, stress, and plant growth-related parameters for the different olive cultivar-pathogen pathotype combinations. We used the following reverse sigmoid function to determine the relationship between disease incidence, disease severity, and intensity of stem vascular colonization:

$$Y_T = f(T) = ak^n/(a^n + T^n)$$

where Y_T is the response of disease-related parameters to soil temperature (T), a determines the maximum asymptote, and k is the half maximum parameter.

For the remaining disease-related parameters and leaf temperature measurements, we used the modified beta function [43]:

$$Y_T = f(T) = G\left[(T - T_{min})/(T_{opt} - T_{min})\right]^{[H(Topt-Tmin)/(Tmax-Topt)]} \left[(Tmax - T)/(Tmax - Topt)^H\right]$$

where Y_T is the response of disease-related parameters to soil temperature, and T_{max} and T_{min} are 36 and 8°C, respectively, which are known maximum and minimum temperatures for growth of *V. dahliae* isolates [16,18,19]. The shape parameter (*H*) determines the soil temperature range near the optimal soil temperature (T_{opt}) at which the response values are close to the maximum response (*G*). For the remaining parameters, we used a Type I combined exponential and power model:

$$Y_T = f(T) = aT^b c^T$$

where Y_T is the response of the measured variable to soil temperature, and *a, b* and *c* are parameters that closely interact to control the shape of the curve [44].

All regression analyses were conducted using the Marquardt nonlinear least-squares iterative procedure for nonlinear models (NLIN) of SAS software (version 9.3; SAS Institute, Cary, NC, USA). The coefficient of determination (R^2), mean square error, standard errors associated with the parameter estimates, confident intervals of predicted values, and pattern of standardized residuals plotted against either predicted values or the response variable were used to evaluate the appropriateness of models to describe the data [45].

3.5. Relationships between experimental treatments and disease, stress, and plant growth-related parameters. The overall response of experimental treatment combinations to disease, stress, and plant growth-related parameters was first explored by cluster analyses. To establish functional groups of correlated experimental treatments, agglomerative clustering based on the Spearman correlation matrix was performed using the Ward clustering method [46]. The optimal number of clusters was estimated on the basis of the average silhouette width according to the Mantel statistic. The number of clusters in which the within-group mean intensity of the link between the objects (i.e., experimental treatments) and their groups was highest (i.e., with the largest average silhouette width) indicated the optimal cluster number. A dendrogram was then produced representing the treatment groups identified [46]. A heat map was developed to visualize the values of the different treatments and parameters used in the analysis. All cluster analysis calculations were performed using R software, version 3.0.2 (R Foundation for Statistical Computing, http://www.R-project.org/) with the *cluster* [47], *gplots* [48] and *vegan* [49] packages.

3.6. Relationship between disease severity classes and stress-related parameters. Two approaches were used to assess the ability to discriminate among disease severity classes and stress-related parameters: logistic regression models and classification trees. Logistic models are direct probability models that are stated in terms of the probability of occurrence of an event (i.e., disease severity class) under a given set of conditions (i.e., stress parameters) [50]. In this study, a multinomial logistic regression model was fitted to each stress parameter as an independent-explanatory variable and disease severity class as the dependent-response variable, using healthy plants as the reference category. Multinomial models with ordinal responses are an extension of standard (i.e., binary) logistic regression to regression with three or more ordered categories [51]. To assess the statistical significance of each independent variable, each model was compared to the null model using a likelihood ratio test. The proportion of the variance explained by each model was evaluated using the maximum rescaled R^2 determination coefficient, and classification accuracy. This was done by using the LOGISTIC procedure of SAS software. To assess the combined effects of all stress-related variables, a multiple logistic regression model was fitted using the stepwise procedure. The developed model was externally validated by partitioning of individuals into two samples: the training sample containing 80% the data of each severity class selected at random and the testing or validation sample with the remaining 20%. The logistic model was fitted using the training sample and externally validated by using the testing sample to assess its classification accuracy.

Classification trees were used to determine the thresholds of stress parameters that discriminated between disease severity classes. The decision tree was obtained by recursive data partitioning, thereby splitting the data set into increasingly smaller subsets based on the predictive variables. The optimal tree was determined using the minimal estimate of cross-validated prediction error for different numbers of splits [52]. Residual mean deviance and misclassification error rates were used as a measure of goodness of fit of the selected tree. The analysis was conducted using the *rpart* package [53] in the R environment. The selected tree was validated by dividing the full data set into two parts and testing for classification accuracy, as described above for the logistic regression analyses.

Results

1. Verticillium wilt development

Treatment effects. Soil temperature, olive cultivar, and pathotype of *V. dahliae* were found to influence the development of Verticillium wilt in olive. Plants grown in soil infested with the D pathotype showed typical symptoms of the defoliating syndrome in the full range of soil temperatures tested, irrespective of the olive cultivar. Disease incidence, disease severity, and intensity of stem vascular colonization decreased with increasing soil temperature according to a reverse sigmoid model with asymptotic optimal values in the range of 16 to 28°C and 16 to 24°C for the interaction of the D pathotype with cvs. Picual and Arbequina, respectively (Fig. 1A–C). In these two cultivars, no significant differences ($P \geq 0.05$) existed between temperatures within the optimal range regarding levels of disease incidence or intensity of stem vascular colonization caused by the D pathotype (Fig. 1A–B). At 32°C, the three disease parameters (i.e., disease incidence, disease severity, and intensity of stem vascular colonization) decreased markedly in both cultivars, reaching 40.0%, 1.00 (on a 0–4 scale), and 44.2% in cv. Picual and 15%, 0.23, and 13.3% in cv. Arbequina, respectively (Fig. 1A–C). Disease was scarce in cv. Picual plants grown in soil infested with the ND pathotype and incubated at 16 to 28°C, and no symptoms were observed at 32°C (Fig. 1A–C). In this combination, disease incidence and intensity of stem vascular colonization was highest at 16°C (i.e., 46.7 and 45%, respectively) and decreased steadily with increasing soil temperature to 5% of stem vascular colonization at 32°C (Fig. 1C).

Time to symptom expression in olive plants was shortest at 24°C in both cultivars grown in soil infested with D *V. dahliae*. Symptoms started to develop 27 to 29 days after planting in cv. Picual and about 3 days later in cv. Arbequina; yet, symptom appearance was delayed the most at 20°C, followed by 16, 28, and 32°C (Fig. 1D). Disease development over time was adequately described by the Gompertz model ($R^2 > 0.97$; RMSE < 0.7631) for all cultivar-soil temperature combinations involving the D pathotype (Fig. 2). The increase in disease severity (DS) became asymptotic (i.e., DS>3 on a 0–4 scale) at all soil temperatures except for cv. Picual (DS = 1.0) at 32°C and cv. Arbequina at 28 and 32°C (DS = 0.77 and 0.23, respectively). Nevertheless, disease severity was always significantly greater ($P < 0.05$) in cv. Picual than in cv. Arbequina regardless of soil temperature (Fig. 1C, Fig. 2).

The beta function adequately described the effects of soil temperature on the reciprocal of the incubation period, the standardized area under the disease severity progress curve (SAUDPC), and the intrinsic rate of disease progression (*rho* parameter of the Gompertz model fitted to temporal disease severity progress). Those three disease-related parameters increased within the range of 16 to 24°C and rapidly started to decrease at 28 and 32°C (Fig. 1D–F). Values of those parameters were significantly lower ($P < 0.05$) in Arbequina plants compared to Picual plants infected with D *V. dahliae* at all soil temperatures, except for the incubation period at the extreme temperatures of 16 and 32°C, for which values were similar in the two cultivars (Fig. 1D–E).

2. Relationship between leaf-level measurements of stress-related parameters and soil temperature

Similarly to the disease-related parameters described above and for all olive cultivar-*V. dahliae* pathotype combinations, leaf temperature (estimated as the difference between leaf and mean air temperature) increased with soil temperature according to a beta model. Leaf temperature increased in the range of 16 to 28°C soil temperature and decreased at a soil temperature of 32°C (Fig. 3A). Leaf temperature in Arbequina plants infected with the D pathotype was 0.03 to 0.64°C higher than that of non-inoculated control plants at 20 to 32°C soil temperature (Fig. 3A). Conversely, there were minor differences in the leaf temperature of Picual plants infected with D or ND *V. dahliae*, although leaf temperature was higher in the most susceptible interaction – cv. Picual/D pathotype (1.47 to 2.29°C) –, decreased in plants infected with the ND pathotype (1.51 to 2.05°C), and was the lowest in control plants (Fig. 3B). Progression of the remaining four stress-related parameters with soil temperature was well described by a Type I combined exponential and power function. The photochemical reflectance index showed different relationships with soil temperature depending on the olive cultivar infected with D *V. dahliae* (Fig. 3C–D). Specifically, minimum values were measured in Arbequina plants grown at 24 and 28°C soil temperature, while the opposite occurred in Picual plants. Overall, however, both cultivars exhibited a similar range of PRI values: 0.03 to 0.04 (Fig. 3C–D). In the remaining three experimental treatment combinations, PRI values were not greatly modified by soil temperature. Values were highest in Arbequina control plants (in a range of 0.031 to 0.034), decreased to a range of 0.025 to 0.031 in the cv. Picual/ND pathotype interaction, and were lowest (0.024 to 0.026) in Picual control plants (Fig. 3C–D). Steady-state fluorescence also exhibited a different relationship with soil temperature depending on the olive cultivar. In cv. Arbequina, steady-state chlorophyll fluorescence values increased with higher soil temperatures ranging from 16 to 24°C and decreased at 28 and 32°C in both D-pathotype-infected and control plants. Yet, steady-state chlorophyll fluorescence values were lower in the control treatment group regardless of soil temperature (Fig. 3D–E). In cv. Picual, the highest steady-state chlorophyll fluorescence values were reached at the extreme 16 and 32°C soil temperatures. In this cultivar, at each soil temperature, steady-state chlorophyll fluorescence values were highest in control plants and tended to decrease in plants infected with *V. dahliae* regardless of the pathotype (Fig. 3D–E). Infection with the D pathotype had a strong effect on chlorophyll content estimated with SPAD readings. Specifically, in both olive cultivars, chlorophyll content was lowest in D *V. dahliae*-infected plants grown at 16 to 24°C (42.4 to 48.0 SPAD units), increased at 28°C (51.3 to 51.8 SPAD units), and reached the highest levels at 32°C (60.9 to 61.5 SPAD units). Chlorophyll content values exhibited minor differences between control plants of both cultivars and Picual plants infected with the ND pathotype at soil temperatures ranging from 20 to 32°C (59.7 to 66.0 SPAD units), but slightly lower values were recorded at 16°C (56.7 to 60.4 SPAD units) (Fig. 4A–B). Similarly, high ethylene production was detected mostly in olive plants infected with the D pathotype and was particularly higher in plants grown at 20 to 24°C, and in Picual plants (12.8 to 25.2 pmol g^{-1} root fresh weight h^{-1}) compared to Arbequina plants. Ethylene production was always lower and almost constant irrespective of soil temperature for the remaining treatment combinations, ranging from 7.4 to 10.4 pmol g^{-1} root fresh weight h^{-1} in the cv. Picual/D pathotype combination and from 4.2 to 9.8 pmol g^{-1} root fresh weight h^{-1} in control plants of both cultivars (Fig. 4C–D).

Plant growth, estimated by the relative rate of canopy growth and dry canopy weight, was also strongly affected by the experimental treatments. The lowest values for both growth measures were observed in plants infected with the D pathotype of *V. dahliae* at a soil temperature ranging from 16 to 24°C. At these soil temperatures, based on measurements of canopy height at the beginning and the end of the experiments, Arbequina plants exhibited a 1.5 to 1.7 rate of canopy growth. The rate of canopy growth of Picual plants was 1.3 to 1.4 times higher than on Arbequina plants (Fig. 5A, C). At the same soil temperature levels, canopy dry weight values exhibited few differences, ranging from 2.4 to 2.8 g in cv. Arbequina and from 2.3 to 2.5 g in Picual plants; this parameter was about 75 to 82% lower than in non-infected control plants (Fig. 5B, D). At a soil temperature of 28°C, both growth parameters increased by 3.4 cm/day and 3.4 g/day in cv. Arbequina, and by 2.4 cm/day and 3.4 g/day in cv. Picual.

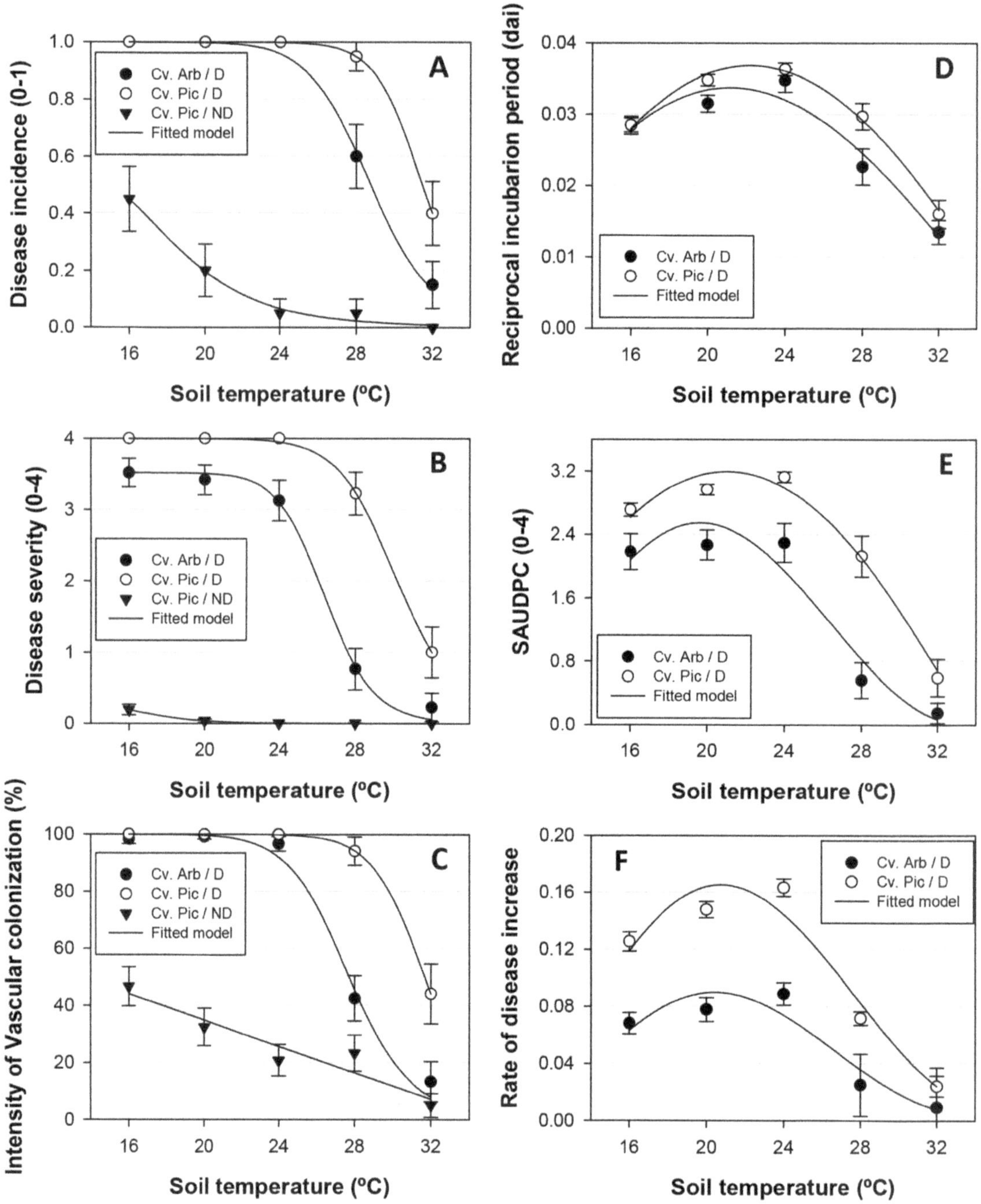

Figure 1. Relationship between Verticillium wilt-related parameters and soil temperature. Relationship between Verticillium wilt-related parameters and soil temperature in olive cvs. Arbequina (Arb) and Picual (Pic) grown in soil infested with the defoliating (D) or non-defoliating (ND) pathotype of *Verticillium dahliae*. **A.** Final disease incidence; **B.** Overall disease severity (0–4 scale: 0 = no symptoms, 4 = dead plant); **C.** Intensity of stem vascular colonization determined at the end of the experiment by isolation in growth media; **D.** Reciprocal incubation period (time to initial symptoms); **E.** Standardized area under the disease severity progress curve (SAUDPC); **F.** Rate of disease increase parameter of the Gompertz function fitted to disease progress data. Each point represents the mean of data from two repeated experiments, each comprising 10 pots with one plant per pot. Vertical bars represent the standard error of the mean. Lines represent the predicted model calculated with a reverse sigmoid function (left panels) or a Beta function (right panels).

At 32°C, maximum growth values were observed in all treatments and both cultivars. Plant growth of Picual plants infected with the ND pathotype was about 24 to 36% lower than that of control plants at a soil temperature of 16 and 20°C but minor differences were observed when plants grew at a soil temperature range of 24 to 32°C. In control treatments, plant growth tended to increase with the increase in soil temperatures but did so at a lower rate than in infected plants. Specifically, soil temperature reduced plant growth only in the lower soil temperature range of 16 and 20°C, and optimal maximum growth took place at 24 to 32°C (Fig. 5).

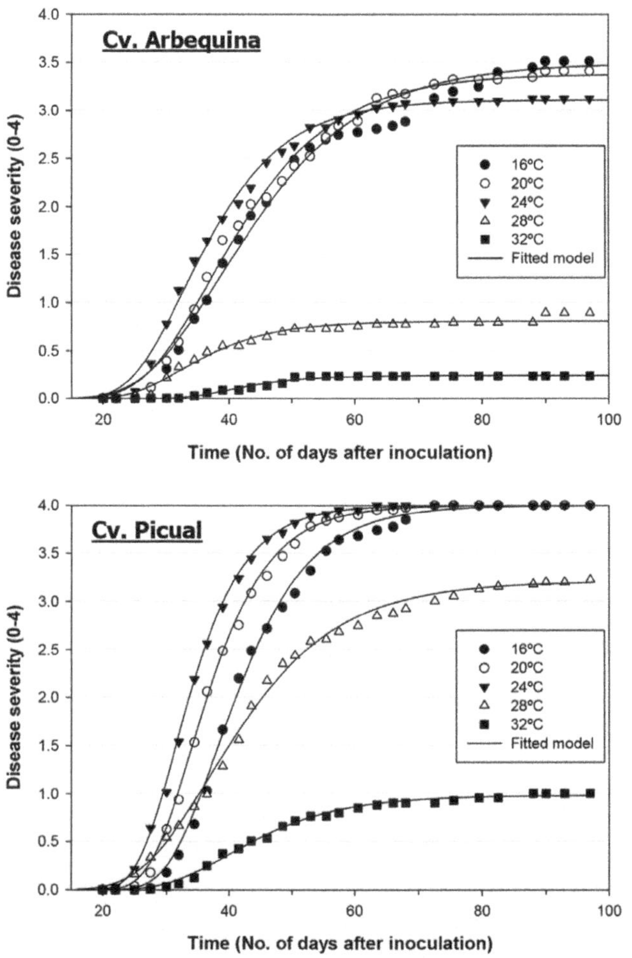

Figure 2. Verticillium wilt disease progress at different soil temperatures. Verticillium wilt disease progression in olive cvs. Arbequina and Picual grown in soil infested with the defoliating pathotype of *Verticillium dahliae* and incubated at different soil temperatures. Each point represents the mean disease severity (0–4 scale: 0 healthy, 4 = dead plant) of data from two repeated experiments, each comprising 10 pots with one plant per pot, at 2-to-3-day intervals. Solid lines represent the predicted disease progress curve calculated with the Gompertz function.

3. Relationships between experimental treatments and parameters related to disease, stress, and plant growth

To further analyze the interactions between *V. dahliae* pathotype, olive cultivar, and soil temperature, we performed a multivariate hierarchical cluster analysis, including all 12 parameters related to disease, stress, and growth as response variables. This analysis led to the creation of four functional groups – A to D – among the 25 experimental treatment combinations (Fig. 6). Group A included seven experimental treatments with a severe disease reaction, including plants infected with the highly virulent D pathotype grown at optimal soil temperature for Verticillium wilt development (i.e., 16 to 28°C for cv. Picual and of 16 to 24°C for cv. Arbequina). Overall, high values of disease severity-related parameters were associated with high levels of leaf temperature and ethylene production and low levels of chlorophyll content and growth-related parameters. In this group, Arbequina plants also exhibited high and low levels in PRI and steady-state chlorophyll fluorescence parameters, respectively; Picual plants exhibited the opposite pattern. Group B comprised four experimental treatments with a moderate disease reaction. It included Arbequina and Picual plants infected with the D pathotype and grown at 28°C and 32°C, respectively, and plants in the cv. Picual/ND pathotype combination grown at 16 or 20°C. This group of treatment combinations exhibited intermediate values in most of the parameters included in the study. Group C comprised three experimental treatments associated with a low level of disease. Specifically, plants in the cv. Arbequina/D pathotype combination grown at 32°C, and plants in the cv. Picual/ND pathotype combination grown at 24 or 28°C. The low Verticillium wilt development observed in treatment combinations within this group was associated with moderate values of stress and plant growth parameters. Group D comprised the remaining eleven treatments with healthy plants, including asymptomatic Picual plants inoculated with the ND pathotype grown at 32°C and all non-inoculated control plants of both cultivars at all soil temperature levels tested (Fig. 6).

4. Relationship between stress-related parameters and Verticillium wilt severity classes

A multinomial logistic regression analysis was performed to determine the relationship between stress-related parameters and Verticillium wilt severity classes. Logistic regression models fitted for each variable (Table 1, Fig. 7) exhibited significant differences between Verticillium wilt severity classes in all stress-related parameters. When fitted individually, chlorophyll content, ethylene production, and leaf temperature were the most explanatory parameters and had the highest correct classification rate (i.e., the Verticillium wilt severity class matched the assigned class with the highest probability) (Table 1, Fig. 7A–E). Although the PRI and steady-state chlorophyll fluorescence values were also statistically significant ($P<0.011$), their explanatory power and correct classification rate was lower (Table 1, Fig. 7E). In addition, we fitted a multiple logistic model using a stepwise procedure, selecting a model that included the four stress parameters with the highest explanatory power (i.e., chlorophyll content, ethylene production, leaf temperature, and steady-state chlorophyll fluorescence) but did not include the PRI. The model explained 98.75% of the total variability and correctly classified 73.40% of the cases (Table 1, Fig. 7F–I). Interestingly, 76 (91.6%) of the 83 healthy plants were correctly classified; of the remaining seven plants, three were classified as asymptomatic and four were classified as plants with only minor symptoms (*data not shown*), demonstrating the ability of the model to discriminate between healthy and diseased plants. Overall, in the selected model the probability of increase of Verticillium wilt symptom severity increased with growing leaf temperature and ethylene production, but the opposite occurred for chlorophyll content and steady-state chlorophyll fluorescence values. Figures 7F to I show the predicted probability distribution curves for the stress parameters included in the model keeping the remaining three stress parameters constant. The predicted probability distribution curves corresponding to Verticillium wilt severity classes from healthy to severely affected were distinct in all the parameters (Fig. 7). Moreover, chlorophyll content values showed distinct curves for all five severity classes (Fig. 7F), and ethylene production showed distinct probability curves for all Verticillium wilt severity classes except for asymptomatic and low symptom severity, which overlapped (Fig. 7G). Finally, leaf temperature (Fig. 7H) and steady-state chlorophyll fluorescence (Fig. 7I) showed distinct probability curves for healthy, asymptomatic, and low symptom severity classes, but overlapping curves for moderately or severely

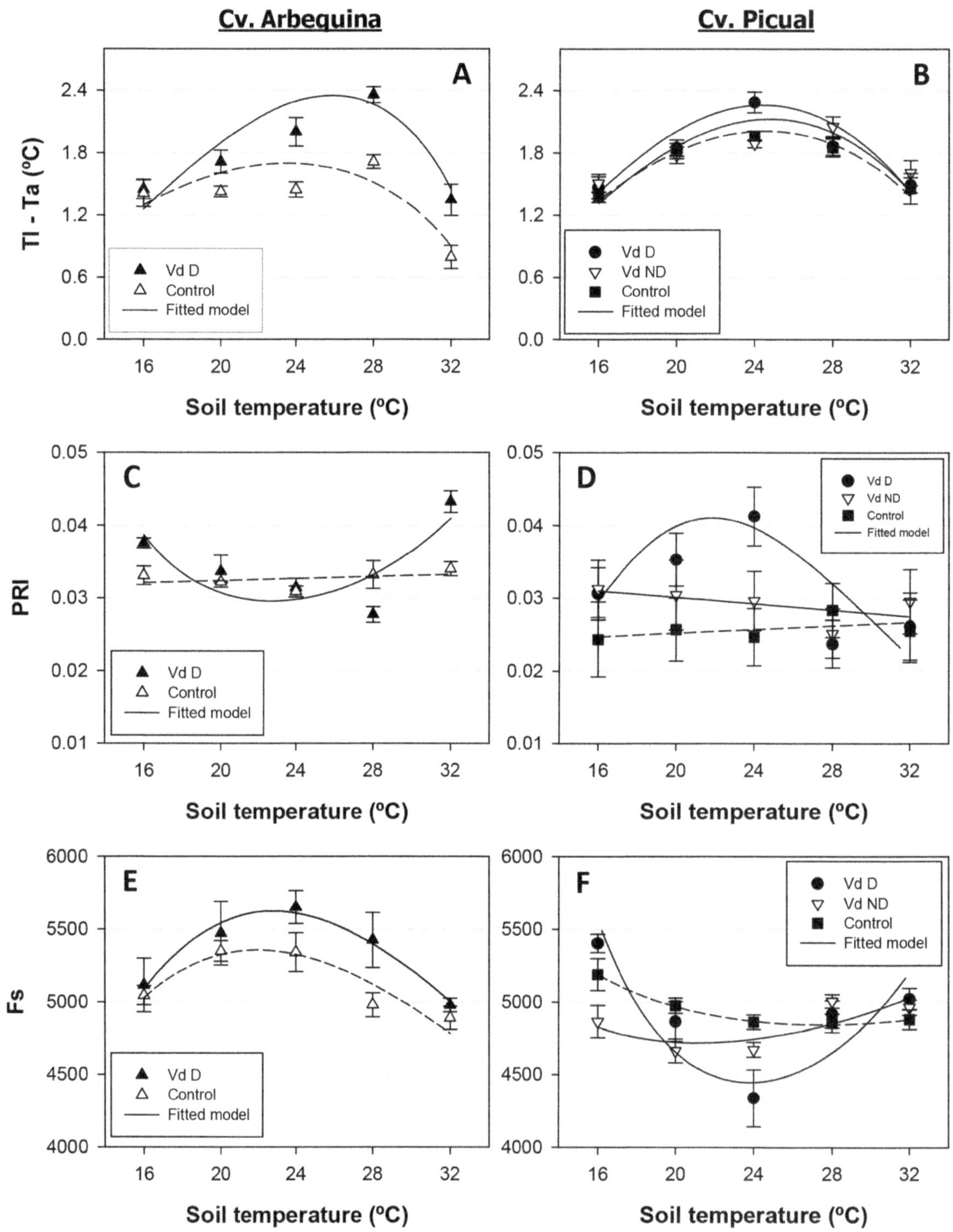

Figure 3. Relationship between stress-related parameters and soil temperature. Relationship between stress-related parameters and soil temperature in olive cvs. Arbequina (Arb) (left panels) and Picual (Pic) (right panels) grown in sterilized soil (control) or in soil infested with the defoliating (D) or the non-defoliating (ND) pathotype of *Verticillium dahliae*. **A.** Leaf temperature minus air temperature (Tl-Ta); **B.** Photochemical reflectance index (PRI); **C.** Steady-state chlorophyll fluorescence (Fs). Each point represents the mean of data from two repeated experiments, each comprising 6 pots with one plant per pot, at 2-week intervals. For each parameter, the average daily increase was calculated as the standardized area under the parameter value progress curve over the observation period. Vertical bars represent the standard error of the mean. Lines represent the predicted model calculated with a Beta function (leaf temperature minus air temperature) or a Type I combined exponential and power function.

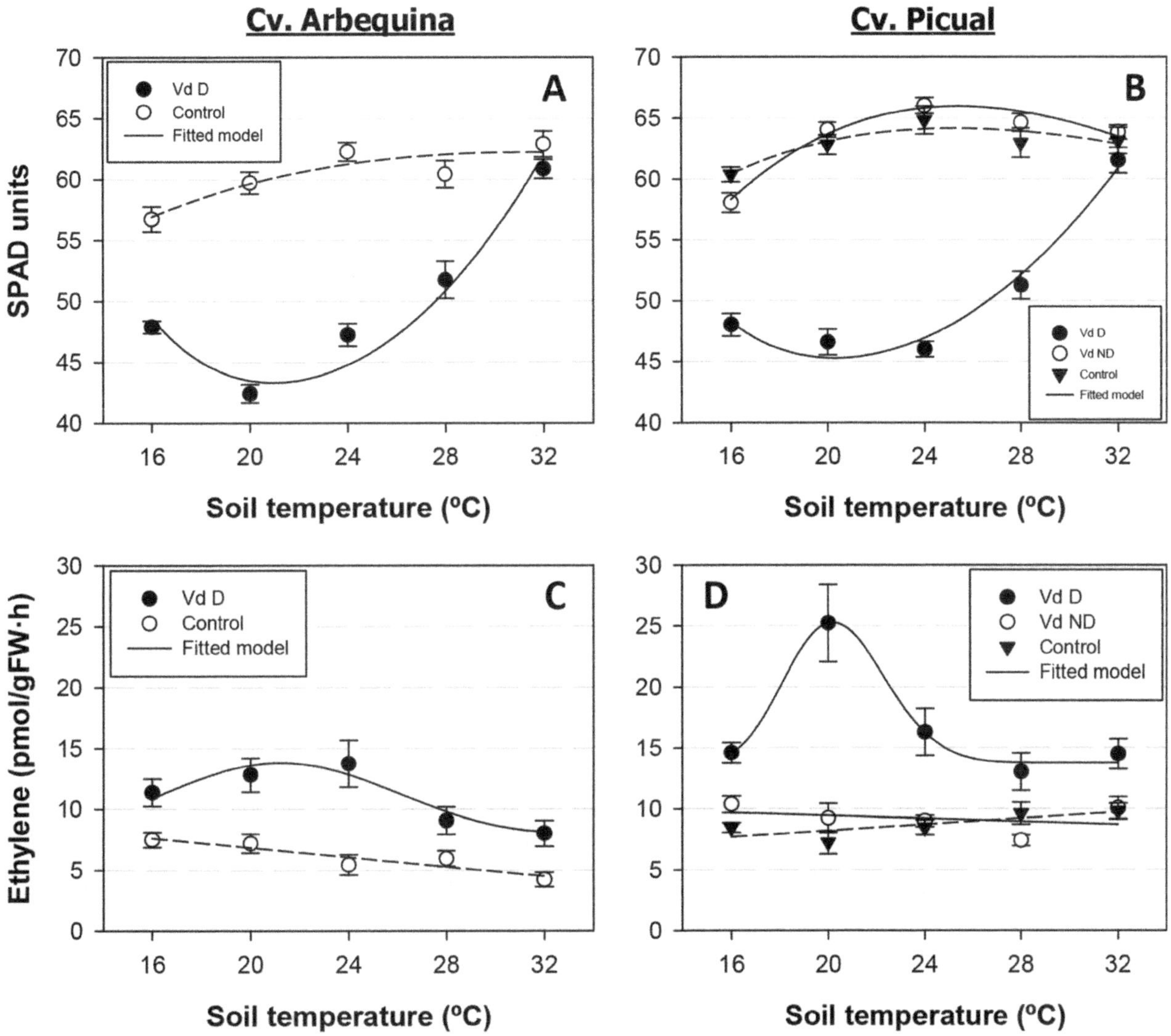

Figure 4. Relationship between stress-related parameters and soil temperature. Relationship between stress-related parameters and soil temperature in olive cvs. Arbequina (Arb) (left panels) and Picual (Pic) (right panels) grown in sterilized soil (control) or in soil infested with the defoliating (D) or the non-defoliating (ND) pathotype of *Verticillium dahliae*. **A.** Chlorophyll content (SPAD readings); **B.** Ethylene production. Each point represents the mean of data from two repeated experiments, each comprising six pots with one plant per pot, at 2-week intervals. For each parameter, the average daily increase was calculated as the standardized area under the parameter value progress curve over the observation period. Vertical bars represent the standard error of the mean. Lines represent the predicted model calculated with a Type I combined exponential and power exponential function.

affected classes. The model was validated using a test data set containing 20% of the data of the original set that had a correct classification rate of 76.60% and explained 99.81% of the variation.

5. Identification of stress-related parameter thresholds

The optimal classification tree fitted to the data had seven terminal nodes and a discrimination ability of 76.6% (Fig. 8). Chlorophyll content (44.7%), ethylene production (37.3%), and steady-state chlorophyll fluorescence (30.4%) were the most important parameters in the construction of the classification tree. Leaf temperature and the PRI accounted for a much lower importance (i.e., 18.5 and 17.7%, respectively). Chlorophyll content was the main factor (i.e., first splitting stress parameter) that differentiated between plants in the healthy and low symptom severity class and those in moderate to severely affected classes with a chlorophyll content value <53.8 SPAD units. At the second level, ethylene and leaf temperature differentiated between both groups. Specifically, healthy plants were separated from asymptomatic plants by an ethylene production threshold of 11.5 pmol g^{-1} root fresh weight h^{-1}. Plants below this threshold with a steady-state chlorophyll fluorescence >4658 were healthy, while plants with a lower fluorescence value were asymptomatic. In addition, plants with ethylene production above 11.5 pmol g^{-1} root fresh weight h^{-1} and a chlorophyll content value >58.5 SPAD units were asymptomatic and those with a lower chlorophyll content value exhibited mild symptoms. Moderate and severe Verticillium wilt classes separated by a chlorophyll content value <53.8 SPAD units were divided into two groups by leaf temperature: plants with a leaf temperature >1.9°C, which exhibited severe Verticillium wilt symptoms, and plants with a lower leaf

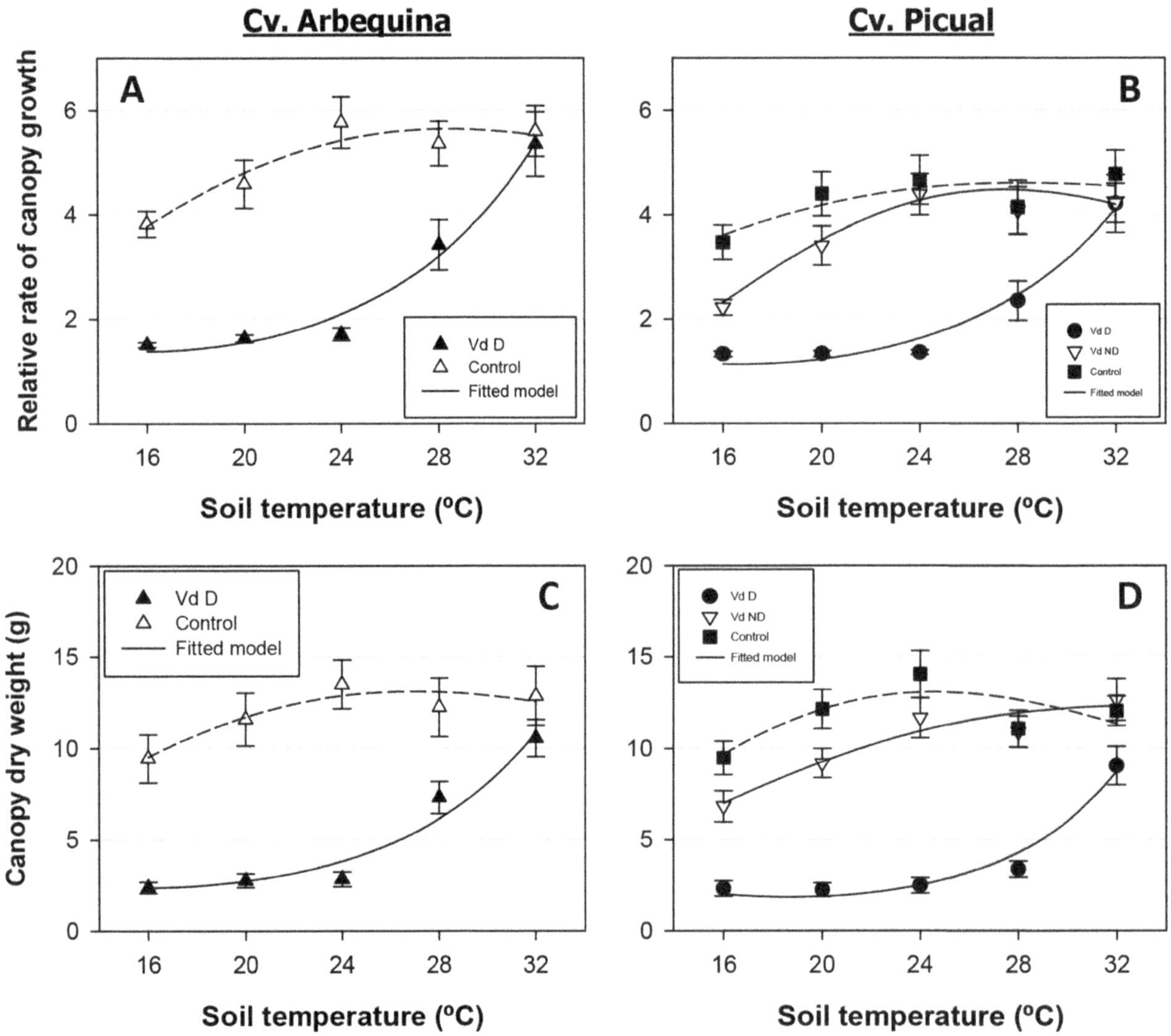

Figure 5. Relationship between plant growth-related parameters and soil temperature. Relationship between plant growth-related parameters and soil temperature in olive cvs. Arbequina (Arb) (left panels) and Picual (Pic) (right panels) grown in sterilized soil (control) or in soil infested with the defoliating (D) or the non-defoliating (ND) pathotype of *Verticillium dahliae*. **A.** Relative rate of canopy growth; **B.** Dry weight of the canopy. Each point represents the mean of data from two repeated experiments, each comprising six pots with one plant per pot, at the end of the experiments. Vertical bars represent the standard error of the mean. Lines represent the predicted model calculated with a Type I power exponential function.

temperature, which were in turn split in two new groups (i.e., plants with moderate Verticillium wilt symptoms and a steady-state chlorophyll fluorescence value <5670 and plants with severe Verticillium wilt symptoms and a steady-state chlorophyll fluorescence value above this threshold). The selected tree was validated using a test data set containing 20% of the data of the original set that was found to have a correct classification rate of 85.11%.

Discussion

The development of Verticillium wilt in various crops has been related to environmental and soil conditions [16]. Soil temperature has been found to affect the development of diseases caused by *V. dahliae* in different crops [16]. However, no studies have explored the relationship between Verticillium wilt development and soil temperature in olive trees. The main objective of this study was to explore the relationship between soil temperature and Verticillium wilt symptom development, considering the potential influence of the virulence of *V. dahliae* pathotypes and the susceptibility of the host cultivar. We also measured several parameters at leaf level to assess the stress caused by the pathogen in olive plants at different soil temperatures and the relationship between such parameters and disease severity.

In our study, the most favorable soil temperature for Verticillium wilt development caused by the *V. dahliae* D pathotype was 24°C in both olive cultivars, showing the highest values in the five disease-related parameters measured (i.e., incubation period, final disease incidence and severity, standardized area under the disease severity progress curve, intrinsic rate of disease increase, and intensity of stem vascular colonization). At 16 and 20°C soil temperatures, high values of these parameters were also observed, although they were lower than those reached at 24°C. At 28 and 32°C, such disease parameters decreased

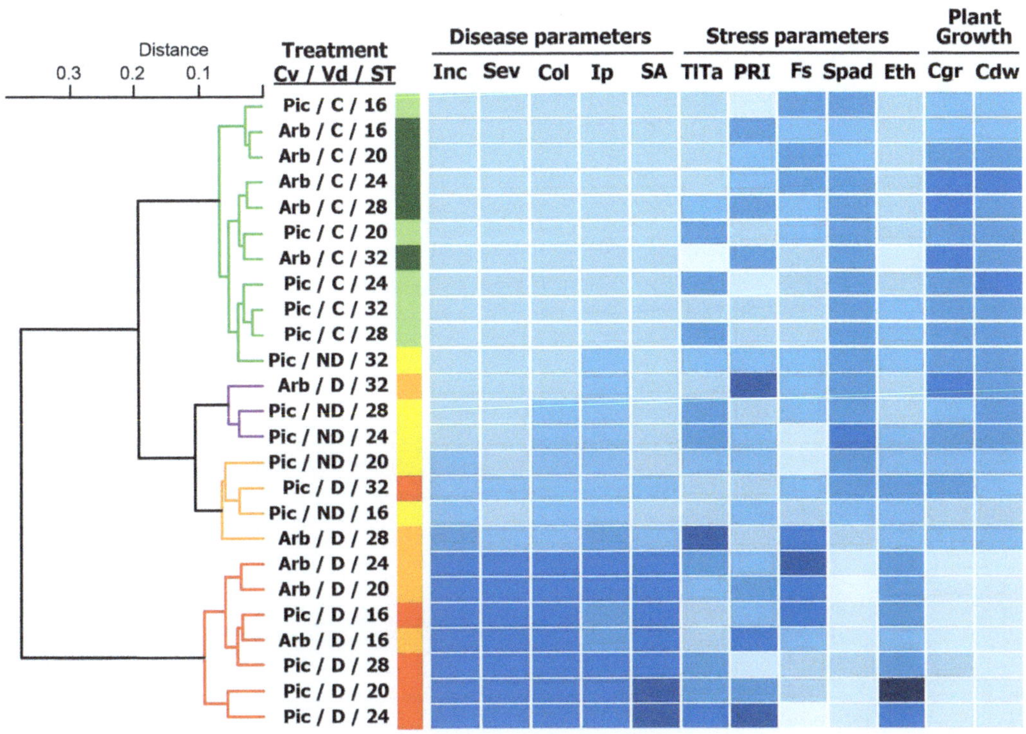

Figure 6. Dendrogram showing results of cluster analyses and heat map representation of disease, stress, and plant growth parameters. The 12 parameters selected for the heat map representation were related to Verticillium wilt reaction (5 parameters): Final disease incidence (Inc), Overall disease severity (Sev), Standardized area under disease severity progress curve (SA), Time to initial symptoms (Ip), and Intensity of stem vascular colonization (Col); related to stress (five parameters): Leaf temperature minus air temperature (TlTa), Steady-state chlorophyll fluorescence (Fs), Photochemical reflectance index (PRI), Chlorophyll content (Spad), and Ethylene production (Eth); and related to plant growth (two parameters): Relative rate of canopy growth (Cgr) and Dry canopy weight (Cdw). Agglomerative cluster analyses were performed based on the Spearman correlation matrix calculated from values of the different parameters using the Ward method. Cluster groupings of experimental treatment combinations represented in different colors were estimated on the basis of the average silhouette width according to the Mantel statistic. In the heat map, for each column, cells represent the relative value of each parameter for each experimental treatment combination of soil temperature, olive cultivar, and non-inoculated control and *Verticillium dahliae* pathotypes of the study from two repeated experiments.

dramatically, reaching the lowest values at 32°C. Our results are in close agreement with most studies on the effect of temperature on disease caused by *V. dahliae* in various host plants [16]. For example, studies conducted in cotton, tomato, and pepper infected by *V. dahliae* revealed that the optimal soil temperatures for *V. dahliae* growth were close to 20°C and temperatures higher than 28°C greatly reduced the development of the disease [16]. Temperature also interacts with the expression level of resistance of host cultivars. Specifically, severe disease symptoms caused by a highly virulent isolate can be modified by a temperature of 28°C

Table 1. Results of the multinomial logistic regression models fitted for each variable separately and the multivariate multinomial logistic regression model fitted with a stepwise method.

	LRT test[b]			
Predictor[a]	Chi-square	P> Chi-square	Max-rescaled R^2 [b]	Correct classification (%)[b]
SPAD	423.05	<0.0001	0.8964	59.57
Ethylene	122.30	<0.0001	0.4854	56.38
Tl-Ta	26.65	<0.0001	0.1358	49.47
Fs	16.78	0.0021	0.0869	47.74
PRI	13.17	0.0105	0.0888	46.28
SPAD, Ethylene, Tl-Ta, Fs	819.35	<0.0001	0.9875	73.40

[a]A multinomial logistic regression model was fitted to each of the stress parameters as an independent variable (predictor) and disease severity class as the dependent variable, using healthy plants as the reference category. To assess the combined effect of all stress-related variables, a multiple logistic regression model was fitted using the stepwise procedure. SPAD: Chlorophyll content; Ethylene: Ethylene production; Tl-Ta: Leaf temperature minus air temperature; Fs: Steady-state chlorophyll fluorescence.

[b]The likelihood ratio test (LRT), maximum rescaled R^2 determination coefficient, and correct classification rate were obtained when using the models for prediction.

Predicted probabilities for single predictor effect

Predicted probabilities for combined effects of predictors for each predictor

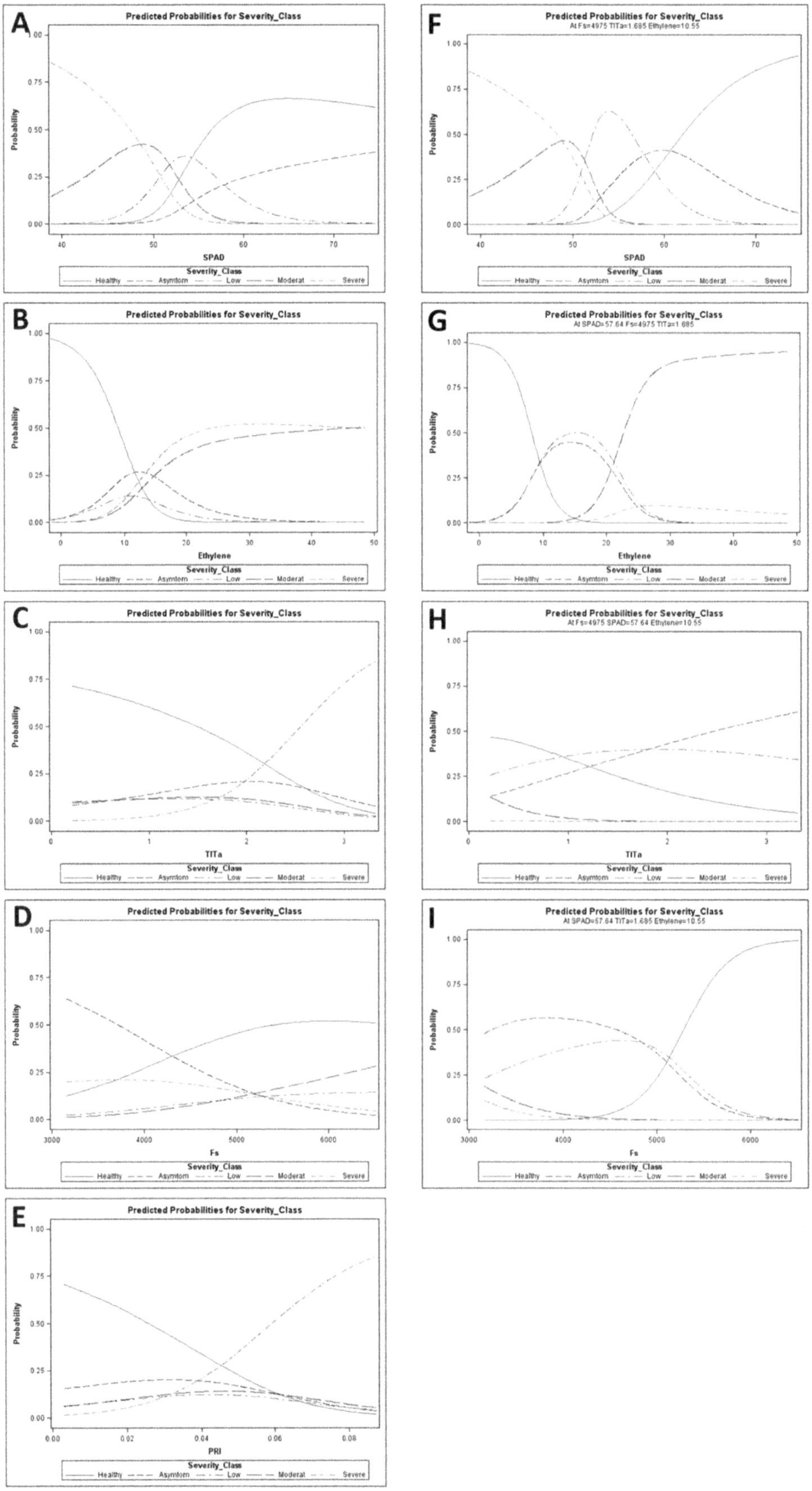

Figure 7. Predicted probabilities according to the multinomial logistic regression model with Verticillium wilt severity as the response variable and stress-related parameters as explanatory variables using healthy plants as the reference category. Left panels (**A to D**) represent the predicted probability distribution curves for each stress parameter fitted separately. Right panels (**E to I**) represent the predicted probability for each of the four stress parameters when the other three are fixed. **A, F.** Chlorophyll content (SPAD value); **B, G.** Ethylene production; **C, H.** Steady-state chlorophyll fluorescence (Fs); **D.** Photochemical reflectance index (PRI); **E, I.** Leaf temperature minus air temperature (Tl-Ta). Severity class indicates the severity of Verticillium wilt symptoms from Healthy control plants to Severe symptom development. Data include a training set of 188 plants selected at random from a total set of 235 plants in two repeated experiments comprising all experimental combinations of soil temperature, olive cultivars, and non-inoculated control and *Verticillium dahliae* pathotypes of the study.

and above to reach those of a moderately susceptible cotton cultivar [16], as occurred in our experiments with Picual plants infected with D *V. dahliae*. High temperatures delay germination of microsclerotia and therefore impair the ability of *V. dahliae* to penetrate the plant and cause disease [54,55]. Values reached by disease-related parameters were higher in cv. Picual than in cv. Arbequina, as previously found by López-Escudero et al. [13], who reported that cv. Picual was more susceptible than cv. Arbequina to the D pathotype. Moreover, stress-related parameters showed the highest differences between D-*V. dahliae*-infected plants and control plants at 24°C. The reason for this was probably that 24°C was the most favorable soil temperature for Verticillium wilt development and consequently that plants at this soil temperature suffered the highest level of stress. Leaf temperature was higher in infected plants than in control plants, and the lowest differences them were recorded at 16 and 32°C soil temperatures. The higher leaf temperatures observed in infected plants than in control plants are consistent with the results of similar studies in oilseed rape plants infected by *V. dahliae* [31] and other soil-borne pathogens [26–31]. In potato, infection with *V. dahliae* has been found to cause lower stomatal conductance, lower transpiration, and therefore higher leaf temperature [56]. In sunflower plants, however, Sadras et al. [57] did not observe changes in leaf temperature or stomatal conductance due to infection with *V. dahliae*.

The photochemical reflectance index was higher in infected plants than in control plants in the cv. Picual/D pathotype combination; by contrast, this only occurred at a soil temperature of 16 and 32°C in the cv. Arbequina/D pathotype combination. These results are consistent with those obtained by Suárez et al. [37,38] in trees subjected to water stress. Leaf chlorophyll fluorescence measurements estimated from steady-state chlorophyll fluorescence showed lower values in D pathotype infected plants of cv. Picual; in the cv. Arbequina/D pathotype combination, however, steady-state chlorophyll fluorescence values were higher in infected plants. The opposite trends observed in the two cultivars for the photochemical reflectance index and steady state fluorescence to soil temperature showed on Fig. 4C to F could be due to the differential response of both olive cultivars to infection by *V. dahliae* and soil temperature. Specifically, maximum disease severity and symptom expression is reached when Picual plants are grown at 20 to 28°C, while at these same soil temperatures disease severity and symptom expression on Arbequina plants are moderate or low. This is also supported by the minor differences showed between cultivars for both stress parameters at extreme soil temperatures (i.e., 16 and 32°C) where disease symptoms are low. The decrease in steady-state chlorophyll fluorescence in cv. Picual/D pathotype infected plants could be also expected according to previous results obtained with trees under water stress [39,58]. Depression in photosynthetic activity mainly due to drought in plants inoculated with wilting fungi has been described in several pathosystems, including potato infected with *V. dahliae* [56], tomato infected with *Verticillium albo-atrum* [59], and *Quercus ilex* infected with *Cryphonectria parasitica* [60]. Chlorophyll content in leaves estimated by SPAD readings was inversely correlated with disease severity, as previously found in potato plants infected with *V. dahliae* [61]. Similarly, SPAD values related to chlorophyll content levels were able to indicate a reduction in photosynthetic activity in tomato plants [25] and grapevines [24] under water stress.

Ethylene is a plant hormone that acts as a signaling molecule in basal plant defense responses [62] and is known to increase rapidly upon *V. albo-atrum* infection [63]. In our study, ethylene production in leaf petioles was greater in infected plants of susceptible cv. Picual than in those of moderately susceptible cv. Arbequina. This agrees with the results of Birem et al. [64], who reported higher levels of ethylene in cv. Picual than in resistant cv. Frantoio plants due to infection by D *V. dahliae*. These findings are also consistent with those of Pegg and Cronshaw [63], who reported a significant production of ethylene in internodes of a susceptible cultivar, but low or no ethylene production in a resistant tomato cultivar following infection with *V. albo-atrum*.

The most favorable soil temperature for Verticillium wilt development of the *V. dahliae* ND pathotype in cv. Picual was 16°C. From that temperature upwards, values of disease incidence and severity and intensity of stem vascular colonization progressively decreased. These values were much lower than those reached by the D pathotype, as previously found by López-Escudero et al. [13], who reported that cv. Picual was more susceptible to the D pathotype than to the ND pathotype. In agreement with such findings, the greatest differences in stress-related parameters between ND pathotype infected plants and control plants were found at 16°C and decreased with rising soil temperatures. These differences were lower than those obtained in the D pathotype, confirming that the lower Verticillium wilt development in plants infected with the ND pathotype is related to a lower stress level. In plants infected with the ND pathotype, however, chlorophyll content and ethylene production showed similar values to those of control plants, revealing that photosynthesis and ethylene production were not affected by pathogen infection. In consequence, plant growth parameters of ND-pathotype-infected plants showed almost no differences when compared to those of control plants.

According to our results, the optimal soil temperature for D pathotype infection was in a range of 16 to 24°C in cv. Picual plants and 20 to 24°C in cv. Arbequina plants. The optimal soil temperature for ND pathotype development was in a range of 16 to 20°C. Differences found in the optimal temperature range for disease development for both *V. dahliae* pathotypes are in agreement with the optimal mycelial growth for isolates of the D and ND pathotypes. Indeed, in *V. dahliae* cotton isolates from southern Spain, Bejarano-Alcázar et al. [18] estimated that the optimal temperature for in vitro growth (over a 21 to 30°C range) was 24 to 27°C for isolates of the D pathotype compared to 21 to 24°C for the ND isolates. In addition, the lower temperature optimum for the ND pathotype may explain why this *V. dahliae* pathotype tends to be geographically restricted to cooler areas of southern Spain such as Granada and Huelva provinces, whereas the D pathotype is present in most olive growing areas but is

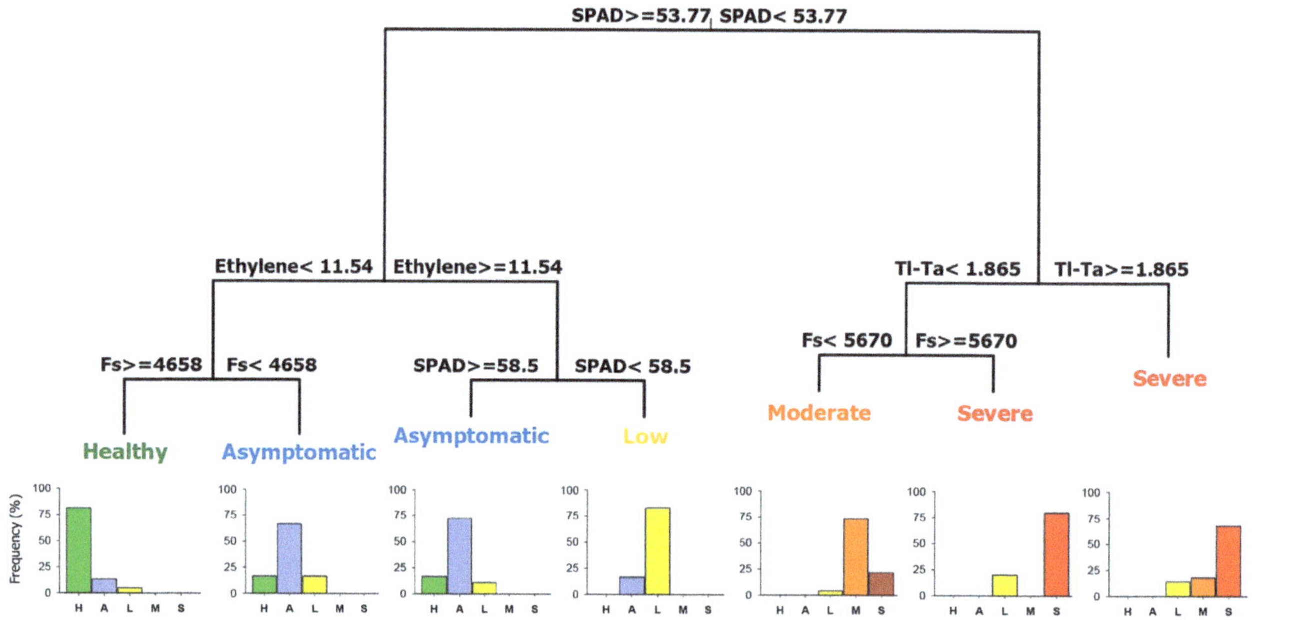

Figure 8. Classification tree to discriminate among Verticillium wilt severity classes based on stress-related parameters. Five stress-related parameters were used: Chlorophyll content (SPAD value), Ethylene production, Steady-state chlorophyll fluorescence (Fs), Leaf temperature minus air temperature (TlTa), and Photochemical reflectance index (PRI). Severity class indicates the severity of Verticillium wilt symptoms from Healthy control plants to Severe symptom development. For each terminal node the most prevalent Verticillium wilt severity class is indicated. The histogram for each terminal node represents the percentage of plants in each severity class (where H = Healthy, A = Asymptomatic, L = Low, M = Moderate and S = Severe symptoms). Data include a training set of 188 plants selected at random from a total set of 235 plants in two repeated experiments comprising all experimental combinations of soil temperature, olive cultivars, and non-inoculated control and *Verticillium dahliae* pathotypes of the study.

particularly prevalent in the warmer areas of the Guadalquivir Valley [5].

The relationship between stress-related parameters and Verticillium wilt severity classes was determined by a multinomial logistic regression analysis and a tree classification. According to the multinomial regression models fitted for each parameter separately, significant differences were found among Verticillium wilt severity classes in all stress-related parameters. The parameters with the highest explanatory power were chlorophyll content, ethylene production, leaf temperature, and steady-state chlorophyll fluorescence. Predicted probability distribution curves corresponding to chlorophyll content values were different for all Verticillium wilt severity classes, which made it possible to discriminate between each disease severity level, even at early stages of *V. dahliae* infection. However, ethylene production showed similar probability curves for asymptomatic and initial-low symptom development classes, while leaf temperature and steady-state chlorophyll fluorescence were similar in moderately and severely affected plants. In fact, ethylene production of plants at early stages of Verticillium wilt development did not differ significantly from that of plants in the asymptomatic class, while leaf temperature and steady-state chlorophyll fluorescence showed significant differences between plants in such classes. Based on the results of the multinomial regression models fitted for each parameter separately, we fitted a multiple logistic regression model including the four best stress parameters: chlorophyll content, ethylene production, leaf temperature, and steady-state chlorophyll fluorescence. The model explained 98.75% of the variance and correctly classified 73.40% of the cases. The optimal classification tree had a similar discrimination ability (76.6%), revealing that chlorophyll content, ethylene production, steady-state chlorophyll fluorescence, and leaf temperature were the most important parameters. As shown by the logistic regression analysis, chlorophyll content was the main factor that differentiated between asymptomatic and low Verticillium wilt severity classes, while ethylene production did not distinguish between them. Nevertheless, leaf temperature and steady-state chlorophyll fluorescence differentiated between moderate and severe Verticillium wilt classes; by contrast, in the logistic regression analysis, these two parameters distinguished between all Verticillium wilt severity classes except moderate and severe classes. The PRI was the parameter with the lowest explanatory power and classification rate in both classification methods. These results obtained at leaf level under controlled conditions confirmed those obtained at leaf and canopy levels under field conditions by Calderón et al. [32], who proved the potential for early detection of *V. dahliae* infection and discrimination among Verticillium wilt severity levels in olive crops using thermal, multispectral, and hyperspectral imagery acquired with an unmanned aerial vehicle. In that study, temperature and chlorophyll fluorescence were identified as the best indicators to detect Verticillium wilt at initial stages of disease development, while the photochemical reflectance and chlorophyll indices proved to be good indicators to detect moderate and severe Verticillium wilt severity classes under field conditions.

In conclusion, the optimal soil temperatures for D and ND pathotype development were 20–24°C and 16–20°C, respectively, with a drastic reduction of Verticillium wilt symptom development at soil temperatures higher than 28°C. Cv. Picual plants were more susceptible to the D than to the ND pathotype and were more affected by the D pathotype than Arbequina plants. Stress-related parameters were able to detect the effects of *V. dahliae* infection and colonization on water flow that eventually cause water stress effects. Results demonstrated that leaf temperature, physiological indices (i.e., photochemical reflectance, steady-state chlorophyll fluorescence, and chlorophyll content), and ethylene production are related to physiological stress caused by Verticillium wilt. Chlorophyll content, steady-state chlorophyll fluorescence, and leaf temperature were identified as the best indicators to detect Verticillium wilt at early stages of disease development, while ethylene production and the photochemical reflectance index were good indicators to detect Verticillium wilt at advanced stages. In addition, chlorophyll content was the parameter with the highest explanatory power and correct classification rate in the classification models used in this study, followed by ethylene production, steady-state chlorophyll fluorescence, and leaf temperature. These results will be useful to better understand the differential geographic distribution of *V. dahliae* pathotypes in southern Spain found by Jiménez-Díaz et al. [5] and to assess the potential effect of climate change on the development of Verticillium wilt of olive under different future climate change scenarios.

Acknowledgments

M. Medina is acknowledged for her support in remote sensing parameter measurements. We thank R.M. Jiménez-Díaz and B.B. Landa from IAS-CSIC for critically reading the manuscript and making valuable suggestions prior to submission.

Author Contributions

Conceived and designed the experiments: RC CL JLTC PJZT JANC. Performed the experiments: RC CL JLTC PJZT JANC. Analyzed the data: RC JANC. Contributed reagents/materials/analysis tools: JANC PJZT. Wrote the paper: RC PJZT JANC.

References

1. Jiménez-Díaz RM, Cirulli M, Bubici G, Jiménez-Gasco LM, Antoniou PP, et al. (2012) Verticillium wilt, a major threat to olive production: Current status and future prospects for its management. Plant Dis 96: 304–329.
2. Tsror L (Lahkim) (2011) Review: Epidemiology and control of Verticillium wilt on olive. Israel J Plant Sci 59: 59–69.
3. Levin AG, Lavee S, Tsror L (2003) Epidemiology of *Verticillium dahliae* on olive (cv. Picual) and its effects on yield under saline conditions. Plant Pathol 52: 212–218.
4. Ruggieri G (1946) A new disease of olive. L'Italia Agricola 83: 369–372.
5. Jiménez-Díaz RM, Olivares-García C, Landa BB, Jiménez-Gasco MM, Navas-Cortés JA (2011) Region-wide analysis of genetic diversity in *Verticillium dahliae* populations infecting olive in southern Spain and agricultural factors influencing the distribution and prevalence of vegetative compatibility groups and pathotypes. Phytopathology 101: 304–315.
6. Villalobos FJ, Testi L, Hidalgo J, Pastor M, Orgaz F (2006) Modelling potential growth and yield of olive (*Olea europaea* L.) canopies. Eur J Agron 24: 296–303.
7. Wilhelm S (1955) Longevity of the Verticillium wilt fungus in the laboratory and field. Phytopathology 45: 180–181.
8. Schreiber LR, Green RJ Jr (1963) Effect of root exudates on germination of conidia and microsclerotia of *Verticillium albo-atrum* inhibited by the soil fungistatic principle. Phytopathology 53: 260–264.
9. Talboys PW (1962) Systemic movement of some vascular pathogens. Trans Br Mycol Soc 45: 280–281.
10. Ayres PG (1978) Water relations of diseased plants. In: Kozlowski TT, editor. Water Deficits and Plant Growth, vol. 5. London, United Kingdom: Academic Press. pp. 1–60.
11. Navas-Cortés JA, Landa BB, Mercado-Blanco J, Trapero-Casas JL, Rodríguez-Jurado D, et al. (2008) Spatiotemporal analysis of spread of *Verticillium dahliae* pathotypes within a high tree density olive orchard in southern Spain. Phytopathology 98: 167–180.
12. Jiménez-Díaz RM, Tjamos EC, Cirulli M (1998) Verticillium wilt of major tree hosts: Olive. In: Hiemstra JA, Harris DC, editors. A Compendium of Verticillium Wilt in Tree Species. Wageningen, The Netherlands: Ponsen and Looijen. pp. 13–16.
13. López-Escudero FJ, del Río C, Caballero JM, Blanco-López MA (2004) Evaluation of olive cultivars for resistance to *Verticillium dahliae*. Eur J Plant Pathol 110: 79–85.
14. López-Escudero FJ, Blanco-López MA (2007) Relationship between the inoculum density of *Verticillium dahliae* and the progress of Verticillium wilt of olive. Plant Dis 91: 1372–1378.
15. DeVay JE, Forrester LL, Garber RH, Butterfield EJ (1974) Characteristics and concentration of propagules of *Verticillium dahliae* in air-dried field soils in relation to the prevalence of Verticillium wilt in cotton. Phytopathology 64: 22–29.
16. Pegg GF, Brady BL (2002) Verticillium Wilts, Wallingford: CABI Publishing. 432 p.
17. McKeen CD (1943) A study of some factors affecting the pathogenicity of *Verticillium albo-atrum* R. & B. Can J Res 21: 95–117.
18. Bejarano-Alcázar J, Blanco-López MA, Melero-Vara JM, Jiménez-Díaz RM (1996) Etiology, importance, and distribution of Verticillium wilt of cotton in southern Spain. Plant Dis 80: 1233–1238.
19. Xu F, Yang L, Zhang J, Guo X, Zhang X, et al. (2012) Effect of temperature on conidial germination, mycelial growth and aggressiveness of the defoliating and nondefoliating pathotypes of *Verticillium dahliae* from cotton in China. Phytoparasitica 40: 319–327.
20. Jackson RD, Pinter PJ Jr (1981) Detection of water stress in wheat by measurement of reflected solar and emitted thermal IR radiation. In: Spectral signatures of objects in remote sensing. Versalles, France: Institut National de la Reserche Agronomique. pp. 399–406.
21. Gamon JA, Peñuelas J, Field CB (1992) A narrow-wave band spectral index that tracks diurnal changes in photosynthetic efficiency. Remote Sens Environ 41: 35–44.
22. Thenot F, Méthy M, Winkel T (2002) The photochemical reflectance index (PRI) as a water-stress index. Int J Remote Sens 23: 5135–5139.
23. Papageorgiu G (1975) Chlorophyll fluorescence; an intrinsic probe of photosynthesis. In: Govindjee, editor. Bioenergetics of Photosynthesis. New York, USA: Academic Press. pp. 319–371.
24. Flexas J, Briantais JM, Cerovic Z, Medrano H, Moya I (2000) Steady-state and maximum chlorophyll fluorescence responses to water stress in grapevine leaves: A new remote sensing system. Remote Sens Environ 73: 282–297.
25. Fanizza G, Ricciardi L, Bagnulo C (1991) Leaf greenness measurements to evaluate water stressed genotypes in *Vitis vinifera*. Euphytica 55: 27–31.
26. Hayat S, Hasan SA, Fariduddin Q, Ahmad A (2008) Growth of tomato (*Lycopersicon esculentum*) in response to salicylic acid under water stress. J Plant Interact 3: 297–304.
27. Pinter PJ, Stanghellini ME, Reginato RJ, Idso SB, Jenkins AD, et al. (1979) Remote detection of biological stresses in plants with infrared thermometry. Science 205: 585–587.
28. Tu JC, Tan CS (1985) Infrared thermometry for determination of root rot severity in bean. Phytopathology 75: 840–844.
29. Mengistu A, Tachibana H, Epstein AH, Bidne KG, Hatfield JD (1987) Use of leaf temperature to measure the effect of brown stem rot and soil moisture stress and its relation to yields of soybeans. Plant Dis 71: 632–634.
30. Nilsson HE (1991) Hand-held radiometry and IR-thermography of plant diseases in field plot experiments. Int J Remote Sens 12: 545–557.
31. Nilsson HE (1995) Remote sensing and image analysis in plant pathology. Annu Rev Phytopathol 15: 489–527.
32. Calderón R, Navas-Cortés JA, Lucena C, Zarco-Tejada PJ (2013) High-resolution airborne hyperspectral and thermal imagery for early detection of *Verticillium* wilt of olive using fluorescence, temperature and narrow-band spectral indices. Remote Sens Environ 139: 231–245.
33. Mercado-Blanco J, Rodríguez-Jurado D, Parrilla-Araujo S, Jiménez-Díaz RM (2003) Simultaneous detection of the defoliating and nondefoliating *Verticillium dahliae* pathotypes in infected olive plants by duplex, nested polymerase chain reaction. Plant Dis 87: 1487–149.
34. Barranco D, Fernández-Escobar R, Rallo L (2004) El cultivo del olivo (The olive tree crop), 5th edition. Madrid, Spain: Ediciones Mundi-Prensa. 800 p.
35. Hoagland DR, Arnon DI (1950) The water culture method for growing plants without soil. California Agricultural Experiment Station, Circular No. 347.
36. Simko I, Piepho H-P (2011) The area under the disease progress stairs: calculation, advantage, and application. Phytopathology 102: 381–389.
37. Suárez L, Zarco-Tejada PJ, Sepulcre-Cantó G, Pérez-Priego O, Miller JR, et al. (2008) Assessing canopy PRI for water stress detection with diurnal airborne imagery. Remote Sens Environ 112: 560–575.
38. Suárez L, Zarco-Tejada PJ, Berni JAJ, González-Dugo V, Fereres E (2009) Modelling PRI for water stress detection using radiative transfer models. Remote Sens Environ 113: 730–740.
39. Zarco-Tejada PJ, González-Dugo V, Berni JAJ (2012) Fluorescence, temperature and narrow-band indices acquired from a UAV for water stress detection using a hyperspectral imager and a thermal camera. Remote Sens Environ 117: 322–337.
40. Yadava UL (1986) A rapid and nondestructive method to determine chlorophyll in intact leaves. HortScience 21: 1449–1450.
41. Marquard RD, Tipton JL (1987) Relationship between extractable chlorophyll and an in situ method to estimate leaf greenness. HortScience 22: 1327.
42. Romera FJ, Alcántara E, De la Guardia MD (1999) Ethylene production by Fe-deficient roots and its involvement in the regulation of Fe-deficiency stress responses by strategy I plants. Ann Bot 83: 51–55.

43. Hau B (1988) Ein erweitertes analytisches modell für epidemien von pflanzenkrankheiten (An extended analytical model for epidemics of plant diseases). Giessen, Germany: Justus-Liebig-Universität. 183 p.
44. Sit V, Poulin-Costello M (1994) Catalogue of curves for curve fitting. Biometrics Information Handbook Series, 4. Victoria, Canada: Forest Science Research Branch, Ministry of Forests. 116 p.
45. Madden LV, Hughes G, van den Bosch F (2007) The Study of Plant Disease Epidemics. St. Paul, MN, USA: APS Press, The American Phytopathological Society. 421 p.
46. Borcard D, Gillet F, Legendre P (2011) Numerical Ecology with R. New York, USA: Springer. 306 p.
47. Maechler M, Rousseeuw P, Struyf A, Hubert M, Hornik K (2013) cluster: Cluster Analysis Basics and Extensions. R package version 1.14.4.
48. Warnes GR, Bolker B, Bonebakker L, Gentleman R, Liaw WHA, et al. (2013) gplots: Various R programming tools for plotting data. R package version 2.12.1.
49. Oksanen J, Blanchet FG, Kindt R, Legendre P, Minchin PR, et al. (2013) vegan: Community Ecology Package. R package version 2.0–10.
50. Hosmer DW, Lemeshow S (2000) Applied Logistic Regression. 2nd ed. New York, USA: John-Wiley & Sons, Inc. 416 p.
51. Agresti A (2007) An Introduction to Categorical Data Analysis 2nd ed. Hoboken, NJ, USA: John Wiley & Sons, Inc. 372 p.
52. Everitt BS, Hothorn T (2010) A Handbook of Statistical Analysis Using R 2nd ed. Boca Raton, FL: Chapman and Hall/CRC Press. 348 p.
53. Therneau T, Atkinson B, Ripley B (2013) rpart: Recursive Partitioning. R package version 4.1–4.
54. Pullman GS, DeVay JE, Garber RH, Weinhold AR (1981) Soil solarization: effects on Verticillium wilt of cotton and soilborne populations of *Verticillium dahliae*, *Pythium* spp., *Rhizoctonia solani*, and *Thielaviopsis basicola*. Phytopathology 71: 954–959.
55. Tjamos EC, Fravel DR (1995) Detrimental effects of sublethal heating and *Talaromyces flavus* on microsclerotia of *Verticillium dahliae*. Phytopathology 85: 388–392.
56. Bowden RL, Rouse DL (1991) Effects of *Verticillium dahliae* on gas exchange of potato. Phytopathology 81: 293–301.
57. Sadras VO, Quiroz F, Echarte L, Escande A, Pereyra VR (2000) Effect of *Verticillium dahliae* on photosynthesis, leaf expansion and senescence of field-grown sunflower. Ann Bot 86: 1007–1015.
58. Zarco-Tejada PJ, Berni JAJ, Suárez L, Sepulcre-Cantó G, Morales F, et al. (2009) Imaging chlorophyll fluorescence from an airborne narrow-band multispectral camera for vegetation stress detection. Remote Sens Environ 113: 1262–1275.
59. Lorenzini G, Guidi L, Nali C, Ciompi S, Soldatini GF (1997) Photosynthetic response of tomato plants to vascular wilt diseases. Plant Sci 124: 143–152.
60. El Omari B, Fleck I, Aranda X, Moret A, Nadal M (2001) Effect of fungal infection on leaf gas-exchange and chlorophyll fluorescence in *Quercus ilex*. Ann For Sci 58: 165–174.
61. Gamliel A, Grinstein A, Peretz Y, Klein L, Nachmias A, et al. (1997) Reduced dosage of methyl bromide for controlling Verticillium wilt of potato in experimental and commercial plots. Plant Dis 81: 469–474.
62. Fradin EF, Thomma BPHJ (2006) Physiology and molecular aspects of Verticillium wilt diseases caused by *V. dahliae* and *V. albo-atrum*. Mol Plant Pathol 7: 71–86.
63. Pegg GF, Cronshaw DK (1976) Ethylene production in tomato plants infected by *Verticillium albo-atrum*. Physiol Plant Pathol 8: 279–295.
64. Birem F, Alcántara E, Blanco-López MA, López-Escudero FJ (2009) Physiologycal differences expressed by susceptible and resistant olive cultivars inoculated with *Verticillium dahliae*. In: 10th International Verticillium Symposium, Book of Abstracts. Corfu Island, Hellas, p. 71.

3

Copper-Deficiency in *Brassica napus* Induces Copper Remobilization, Molybdenum Accumulation and Modification of the Expression of Chloroplastic Proteins

Vincent Billard[1,2,3], Alain Ourry[1,2,3], Anne Maillard[1,2,3], Maria Garnica[4], Laurent Coquet[5], Thierry Jouenne[5], Florence Cruz[6], José-Maria Garcia-Mina[4], Jean-Claude Yvin[6], Philippe Etienne[1,2,3]*

1 Normandie Université, Caen, France, **2** UNICAEN, UMR 950 Ecophysiologie Végétale, Agronomie et nutritions N, C, S, Caen, France, **3** INRA, UMR 950 Ecophysiologie Végétale, Agronomie et nutritions N, C, S, Caen, France, **4** Timac Agro Spain, Poligono de Arazuri-Orcoyen, Orcoyen, Spain, **5** Plateforme de protéomique PISSARO, UMR6270 CNRS Faculté des Sciences de Rouen, Mont-Saint-Aignan, France, **6** Centre de Recherche International en Agroscience, CRIAS-TAI, Groupe Roullier, Dinard, France

Abstract

During the last 40 years, crop breeding has strongly increased yields but has had adverse effects on the content of micronutrients, such as Fe, Mg, Zn and Cu, in edible products despite their sufficient supply in most soils. This suggests that micronutrient remobilization to edible tissues has been negatively selected. As a consequence, the aim of this work was to quantify the remobilization of Cu in leaves of *Brassica napus* L. during Cu deficiency and to identify the main metabolic processes that were affected so that improvements can be achieved in the future. While Cu deficiency reduced oilseed rape growth by less than 19% compared to control plants, Cu content in old leaves decreased by 61.4%, thus demonstrating a remobilization process between leaves. Cu deficiency also triggered an increase in Cu transporter expression in roots (*COPT2*) and leaves (*HMA1*), and more surprisingly, the induction of the *MOT1* gene encoding a molybdenum transporter associated with a strong increase in molybdenum (Mo) uptake. Proteomic analysis of leaves revealed 33 proteins differentially regulated by Cu deficiency, among which more than half were located in chloroplasts. Eleven differentially expressed proteins are known to require Cu for their synthesis and/or activity. Enzymes that were located directly upstream or downstream of Cu-dependent enzymes were also differentially expressed. The overall results are then discussed in relation to remobilization of Cu, the interaction between Mo and Cu that occurs through the synthesis pathway of Mo cofactor, and finally their putative regulation within the Calvin cycle and the chloroplastic electron transport chain.

Editor: Paul A. Cobine, Auburn University, United States of America

Funding: VB's PhD grant was funded by the Regional Council of Basse-Normandie (http://www.region-basse-normandie.fr/) and Timac Agro International (http://www.timacagro.fr/). Regional Council of Basse-Normandie had no role in study design, data collection and analysis, decision to publish, or preparation of the manuscript. Timac Agro Internationnal had a role in study design and analysis.

Competing Interests: The authors have declared that no competing interests exist.

* Email: philippe.etienne@unicaen.fr

Introduction

While N, P, K and more recently S fertilizations are common practices in agriculture, the supply of micronutrients is less frequently considered. However, according to the World Health Organization [1], two billion people around the world suffer from micronutrient deficiencies (mostly Fe, Cu, Zn and Mg), causing 7.3% of disease burden. Among the micronutrients involved in human metabolism, Cu deficiency, for example, causes symptoms of diseases such as immune defects or anaemia [2]. Without stock or sufficient intake, undernourished children are the first affected [3], however Cu deficiency is also observed in developed countries [4,5]. In the human diet, Cu is mainly present in meat (especially liver and offal), fish and nuts [6]. However, these food supplies are not available to most populations suffering from micronutrients deficiencies. As a consequence, staple foods (cereals, tubers and oils) must be fortified in order to provide the minimum dose of essential micronutrients, such as Cu, for these populations [7].

Over the past 60 years, the micronutrient content (mostly Fe, Zn, Mg and Cu) has been reduced in edible products despite sufficient micronutrient availability in most cultivated soils, and this is the result of varietal selection that aimed to achieve higher yields [8,9]. Consequently, whether a genetic improvement of micronutrient uptake by roots would increase the edible plant content remains questionable. Alternatively, improving the transfer of such micronutrients in edible parts from remobilization of vegetative tissue could be an alternative for breeding if we assume that the recycling (i.e. remobilization) of such nutrients is a significant process. In addition, it has been shown that some soils can lack Cu resulting in dramatic agricultural effects [10]. Thus, Cu deficiency in plants affects human health either directly (decrease in yield) or indirectly (symptoms induced by a lack of micronutrients) [11,12].

As a transition metal [13], Cu is involved in numerous processes in plants. Most of Cu's functions in plants rely on enzymatically bound Cu (more than one hundred proteins identified so far [14]), mostly catalysing redox reactions. As a consequence, only 2% of plant Cu occurs in its free form. Chloroplasts contain about half of the total plant Cu, where it is especially associated with proteins involved in the electron transport chain such as plastocyanin [15]. Three Cu ions are also required for assembly of the active cytochrome c oxydase complex in the mitochondrial electron transport chain. Additionally, numerous proteins known to bind Cu or to be regulated by Cu are involved in the Calvin cycle or in the Tricarboxylic Acid Cycle (TCA), for example Fructose 1–6 Bis Phosphatase (FBPase), Glyceraldehyde 3 Phosphate DeHydrogenase (G3PDH) [16] and PhosphoGlycerate Kinase (PGK) [17]. Cu is also important for enzymes involved in cell detoxification, such as glutathione-S-transferase [18] or Cu-Zn superoxide dismutase [19], or in cell wall metabolism such as ascorbate and polyphenol oxidases [20].

From a genetic point of view, some studies have reported a high variation in Cu content between accessions of the same species such as spinach, pea [21] and cassava [22]. While it has been suggested that Cu management should be easily improved by breeding [4], mechanisms that could increase the Cu content in edible tissues, such as remobilization, have been poorly described in the literature.

An important agricultural crop, such as *Brassica napus*, usually requires high levels of nutrients and hence fertilization to reach an optimal yield with preserved quality. Indeed, winter oilseed rape is highly sensitive to sulphur (S) and nitrogen (N) deficiencies with negative consequences for yield and seed quality [23] and therefore, requires high doses of N and S fertilizers. These strong needs for fertilizers are partly due to the low (macro)-nutrient use efficiency (defined by the ratio of seed to plant nutrient contents) of oilseed rape. For example, it has been reported that its low N use efficiency results from a relatively inefficient endogenous N mobilization [24,25] that occurs mostly during leaf senescence [26]. In contrast, S remobilization from leaves to the seeds (linked to induction of tonoplastic sulphate transporters allowing the mobilization of sulphate previously stored in leaves [27]) is not triggered by leaf senescence but rather by S deficiency. Therefore, without S deficiency, the S content in fallen leaves remains high, up to 2% of dry weight. According to these studies, *Brassica napus* constitutes an important model of defective agro-environmental interactions in studies of nutrient remobilization from leaves. Moreover, it constitutes a relevant plant species model because its genetic proximity to *Arabidospis thaliana* provides easy access to molecular tools while maintaining an agronomic relevance.

To date, deciphering of the plant response to Cu deficiency has been performed through Fe and Zn interactions, identification and characterization of genes encoding transporters (such as *COPT* and *HMA* families), and monitoring target proteins and metabolites [28–31]. However, to our knowledge, the effect of Cu deficiency on growth and plant metabolism is currently poorly reported. Indeed, only one article [32] describes the use of a microarray approach to study modifications of the *Arabidopsis thaliana* transcriptome in response to Cu deficiency.

The first aim of this study was to quantify the effect of Cu deficiency on growth of *Brassica napus* and to assess the mobility of Cu between plant tissues (i.e. Cu remobilization). Moreover, a wider analysis of plant nutrients has been used to monitor the effect of Cu-deficiency on the uptake of some macro (Ca, K, Mg, N, P and S) and micronutrients (B, Fe, Mn, Mo, Na and Zn). Finally, a molecular approach combining proteomic and targeted transcriptomic studies has been used to identify the main metabolic pathways affected by Cu deficiency.

Materials and Methods

Growth conditions

Seeds of *Brassica napus* var. Boheme were surface-sterilized by exposure to 80% ethanol for 30 s followed by 20% sodium hypochlorite for 10 min. After 10 washes with demineralised water, seeds were germinated on perlite over demineralised water for 2 days in the dark and 1 week under natural light in a greenhouse. Just after first leaf emergence, seedlings were transferred to a 20 L tank containing the following nutrient solution: KNO_3 1.25 mM, $Ca(NO_3)_2$ 1.25 mM, KH_2PO_4 0.25 mM, $MgSO_4$ 0.5 mM, EDTA, 2 NaFe 0.2 mM, H_3BO_3 0.01 mM, $MnSO_4$ 5 µM, $ZnSO_4$ 3 µM, $(NH_4)_6Mo_7O_{24}$ 0.7 µM, $CoCl_2$ 0.1 µM, $NiCl_2$ 0.04 µM, SiO_2 0.1 mM, $CaCl_2$ 20 1.25 mM, KCl 0.25 mM. Control nutrient solution also contained NaOH 0.1 mM and $CuSO_4$ 0.7 µM while Na_2SO_4 1 µM was added to Cu-depleted nutrient solution. These nutrient solutions were renewed every two days. Plants were grown under greenhouse conditions with a thermoperiod of 20°C/17°C day/night and a photoperiod of 16 h. Natural light was supplemented with high pressure sodium lamps (Philips, MASTER GreenPower T400W) supplying an average photosynthetically active radiation of 280 µmol photons.$m^{-2}.s^{-1}$ at canopy height. After one week of growth, plants were separated into 2 sets: control plants receiving normal nutrient solution and Cu-deficient plants receiving the Cu-depleted nutrient solution (for details, see above) over 25 days. Four independent samples, each consisting of three plants, were harvested at the beginning of Cu depletion (t = 0) and after 25 days with (control) or without Cu (–Cu). Leaves and petioles present at the beginning of Cu depletion (referred as "old leaves" and "old petioles", respectively) were distinguished from leaves appearing during Cu depletion (referred as "young leaves" and "young petioles", respectively). At each date of harvest (t = 0 and t = 25 days), whole roots from control and Cu-depleted plants were collected. An aliquot of each tissue was weighed and dried in an oven (60°C) for dry weight (DW) determination and ground to fine powder for IRMS and ICP-OES analysis. Thereafter, the remaining fresh tissues were frozen in liquid nitrogen and stored at –80°C for transcriptomic and proteomic analyses.

Every five days throughout the culture period, non-destructive determination of chlorophyll content of young and old leaves was performed using a SPAD chlorophyll meter (SPAD-502 model, Minolta, Tokyo, Japan). The determination was carried using three replicates of ten measurements performed on independent leaves.

Analysis of nutrients in plant tissues

For the analysis of total N and S contents, an aliquot of around 4 mg DW of each plant organ sample was placed in tin capsules for total N and S analysis using an IRMS spectrometer (Isoprime, GV Instrument, Manchester, UK) linked to a C/N/S analyser (EA3000, Euro Vector, Milan, Italy). The total amount of N or S (N_{tot} or S_{tot}) in a tissue "i" at a given time "t" is calculated:

$$N_{tot}(\text{or } S_{tot}) = \%\ N_{i,t}(\text{or } S_{i,t}) \times DWi,t/100$$

Other nutrients (K, Ca, S, P, Mg, Fe, Na, Mn, B, Si, Zn and Cu) were analysed by Inductively Coupled Plasma Optical Emission Spectrometry (ICP-OES, Thermo Elemental Co. Iris Intrepid II XDL) with prior microwave acid sample digestion (8 mL of concentrated HNO_3 and 2 mL of H_2O_2 for 0.5 g DW), using a protocol previously described by Mora *et al.* [33].

RNA extraction

Total RNA was extracted from 200 mg of frozen samples ground to a powder with a pestle in a mortar containing liquid nitrogen. The resulting powder was suspended in 750 µl of extraction buffer [0.1 M TRIS, 0.1 M LiCl, 0.01 M EDTA, 1% SDS (w/v), pH 8] and 750 µl of hot phenol (80°C, pH 4). This mixture was vortexed for 30 s and, after addition of 750 µl of chloroform/isoamylalcohol (24:1, v/v), the homogenate was centrifuged (15 000 g, 5 min, 4°C). The supernatant was transferred into a 4 M LiCl solution and incubated overnight at 4°C. After centrifugation (15 000 g, 30 min, 4°C), the pellet was suspended in 100 µl of sterile water. RNA was then purified with an RNeasy mini kit according to the manufacturer's protocol (Qiagen, Courtaboeuf, France). Quantification of total RNA was performed by spectrophotometry at 260 nm (BioPhotometer, Eppendorf, Le Pecq, France) before Reverse Transcription (RT) and real time Quantitative Polymerase Chain Reaction (Q-PCR) analysis.

Reverse transcription (RT) and Q-PCR analysis

For RT, 1 µg of total RNA was converted to cDNA with an "iScript cDNA synthesis kit" according to the manufacturer's protocol (Bio-Rad, Marne-la-Coquette, France).

Q-PCR amplifications were performed using specific primers for each housekeeping gene (*EF1-α* and *18S rRNA*) and target genes (Table 1). Q-PCRs were performed with 4 µl of 200x diluted cDNA, 500 nM of primers, and 1x SYBR Green PCR Master Mix (Bio–Rad, Marne–la–Coquette, France) in a real-time thermocycler (CFX96 Real Time System, Bio–Rad, Marne–la–Coquette, France). A 2 step program, composed of 42 cycles of a denaturing step at 95°C for 15 s followed by an annealing and extending step at 60°C for 40 s, was used for all pairs of primers (Table 1) except for *COPT2*, for which we used a 3 step program. In this case the denaturing step was as described previously. Annealing was at 58.6°C for 10 s and was followed by an extending step at 72°C for 30 s. For each pair of primers, a threshold value and PCR efficiency were determined using a cDNA preparation diluted >10-fold. For all pairs of primers, PCR efficiency was around 100%. The specificity of PCR amplification was examined by monitoring the presence of the single peak in the melting curves after Q-PCRs and by sequencing the Q-PCR product to confirm that the correct amplicon was produced from each pair of primers (Eurofins, Ebersberg, Germany). The relative expression of the genes in each sample was compared with the control sample [corresponding to untreated plants at the same time of harvest] and was determined with the delta delta Ct (ΔΔCt) method using the following equation [34]: Relative expression $= 2^{-\Delta\Delta Ct}$, with $\Delta\Delta Ct = \Delta Ct_{treated} \text{-} \Delta Ct_{control}$ and with $\Delta Ct = Ct_{target\ gene} - [\sqrt{(Ct_{EF1-\alpha} \times Ct_{18S\ rRNA})}]$, where Ct refers to the threshold cycle determined for each gene in the exponential phase of PCR amplification and $[\sqrt{(Ct_{EF1-\alpha} \times Ct_{18S\ rRNA})}]$ corresponds to the geometric average of Ct of the reference genes. Using this analysis method, the relative expression of the different genes in the control sample of the experiment was equal to 1 [34], and on this basis the relative expression of other treatments was then compared with the control. For *COPT2*, because no transcript was detected in control plants (*i.e.* $Ct_{control}$ is undetermined), values are not expressed relatively to control but as $2^{-\Delta Ct}$.

Extraction and determination of total proteins

Two hundred mg of fresh matter from old leaf samples were ground to a fine powder in liquid nitrogen in the presence of 50 mg of poly(vinylpolypyrrolidone) (PVPP). The addition of PVPP is used to fix plant polyphenols that might interfere with the quantification of proteins or during separation of proteins by electrophoresis. The ground material was dissolved in 1.75 mL of TCA/acetone solution (10% TCA (w/v) prepared in acetone). After centrifugation (3 min, 16 000 g, 4°C), the protein pellet was purified according to the protocol adapted from Wang *et al.* [35]. The protein pellet obtained after precipitation with TCA/acetone (10% TCA (w/v) was suspended in 1.75 mL of 0.1 M ammonium acetate dissolved in 80% methanol. After homogenization and centrifugation (16 000 g, 3 min, 4°C), the pellet was washed with 1.75 mL of 80% acetone and centrifuged again (16 000 g, 3 min, 4°C). The supernatant was removed and the pellet was dried under vacuum (Speedvac Concentrator 5301, Eppendorf, France) for 5 min at 50°C and then suspended with 0.8 mL of phenol at pH 7.9 and in 0.8 mL of dense SDS buffer (30% sucrose, 2% SDS, 0.1 M Tris–HCl, pH 8.0, 0.5% 2–mercaptoethanol). After 5 min incubation at 4°C and centrifugation (16 000 g, 3 min, 4°C), the phenol phase was transferred to a new tube and supplemented with 1.75 mL of 0.1 M ammonium acetate and stored at –20°C overnight. Afterwards, ammonium acetate was used to precipitate proteins to enable their collection by centrifugation (16 000 g, 5 min, 4°C). The protein pellet was then washed with 1.75 mL of 100% methanol and again with 1.75 mL of 80% acetone. Residual acetone was removed by vacuum evaporation over a few minutes. The pellet was resuspended in 400 µL of R2D2 rehydration buffer [5 M urea, 2 M thiourea, 2% CHAPS, 2% N–decyl–N,N–dimethyl–3–ammonio–1–propanesulfonate, 20 mM dithiothreitol, 5 mM Tris (2–carboxy– ethyl) phosphine, 0.5% IPG buffer] (GE Healthcare, Saclay, France), pH 4 to 7 [36]. The total protein concentration was determined by the method previously described by Bradford [37] using bovine serum albumin as standard.

Table 1. Q-PCR primer sets.

Gene	Accession number	Forward	Reverse	Gene Function
EF1-α	DQ312264	5'-gcctggtatggttgtgacct-3'	5'-gaagttagcagcacccttgg-3'	
18S rRNA	GQ380689	5'-cggataaccgtagtaattctag-3'	5'-gtactcattccaattaccagac-3'	
COPT2	NM_114557	5'-tgcacatgaccttcttttgg-3'	5'-gtcatcggagggtttgttga-3'	Cu uptake
HMA1	NM_119890.6	5'-gtacagctgaccgaggaagc-3'	5'tgcccataaatgggttcaat-3'	Cu allocation to chloroplast
MOT1	NM_128127	5'-ctcgccaggatttggactta-3'	5'agatccccaacacgaacaag-3'	Mo uptake

EF1α and *18S rRNA* were housekeeping genes used for relative gene expressions by Q-PCR analysis.

Two-dimensional electrophoresis (2-DE) and image analysis

2-DE was performed according to the protocol detailed by Desclos *et al.* [25]. Gels were stained using the silver-staining procedure described by Blum *et al.* [38] and scanned with a ProXPRESS 2D proteomic imaging system (Perkin-Elmer, Courtaboeuf, France). Images of the 2-DE gels were analysed using the Progenesis SameSpots software v3.0 (Nonlinear Dynamics, Newcastle upon Tyne, UK) according to the manufacturer's protocol. Gels from three independent biological replicates were used. Spot detection, warping, and matching were performed automatically by the software and manually validated. Artefacts due to non-specific silver nitrate staining or spots that could not be confidently verified as true matches were disregarded rather than manually edited, and misalignments were corrected by manual warping when appropriate. Mw and *pI* were calculated using SameSpots software calibrated with commercial molecular mass standards (prestained precision protein standards; Bio-Rad, Marne-la-Coquette, France) run in a separate marker lane on 2-DE gel.

RuBisCO relative quantification

In order to observe maximum spots, silver staining was performed on 2-DE with a saturated RuBisCO signal. The amount of RuBisCO was then measured by Experion Pro260 Analysis Kit (Bio-Rad, Marne-la-Coquette, France). The Experion automated electrophoresis system applies a combination of microfluidic separation technology and sensitive fluorescent detection of proteins [39]. The protein extracts were treated with the Experion reagents and then separated on the Experion automated electrophoresis station according to the manufacturer's instruction. Peak areas corresponding to large and small RuBisCO subunits were then compared using software given by manufacturer.

Protein Identification by ESI LC-MS/MS

Spots of interest were excised from 2D gels and washed several times with water and dried for a few minutes. Trypsin digestion was performed overnight with a dedicated automated system (MultiPROBE II, Perkin-Elmer). The gel fragments were subsequently incubated twice for 15 min in a 0.1% CH_3CN solution in water to allow extraction of peptides from the gel pieces. Peptide extracts were then dried and dissolved in starting buffer for chromatographic elution, consisting of 3% CH_3CN and 0.1% HCOOH in water. Peptides were enriched and separated using lab-on-a-chip technology (Agilent, Massy, France) and fragmented using an on-line XCT mass spectrometer (Agilent). The ESI LC-MS/MS data were converted into DTA- format files that were further searched for proteins with MASCOT Daemon (Matrix Science [40]). For protein identification, two strategies were employed to mine the maximum data. Measured peptides were searched in the NCBI nr-protein sequence database, *viridiplantae* (green plants [41]), and in the *Brassica* EST database (*Brassica* Genome Gateway 2007, [42]). Proteins with two or more unique peptides matching the protein sequence with a score >53, as defined by MASCOT, were considered as a positive identification. The spectra of each peptide were verified manually. In cases where protein identification data were lacking, BLAST analysis was performed [43].

Data and statistical analysis

Regarding the growth, ICP-OES and IRMS analyses, all experiments were conducted with 4 independent biological replicates each corresponding to 3 plants. Proteomics and Q-PCR analyses were performed on 3 independent biological replicates each corresponding to 3 plants. Data are given as mean ± SE for n = 3 or 4. For Q-PCR, SE and Student's T-test were based on ΔΔCt (or ΔCt for *COPT2*, see above for more information), when the relation between values is still linear. According to Yuan *et al.* 2006 [44], a confidence interval of each mean value (with min and max values) can be calculated and has been presented in the results section as [mean of relative expression] (Min value-Max value) with Min value = $2^{-(\Delta\Delta Ct+SE)}$ and Max value = $2^{-(\Delta\Delta Ct-SE)}$ (Min value = $2^{-(\Delta Ct+SE)}$ and Max value = $2^{-(\Delta Ct-SE)}$ for *COPT2*). Conversion of the results of an exponential process into a linear comparison of amounts leads to an asymmetric distribution [34].

All data were analysed with Student's T-test ($p<0.05$) and marked by an asterisk (*) or cross (†) when significantly different.

Results

Cu depletion affects the growth and Cu content of *B. napus*

After 25 days of culture, control whole plant DW increased from 3.23±0.24 g to 39.67±1.26 g (table 2). The total DW of Cu-depleted plants was decreased by 18.4% relative to control plants (table 2, from 39.67 to 32.37 g $DW.plant^{-1}$). This was the result of a significant reduction in root DW (from 4.11±0.23 in control plants to 3.73±0.06 g $DW.plant^{-1}$ in Cu-depleted plants) and secondly, of old petioles DW (from 8.07±0.60 to 6.31±0.49 g $DW.plant^{-1}$). No significant differences in growth rates were found for other organs. Control plants accumulated 251.5 µg of Cu while the Cu content in Cu-depleted plants only rose from 29.0±4.0 µg to 74.8±3.0 µg. The slight increase in total Cu in Cu-depleted plants was the result of a trace of Cu found in the mineralized water used for the nutrient solution (0.47±0.00 µM, data not shown). In roots of control plants, the amount of Cu increased from 14.7±3.0 µg to 155.2±13.2 µg but only reached 43.4±1.5 µg in Cu deficient plants. In aerial tissues that developed under Cu depletion, the Cu amount was reduced from 11.8±1.9 µg to 2.6±0.7 µg and from 76.2±4.8 µg to 16.2±2.3 µg in petioles and leaves, respectively. In petioles already present at the time of the Cu depletion, the amount of Cu rose from 2.1±0.2 µg to 8.0±0.7 µg during depletion while it reached 20.3±2.4 µg in control plants at the end of the experiment. The amount of Cu in old leaves of control plants did not vary significantly during the experiment, while in Cu-depleted plants, it was reduced from 12.2±1.1 µg to 4.7±1.0 µg indicating a remobilization of 61.4±4.2% of the Cu initially present in these leaves.

No significant difference was observed for RuBisCO by *Experion* method (Figure 1.A). During the whole experiment, the leaf chlorophyll content (SPAD values, Figure 1.B) were not significantly different between control and Cu-depleted plants. At the end of the experiment, this was confirmed in old leaves (Figure 1.C). Also, N content in old leaves didn't present any difference between the two conditions (Figure 1.D).

Cu depletion increases expression of Cu transporters in *Brassica napus*

The relative expression of genes involved in Cu transport (*HMA1, COPT2*) is given in Figure 1. Expression of the *HMA1* gene encoding a chloroplastic Cu transporter was monitored in leaves, whereas *COPT2*, which encodes a transporter involved in Cu uptake, was monitored in roots. The relative number of *HMA1* transcripts increased 2.68 fold (2.23-3.24) after 25d of Cu

Table 2. Biomass and Cu content in *Brassica napus* L. at t = 0 and after 25 days of culture with (control) or without Cu (−Cu).

		t=0	Control (25d)	-Cu (25d)
Old leaves	DW (g)	2.13±0.17	4.78±0.79	4.83±0.93
	Cu amount (μg)	12.2±1.1	17.0±2.5	4.7±1.0*†
	% remobilized	NA	NA	60.36±4,15
Old petioles	DW (g)	0.63±0.04	8.07±0.60	6.31±0.49*†
	Cu amount (μg)	2.1±0.2	20.3±2.4*	8.0±0.7*†
Young leaves	DW (g)	NA	12.7±0.94	8.83±1.64
	Cu amount (μg)	NA	76.2±4.8	16.2±2.3 †
Young petioles	DW (g)	NA	10.02±0.78	8.66±0.55
	Cu amount (μg)	NA	11.8±1.9	2.6±0.7 †
Roots	DW (g)	0.48±0.03	4.11±0.23*	3.73±0.06*
	Cu amount (μg)	14.7±3.0	155.2±13.2*	43.4±1.5*†
Total	DW (g)	3.23±0.24	39.67±1.26*	32.37±1.71*†
	Cu amount (μg)	29.0±4.0	280.5±13.5*	74.8±3.0*†

* and † represent significant differences at $p<0.05$ compared to t = 0 and control, respectively. NA: Not Applicable.

depletion (Figure 2.A). Transcripts of *COPT2* were detected only in roots from Cu-depleted plants suggesting that the expression of the *COPT2* transporter was up-regulated and/or de-repressed in *Brassica napus* roots in response to Cu deficiency (Figure 2.B).

Cu depletion affects uptake of some mineral nutrients

Figure 3 shows that Cu deprivation decreased Cu uptake by 81.8±2.0%, but also affected the uptake of Na (−41.0±3.8%), B (−31.6±2.1%) and Mg (−28.5±2.8%). while plant dry weight was reduced by 18.4±2.5% (Table 2). Cu deficiency had no effect on total plant uptake for nutrients such as N, Ca, K, S, P, B, Fe, Mn and Zn, despite the difference of biomass. In contrast, uptake of Mo was increased by 121.8% despite the reduced plant growth and was associated with a very strong up-regulation of the molybdenum transporter 1 gene (*MOT1,* Figure 2.C), having an important role in efficient Mo uptake [45,46]. Indeed, in Cu-depleted plants, the number of *MOT1* transcripts was found to be 159.61 fold (127–200) higher than in control plants.

Cu depletion induces proteomic modifications

Leaves already present at the beginning of Cu depletion were the most affected tissues, not only due to their decreased Cu content and hence the mobilization of Cu (by about 60%) to other tissues, but also according to their decrease of Mg (−42.1±11.0%), Na (−42.6±13.6%), Ca (−35.8±10.0%) and Fe (−27.9±13.5%) contents. Therefore, leaves were selected for a comparative proteomic approach in order to highlight the main metabolic modifications induced by Cu depletion. Representative 2D gels corresponding to protein extracts from leaves of control and Cu-depleted plants are shown in Figure 4. On each gel, 693 proteins were identified. Their comparative analysis revealed 48 proteins significantly up- or down-regulated by Cu depletion. Because different spots correspond to different isoforms of the same protein, only 33 different proteins were identified. Among them, 25 proteins were accumulated in Cu-depleted old leaves. Transketolase-like protein (#7 to 11, Figure 4 and Table 3) was the strongest protein induced with four isoforms that were accumulated more than fourfold compared to control. On the other hand, repression was significant in only 8 of the 33 identified

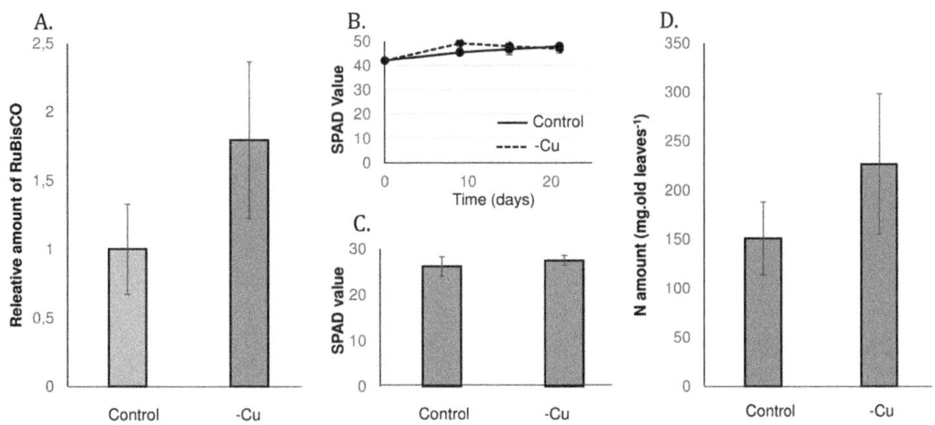

Figure 1. Parameters associated with senescence. (A.) Relative amount of RuBisCO in old leaves. Relative chlorophyll content (SPAD value) during (B.) the whole experiment at whole plant level and (C.) after 25 days of depletion in old leaves (D.) N content in old leaves.

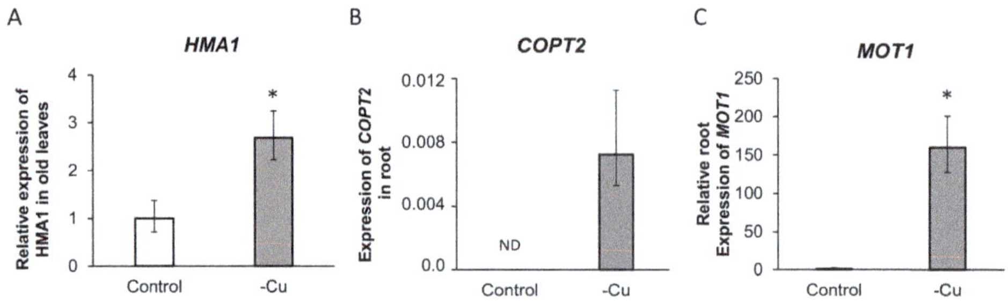

Figure 2. Relative accumulation of transcript of Cu transporters (A. *HMA1* in old leaves and B. *COPT2* in roots) and Molybdenum transporter 1 (C. *MOT1* in roots) in *Brassica napus* L after 25 days of culture with (control) or without Cu (−Cu). Control is represented by white bars and -Cu by grey bars. *MOT1* and *HMA1* values are expressed as $2^{-\Delta\Delta Ct}$ while *COPT2 value* is expressed as $2^{-\Delta Ct}$ (for details, see "Material and Methods"). * represents significant differences at $p<0.05$. ND: not detected.

proteins, with the Oxygen Evolving Enhancer (OEE, #33 to 37) having the greatest repression in our study (−2.3 to −3.93). Among all proteins (Table 3), five were involved in stress responses (disease/defence or defence/secondary metabolite), for example glutathione reductase (#18), and seventeen proteins were involved in energy processes, for example aldolase (#27) and ATP synthase (#17). Protein annotations in databases also indicated that chloroplasts were the main locations of the identified proteins with 17 references, including the most repressed, OEE (#33 to 37), and the most induced, Transketolase-like protein (#7 to 11).

Considering some of the characteristic Cu proteins reported in the literature [47], Plastocyanin was too small (around 10 kDa) to be found on our gels while Cu-Zn superoxide dismutase and Polyphenol oxidase were not differentially expressed and/or detected.

Discussion

Cu remobilization from old leaves is induced by Cu deficiency

In this study, the effectiveness of Cu deficiency can be attested by the low Cu concentration found in old and young leaves of Cu depleted plants (0.97±0.03 $\mu g.g^{-1}$ DW and 1.66±0.40 $\mu g.g^{-1}$ DW, respectively) which was lower than the usually admitted minimum concentration (5 $\mu g.g^{-1}$ DW) [48]. Using Cu deficiency we were able to demonstrate a decrease in the Cu content of leaves by 61.4±4.2% (Table 2). To our knowledge, alongside work on *Arabidopsis thaliana* [49], this is one of the few studies that shows that Cu can be recycled between leaves. However, the Cu remobilization could be underestimated in our conditions because some traces of Cu were found in the nutrient solution, as suggested by the slight increase in the overall Cu content in Cu-depleted plants (Table 2). A precise estimation of Cu remobilization would require the use of stable isotopes. This remobilization of Cu in leaves was concomitant with an increase in *COPT2* and *HMA1* expression. Despite a tendency of an increase of RuBisCO and N amounts, the RuBisCO, N amounts and chlorophyll contents in old leaves were not significantly differentially affected between control and Cu-depleted plants (Figure 1). Then we could assume that Cu remobilization was not associated with leaf senescence. This is also consistent with the down-regulation of senescence-associated proteases such as FtsH2 (#14, Table 3), a protease involved in the degradation of photosystem II [26,50]. Thus, this study suggests that Cu depletion induces a Cu mobilization independent of the distinguishing marks of senescence. This is different to what has been found for N for example, whose deficiency is known to increase leaf senescence in *Brassica napus* [25].

Effect of Cu deficiency on metabolic pathway in chloroplasts from old leaves of *B. napus*

Because the spots of our 2D–E gels correspond to the most abundant proteins in old leaves, numerous proteins (especially transporters and transcription factors) corresponding to transcripts

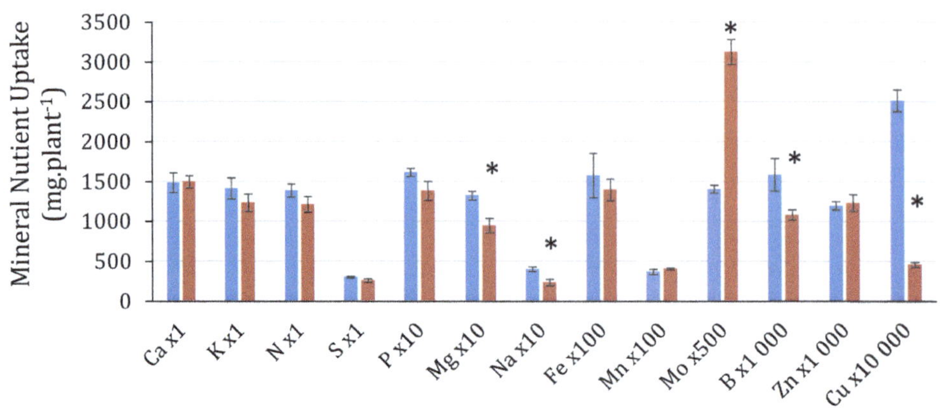

Figure 3. Mineral nutrient uptake in control plants (blue) and Cu-depleted plants over 25 Days (red). Due to large differences between nutrients, factors have been attributed to each nutrient to allow representation on the same graph. * indicates significant difference between control and Cu depleted plants at $p=0.05$.

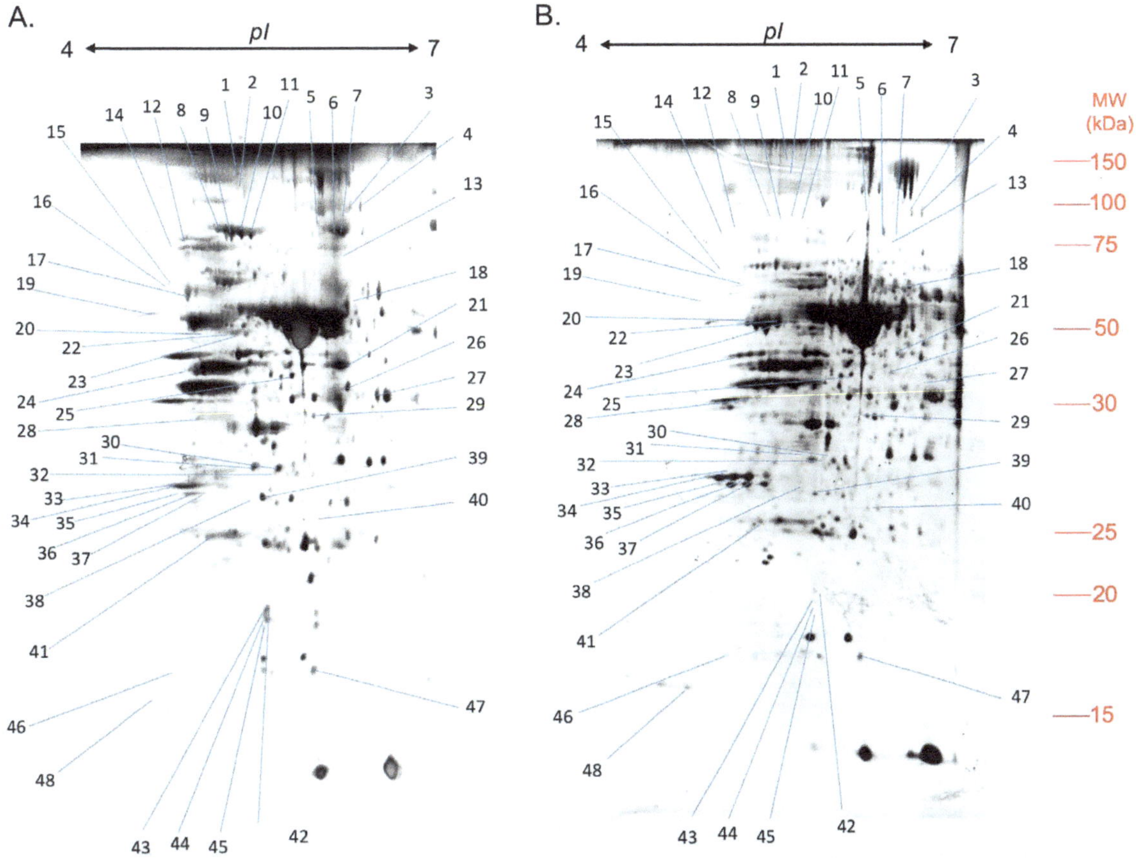

Figure 4. Representative Silver-stained 2-DE gels of proteins from old leaves of *Brassica napus* L. after 25 days of culture with (control, A) or without Cu (−Cu, B). A total of 150 µg of total proteins was loaded on an IEF strip forming an immobilized pH linear gradient from 4 to 7 (for details, see "Materials and Methods"). Numbers indicate sequenced and identified proteins presented in table 3. Molecular mass markers (MW in kDa) are listed on the right side of the gel.

identified in previous studies as regulated by Cu depletion could not be observed in the present work. Thus, this study complement previous transcriptomic data on Cu remobilisation [51,52]. This proteomic analysis of Cu-deficient old leaves showed that more than half of the up or down-regulated proteins were localised in chloroplasts (Table 3) indicating potential disturbance, despite an absence of chlorophyll content variation (Figure 1.B, C). Five of these proteins are involved in the Calvin cycle and 5 are involved in the thylakoid electron transport chain (Figure 5). From a general point of view, among the 33 proteins that were differentially expressed relatively to control plants (Table 3), 11 of them are known to rely on the presence of Cu for their synthesis and/or activity, and 9 were over expressed (#15, 16, 17, 19, 27, 29, 39 and 43). As a consequence, it could be suggested that the lack of Cu reduced their activity which may explain their up-regulation (or alternatively reduced their down-regulation). A similar regulation would also occur for enzymes that are located upstream or downstream of such copper deficiency up-regulated proteins. For example, this is the case (Figure 5) within the Calvin cycle for Transketolases (#6–11) or TPI (#41), which are located upstream and downstream of Aldolase (#27), a copper dependent protein that was up-regulated by Cu-deficiency.

Among Cu binding or regulated enzymes within the Calvin cycle (FBPase [53], GAPDH, Aldolase [16], RuBisCO [18], PGK [17]), only Aldolase (#27) and PGK (#21) were found to be over represented (Table 3 and Figure 5). However, whilst RuBisCO was not significantly affected (Figure 1), its main regulator RuBisCO activase (#24) was increased by about 3 folds (Table 3). It should be noted that PGK, Aldolase, Transketolase and TPI are also involved in glycolysis (in the cytosol or plastids [54]) and their regulation should therefore induce mitochondrial defects.

GAP is a crosspoint between numerous pathways, with B6 vitamin biosynthesis being one of them [55]. However, under an impaired Calvin cycle, more GAP would be available for this cycle (Figure 5). Vitamin B6 can also be synthesized through Pyridoxin biosynthesis protein (#32, Table 3) which is over represented in old leaves of Cu-depleted *B. napus*. The B6 dependent protein Glutamate-1-semialdehyde aminomutase (GSA, #25) [56] with a role in chlorophyll biosynthesis also accumulated (Figure 5).

Differentially expressed proteins located in thylakoids (Figure 5) could be subdivided into two groups: the first one including OEE (#33 to 37), NADPH subunit 48 (#20) and ATP synthase (#17) were down-regulated, which is consistent with previous work reporting a down-regulation of photosynthesis-related genes in response to Cu deficiency [51]. The second group corresponded to [Fe-S] cluster containing proteins such as Ferredoxin (#30 and 31) and Rieske domain protein (#47), which were over expressed in response to Cu depletion. We suggest that overall the electron transport chain was down-regulated, but an impairment of [Fe-S] cluster biosynthesis could be compensated by an up-regulation of *de novo* synthesis of [Fe-S] proteins. This stimulation of [Fe-S] cluster proteins can be linked to the accumulation of Rhodanese

Table 3. Abundance of identified proteins over (positive value) or under (negative value) accumulated in old leaves of Cu-depleted *B. napus* relative to control plants.

Spot no.	Relative Abundance	Exp. pI/ Mw	Theo. pI/Mw	PM	SC (%)	Protein name/Organism/NCBI accession no.	Functional classification/Sub-Cellular localization	Note
1	3.12	5.7/115	5.39/104.3	30	20	Myrosinase binding protein [*Brassica napus*]/gi|1711296	11. Disease/Defence	
2	2.11	6.3/99	6.59/102.2	117	40	Chaperone protein ClpC, chloroplastic-like [*Fragaria vesca* subsp. vesca]/gi|470122943	08. Intracellular traffic/Chloroplast	
3	1.75	6.64/90	6.12/84.3	58	24	Cobalamin-independent methionine synthase [*Arabidopsis lyrata* subsp. lyrata]/gi|297807807	01. Metabolism	**Zn** [73]
4	2.78	6.53/91	6.12/84.3	29	19	Cobalamin-independent methionine synthase [*Arabidopsis lyrata* subsp. lyrata]/gi|297807807	01. Metabolism	**Zn** [73]
5	4.16	6.26/80	6.16/81.2	13	13	NADH Ubiquinone oxydoreductase 75 kD subunit [*Arabidopsis thaliana*]/gi|30693102	02. Energy/Chloroplast-Mitochondria	[Fe-S] cluster
6	5.23	6.5/81	5.80/81.5	15	18	Transketolase-like protein [*Arabidopsis thaliana*]/gi|7329685	02. Energy/Chloroplast-Mitochondria	
7	2.22	6.41/80	6.12/79.9	13	10	Transketolase [*Arabidopsis thaliana*]/gi|30689983	02. Energy/Chloroplast-Mitochondria	
8	4.79	5.74/75	6.16/81.2	46	24	Transketolase-like protein [*Arabidopsis thaliana*]/gi|7329685	02. Energy/Chloroplast-Mitochondria	
9	4.17	5.78/75	6.16/81.2	61	27	Transketolase-like protein [*Arabidopsis thaliana*]/gi|7329685	02. Energy/Chloroplast-Mitochondria	
10	4.46	5.67/76	6.16/81.2	62	29	Transketolase-like protein [*Arabidopsis thaliana*]/gi|7329685	02. Energy/Chloroplast-Mitochondria	
11	4.23	5.54/74	6.16/81.2	44	26	Transketolase-like protein [*Arabidopsis thaliana*]/gi|7329685	02. Energy/Chloroplast-Mitochondria	
12	2.06	5.25/73	5.08/70.7	49	37	Heat shock cognate protein HSC70 [*Brassica napus*]/gi|2655420	08, intracellular traffic/Chloroplast	*Zn* [16]
13	−2.07	6.48/70	5.56/63.4	10	11	Phosphoglucomutase [*Arabidopsis thaliana*]/gi|15223226	02. Energy/Chloroplast	
14	−2.03	5.17/69	5.62/72.0	25	16	ATP-dependent zinc metalloprotease FTSH 2, chloroplastic-like [*Brachypodium distachyon*]/gi|357123383	06. Prot. Dest. & Storage/Chloroplast	***Zn*** [74]
15	1.54	6.14/64	5.22/49.4	11	7	Thylakoid rhodanese-like protein [*Arabidopsis thaliana*]/gi|18411523	02. Energy/Chloroplast	***Cu*** and ***Zn*** [75]
16	1.4	6.08/62	5.66/47.6	5	4	Thylakoid rhodanese-like protein [*Arabidopsis thaliana*]/gi|18411523	02. Energy/Chloroplast	***Cu*** and ***Zn*** [75]
17	2.63	5.25/62	5.97/47.0	85	42	ATPase subunit I [*Arabidopsis lyrata* subsp. lyrata]/gi|297837977	02. Energy/Chloroplast-Mitochondria	*Cu* and *Zn* [16]
18	−2.67	6.52/61	8.03/60.6	15	19	Glutathione reductase [*Brassica juncea*]/gi|13448672	20. Defence & Secondary metabolism/Vacuole	*Cu* [16]
19	2.09	4.92/55	5.23/50.6	7	5	Rhodanese-like domain-containing protein 4, chloroplastic-like [*Cicer arietinum*]/gi|502179307	01. Metabolism/Chloroplast	***Cu*** [75]
20	−2.27	5.45/50	6.49/51.0	19	16	NAD(P)H dehydrogenase subunit 48 [*Arabidopsis thaliana*] gi|18394307	02. Energy/Chloroplast	
21	1.52	6.44/47	5.91/50.1	25	21	Phosphoglycerate kinase 1 [*Arabidopsis thaliana*]/gi|15230595	02. Energy/Chloroplast-Mitochondria	*Cu* [17]
22	1.92	5.58/53	6.06/54.2	19	19	Mitochondrial processing peptidase alpha subunit, putative [*Arabidopsis thaliana*] gi|21594004	06. Prot. Dest. & Storage/ Mitochondria	
23	2.81	5.53/52	6.06/54.2	7	11	Mitochondrial processing peptidase alpha subunit, putative [*Arabidopsis thaliana*] gi|21594004	06. Prot. Dest. & Storage/ Mitochondria	
24	2.97	5.11/46	6.78/47.9	61	46	Chloroplast ribulose-1,5-bisphosphate carboxylase/oxygenase activase [*Brassica oleracea*]/gi|383470439	02. Energy/Chloroplast	

Table 3. Cont.

Spot no.	Relative Abundance	Exp. pI/Mw	Theo. pI/Mw	PM	SC (%)	Protein name/Organism/NCBI accession no.	Functional classification/Sub-Cellular localization	Note
25	1.7	6.03/44	6.43/50.2	44	32	Glutamate-1-semialdehyde 2,1-aminomutase 2 [*Arabidopsis thaliana*]/gi|15229018	01. Metabolism	B6 dependent [56]
26	1.42	6.44/42	5.91/15.1	16	45	Cinnamyl alcohol dehydrogenase [*Brassica napus*]/gi|6683959	01. Metabolism	
27	1.46	6.64/40	6.22/38.4	19	37	Fructose-bisphosphate aldolase [*Arabidopsis thaliana*]/gi|15231715	02. Energy/Chloroplast-Mitochondria	*Cu* [16] and ***Zn*** [76]
28	1.47	5.24/37	5.14/32.0	9	35	40S ribosomal protein [*Brassica napus*]/gi|15214300	05. Protein synthesis/Ribosome	
29	1.33	6.16/37	8.54/35.8	29	31	Mitochondrial NAD-dependent malate dehydrogenase [*Arabidopsis thaliana*]/gi|21592905	02. Energy/Mitochondria	*Cu* [16] and ***Zn*** [77]
30	1.72	5.7/32	8.32/40.3	49	34	Ferredoxin-NADP(+)-oxidoreductase 1 [*Arabidopsis thaliana*]/gi|15239282	02. Energy/Chloroplast	[Fe-S] cluster
31	1.56	5.8/32	8.66/40.1	34	29	Ferredoxin-NADP+ reductase [*Arabidopsis thaliana*]/gi|5730139	02. Energy/Chloroplast	[Fe-S] cluster
32	2.1	5.84/32	6.12/33.0	13	25	Pyridoxine biosynthesis protein [*Lotus japonicus*]/gi|72256517	11. Disease/Defence/Cytosol	B6 biosynthesis [55]
33	−3.93	4.88/31	5.92/35.0	5	15	Oxygen-evolving enhancer protein 1–2 [*Arabidopsis thaliana*]/gi|15230324	02. Energy/Chloroplast	***Mn*** [78]
34	−2.3	4.92/30	5.92/35.0	81	54	Oxygen-evolving enhancer protein 1–2 [*Arabidopsis thaliana*]/gi|15230324	02. Energy/Chloroplast	***Mn*** [78]
35	−2.56	4.94/30	5.93/35.1	36	48	Photosystem II subunit O-2 [*Arabidopsis lyrata* subsp. lyrata]/gi|297819782	02. Energy/Chloroplast	***Mn*** [78]
36	−2.95	5.03/30	5.93/35.1	67	52	Photosystem II subunit O-2 [*Arabidopsis lyrata* subsp. lyrata]/gi|297819782	02. Energy/Chloroplast	***Mn*** [78]
37	−2.8	5.32/30	5.92/35.0	84	55	Oxygen-evolving enhancer protein 1–2 [*Arabidopsis thaliana*]/gi|15230324	02. Energy/Chloroplast	***Mn*** [78]
38	1.69	5.61/29	8.65/33.7	13	11	Thioredoxin-like protein CDSP32 [*Arabidopsis thaliana*]/gi|15222954	11. Disease/Defence/Cytosol	
39	3.81	5.73/29	5.54/29.5	23	17	Carbonic anhydrase 1 [*Arabidopsis thaliana*]/gi|30678347	01. Metabolism/Chloroplast	***Cu*** and ***Zn*** [79]
40	−1.81	6.17/27	6.28/27.6	29	34	Mitochondrial F1F0-ATP synthase subunit Fad [*Arabidopsis thaliana*]/gi|15227104	02. Energy/Mitochondria	Cu [18] and *Zn* [16]
41	1.55	5.28/26	7.67/33.3	10	16	Triosephosphate isomerase [*Arabidopsis thaliana*]/gi|15226479	02. Energy/Chloroplast-Mitochondria	
42	2.41	5.73/20	5.37/19.9	9	26	Uncharacterized protein [*Arabidopsis thaliana*]/gi|18391006	13. Unclassified	
43	2.79	5.7/19	5.47/35.7	41	38	Chloroplast beta-carbonic anhydrase [*Brassica napus*]/gi|297787439	01. Metabolism/Chloroplast	***Cu*** and ***Zn*** [79]
44	2.6	5.75/19	6.81/21.8	6	17	Germin-like protein [*Arabidopsis thaliana*]/gi|1755154	12. Unclear Classification	
45	1.99	5.7/19	6.27/22.0	7	23	RecName: Full = Germin-like protein 1; Flags: Precursor [*Sinapis alba*]/gi|1169944	12. Unclear Classification	
46	−2.6	4.99/17	9.12/24.7	18	31	Peroxiredoxin-2E [*Arabidopsis thaliana*]/gi|15231718	11. Disease/Defence/Cytosol	
47	2.74	6.11/17	8.80/24.3	22	22	Rieske FeS protein [*Arabidopsis thaliana*]/gi|9843639	02. Energy/Chloroplast	[Fe-S] cluster
48	−1.76	4.58/16	5.24/17.9	6	15	Probable glycine cleavage system H protein 2 [*Arabidopsis thaliana*] gi|15223217	02. Energy/Chloroplast	

Experimental and theoretical pI/Mw, the number of LC–MS/MS matched peptides (PM), the SCORE and the percentage of sequence coverage (SC) obtained are also indicated. For each protein, the assigned best–matched protein is listed with the organism in which it was identified and its GenBank protein accession number is indicated. Elements given in notes correspond to ligand (in italics) or regulators (in bold) of the corresponding protein described in previous studies (indicated in brackets).

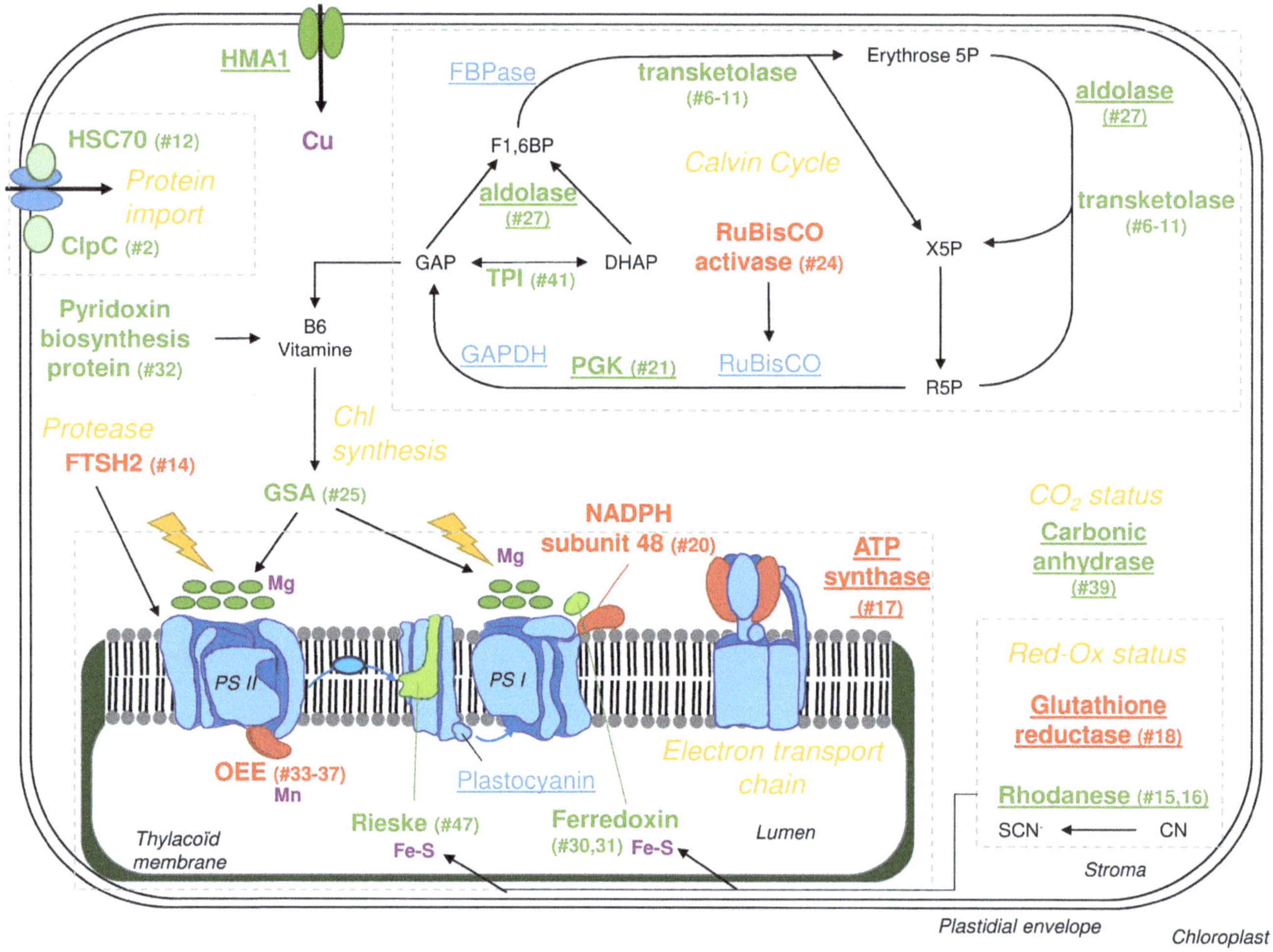

Figure 5. Effect of Cu deficiency on metabolic pathwayspathway in chloroplasts from old leaves of *B. napus*. Bold red indicates a decrease in the amount of protein while bold green indicates an increase. Spot number(s) corresponding to proteins identified in table 3 are indicated in parentheses. Intermediate metabolites are in black, proteins not identified in 2D-E analysis are in blue. Pathways are encircled by grey dashed lines and named in orange italics. Underlines indicate Cu binding and/or known Cu regulation. Other elements known to bind and/or regulate proteins are indicated in purple. PS I: Photosystem I; PS II Photosystem II; R5P: ribulose-5-phosphate; X5P: xylose-5-phosphate; F1,6BP: Fructose-1,6-bisphosphate; GAPDH: GlycerAldehyde Phosphate DeHydrogenase; PGK: PhosphoGlucoKinase; GAP: GlycerAldehyde-3-Phosphate; FBPase: Fructose-1,6-BisPhosphatase; TPI: Triose Phosphate Isomerase; GSA: Glutamate-1-Semialdehyde 2,1-Aminomutase; OEE: Oxygen Evolving Enhancer; Chl: Chlorophyll; CN: cyanate; SCN: Thiocyanate.

family members (#15 to 16), previously described as protecting [Fe-S] clusters [57] in addition to their role in cyanate detoxification.

Previous studies [52,58] highlighted the transcription factor SPL7 (SQUAMOSA-promoter binding link protein 7) which seems to coordinate the early response to Cu deficiency by targeting a Cu response element (CuRE) in the promoter of genes differentially expressed during Cu deficiency. Unfortunately, we couldn't performed an analysis on promoters of genes corresponding to the proteins presented in table 3 as *B. napus* genome sequence is not yet available. However, regulation seems to be complex as Cu deficiency induces a down-regulation at transcript level of carbonic anhydrase [52], but it was up-regulated at protein level in our study (#39). It should be noted that the Carbonic anhydrase regulation is independent from SPL7 [52] and thus its regulation should not be directly induced by Cu deficiency. Moreover, previous studies on Cu deficiency also reported a variation of microRNAs [51,52] or proteases ([59] as well as FTSH2 #14, Table 3 and Figure 5) that could impact on the translation and the proteolysis and thus lead to differences between transcriptomic and proteomic results.

Crosstalk between Cu, [Fe-S] and Mo

Surprisingly, among all quantified mineral nutrients, only Mo uptake was strongly increased in Cu-depleted plants (Figure 1) and this was further linked to strong up-regulation of *MOT1* gene in roots (Figure 2).

The relation between Mo and Cu homeostasis has been reported several times but remains poorly explained [60–63]. As an example, in lesser yam (*Dioscorea esculenta*) [63], molybdenum reached a higher concentration in Cu depleted plants than in plants submitted to any other nutrient deficiencies.

The crosstalk between Cu-Mo and Cu-[Fe-S] established from the literature and the present study are summarised in Figure 6 and can explain the up-regulation of Mo uptake (Figure 1) and expression of [Fe-S] proteins (Rieske #47 and Ferredoxin #30, 31, table 3) by Cu deficiency. Indeed, the chloroplast produces its own [Fe-S] clusters through the SUF (sulfur assimilation proteins)

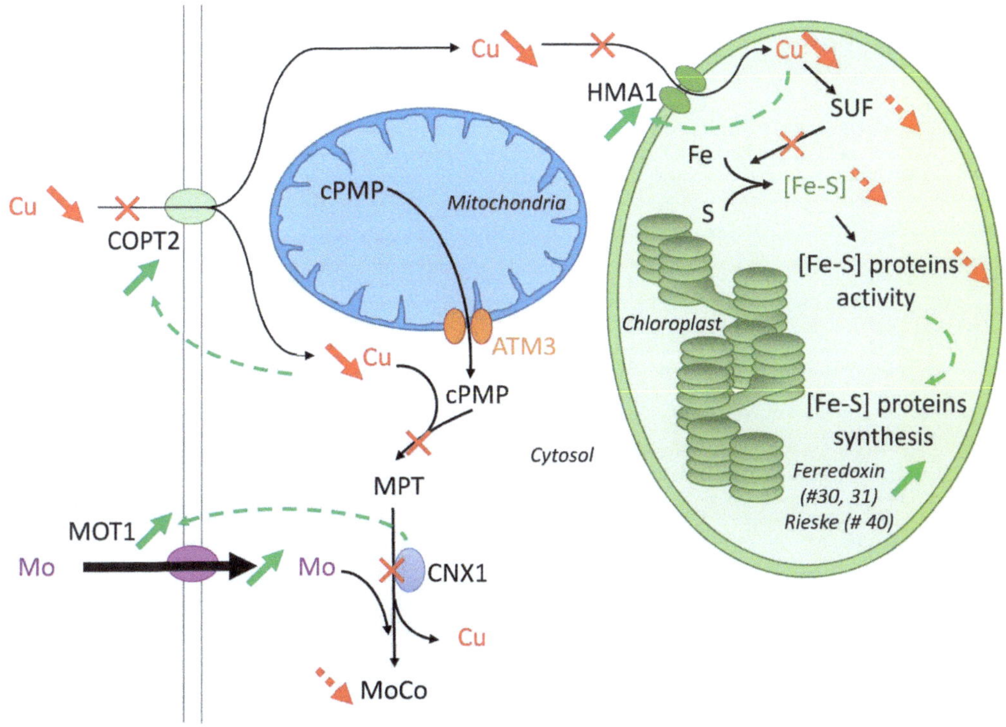

Figure 6. Crosstalk between Cu, [Fe-S] and Mo. Bold arrows indicate variation (green: increase; red: decrease). Full arrows represent observations from this study while dashed arrows have been deduced from the literature. The synthesis of chloroplastic [Fe-S] clusters is controlled by SUF (Sulphur Assimilation Protein), itself controlled by the level of Cu. On the other hand, in the mitochondria, the MoCo precursor cPMP, is exported through mitochondrial membrane by support of the ATM3 transporters. In the cytosol, the second and third steps of MoCo biosynthesis are highly dependent on Cu levels. MoCo is then incorporated into Mo proteins such as Nitrate reductase and into Mo-[Fe-S] proteins such as Xanthine dehydrogenase. The up-regulation of Copt2, HMA1 and Mot1 could be explained by feedback signals due to a low [Cu] in cell, in cytosol and an impairment of CNX1 activity respectively. cPMP: cyclic Pyranopterin MonoPhosphate; MPT: MolybdoPTerin; ATM: ATP-binding cassette Transporters of Mitochondria.

machinery [64], which may be inhibited by Cu deficiency, as previously described in *Bacillus subtilis* [65]. As previously discussed, our leaf proteomic analysis revealed two identified proteins that were over-expressed during Cu deficiency, Ferredoxin and Rieske protein which contain [Fe-S] clusters (#30, 31, 47, Table 3, Figure 6). We may hypothesized that their activity was lowered due to [Fe-S] cluster destabilization by Cu-depletion. This should lead to the observed up-regulation of both proteins to compensate for a possible loss of activity. To minimize the effect of Cu deficiency in chloroplast, the Cu transport across Chloroplast membrane is stimulated through the up regulation of *HMA1* (Figure 2.C, 5, 6).

MOT1 was previously identified as a S transporter (*SULTR5;2*) and it was identified very recently as a high affinity transporter of Mo [45]. Its localization remains unclear (mitochondrial membrane [46] or plasma membrane [45]). Its precise function has not been reported yet, but its importance for an efficient Mo uptake is acknowledged [66]. At cellular level, Mo is essentially used by the Molybdenum Cofactor (MoCo) which is essential for key enzymes such as nitrate reductase and aldehyde oxidase [66]. The first step of the MoCo biosynthesis occurs in the mitochondria and produces cPMP (Figure 6, [67]). cPMP is then exported across the mitochondria membrane by support of the ATM3 transporter [68]. In the cytosol, it must bind Cu in order to be metabolised into MPT representing a step that should be affected in Cu depleted plants. Then, the CNX1 multidomain protein releases Cu and inserts Mo from MPT to give a functional MoCo molecule [69]. It has been reported that CNX1 binds to the cytoskeleton [70] and could have a role as a Mo sensor, interacting with MOT1 transporter to regulates Mo concentration in the cell [71]. Overall, it would explain how the Cu deficiency indirectly up-regulates the Mo uptake through the increase of MOT1 expression. On the other side, a transcriptomic analysis of *Arabidopsis thaliana* WT and *mot1* mutants submitted to Mo deficiency [72] did not reveal a modification of the expression of genes involved in Cu metabolism.

The impairment of Mo metabolism could also be the cause of Ferredoxin and Rieske up-regulation. Indeed, a strong crosstalk exists between Fe and Me metabolisms: some proteins involved in MoCo biosynethesis are Fe-S (but these Fe-S clusters assembly are independent from the SUF machinery) [66], and Ferredoxin transcripts were up-regulated during an Mo deficiency in *mot1* mutants [72].

Acknowledgments

We are most grateful to the PLATIN' (Plateau d'Isotopie de Normandie) core facility for all element analysis used in this study and the authors deeply appreciate the technical help of Josette Bonnefoy and Marie-Paule Bataillé. The authors thank Laurence Cantrill for improving the English in the manuscript. The authors deeply thank the reviewers for greatly improving the quality of this manuscript with their comments.

Author Contributions

Conceived and designed the experiments: VB AO PE JCY FC. Performed the experiments: VB AM MG JMGM TJ LC. Analyzed the data: VB AM

MG LC. Contributed reagents/materials/analysis tools: MG JMGM TJ LC. Contributed to the writing of the manuscript: VB AO PE.

References

1. Allen L, World Health Organization, Food and Agriculture Organization of the United Nations (2006) Guidelines on food fortification with micronutrients. Geneva; Rome: World Health Organization; Food and Agriculture Organization of the United Nations.
2. Collins JF, Klevay LM (2011) Copper. Adv Nutr 2: 520–522. doi:10.3945/an.111.001222.
3. Steinnes E (2009) Soils and geomedicine. Environ Geochem Health 31: 523–535. doi:10.1007/s10653-009-9257-2.
4. White PJ, Broadley MR (2005) Biofortifying crops with essential mineral elements. Trends Plant Sci 10: 586–593. doi:10.1016/j.tplants.2005.10.001.
5. Ruston D, Great Britain. Office for National Statistics. Social Survey Division, Medical Research Council. Human Nutrition Research, Great Britain. Food Standards Agency (2004) The national diet & nutrition survey: adults aged 19 to 64 years: volume 4: nutritional status (anthropometry and blood analytes), blood pressure and physical activity. London: TSO.
6. ANSES (2012) Anses Table Ciqual 2012 Composition nutritionnelle des aliments. Ciqual Fr Food Comp Table Version 2012. Available: http://www.ansespro.fr/TableCIQUAL/index.htm. Accessed 11 February 2014.
7. Bouis I (2012) Biofortification: Leveraging agriculture to reduce hidden hunger. Ed Shenggen Fan Rajul Pandya-Lorch: 83.
8. Fan M-S, Zhao F-J, Fairweather-Tait SJ, Poulton PR, Dunham SJ, et al. (2008) Evidence of decreasing mineral density in wheat grain over the last 160 years. J Trace Elem Med Biol Organ Soc Miner Trace Elem GMS 22: 315–324. doi:10.1016/j.jtemb.2008.07.002.
9. Garvin DF, Welch RM, Finley JW (2006) Historical shifts in the seed mineral micronutrient concentration of US hard red winter wheat germplasm. J Sci Food Agric 86: 2213–2220. doi:10.1002/jsfa.2601.
10. Peverill KI, Sparrow LA, Reuter DJ (1999) Soil Analysis: An Interpretation Manual. Csiro Publishing. 427 p.
11. Alloway BJ (2013) Bioavailability of Elements in Soil. In: Selinus O, editor. Essentials of Medical Geology. Springer Netherlands. 351–373. Available: http://link.springer.com/chapter/10.1007/978-94-007-4375-5_15. Accessed 11 February 2014.
12. Andersen P (2007) A Review of Micronutrient Problems in the Cultivated Soil of Nepal. Mt Res Dev 27: 331–335. doi:10.1659/mrd.0915.
13. Yruela I (2013) Transition metals in plant photosynthesis. Metallomics 5: 1090–1109. doi:10.1039/C3MT00086A.
14. Yruela I (2009) Copper in plants: acquisition, transport and interactions. Funct Plant Biol 36: 409. doi:10.1071/FP08288.
15. Hänsch R, Mendel RR (2009) Physiological functions of mineral micronutrients (Cu, Zn, Mn, Fe, Ni, Mo, B, Cl). Curr Opin Plant Biol 12: 259–266. doi:10.1016/j.pbi.2009.05.006.
16. Tan Y-F, O'Toole N, Taylor NL, Millar AH (2010) Divalent Metal Ions in Plant Mitochondria and Their Role in Interactions with Proteins and Oxidative Stress-Induced Damage to Respiratory Function. Plant Physiol 152: 747–761. doi:10.1104/pp.109.147942.
17. Bailey E, Stirpe F, Taylor CB (1968) Regulation of rat liver pyruvate kinase. The effect of preincubation, pH, copper ions, fructose 1,6-diphosphate and dietary changes on enzyme activity. Biochem J 108: 427–436.
18. Kung C-CS, Huang W-N, Huang Y-C, Yeh K-C (2006) Proteomic survey of copper-binding proteins in Arabidopsis roots by immobilized metal affinity chromatography and mass spectrometry. Proteomics 6: 2746–2758. doi:10.1002/pmic.200500108.
19. Tainer JA, Getzoff ED, Richardson JS, Richardson DC (1983) Structure and mechanism of copper, zinc superoxide dismutase. Nature 306: 284–287. doi:10.1038/306284a0.
20. Broadley M, Brown P, Cakmak I, Rengel Z, Zhao F (2012) Chapter 7 - Function of Nutrients: Micronutrients. In: Marschner P, editor. Marschner's Mineral Nutrition of Higher Plants (Third Edition). San Diego: Academic Press. 191–248. Available: http://www.sciencedirect.com/science/article/pii/B9780123849052000078. Accessed 14 April 2014.
21. Grusak MA, Cakmak I (2005) 12 Methods to improve the crop-delivery of minerals to humans and livestock. Plant Nutr Genomics: 265.
22. Chavez AL, Bedoya JM, Sánchez T, Iglesias C, Ceballos H, et al. (2000) Iron, carotene, and ascorbic acid in cassava roots and leaves. Food Nutr Bull 21: 410–413.
23. Dubousset L, Etienne P, Avice JC (2010) Is the remobilization of S and N reserves for seed filling of winter oilseed rape modulated by sulphate restrictions occurring at different growth stages? J Exp Bot 61: 4313–4324. doi:10.1093/jxb/erq233.
24. Etienne P, Desclos M, Le Gou L, Gombert J, Bonnefoy J, et al. (2007) N-protein mobilisation associated with the leaf senescence process in oilseed rape is concomitant with the disappearance of trypsin inhibitor activity. Funct Plant Biol 34: 895–906.
25. Desclos M, Dubousset L, Etienne P, Le Caherec F, Satoh H, et al. (2008) A proteomic profiling approach to reveal a novel role of Brassica napus drought 22 kD/water-soluble chlorophyll-binding protein in young leaves during nitrogen remobilization induced by stressful conditions. Plant Physiol 147: 1830–1844. doi:10.1104/pp.108.116905.
26. Avice J-C, Etienne P (2014) Leaf senescence and Nitrogen Remobilization Efficiency in oilseed rape (Brassica napus L.). J Exp Bot in press. doi:JEXBOT/2014/119172, IF: 5,54.
27. Abdallah M, Dubousset L, Meuriot F, Etienne P, Avice J-C, et al. (2010) Effect of mineral sulphur availability on nitrogen and sulphur uptake and remobilization during the vegetative growth of Brassica napus L. J Exp Bot 61: 2635–2646. doi:10.1093/jxb/erq096.
28. Waters BM, Armbrust LC (2013) Optimal copper supply is required for normal plant iron deficiency responses. Plant Signal Behav 8.
29. Garcia-Molina A, Andrés-Colás N, Perea-García A, Neumann U, Dodani SC, et al. (2013) The Arabidopsis COPT6 transport protein functions in copper distribution under copper-deficient conditions. Plant Cell Physiol 54: 1378–1390. doi:10.1093/pcp/pct088.
30. Deng F, Yamaji N, Xia J, Ma JF (2013) A member of the heavy metal P-type ATPase OsHMA5 is involved in xylem loading of copper in rice. Plant Physiol 163: 1353–1362. doi:10.1104/pp.113.226225.
31. Higashi Y, Takechi K, Takano H, Takio S (2013) Involvement of microRNA in copper deficiency-induced repression of chloroplastic CuZn-superoxide dismutase genes in the moss Physcomitrella patens. Plant Cell Physiol 54: 1345–1355. doi:10.1093/pcp/pct084.
32. Wintz H, Fox T, Wu Y-Y, Feng V, Chen W, et al. (2003) Expression Profiles of Arabidopsis thaliana in Mineral Deficiencies Reveal Novel Transporters Involved in Metal Homeostasis. J Biol Chem 278: 47644–47653. doi:10.1074/jbc.M309338200.
33. Mora V, Bacaicoa E, Zamarreño A-M, Aguirre E, Garnica M, et al. (2010) Action of humic acid on promotion of cucumber shoot growth involves nitrate-related changes associated with the root-to-shoot distribution of cytokinins, polyamines and mineral nutrients. J Plant Physiol 167: 633–642. doi:10.1016/j.jplph.2009.11.018.
34. Livak KJ, Schmittgen TD (2001) Analysis of Relative Gene Expression Data Using Real-Time Quantitative PCR and the 2−ΔΔCT Method. Methods 25: 402–408. doi:10.1006/meth.2001.1262.
35. Wang W, Scali M, Vignani R, Spadafora A, Sensi E, et al. (2003) Protein extraction for two-dimensional electrophoresis from olive leaf, a plant tissue containing high levels of interfering compounds. ELECTROPHORESIS 24: 2369–2375. doi:10.1002/elps.200305500.
36. Méchin V, Consoli L, Le Guilloux M, Damerval C (2003) An efficient solubilization buffer for plant proteins focused in immobilized pH gradients. Proteomics 3: 1299–1302. doi:10.1002/pmic.200300450.
37. Bradford MM (1976) A rapid and sensitive method for the quantitation of microgram quantities of protein utilizing the principle of protein-dye binding. Anal Biochem 72: 248–254.
38. Blum H, Beier H, Gross HJ (1987) Improved silver staining of plant proteins, RNA and DNA in polyacrylamide gels. ELECTROPHORESIS 8: 93–99. doi:10.1002/elps.1150080203.
39. Bradová J, Matějová E (2008) Comparison of the Results of SDS PAGE and Chip Electrophoresis of Wheat Storage Proteins. Chromatographia 67: 83–88. doi:10.1365/s10337-008-0545-2.
40. Matrix science (n.d.) Mascot search engine | Protein identification software for mass spec data. Available: http://www.matrixscience.com/. Accessed 11 February 2014.
41. RefSeq: NCBI Reference Sequence Database (n.d.). Available: http://www.ncbi.nlm.nih.gov/refseq/. Accessed 22 April 2014.
42. Brassica Genome Gateway (n.d.). Available: http://brassica.nbi.ac.uk/. Accessed 11 February 2014.
43. BLAST: Basic Local Alignment Search Tool (n.d.). Available: http://blast.ncbi.nlm.nih.gov/Blast.cgi?CMD=Web&PAGE_TYPE=BlastHome. Accessed 14 February 2014.
44. Yuan JS, Reed A, Chen F, Stewart CN (2006) Statistical analysis of real-time PCR data. BMC Bioinformatics 7: 85. doi:10.1186/1471-2105-7-85.
45. Tomatsu H, Takano J, Takahashi H, Watanabe-Takahashi A, Shibagaki N, et al. (2007) An Arabidopsis thaliana high-affinity molybdate transporter required for efficient uptake of molybdate from soil. Proc Natl Acad Sci 104: 18807–18812. doi:10.1073/pnas.0706373104.
46. Baxter I, Muthukumar B, Park HC, Buchner P, Lahner B, et al. (2008) Variation in molybdenum content across broadly distributed populations of Arabidopsis thaliana is controlled by a mitochondrial molybdenum transporter (MOT1). PLoS Genet 4: e1000004.
47. Ravet K, Danford FL, Dihle A, Pittarello M, Pilon M (2011) Spatiotemporal analysis of copper homeostasis in Populus trichocarpa reveals an integrated molecular remodeling for a preferential allocation of copper to plastocyanin in the chloroplasts of developing leaves. Plant Physiol 157: 1300–1312. doi:10.1104/pp.111.183350.
48. Marschner P (2012) Marschner's Mineral Nutrition of Higher Plants. Academic Press. 668 p.

49. Del Pozo T, Cambiazo V, González M (2010) Gene expression profiling analysis of copper homeostasis in Arabidopsis thaliana. Biochem Biophys Res Commun 393: 248–252. doi:10.1016/j.bbrc.2010.01.111.
50. Burda K, Kruk J, Schmid GH, Strzalka K (2003) Inhibition of oxygen evolution in Photosystem II by Cu(II) ions is associated with oxidation of cytochrome b559. Biochem J 371: 597–601. doi:10.1042/BJ20021265.
51. Bernal M, Casero D, Singh V, Wilson GT, Grande A, et al. (2012) Transcriptome sequencing identifies SPL7-regulated copper acquisition genes FRO4/FRO5 and the copper dependence of iron homeostasis in Arabidopsis. Plant Cell Online 24: 738–761.
52. Andrés-Colás N, Perea-García A, de Andrés SM, Garcia-Molina A, Dorcey E, et al. (2013) Comparison of global responses to mild deficiency and excess copper levels in Arabidopsis seedlings. Metallomics 5: 1234–1246.
53. Slovacek RE, Monahan BC (1983) Reductive activation of fructose-1,6-bisphosphatase and the peroxide effect on chloroplast photosynthesis. Arch Biochem Biophys 224: 310–318.
54. Munoz-Bertomeu J, Cascales-Minana B, Mulet JM, Baroja-Fernandez E, Pozueta-Romero J, et al. (2009) Plastidial Glyceraldehyde-3-Phosphate Dehydrogenase Deficiency Leads to Altered Root Development and Affects the Sugar and Amino Acid Balance in Arabidopsis. Plant Physiol 151: 541–558. doi:10.1104/pp.109.143701.
55. Mooney S, Hellmann H (2010) Vitamin B6: Killing two birds with one stone? Phytochemistry 71: 495–501. doi:10.1016/j.phytochem.2009.12.015.
56. Hennig M, Grimm B, Contestabile R, John RA, Jansonius JN (1997) Crystal structure of glutamate-1-semialdehyde aminomutase: an alpha2-dimeric vitamin B6-dependent enzyme with asymmetry in structure and active site reactivity. Proc Natl Acad Sci U S A 94: 4866–4871.
57. Cereda A, Carpen A, Picariello G, Tedeschi G, Pagani S (2009) The lack of rhodanese RhdA affects the sensitivity of Azotobacter vinelandii to oxidative events. Biochem J 418: 135. doi:10.1042/BJ20081218.
58. Yamasaki H, Hayashi M, Fukazawa M, Kobayashi Y, Shikanai T (2009) SQUAMOSA promoter binding protein–like7 is a central regulator for copper homeostasis in Arabidopsis. Plant Cell Online 21: 347–361.
59. Castruita M, Casero D, Karpowicz SJ, Kropat J, Vieler A, et al. (2011) Systems Biology Approach in Chlamydomonas Reveals Connections between Copper Nutrition and Multiple Metabolic Steps. Plant Cell Online 23: 1273–1292. doi:10.1105/tpc.111.084400.
60. Lidon FC, Henriques FS (1993) Effects of copper toxicity on growth and the uptake and translocation of metals in rice plants. J Plant Nutr 16: 1449–1464. doi:10.1080/01904169309364626.
61. Wallace A, Romney EM, Alexander GV, Kinnear J (1977) Phytotoxicity and some interactions of the essential trace metals iron, manganese, molybdenum, zinc, copper, and boron. Commun Soil Sci Plant Anal 8: 741–750. doi:10.1080/00103627709366768.
62. Basak A, Mandal LN, Haldar M (1982) Interaction of phosphorus and molybdenum in relation to uptake and utilization of molybdenum, phosphorus, zinc, copper and manganese by rice. Plant Soil 68: 261–269.
63. O'Sullivan JN, Ernest J (2007) Nutrient deficiencies in lesser yam (Dioscorea esculenta) characterized using constant-water table sand culture. J Plant Nutr Soil Sci 170: 273–282. doi:10.1002/jpln.200625049.
64. Pilon M, Abdel-Ghany SE, Van Hoewyk D, Ye H, Pilon-Smits EAH (2006) Biogenesis of iron-sulfur cluster proteins in plastids. Genet Eng (N Y) 27: 101–117.
65. Chillappagari S, Seubert A, Trip H, Kuipers OP, Marahiel MA, et al. (2010) Copper Stress Affects Iron Homeostasis by Destabilizing Iron-Sulfur Cluster Formation in Bacillus subtilis. J Bacteriol 192: 2512–2524. doi:10.1128/JB.00058-10.
66. Bittner F (2014) Molybdenum metabolism in plants and crosstalk to iron. Front Plant Sci 5. Available: http://www.ncbi.nlm.nih.gov/pmc/articles/PMC3916724/. Accessed 18 August 2014.
67. Mendel RR (2013) The Molybdenum Cofactor. J Biol Chem 288: 13165–13172. doi:10.1074/jbc.R113.455311.
68. Teschner J, Lachmann N, Schulze J, Geisler M, Selbach K, et al. (2010) A Novel Role for Arabidopsis Mitochondrial ABC Transporter ATM3 in Molybdenum Cofactor Biosynthesis. Plant Cell Online 22: 468–480. doi:10.1105/tpc.109.068478.
69. Kuper J, Llamas A, Hecht H-J, Mendel RR, Schwarz G (2004) Structure of the molybdopterin-bound Cnx1G domain links molybdenum and copper metabolism. Nature 430: 803–806. doi:10.1038/nature02681.
70. Schwarz G, Schulze J, Bittner F, Eilers T, Kuper J, et al. (2000) The Molybdenum Cofactor Biosynthetic Protein Cnx1 Complements Molybdate-Repairable Mutants, Transfers Molybdenum to the Metal Binding Pterin, and Is Associated with the Cytoskeleton. Plant Cell Online 12: 2455–2471. doi:10.1105/tpc.12.12.2455.
71. Tejada-Jiménez M, Galván A, Fernández E, Llamas Á (2009) Homeostasis of the micronutrients Ni, Mo and Cl with specific biochemical functions. Curr Opin Plant Biol 12: 358–363. doi:10.1016/j.pbi.2009.04.012.
72. Ide Y, Kusano M, Oikawa A, Fukushima A, Tomatsu H, et al. (2011) Effects of molybdenum deficiency and defects in molybdate transporter MOT1 on transcript accumulation and nitrogen/sulphur metabolism in Arabidopsis thaliana. J Exp Bot 62: 1483–1497. doi:10.1093/jxb/erq345.
73. Ferrer J-L, Ravanel S, Robert M, Dumas R (2004) Crystal Structures of Cobalamin-independent Methionine Synthase Complexed with Zinc, Homocysteine, and Methyltetrahydrofolate. J Biol Chem 279: 44235–44238. doi:10.1074/jbc.C400325200.
74. Komenda J, Knoppová J, Krynická V, Nixon PJ, Tichý M (2010) Role of FtsH2 in the repair of Photosystem II in mutants of the cyanobacterium Synechocystis PCC 6803 with impaired assembly or stability of the CaMn4 cluster. Biochim Biophys Acta BBA - Bioenerg 1797: 566–575. doi:10.1016/j.bbabio.2010.02.006.
75. Volini M, Van Sweringen B, Chen F-S (1978) Effects of metal-ion binding on rhodanese activity. Arch Biochem Biophys 191: 205–215. doi:10.1016/0003-9861(78)90083-8.
76. Berry A, Marshall KE (1993) Identification of zinc-binding ligands in the class II fructose-1,6-bisphosphate aldolase of Escherichia coli. FEBS Lett 318: 11–16.
77. Mahmoud YA, el Souod SM, Niehaus WG (1995) Purification and characterization of malate dehydrogenase from Cryptococcus neoformans. Arch Biochem Biophys 322: 69–75.
78. Ikeuchi M, Inoue Y (1986) Characterization of O2 evolution by a wheat photosystem II reaction center complex isolated by a simplified method: disjunction of secondary acceptor quinone and enhanced Ca2+ demand. Arch Biochem Biophys 247: 97–107.
79. Giulia M, Caricato R, Elena M, Erroi E, Schettino T (2012) Carbonic Anhydrase and Heavy Metals. In: Ekinci D, editor. Biochemistry. InTech. Available: http://www.intechopen.com/books/biochemistry/carbonic-anhydrase-and-heavy-metals. Accessed 19 February 2014.

Latitudinal Environmental Niches and Riverine Barriers Shaped the Phylogeography of the Central Chilean Endemic *Dioscorea humilis* (Dioscoreaceae)

Juan Viruel[1], Pilar Catalán[1,2], José Gabriel Segarra-Moragues[3]*

1 Departamento de Agricultura y Economía Agraria, Escuela Politécnica Superior de Huesca, Universidad de Zaragoza, Huesca, Spain, **2** Department of Botany, Institute of Biology, Tomsk State University, Tomsk, Russia, **3** Departamento de Ecología, Centro de Investigaciones sobre Desertificación (CIDE), Consejo Superior de Investigaciones Científicas (CSIC), Moncada, Valencia, Spain

Abstract

The effects of Pleistocene glaciations and geographical barriers on the phylogeographic patterns of lowland plant species in Mediterranean-climate areas of Central Chile are poorly understood. We used *Dioscorea humilis* (Dioscoreaceae), a dioecious geophyte extending 530 km from the Valparaíso to the Bío-Bío Regions, as a case study to disentangle the spatio-temporal evolution of populations in conjunction with latitudinal environmental changes since the Last Inter-Glacial (LIG) to the present. We used nuclear microsatellite loci, chloroplast (cpDNA) sequences and environmental niche modelling (ENM) to construct current and past scenarios from bioclimatic and geographical variables and to infer the evolutionary history of the taxa. We found strong genetic differentiation at nuclear microsatellite loci between the two subspecies of *D. humilis*, probably predating the LIG. Bayesian analyses of population structure revealed strong genetic differentiation of the widespread *D. humilis* subsp. *humilis* into northern and southern population groups, separated by the Maipo river. ENM revealed that the ecological niche differentiation of both groups have been maintained up to present times although their respective geographical distributions apparently fluctuated in concert with the climatic oscillations of the Last Glacial Maximum (LGM) and the Holocene. Genetic data revealed signatures of eastern and western postglacial expansion of the northern populations from the central Chilean depression, whereas the southern ones experienced a rapid southward expansion after the LGM. This study describes the complex evolutionary histories of lowland Mediterranean Chilean plants mediated by the summed effects of spatial isolation caused by riverine geographical barriers and the climatic changes of the Quaternary.

Editor: Tzen-Yuh Chiang, National Cheng-Kung University, Taiwan

Funding: Financial support for this study was provided by a Fundación BBVA BIOCON 05-093/06 project grant to PC and JGSM. JV was supported by a Fundación BBVA Ph.D. grant. JGSM was supported by two consecutive Spanish Aragón Government "Araid" and Spanish Ministry of Science and Innovation "Ramón y Cajal" postdoctoral contracts. PC was partially funded by a Bioflora (http://bifi.es/bioflora/) research team grant co-funded by the Spanish Aragón Government and the European Social Fund. The funders had no role in study design, data collection and analysis, decision to publish, or preparation of the manuscript.

Competing Interests: The authors have declared that no competing interests exist.

* Email: jogasemo@gmail.com

Introduction

Historical, geographical and climatic events have a strong influence on the genetic diversity of species [1]. In South America, most of the biogeographical studies of plants have focused on the effects of Pleistocene glaciations and postglacial climatic fluctuations on Andean species, having identified several lowland refugia [2–4]. However, the phylogeography of lowland species inhabiting ice-free areas during glaciations remains scarcely documented. Population genetic diversity and structure of lowland taxa are not expected to have been severely impacted by the direct effect of glaciations because of the absence of ice sheets in the central Chilean depression [5–6] and the North-to-South arrangement of the Andes, which allowed latitudinal migration [7]. Additionally, the central Chilean depression and its surrounding coastal areas provided the most suitable and stable environments for the establishment of plant and animal populations during Quaternary glaciations [2,4,8–9]. Unlike the high Andean regions, the areas currently occupied by lowland species likely allowed *in situ* survival during glaciations; however, global temperature cooling during the glaciations could have also contributed to narrowing their geographical ranges to warmer areas.

During the Last Glacial Maximum (LGM, 25000–15000 years ago), ice sheets extended from 56°S to 35°S along the Andes [5–6]. These extensive glaciations affected the central Chilean valleys of Maipo and Aconcagua [10]. Although Quaternary glaciers reached down to 1200–2800 m.a.s.l. [10–11], their occurrence was coupled with a decrease in temperature and an increase in the precipitation rates at lower altitudes [12–13].

In addition to the West-to-East barriers imposed by the Coastal Cordillera and the Andean mountains, it has been proposed that large rivers (e.g. Aconcagua, Maipo) that completely cross Chile may contribute to within-species differentiation [14]. Water volume carried by those rivers fluctuated concomitantly with

Pleistocene glaciations, increasing considerably due to ice-melting from the Andes. Accordingly, their potential barrier effect to species migration was stronger during the glacial periods than during the interglacials [15]. The genetic structure of central Chilean lowland species during these glaciations may have been affected by an East-to-West contraction of their distribution ranges towards the central Chilean depression and by their dispersal ability to bypass the transversal river barriers during latitudinal migration.

The *Epipetrum* group of *Dioscorea* is a small evolutionary lineage of the Dioscoreaceae including two species, *D. humilis* Colla and *D. biloba* (Phil.) Caddick & Wilkin, with two subspecies in each [16] that probably originated in the late Miocene (Viruel *et al.*, unpublished data). The diversification of this small group followed the retreat of the marine transgressions of the middle Miocene (15-11 Ma) which covered central Chile, providing new lands available for plant colonization from the late Miocene onwards [17]. *Dioscorea humilis* is a dioecious, diploid ($2n = 14$), dwarf geophyte with a widespread distribution spanning five central Chilean regions (530 km), from its northernmost limit in Valparaíso to its southernmost limit in Bío-Bío [16,18] (Fig. 1, Table 1). Its current distribution range is included within the Mediterranean-type bioclimatic region of Chile [19], which is bounded northwards by the Atacama Desert and southwards by temperate forests [8]. This North-to-South range covers three different climatic environments (Fig. 1): semi-arid, sub-humid and humid Mediterranean climates [20]. *Dioscorea humilis* occurs in the lowland depression between the coastal mountain range and the Andes. It includes two subspecies, the widespread *D. humilis* subsp. *humilis* and the narrow parapatric Maule coastal endemic *D. humilis* subsp. *polyanthes* (F. Phil.) Viruel, Segarra-Moragues & Villar [16] (Fig. 1).

Dioscorea humilis has a sprawling habit with shoots creeping among rock crevices. Flowers are tiny and inconspicuous; those of males are produced in pauciflorate racemes, and those of females are generally solitary. The pollination mechanisms are unknown, but flower morphology suggests the implication of a small-sized insect. The wingless seeds are produced in capsules which are sustained by spirally curled peduncles that attach capsules close to the ground or inside rock crevices, suggesting extremely short-distance seed dispersal [16].

We used nuclear microsatellite markers and cpDNA sequences to document the current patterns of population genetic diversity and structure in *D. humilis*. Additionally, Environmental Niche Modelling (ENM) was estimated on the current range extension of the infraspecific genetic groups and projected to two past scenarios, the Last Glacial Maximum (LGM) and Last Inter-Glacial (LIG). Phylogeographical patterns obtained from molecular markers, together with the estimated past variation in range extension, were investigated to elucidate the effect of Pleistocene glaciations and geological and hydrological barriers in the evolutionary history of lowland central Chilean species with limited dispersal abilities like *D. humilis*.

Materials and Methods

Ethic Statement

Necessary permits for fieldwork and sampling were obtained from the Corporación Nacional Forestal (CONAF-Chile).

Plant Sampling, DNA Extraction and Microsatellite Amplification

Fresh leaves from a total 558 individuals from 17 populations of *D. humilis* were collected throughout its entire distribution range. Fifteen populations (Dhh01-Dhh15) corresponded to *D. humilis* subsp. *humilis* and two populations (Dhp01-Dhp02) to *D. humilis* subsp. *polyanthes* (Table 1, Fig. 1). Eight populations of *D. humilis* subsp. *humilis* (Dhh01-Dhh08) were located North of the Maipo river basin, growing in semi-arid Mediterranean-type climate areas, whereas the other seven populations (Dhh09-Dhh15) were located South of it (Table 1, Fig. 1). Five of these (Dhh09-Dhh14) were growing in sub-humid Mediterranean-type climate areas, and the southernmost population (Dhh15), together with two populations of *D. humilis* subsp. *polyanthes* (Dhp01-Dhp02), were growing in humid Mediterranean-type climate areas [20]. DNA extraction followed the procedure described in [21]. Individuals were genotyped for eight unlinked microsatellite following [22].

Plastid DNA Amplification and Sequencing

Two plastid regions, *trn*T-L and *trn*L-F [23] were amplified and sequenced in up to six individuals per population following [21]. Sequences were deposited in Genbank under the accession numbers KF357945-KF357955. A combined matrix of individual sequences of both plastid regions totalling 54 sequences was used in subsequent analyses.

Microsatellite Analysis

Allele frequencies and genetic diversity indices were calculated in all populations using GENETIX 4.05 [24]. Deviations from Hardy-Weinberg equilibrium were tested in all populations using GENEPOP v. 4.0 [25]. Different taxonomic and geographical population groups were compared to reveal differences in average values of allelic richness (A^*), observed heterozygosity (H_O), genetic diversity within populations (H_S), inbreeding coefficient (F_{IS}) and population differentiation (F_{ST}) using FSTAT v. 2.9.3.2 [26], and differences were tested for significance with 10,000-permutation tests. Population pairwise differentiation (F_{ST}) was calculated with ARLEQUIN 3.11 [27] and tested for significance using 1000 replicates. ARLEQUIN was also used to generate a matrix of pairwise linearized F_{ST} values (i.e. $F_{ST}/(1-F_{ST})$; [28]), which was correlated to a log-transformed matrix of geographical distances between populations to test for Isolation By Distance (IBD) through Mantel tests. Significance of correlation was tested with 1000 permutations with NTSYSpc 2.11 [29].

Pairwise D_A genetic distances [30] between populations were calculated with POPULATIONS 1.2.3 [31] and used to conduct a Principal Coordinates Analysis (PCO) and a Minimum Spanning Tree (MST) that was superimposed on the PCO plots using NTSYSpc 2.11.

Population genetic structure was investigated by means of Analysis of Molecular Variance (AMOVA) which was performed in ARLEQUIN 3.11 to partition in different population groups according to the taxonomical or geographical membership. The significance of the analyses was tested with 1000 replicates.

Bayesian clustering was also used to infer population genetic structure using STRUCTURE 2.1 [32]. Analyses were based on an admixture ancestry model with correlated allele frequencies, for a range of K genetic clusters from one to 19, with ten replicates for each K. The analyses were performed with a burn-in period and a run length of the Monte Carlo Markov Chain (MCMC) of 7×10^5 and 7×10^6 iterations, respectively. The most likely number of genetic clusters (K) was determined according to Evanno *et al.* [33].

Plastid DNA Data Analyses

Haplotype polymorphism was estimated within populations and within genetic and geographical groups through the analysis of the number of segregating sites (S), the number of haplotypes (h), the

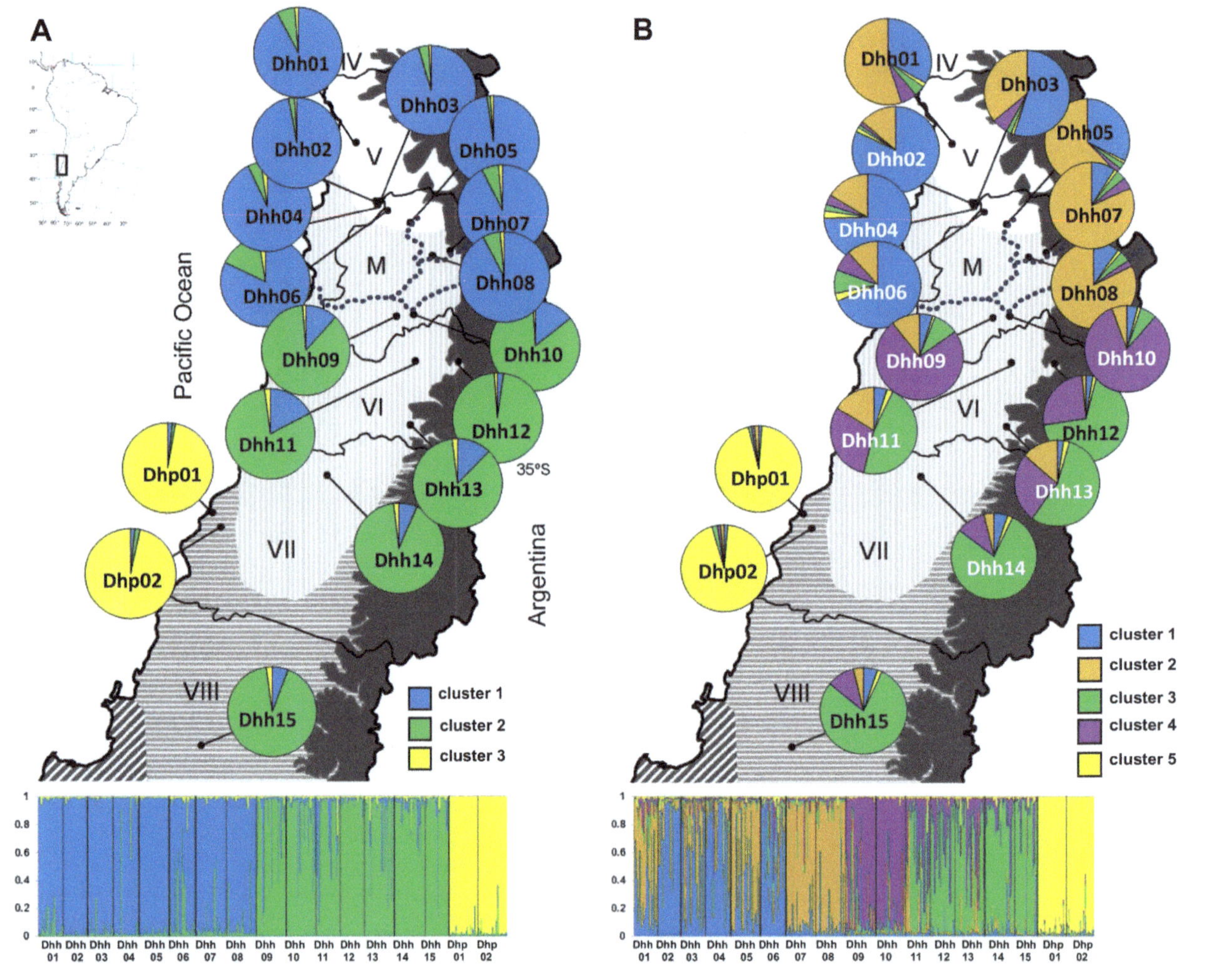

Figure 1. Geographical distribution of sampled populations of *Dioscorea humilis* in Chile (Table 1) and Bayesian analyses of the genetic structure of 15 populations of *D. humilis* subsp. *humilis* and two populations of *D. humilis* subsp. *polyanthes* based on nuclear microsatellite data. The mean proportion of membership of each predefined population to each of the A, three ($K=3$), and B, five ($K=5$), most likely inferred genetic clusters is shown. The dotted line indicates the location of the Maipo river. Chilean administrative regions: IV, Coquimbo, V, Valparaíso, M, Metropolitana, VI, Libertador General Bernardo O'Higgins, VII, Maule, and VIII, Bío-Bío. Geographical ranges of five climatic zones in central Chile from Castillo *et al.*, [20] are superimposed on the maps. From North to South: semiarid Mediterranean (white), sub-humid Mediterranean (vertical shading), humid Mediterranean (horizontal shading), hyper humid Mediterranean (diagonal shading) and eastern Andean Continental (solid grey). Map contour constructed from spatial data retrieved from http://www.diva-gis.org/gdata.

haplotype diversity index (*Hd*) and the average number of pairwise nucleotide differences between DNA sequences ($\theta\pi$) [34] with DnaSP5 [35]. Indels encompassing two to several nucleotides were reduced to single gaps and treated as a fifth nucleotide state for a statistical parsimony haplotype network analysis with TCS v. 1.21 software [36].

Environmental Niche Modelling Analyses

Environmental niche modelling (ENM) was conducted to evaluate the potential distribution of the geographical groups of *D. humilis* under current climatic conditions and under Last Glacial Maximum (LGM) and Last Interglacial (LIG) conditions. A set of 19 bioclimatic variables (Table S1 in Appendix S1) retrieved from WorldClim (www.worldclim.org) plus the altitude were used, and GIS layers with 30 sec resolution were clipped to the extent of central Chilean regions using DIVA-GIS [37]. Correlation among environmental variables was determined by Mantel tests using XLSTAT and tested for significance with 1000 random permutations (Table S1 in Appendix S1). Then we selected a reduced set of nine uncorrelated environmental variables with higher percent contribution (PC) and permutation importance (PI) based on jackknife pseudosampling on the ENM of *D. humilis* (Tables S1 and S2 in Appendix S1): altitude, bio3 (isothermality), bio4 (temperature seasonality), bio6 (minimum temperature of coldest month), bio7 (annual range temperature), bio9 (mean temperature of driest quarter), bio15 (precipitation seasonality), bio18 (precipitation of warmest quarter) and bio19 (precipitation of coldest quarter).

Additionally, we assessed pairwise correlations between all 20 environmental variables studied and pairwise D_A population genetic distances [30], and pairwise population linearized F_{ST} [28], among populations of *D. humilis*, and the correlation between the 20 environmental variables and latitude, using the Mantel test with 1000 random permutations.

The maximum entropy algorithm implemented in MAXENT v. 3.3.3k [38–39] was used to construct the models. Maxent is

Table 1. Population data and genetic diversity indices of 17 Chilean populations of *Dioscorea humilis* for eight nuclear microsatellite loci and overall estimates for taxonomic and geographical population groups.

Code	Locality	Latitude	Longitude	Altitude (m)	N^1	A^1	H_O^1	H_E^1	F_{IS}^1
***Dioscorea humilis* subsp. *Humilis* (Dhh01 to Dhh15)**					488	3.969	0.468	0.425	−0.103***
Northern *D. humilis* subsp. *humilis* (Dhh01 to Dhh08)					257	3.800	0.466	0.410	−0.136***
Dhh01	Valparaíso: Catapilco, rincón de La Mestiza	32° 32′ 33.4″ S	71° 17′ 37.7″ W	90	30	4.125	0.471	0.405	−0.165**
Dhh02	Metropolitana: Granizo, Parque Nacional de La Campana, sector Granizo	32° 58′ 54.9″ S	71° 07′ 58.2″ W	450	28	4.000	0.424	0.386	−0.100**
Dhh03	Metropolitana: Granizo, Parque Nacional de La Campana, La Troya bridge.	32° 59′ 05.3″ S	71° 08′ 17.8″ W	420	30	4.125	0.425	0.371	−0.148**
Dhh04	Metropolitana: Parque Nacional de La Campana, sector Cajón grande	33° 00′ 06.1″ S	71° 07′ 53.5″ W	340	30	3.500	0.350	0.391	+0.107*
Dhh05	Metropolitana: Carretera a Til Til, Entrance to Parcelación de El tranque.	33° 08′ 40.7″ S	70° 52′ 34.8″ W	530	36	3.875	0.438	0.392	+0.036**
Dhh06	Metropolitana: Road from Til Til to Limache, detour to La Vega through el Camarico.	33° 03′ 28.3″ S	71° 02′ 57.2″ W	700	31	3.875	0.510	0.458	+0.019**
Dhh07	Metropolitana: Santiago de Chile, Cerro Manquehue, creek starting from Agua del Palo.	33° 21′ 50.3″ S	70° 34′ 56.9″ W	820	36	3.625	0.523	0.423	−0.241**
Dhh08	Metropolitana: Santiago de Chile, Cerro de Renca.	33° 23′ 43.1″ S	70° 42′ 38.9″ W	540	36	4.500	0.560	0.453	−0.135**
Southern *D. humilis* subsp. *humilis* (Dhh09 to Dhh15)					231	4.163	0.472	0.442	−0.068***
Dhh09	Metropolitana: Maipo, Cerro Cantillana, side of forest track from Rague to Pabellón.	33° 51′ 10.8″ S	70° 58′ 44.5″ W	630	36	4.625	0.460	0.467	+0.108**
Dhh10	Metropolitana : Maipo, Cerros de Aculeo.	33° 50′ 04.4″ S	70° 51′ 14.0″ W	380	36	4.714	0.549	0.521	−0.044^{ns}

Table 1. Cont.

Code	Locality	Latitude	Longitude	Altitude (m)	N^1	A^1	$H_O{}^1$	$H_E{}^1$	$F_{IS}{}^1$
Dhh11	OHiggins: Road from Rancagua to Doñihue.	34° 11′ 28.2″ S	70° 50′ 39.4″ W	430	30	5.000	0.432	0.459	+0.078**
Dhh12	OHiggins: Road from Coya to Pangal.	34° 12′ 03.2″ S	70° 30′ 58.4″ W	850	29	3.250	0.453	0.357	−0.177**
Dhh13	OHiggins: Road from San Fernando to Tinguiririca, next to La Rufina forest track.	34° 40′ 27.0″ S	70° 52′ 56.4″ W	550	36	4.625	0.485	0.444	−0.073**
Dhh14	Maule: Road from Curicó to Sagrada Familia.	35° 03′ 03.0″ S	70° 31′ 03.2″ W	170	36	4.875	0.447	0.413	+0.046ns
Dhh15	Bio-Bio: Between Yumbel and Monteáguila	37° 05′ 00.4″ S	72° 28′ 46.1″ W	110	28	3.875	0.556	0.502	−0.064*
***Dioscorea humilis* subsp. *polyanthes* (Dhp01-Dhp02)**					**70**	4.131	0.574	0.480	−0.195***
Dhp01	Maule: Constitución, just after the river Maule bridge to Putú.	35° 20′ 05.6″ S	72° 23′ 17.2″ W	20	35	4.250	0.598	0.489	−0.222**
Dhp02	Maule: de Constitución a San Javier	35° 26′ 24.1″ S	72° 20′ 00.2″ W	330	35	4.500	0.550	0.475	−0.162ns

[1] *N*, sample size; *A*, mean number of alleles per locus; H_O, H_E, observed and expected heterozygosity, respectively; F_{IS}, inbreeding coefficient. Significance: *, $P<0.05$, **, $P<0.01$, ***, $P<0.001$, ns, not significant.

optimal for ENM when using small sample sizes [40] and when environmental predictions are poorly influenced by the addition of irrelevant bioclimatic variables [41]. The *D. humilis* data fit these requirements since no significant increase in the area under the curve (AUC) values was observed when using all variables compared to those from the reduced set of variables (Table S3 in Appendix S1).

Occurrence data were split into training data (75%) to build the model and test data (25%) to test the accuracy of the model. Fifteen subsample replicates were performed in each run using the default options and 1000 iterations. Model accuracy was assessed with the AUC value of the receiver-operating characteristic curve (ROC) [38]. The contribution of each environmental variable to the ENM was evaluated through a Jackknife pseudosampling (see above). A tenth percentile threshold was applied for all models.

ENM were conducted for the two northern (Dhh01-Dhh08) and southern (Dhh09-Dhh15) population groups of *D. humilis* ssp. *humilis*. The low number of known populations of *D. humilis* subsp. *polyanthes* (Dhp01-02) precluded a confident ENM analysis of this taxon.

ENMs were projected to LGM (c. 21 ka BP), with 2.5 arc-minutes resolution [42], and to LIG (c. 120–140 ka BP), with 30 arc-seconds resolution [43] scenarios. Two palaeoclimatic layers simulated for two general atmospheric circulation models were used for LGM: the Community Climate System Model (CCSM, [44]) and the Model for Interdisciplinary Research on Climate (MIROC, [45]). Both CCSM and MIROC layers were combined following a conservative approach by including their overlapping predicted areas [46]. Current minimum predicted values were used to determine the past minimal predicted areas, assuming that the environmental requirements of *D. humilis* subsp. *humilis* have remained stable during at least since LIG.

A complementary ENM approach was done through a Principal Component Analysis (PCA), which was constructed with the raw data obtained from the 19 climatic variables and the altitude for each population of *D. humilis* using PAST 2.17c [47], Fig. S1 in Appendix S1).

Results

Microsatellite Genetic Diversity in *Dioscorea humilis*

All eight microsatellite loci were polymorphic and amplified a total of 79 alleles in the 17 studied populations of *D. humilis* (Table S4 in Appendix S2). The number of alleles per locus ranged from three (B633) to 24 (H442) with a mean of 9.88±6.38 (±SD) alleles per locus. The mean number of alleles per locus and population ranged from 3.25 (Dhh12) to 5.00 (Dhh11, Table 1). Of the 79 microsatellite alleles scored, 34 (43.04%) were shared by both subspecies of *D. humilis*, while 36 (45.57%) and 9 (11.39%) were exclusive to *D. humilis* subsp. *humilis* and *D. humilis* subsp. *polyanthes*, respectively (Table S4 in Appendix S2).

Observed heterozygosities ranged from 0.350 (Dhh04) to 0.598 (Dhp01), and unbiased expected heterozygosities from 0.357 (Dhh12) to 0.521 (Dhh10) (Table 1). Five of the 17 populations showed HW deviations towards heterozygote deficiency; three, including one population of *D. humilis* subsp. *polyanthes*, showed non-significant departure from HW equilibrium, and the remaining eight populations of *D. humilis* subsp. *humilis* and one of *D. humilis* subsp. *polyanthes* showed a significant heterozygote excess (Table 1).

No significant differences were detected for the tested genetic diversity indices between *D. humilis* subsp. *humilis* and *D. humilis* subsp. *polyanthes*, except for observed heterozygosity (H_o). Surprisingly, the more restricted endemic *D. humilis* subsp. *polyanthes* showed significantly higher ($p = 0.033$) average H_o (Table 1). Similarly, the comparison of northern and southern population groups of *D. humilis* subsp. *humilis* failed to find significant differences at any of the tested indices (Table 1).

Population Structure of *Dioscorea humilis*

Moderate but significant (different from zero; $p<0.05$) levels of population differentiation were observed among populations (results not shown). Higher average F_{ST} values were found between populations of both subspecies (average $F_{ST} = 0.295$) than among populations within subspecies (average Dhh $F_{ST} = 0.145$; Dhp $F_{ST} = 0.011$). Similarly, a higher average differentiation was observed between northern and southern populations groups of *D. humilis* subsp. *humilis* (average $F_{ST} = 0.198$), than among populations within northern populations (Dhh01-Dhh08 $F_{ST} = 0.069$) and southern populations (average Dhh09-Dhh15 $F_{ST} = 0.109$) of *D. humilis* subsp. *humilis* with differentiation between the groups not being statistically significant (Table 2).

Bayesian analysis of population structure showed a maximum $\Delta K = 1598.45$ value for $K = 3$ (Fig. S2 in Appendix S2). In this clustering, individuals of *D. humilis* subsp. *polyanthes* showed a high proportion of membership to cluster 3 and those of *D. humilis* subsp. *humilis* to cluster 1 (populations Dhh01-Dhh08) or to cluster 2 (populations Dhh09-Dhh15; Fig. 1a). Mean F_{ST} values corresponding to the divergence between clusters 1, 2 and 3 and the hypothetical ancestral population were 0.114, 0.201 and 0.266, respectively, indicating that populations showing a higher membership to cluster 1 were less diverged from the ancestral population. A further maximum $\Delta K = 127.90$ value was obtained for $K = 5$ (Fig. S2 in Appendix S2) which separated the populations of *D. humilis* subsp. *humilis* into two additional genetic clusters (clusters 1–4; Fig. 1b).

Non-hierarchical AMOVA attributed 19.03% of the total variation to among populations of *D. humilis s.l.*, and 15.00% of the total variation to among populations of *D. humilis* subsp. *humilis* (Table 2). In hierarchical AMOVA, the largest proportion of variation among groups (21.37%) was obtained for a taxonomical grouping of populations into subspecies. AMOVA based on a geographical grouping of populations attributed 12.15% of the variation to differences between northern and southern groups of *D. humilis* subsp. *humilis* and a lower proportion of variance (7.65%) to differences among populations within groups (Table 3). The grouping of *D. humilis* subsp. *humilis* populations into four genetic clusters did not increase the variance among groups (11.81%) but lowered the proportion of variance among populations within groups (5.34%).

PCO showed results consistent with STRUCTURE (Figs. 1 and 2). Populations of *D. humilis* subsp. *polyanthes* separated at a large distance from populations of *D. humilis* subsp. *humilis* (Fig. 2). Clustering of populations of this latter taxon was consistent with their geographical distribution (Fig. 2). PCO with superimposed MST analysis identified the closer relationship of *D. humilis* subsp. *polyanthes* to the southern populations of *D. humilis* subsp. *humilis* (Fig. 2).

A significant correlation between pairwise geographical distances and linearized F_{ST} values was found in both *D. humilis s.l.* ($r = 0.537$, $p = 0.001$) and *D. humilis* subsp. *humilis* ($r = 0.416$, $p = 0.004$) populations (Fig. 3a), thus showing significant isolation by distance (IBD). However, the pattern of IBD vanished when this analysis was separately conducted within both northern ($r = 0.296$, $p = 0.080$) and southern ($r = 0.005$, $p = 0.420$) geographical groups of *D. humilis* subp. *humilis* (Fig. 3b).

Table 2. Analyses of molecular variance (AMOVA) of *Dioscorea humilis* populations based on microsatellite data.

Source of variation (groups)	Sum of squared deviations (SSD)	d.f.	Variance components	% of the total variance
1. *Dioscorea humilis* s.l.				
Among populations	443.049	16	0.39637	19.03
Within populations	1852.940	1099	1.68602	80.97
2. Taxonomic membership: *humilis* (Dhh01-Dhh15) *vs. polyanthes* (Dhp01. Dhp02)				
Among groups	150.915	1	0.53219	21.37
Among populations within groups	292.314	15	0.27222	10.93
Within populations	1852.940	1099	1.68602	67.70
3. *Dioscorea humilis* subsp. *humilis s.l.*				
Among populations	288.727	14	0.29176	15.00
Within populations	1589.255	961	1.65375	85.00
4. Geographical membership of *D. humilis* subsp. *humilis*: northern (Dhh01–08), *vs.* southern (Dhh09–15)				
Among groups	133.969	1	0.25056	12.15
Among populations within groups	154.758	13	0.15781	7.65
Within populations	1589.255	961	1.65375	80.20
5. Genetic membership (excluding *D. humilis* susbp. *polyanthes*): cluster Dhh01, Dhh05, Dhh07-Dhh08 *vs.* cluster Dhh02-Dhh04, Dhh06 *vs.* Dhh09-Dhh10 *vs.* Dhh11-Dhh15				
Among groups	194.783	3	0.23571	11.81
Among populations within groups	93.944	11	0.10659	5.34
Within populations	1589.255	961	1.65375	82.85

Plastid Haplotype Diversity in *Dioscorea humilis*

The combination of *trn*L–F and *trn*T–L plastid DNA regions produced eight haplotypes (Fig. 4; Table S5 in Appendix S2). Six and one haplotypes were restricted to *D. humilis* subsp. *humilis* and subsp. *polyanthes* respectively, and one haplotype was shared between both taxa. Haplotype IV was widespread in 12 populations of *D. humilis*, including one population of *D. humilis* subsp. *polyanthes*, and had the highest outgroup probability (0.771). Three haplotypes (I, II and III) were restricted to some western populations of northern *D. humilis* subsp. *humilis*, whereas haplotypes V, VI and VII were restricted to some southern populations of *D. humilis* subsp. *humilis* (Fig. 4a; Table S5 in Appendix S2).

Haplotype diversity was higher in *Dioscorea humilis* subsp. *humilis* ($S = 6$, $\theta\pi = 0.789$), than in *D. humilis* subsp. *polyanthes* ($S = 1$, $\theta\pi = 0.500$) (Table 3), as expected for the wider distribution range and population abundance of the former. The northern group of populations of *D. humilis* subsp. *humilis* (Dhh01-Dhh08) showed higher haplotype diversity ($S = 3$, $\theta\pi = 0.873$) compared to the southern group (Dhh09-Dhh15) of populations ($S = 3$, $\theta\pi = 0.534$) (Table 3).

TCS estimated a 95% maximum connection of 17 steps incorporating all eight haplotypes into the network and inferred three missing haplotypes (Fig. 4b). The haplotype network showed a star-like pattern with four of the six derived haplotypes directly connected to the most widespread one (Hap. IV; Fig. 4). Three haplotypes were connected to the central one at a larger number of mutations. Two of them were private to the northernmost population (Hap. I and II) whereas the other (Hap. V) was private to a population from the southern group (Fig. 4).

Environmental Niche Modelling

All rainfall-derived variables (bio12-bio19) and all but three temperature-derived variables (bio6, bio8 and bio11) were significantly correlated to genetic distances (Table S1 in Appendix S1). Also, latitude was highly correlated to all rainfall-derived variables (bio12-bio19), to two temperature-derived variables (bio2-bio3), and to both pairwise populations' D_A and F_{ST} genetic distances (Table S1 in Appendix S1).

According to response curves and Jackknife tests, the most informative variables for the ENM of *D. humilis s.l.* were altitude and three climatic variables derived from rainfall data (bio16, precipitation wettest quarter; bio18, precipitation warmest quarter; and bio19, precipitation coldest quarter). At the subspecies level, the variables with the largest contributions to the ENM of *D. humilis* subsp. *humilis* were altitude and the climatic variables bio8 (mean temperature of wettest quarter), bio9 (mean temperature of driest quarter) and bio15 (precipitation seasonality) (Table S2 in Appendix S1). Independent ENM for northern (Dhh01-Dhh08) and southern (Dhh09-Dhh15) genetic groups of *D. humilis* subsp. *humilis* revealed that the variables bio15 and bio18 were most informative for the northern group, whereas bio8, bio9 and bio15 were most informative for the southern group (Table S2 in Appendix S1). All projections (Figs. 5a–5c) showed excellent predictive success rates, with AUC values higher than 0.9 (Table S3 in Appendix S1). The PCA of environmental variables for the *D. humilis* populations (Fig. S1 in Appendix S1) separated northern from southern genetic groups of *D. humilis* subsp. *humilis*. The southernmost population, Dhh15, showed a distinct set of climatic conditions from the others, according to its separated position, and clustered together with the *D. humilis* subsp. *polyanthes* populations in the bidimensional PCA plot (Fig. S1 in Appendix S1).

Table 3. Plastid combined *trn*TL- *trn*LF haplotype diversity analysis of *D. humilis* populations and geographical/genetic groups.

Population/Group	*N*	*S*	*h*	*Hd*	θπ
D. humilis ssp humilis (Dhh01-15)	45	6	6	0.391	0.789 (0.000–2.404)
Northern range (Dhh01-08)	27	3	3	0.325	0.873 (0.000–2.610)
Dhh01-06	21	3	3	0.400	1.089 (0.000–3.124)
Dhh07-08	6	0	1	0.000	-
Dhh01	5	1	2	0.400	0.397 (0.000–1.800)
Dhh02	2	0	1	0.000	-
Dhh03	2	0	1	0.000	-
Dhh04	4	0	1	0.000	-
Dhh05	4	0	1	0.000	-
Dhh06	4	0	1	0.000	-
Dhh07	3	0	1	0.000	-
Dhh08	3	0	1	0.000	-
Southern range (Dhh09-15)	18	3	4	0.477	0.534 (0.000–1.863)
Dhh09-10	4	1	2	0.667	0.664 (0.000–2.500)
Dhh11-15	14	2	3	0.385	0.406 (0.000–1.560)
Dhh09	2	0	1	0.000	-
Dhh10	2	0	1	0.000	-
Dhh11	2	0	1	0.000	-
Dhh12	2	0	1	0.000	-
Dhh13	3	0	1	0.000	-
Dhh14	4	1	2	0.667	0.673 (0.000–2.500)
Dhh15	3	1	2	0.667	0.656 (0.000–2.667)
D. humilis ssp polyanthes (Dhp01-02)	9	1	2	0.500	0.506 (0.000–1.889)
Dhp01	3	0	1	0.000	-
Dhp02	6	0	1	0.000	-
Total	54	7	7	0.419	0.768 (0.000–2.379)

Population codes, sample size (*N*), and combined *trn*TL- *trn*LF haplotype frequency parameters: number of segregating sites (*S*), number of distinct haplotypes (*h*), and haplotype diversity (*Hd*) and molecular diversity (θπ) estimates (with 95% confidence intervals of θπ generated through 10,000 θ-based simulations under the coalescence model using the program DNAsp v.5 [35].

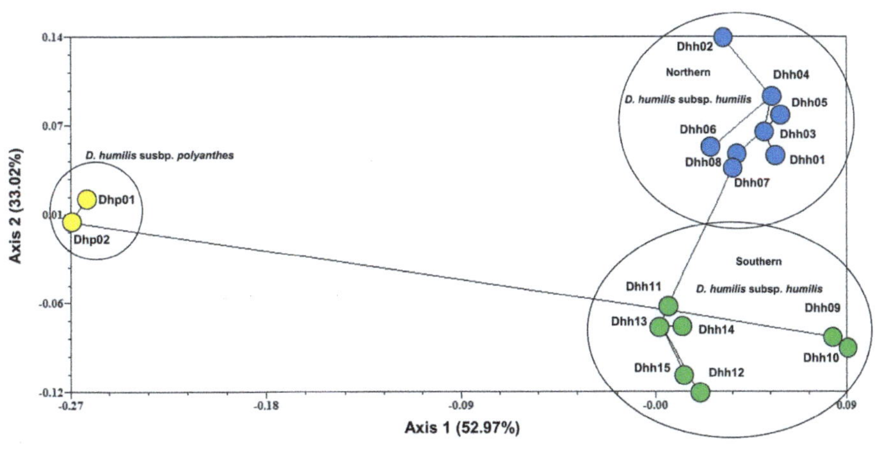

Figure 2. Principal Coordinates Analysis (PCO) showing the genetic relationships among populations of *Dioscorea humilis* based on D_A genetic distance [30]**.** Populations of *D. humilis* subsp. *polyanthes* (Dhp01, Dhp02), yellow circles; northern populations of *Dioscorea humilis* subsp. *humilis* (Dhh01 to Dhh08), blue circles; southern populations of *D. humilis* subsp. *humilis* (Dhh09 to Dhh15), green circles.

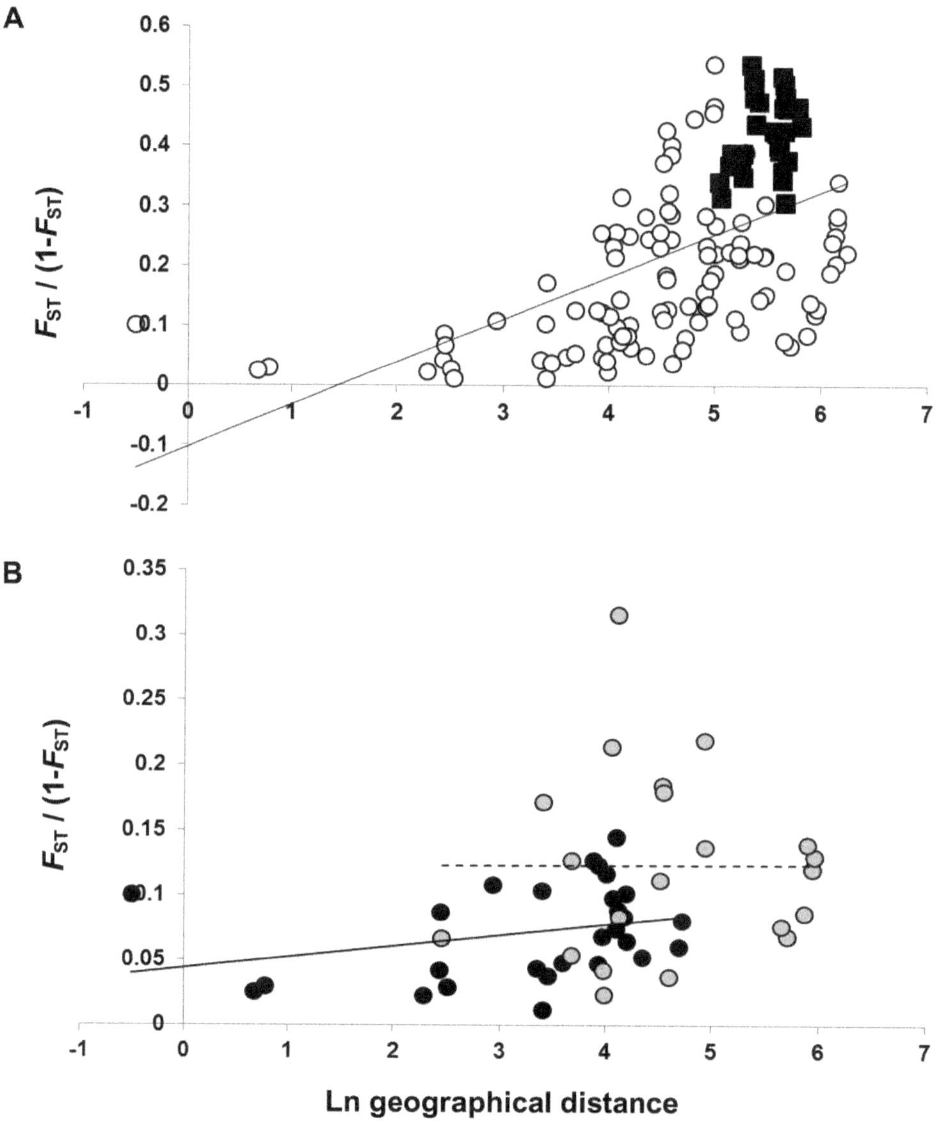

Figure 3. Isolation by distance analyses. Correlation between log-transformed pairwise geographical distances and linearized F_{ST} values [28] among populations of *D. humilis*. A. *D. humilis s.l.* where open circles represent pairwise comparisons among populations of *D. humilis*. subsp. *humilis* and black squares represent pairwise comparisons among populations of *D. humilis*. subsp. *humilis* and *D. humilis* subsp. *polyanthes*. Correlation between matrices was $r=53.74\%$, $p=0.001$ for *D. humilis s.l.* and $r=41.61\%$, $p=0.004$ for *D. humilis* subsp. *humilis* B. IBD analyses within geographical groups of *D. humilis* susbp. *humilis*, where black circles represent pairwise comparisons among populations of the northern group (Dhh01-Dhh08) and grey circles represent pairwise comparisons among populations of the southern group (Dhh09-Dhh15). Correlation between matrices was $r=29.62\%$, $p=0.080$ and $r=0.47\%$, $p=0.420$ for the northern and southern groups, respectively.

The ENM for current environmental conditions was mostly concordant with the current distribution of northern and southern genetic groups of *D. humilis* subsp. *humilis* (Fig. 5c), except for the southernmost population, Dhh15. Potential areas of contact were predicted on the eastern boundary of their distributions. Projections to the LIG (Fig. 5a) predicted a minimal extension area for the potential distribution of the southern group and a larger extension for the northern group. Projections to the LGM (Fig. 5b) predicted a substantial reduction in the southern group and, to a lesser extent, of the northern group. The present and the two historical models predicted areas with unsuitable environmental conditions between the northern and southern groups (Fig. 5), and an increase in the potential areas of contact between the two groups since the LGM to the present.

Discussion

Genetic Diversity, Genetic Structure and Diversification of *Dioscorea humilis*

Genetic diversity and population structure in plant species is determined by various abiotic and biotic factors, some of which have triggered population differentiation [48] and speciation processes [49–50]. Biotic factors have been globally assigned to life-history (e.g. life form), and reproductive traits (e.g. reproductive systems, pollination and seed dispersal mechanisms [51]).

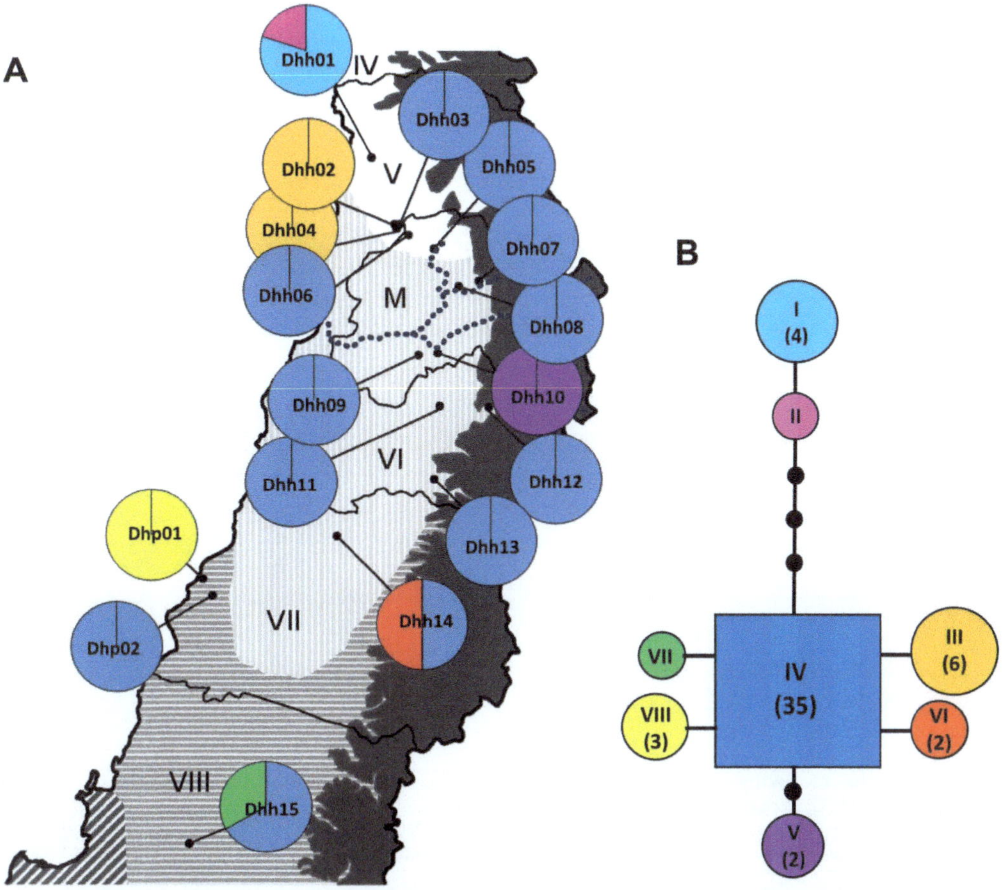

Figure 4. Plastid haplotype diversity in *Dioscorea humilis*. A. Geographical distribution of eight chloroplast haplotypes in 17 populations of *Dioscorea humilis*. Pie charts indicate relative frequencies of each haplotype in each population. The dotted line indicates the location of the Maipo river. Chilean administrative regions and climatic regions are indicated as in Fig. 1. B. Parsimony Network showing the relationships among eight haplotypes. Black dots indicate unsampled or extinct haplotypes. The size of the circles or squares is proportional to the number of sequences representing each haplotype, and is indicated in parentheses when higher than one. Map contour constructed from spatial data retrieved from http://www.diva-gis.org/gdata.

However, closely related taxa, such as the two subspecies of *D. humilis*, show similar biotic parameters. Abiotic factors include, as most relevant, climatic variables and barriers to dispersal. Our results suggest that the conjunction of these two later factors caused the intraspecific split within *D. humilis*.

Our analyses revealed moderate levels of allelic diversity and heterozygosity across populations of *D. humilis* ($A = 3.25$–5.0, $H_O = 0.350$–0.598, $H_E = 0.357$–521, Table 1), which were relatively lower than in the sister species *D. biloba* ($A = 5.14$–7.29, $H_O = 0.345$–0.686, $H_E = 0.458$–0.706, Table S6 in Appendix S3). However, genetic diversity parameters of *D. humilis* were in the range of other yam species with a different combination of life-history (climbers), reproductive (winged seeds) and distribution range (broad range, non-endemic), characteristics that, contrary to *D. humilis* should predispose them to higher levels of genetic diversity (Table S6 in Appendix S3). By contrast, *D. humilis* showed higher genetic diversity than those of species of the *Borderea* group of *Dioscorea* which are comparable in morphological and reproductive traits (Table S6 in Appendix S3), though the Pyrenean *Dioscorea* species differ from *D. humilis* in their even narrower distributions [52−53].

Widespread taxa tend to maintain higher levels of genetic diversity compared to geographically restricted congeners [54]. However, genetic diversity in *D. humilis* subsp. *polyanthes* was not significantly lower than in *D. humilis* subsp. *humilis*, despite its more restricted geographical range (Table 1). This result could suggest equally or more efficient mechanisms buffering against genetic loss in *D. humilis* subsp. *polyanthes*.

Our study also revealed a strong geographical structure of nuclear microsatellite variation throughout the range of *D. humilis* (Fig. 1), with populations of *D. humilis* subsp. *polyanthes* clearly separate from those of *D. humilis* subsp. *humilis* that split into clusters 1 (northern populations Dhh01-Dhh08) and 2 (southern populations Dhh09-Dhh15; Fig. 1a). Clustering analyses (Fig. 2) and AMOVA (Table 3) also found a major differentiation between the two subspecies, in agreement with their morphological distinction [16]. However, plastid DNA haplotype sharing (haplotype IV) between the subspecies and the occurrence of a private plastid haplotype (VIII) in Dhp01, directly derived from the most common haplotype (IV), suggests recent diversification mediated by isolation by distance (Fig. 3a), with incomplete lineage sorting in *D. humilis* subsp. *polyanthes* (Fig. 4) or alternatively, introgression between subspecies.

Bayesian F_{ST} values supported an origin of the species in the northern region and a derived recent origin of *D. humilis* subsp. *polyanthes* from the southern group of *D. humilis* subsp. *humilis*. This was corroborated by the highest haplotypic diversity of the northern group (Table 3) and by the PCO-MST analysis, which

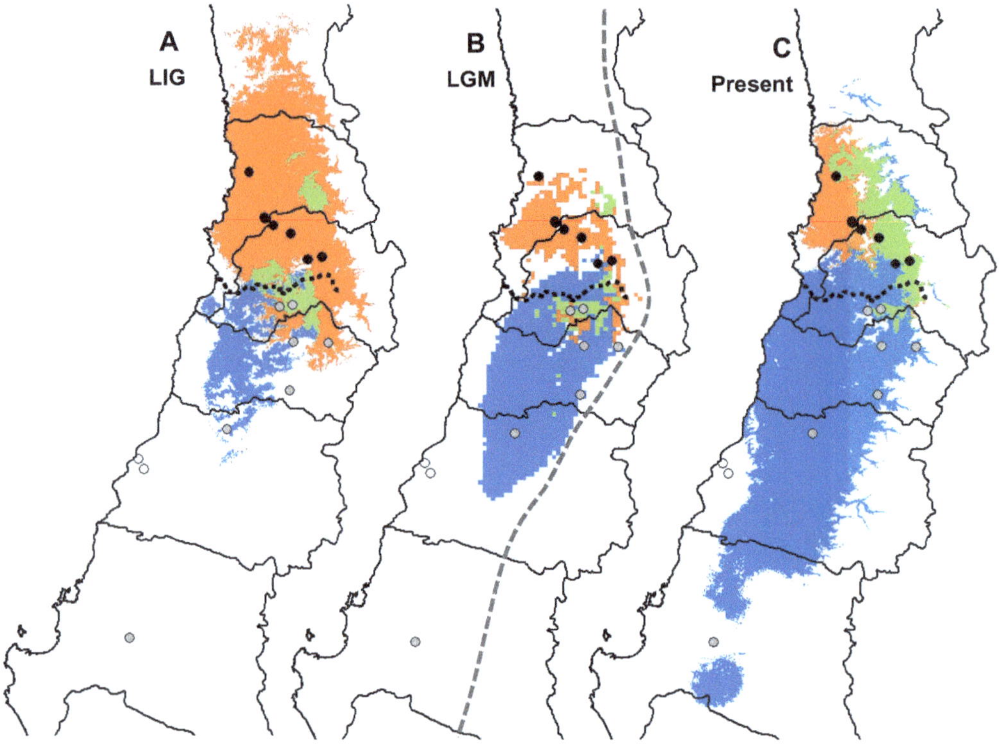

Figure 5. Environmental Niche Modelling (ENM) of *Dioscorea humilis* estimated under Last Interglacial (LIG) (A), Last Glacial Maximum (LGM) (B) and current (C) climate conditions. In orange, northern *D. humilis* subsp. *humilis* genetic group, in blue, southern genetic group and in bright green, overlapping areas among predicted distributions. A tenth percentile threshold was applied. Black circles: northern populations of *D. humilis* subsp. *humilis* (Dhh01 to Dhh08); grey circles: southern populations of *D. humilis* subsp. *humilis* (Dhh09 to Dhh15). The black dashed line indicates the location of the Maipo river. The grey dashed line in B indicates the approximate extent of the ice sheet during the LGM after [5]. Map contours constructed from spatial data retrieved from http://www.diva-gis.org/gdata.

indicated the closeness of *D. humilis* subsp. *polyanthes* to the southern *D. humilis* subsp. *humilis* populations (Fig. 2).

The intraspecific divergence of the southern *D. humilis* subsp. *polyanthes* from the southern group of *D. humilis* subsp. *humilis* could have been a consequence of local environmental adaptation. However, despite the lack of an ENM for the former taxon, the range of values for its 19 bioclimatic variables overlap with that of the latter group and are not significantly different from them (Table S2 in Appendix S1). Thus, the explanation for their divergence is other than a climatically driven speciation process; it may be rather the consequence of geographical isolation and incomplete plastid lineage sorting of *D. humilis* subsp. *polyanthes* from the central Chilean depression southern *D. humilis* subsp. *humilis* group during the last glacial and interglacial phases (Fig. 5).

Influence of Current and Past Latitudinal Climatic Heterogeneity on the Genetic Structure of *Dioscorea humilis* subsp. *humilis*

A noticeable finding was the strong geographical structure detected among *D. humilis* subsp. *humilis* populations, which separated into two North-to-South genetic groups (Fig. 1a). Such spatial patterns are usually driven by the effect of strong geographical or climatic barriers to dispersal, as proposed for *Hordeum chilense* Roem. & Schult., which is similarly distributed along a climatic gradient in Chile [20]. The distribution of the genetic groups of *D. humilis* subsp. *humilis* mostly paralleled those of the main Chilean latitudinal climatic zones (Fig. 1). Mountain chains in the central Chilean depression show lower altitudes and may not significantly contribute to latitudinal isolation of populations. Our study showed that the genetic divergence of the two population groups occurred northwards and southwards of the Maipo river basin (Fig. 1). Indeed, river basins have been identified as efficient barriers to dispersal for seed plants, such as in Chinese populations of *Vitex negundo* L. (Verbenaceae) on opposite shores of the Yangtze river [55]. Specifically, the role of the Maipo river as a geographical barrier has been highlighted for other organisms with potentially higher dispersal capabilities than *D. humilis*, such as the snake *Philodryas chamissonis* Wiegmann [14]. The Maipo river acting as a geographical barrier to gene flow could contribute to explain the IBD pattern across the range of *D. humilis* subsp. *humilis* (Fig. 3a), and the abrupt difference in genetic structure between the two geographical groups (Fig. 1a). However, this IBD pattern does not apply for within-group pairwise population comparisons (Fig. 3b). The absence of IBD within the two geographical areas of *D. humilis* subsp. *humilis* contrasts with life-history and reproductive traits of the species which all point towards extremely short dispersal distances [16]. Therefore, the observed patterns are probably mirroring historical gene flow among populations within ranges preceding range expansions in the Holocene and a relatively rapid postglacial expansion by unknown vectors.

Geographical and historical variations of environmental variables have been demonstrated to greatly influence genetic divergence among populations [20,56]. Our ENM analyses indicate a strong latitudinal ecological differentiation throughout the current range of *D. humilis* subsp. *humilis* into two well defined environmental niches (Fig. 5c). Past projection of niche models indicate that this ecological niche differentiation likely originated earlier than LIG (Fig. 5a), and that ecological conditions have been maintained until present times. The current separation of the groups by the Maipo river basin [14] matches the ecogeographical division of the *D. humilis* range into northern semiarid and southern subhumid Mediterranean areas (Fig. 1, [20]). Ice-cover during the LGM, which reached to 35°S in the Andes [5–6]), together with its northwards influence that extended to approximately 33°S, could have strengthened the barrier effect of the Maipo river. Water volume of this river originating from the Andes was likely higher during LGM than in present times [14], which could account for the allopatric distribution of the two population groups, thereby contributing to the observed genetic differentiation between them (Fig. 5). Bayesian F_{ST} values supported the ancestry of the northern populations, suggesting a likely origin of the species in its northern range, in the overlapping area with its congener *D. biloba* [16], followed by further southwards expansion.

Nonetheless, the predicted extension of the potential distribution areas of the two population groups could have fluctuated both in latitude and longitude during glacial and interglacial episodes, as expected from changes in climatic parameters in those areas following periods of warming (LIG and present) and cooling (LGM, Fig. 5). The predicted distribution area of the northern population group showed a maximum extension during LIG (Fig. 5a), whereas the strong reduction during LGM (Fig. 5b) was maintained until present times (Fig. 5c). Contrastingly, the predicted distribution area of the southern group showed a progressive increase in extension from LIG (Fig. 5a) to present times (Fig. 5c). The potential overlap of distribution areas between the two groups reached its maximum extension during present times (Fig. 5c). However, it was restricted to the eastern range of the present distribution of the species, suggesting that lineage migration and admixture, as denoted by the occurrence of the common plastid haplotype IV, is likely to have occurred only along the eastern boundaries of both distribution areas (Fig. 5). Predicted environmental niche models of the northern and southern groups of *D. humilis* were also consistent with a contraction towards the central Chilean depression during the LGM (Fig. 5b), preceded by broader eastern and western distributions of the potential areas of the northern group during the LIG (Fig. 5a).

Bayesian analyses of nuclear microsatellite variation and of plastid haplotypes also revealed genetic signatures of postglacial population expansion within the northern and southern groups of *D. humilis* subsp. *humilis* (Figs. 1b, 4a). Concerning the northern group, western populations predominantly showed a microsatellite genetic membership to cluster 1, whereas eastern populations showed a predominant genetic membership to cluster 2 (Fig. 1a). This was paralleled to a lesser extent by the slower-evolving cpDNA data (Fig. 4a), where three northwestern populations showed three cpDNA haplotypes that were not represented in eastern populations (Fig. 4a). This would indicate a further isolation of the northwestern populations, which, unlike the eastern ones, did not admix with the southern ones.

Contrastingly, a North-to-South expansion was detected in the southern group of *D. humilis* subsp. *humilis*, supported by a gradual North-to-South decrease in microsatellite genetic membership to cluster 4, and an increase in membership to cluster 3 (Fig. 1b), agreeing with the predicted postglacial southwards expansion (Fig. 5c). Exclusive cpDNA haplotypes were scattered among populations in this range, and were all derived directly from the most common haplotype (Fig. 4b), suggesting recent divergence and dispersal [57–58].

Conclusions

Our study represents a significant contribution to the understanding of the phylogeography of lowland plants from the central Mediterranean area of Chile. Genetic and ENM analyses suggest that *D. humilis* subsp. *polyanthes* diverged from southern populations of *D. humilis* subsp. *humilis* due to local niche adaptation to coastal areas.

The study has also revealed a strong phylogeographical structure within *D. humilis* subsp. *humilis* and identified two highly differentiated genetic groups with distributions that match present latitudinal environmental heterogeneity in the area [20,59]. The genetic differentiation of these two groups could have been triggered by a coupled effect of adaptation to divergent ecological parameters of higher and lower aridity in the northern and southern geographical areas, respectively [60–62], enhanced by the permanent geographical barrier of the Maipo river basin between the two areas [14].

Supporting Information

Appendix S1 Environmental niche model analysis of *Dioscorea humilis*.

Appendix S2 Microsatellite allele frequencies, Bayesian estimation of genetic clusters and plastid haplotype frequencies in populations of *Dioscorea humilis*.

Appendix S3 Comparison of microsatellite genetic diversity in yam species.

Acknowledgments

We thank E. Pérez-Collazos for his help during fieldwork, D. López for her assistance in ENM analyses and Emily Lemonds for revising the English text. Environmental data was retrieved from The international modelling groups and the Laboratoire des Sciences du Climat et de l'Environnement (LSCE). The PMIP 2 Data Archive is supported by CEA, CNRS and the Programme National d'Etude de la Dynamique du Climat (PNEDC).

Author Contributions

Conceived and designed the experiments: PC JGSM. Performed the experiments: JV. Analyzed the data: JV JGSM. Contributed reagents/materials/analysis tools: PC JGSM. Contributed to the writing of the manuscript: JV JGSM PC.

References

1. Schaal BA, Hayworth DA, Olsen KM, Rauscher JT, Smith WA (1998) Phylogeographic studies in plants: problems and prospects. Molecular Ecology, 7, 465–474.
2. Markgraf V, McGlone M, Hope G (1995) Neogene paleoenvironmental and paleoclimatic change in southern temperate ecosystems – a southern perspective. Trends in Ecology and Evolution, 10, 143–147.
3. Villagrán C (1991) Historia de los bosques templados del sur de Chile durante el tardiglacial y postglacial. Revista Chilena de Historia Natural, 64, 447–460.
4. Villagrán C (2001) Un modelo de la historia de la vegetación de la Cordillera de La Costa de Chile central-sur: la hipótesis glacial de Darwin. Revista Chilena de Historia Natural, 74, 783–803.
5. Clapperton C (1994) The Quaternary glaciation of Chile: a review. Revista Chilena de Historia Natural, 67, 369–383.

6. McCulloch RD, Bentley MJ, Purves RS, Hulton NRJ, Sugden DE, et al. (2000) Climatic inferences from glacial and palaeoecological evidence at the last glacial termination, southern South America. Journal of Quaternary Science, 15, 409–417.
7. Armesto J, Arroyo MTK, Hinojosa LF (2007) *The mediterranean Environment of Central Chile*. In: The Physical Geography of South America. Eds Veblen TT, Ypung KR, Orme AR, 184–199. Oxford University Press, New York.
8. Bull-Hereñu K, Martínez EA, Squeo FA (2005) Structure and genetic diversity in *Colliguaja odorifera* Mol. (Euphorbiaceae), a shrub subjected to Pleisto-Holocenic natural perturbations in a Mediterranean South American region. Journal of Biogeography, 32, 1129–1138.
9. Vásquez M, Torres-Pérez F, Lamborot M (2007) Genetic variation within and between four chromosomal races of *Liolaemus monticola* in Chile. Herpetological Journal, 17, 149–160.
10. Caviedes CN, Paskoff R (1975) Quaternary glaciations in the Andes of North-Central Chile. Journal of Glaciology, 14, 155–170.
11. Santana-Aguilar M (1973) La glaciation Quaternaire dans les Andes de Rancagua (Chili central). Bulletin of the Association of Geography France, 406/407, 473–483.
12. Graf K (1994) Discussion of palynological methods and paleoclimatical interpretations in northern Chile and the whole Andes. Revista Chilena de Historia Natural, 67, 405–415.
13. Lamy F, Hebbeln D, Wefer G (1999) High resolution marine record of climatic change in mid-latitude Chile during the last 28,000 years based on terrigenous sediment parameters. Quaternary Research, 51, 83–93.
14. Sallaberry-Pincheira N, Garin CF, González-Acuña D, Sallaberry MA, Vianna JA (2011) Genetic divergence of Chilean long-tailed snake (*Philodryas chamissonis*) across latitudes: conservation threats for different lineages. Diversity and Distributions, 17, 152–162.
15. Lamborot M, Eaton LC (1997) The Maipo River as a geographical barrier to *Liolaemus monticola* (Torpiduridae) in the mountain ranges of central Chile. Journal of Zoological Systematics and Evolutionary Research, 35, 105–111.
16. Viruel J, Segarra-Moragues JG, Pérez-Collazos E, Villar L, Catalán P (2010a) Systematic revision of the *Epipetrum* group of *Dioscorea* (Dioscoreaceae) endemic to Chile. Systematic Botany, 35, 40–63.
17. Donato M (2006) Historical biogeography of the family Tristiridae (Orthoptera: Acridomorpha) applying dispersal-vicariance analyses. Journal of Arid Environments, 66, 421–434.
18. Viruel J, Segarra-Moragues JG, Pérez-Collazos E, Villar L, Catalán P (2008) The diploid nature of the Chilean *Epipetrum* and a new base number in the Dioscoreaceae. New Zealand Journal of Botany, 46, 327–339.
19. Amigo J, Ramírez C (1998) A bioclimatic classification of Chile: woodland communities in the temperate zone. Plant Ecology, 136, 9–26.
20. Castillo A, Dorado G, Feuillet C, Sourdille P, Hernández P (2010) Genetic structure and ecogeographical adaptation in wild barley (*Hordeum chilense* Roemer et Schultes) as revealed by microsatellite markers. BMC Plant Biology, 10, 266–278.
21. Viruel J, Catalán P, Segarra-Moragues JG (2012) Disrupted phylogeographical SSR and cpDNA patterns indicate a vicariance rather than long-distance dispersal origin for the disjunct distribution of the Chilean endemic *Dioscorea biloba* (Dioscoreaceae) around the Atacama Desert. Journal of Biogeography, 39, 1073–1085.
22. Viruel J, Catalán P, Segarra-Moragues JG (2010b) New microsatellite loci in the dwarf yams *Dioscorea* gr. *Epipetrum* (Dioscoreaceae). American Journal of Botany, 97, e121-e12.
23. Taberlet P, Gielly L, Pautou G, Bouvet J (1991) Universal primers for amplification of three non-coding regions of chloroplast DNA. Plant Molecular Biology, 17, 1105–1109.
24. Belkhir K, Borsa P, Chikhi L, Raufaste N, Bonhomme F (2004) *GENETIX 4.05, logiciel sous Windows TM pour la génétique des populations*. Laboratoire Génome, Populations, Interactions, CNRS UMR 5000, Université de Montpellier II, Montpellier (France).
25. Rousset F (2008) GENEPOP'007: a complete re-implementation of the GENEPOP software for Windows and Linux. Molecular Ecology Resources, 8, 103–106.
26. Goudet J (2001) *FSTAT v. 2.9.3.2, a program to estimate and test gene diversities and fixation indices*. Available from: http://www2.unil.ch/popgen/softwares/fstat.htm.
27. Excoffier L, Laval G, Schneider S (2005) Arlequin ver. 3.0: An integrated software package for population genetics data analysis. Evolutionary Bioinformatics Online, 1, 47–50.
28. Slatkin M (1995) A measure of population subdivision based on microsatellite allele frequencies. Genetics, 139, 457–462.
29. Rohlf FJ (2002) *NtSYSpc, Numerical Taxonomy and Multivariate analysis System. Version 2.11a, User guide*. Exeter software, New York, 38 p.
30. Nei M, Tajima F, Tateno Y (1983) Accuracy of estimated phylogenetic trees from molecular data. Journal of Molecular Evolution, 19, 153–170.
31. Langella O (2000) *Populations (Logiciel de genétique des populations)*. CNRS, France.
32. Pritchard JK (2002) *Documentation for STRUCTURE software: version 2*. Available at: http://pritch.bsd.uchicago.edu/software/readme_2_1/readme.html.
33. Evanno G, Regnaut S, Goudet J (2005) Detecting the number of clusters of individuals using the software STRUCTURE: a simulation study. Molecular Ecology, 14, 2611–2620.
34. Tajima F (1983) Evolutionary relationship of DNA sequences in finite population. Genetics, 105, 437–460.
35. Librado P, Rozas J (2009) DnaSP v5: A software for comprehensive analysis of DNA polymorphism data. Bioinformatics, 25, 1451–1452.
36. Clement M, Posada D, Crandall KA (2000) TCS: a computer program to estimate gene genealogies. Molecular Ecology, 9, 1657–1659.
37. Hijmans RJ, Cameron SE, Parra JL, Jones PG, Jarvis A (2005) Very high resolution interpolated climate surfaces for global land areas. International Journal of Climatology, 25, 1965–1978.
38. Phillips SJ, Anderson RP, Schapire RE (2006) Maximum entropy modeling of species geographic distributions. Ecological Modelling, 190, 231–259.
39. Phillips SJ, Dudík M, Elith J, Graham CH, Lehmann A, et al. (2009) Sample selection bias and presence - only distribution models: implications for background and pseudoabsence data. Ecological Applications, 19, 181–197.
40. Petal RG, Raxworthy CJ, Nakamura M, Peterson AT (2007) Predicting species distributions from small numbers of occurrence records: a test case using cryptic geckos in Madagascar. Journal of Biogeography, 34, 102–117.
41. Pease KM, Freedman AH, Pollinger JP, McCormack JE, Buermann W, et al. (2009) Landscape genetics of California mule deer (*Odocoileus hemoinus*): the roles of ecological and historical factors in generating differentiation. Molecular Ecology, 18, 1848–1862.
42. Braconnot P, Otto-Bliesner B, Harrison S, Joussaume S, Peterschmitt J-Y, et al. (2007) Results of PMIP2 coupled simulations of the Mid-Holocene and Last Glacial Maximum - Part 1: experiments and large-scale features. Climate of the Past, 3, 261–277.
43. Otto-Bliesner B-L, Marshall SH, Overpeck JT, Miller GH, Hu A, et al. (2006) Simulating Arctic Climate Warmth and Icefield Retreat in the Last Interglaciation. Science, 24, 1751–1753.
44. Collins WD, Bitz CM, Blackmon ML, Bonan GB, Bretherton CS, et al. (2006) The community climate system model version 3 (CCSM3). Journal of Climate, 19, 2122–2143.
45. Hasumi H, Emori S (2004) *K-1 coupled GMC (MIROC) description* (ed. by H. Hasumi and S. Emori). K-1 Technical Report No. 1, September, 2004. Center for Climate System Research, University of Tokyo, Tokyo.
46. Waltari E, Hijmans RJ, Peterson AT, Nyari AS, Perkins SL, et al. (2007) Locating Pleistocene refugia: comparing phylogeographic and ecological niche model predictions. PLoS ONE, 2, e563.
47. Hammer Ø, Harper DAT, Ryan PD (2001) PAST: Paleontological statistics software package for education and data analysis. Palaeontologia Electronica, 4, 1–9.
48. Zattara EE, Premoli AC (2005) Genetic structuring in Andean landlocked populations of *Galaxias maculatus*: effects of biogeographic history. Journal of Biogeography, 32, 5–14.
49. Dillon MO, Tu T, Xie L, Quipuscoa-Silvestre V, Wen J (2009) Biogeographic diversification in *Nolana* (Solanaceae), a ubiquitous member of the Atacama and Peruvian Deserts along the western coast of South America. Journal of Systematics and Evolution, 47, 457–476.
50. Schmidt-Jabaily R, Sytsma KJ (2010) Phylogenetics of *Puya* (Bromeliaceae): placement, major lineages, and evolution of Chilean species. American Journal of Botany, 97, 337–356.
51. Hamrick JL, Godt MJW (1996) Effects of life history traits on genetic diversity in plant species. Philosophical Transactions of the Royal Society B: Biological Sciences, 351, 1291–1298.
52. Segarra-Moragues JG, Palop-Esteban M, González-Candelas F, Catalán P (2005) On the verge of extinction: genetics of the Critically Endangered Iberian plant species, *Borderea chouardii* (Dioscoreaceae) and implications for conservation management. Molecular Ecology, 14, 969–982.
53. Segarra-Moragues JG, Palop-Esteban M, González-Candelas F, Catalán P (2007) Nunatak survival *vs. tabula rasa* in the Central Pyrenees: a study on the endemic plant species *Borderea pyrenaica* (Dioscoreaceae). Journal of Biogeography, 34, 1893–1906.
54. Gitzendanner MA, Soltis PM (2000) Patterns of genetic variation in rare and widespread plant congeners. American Journal of Botany, 87, 783–792.
55. Zhang Z-Y, Zheng X-M, Ge S (2007) Population genetic structure of *Vitex negundo* (Verbenaceae) in Three-Gorge Area of the Yangtze River: The riverine barrier to seed dispersal in plants. Biochemical Systematics and Ecology, 35, 506–516.
56. Li Y, Fahima T, Korol AB, Peng J, Röder MS, et al. (2000) Microsatellite diversity correlated with ecological-edaphic and genetic factors in three microsites of wild emmer wheat in North Israel. Molecular Biology and Evolution, 17, 851–862.
57. Canestrelli D, Nascetti G (2008) Phylogeography of the pool frog *Rana (Pelophylax) lessonae* in the Italian peninsula and Sicily: multiple refugia, glacial expansions and nuclear–mitochondrial discordance. Journal of Biogeography, 35, 1923–1936.
58. Grill A, Amori G, Aloise G, Lisi I, Tosi G, et al. (2009) Molecular phylogeography of European *Sciurus vulgaris*: refuge within refugia? Molecular Ecology, 18, 2687–2699.
59. Pezoa LS (2003) *Recopilación y análisis de la variación de las temperaturas (período 1965–2001) y las precipitaciones (período 1931–2001) a partir de la información de estaciones meteorológicas de Chile entre los 33° y 53° de latitud*

sur. Tesis de grado Ingeniería Forestal, Facultad de Ciencias Forestales, Universidad Austral de Chile, Valdivia, Chile. 99 p.
60. Van Hulsen C (1967) Klimagliederung in Chile auf der Basis von Häufigkeitsverteilungen der Niederschlagssummen. Freiburger Geographische, 4, 1–109.
61. Veit H (1996) Southern Westerlies during the Holocene deduced from geomorphological and pedological studies in the Norte Chico, Northern Chile (27–33°S). Palaeogeography, Palaeoclimatology, Palaeoecology, 123, 107–119.
62. Bonilla CA, Vidal KL (2011) Rainfall erosivity in Central Chile. Journal of Hydrology, 410, 126–133.

5

Cadmium Telluride Quantum Dots (CdTe-QDs) and Enhanced Ultraviolet-B (UV-B) Radiation Trigger Antioxidant Enzyme Metabolism and Programmed Cell Death in Wheat Seedlings

Huize Chen[1,2]♥, Yan Gong[1,3]♥, Rong Han[1,2]*

1 Higher Education Key Laboratory of Plant Molecular and Environmental Stress Response (Shanxi Normal University) in Shanxi Province, Linfen, China, 2 School of Life Science, Shanxi Normal University, Linfen, China, 3 School of Chemistry and Material Science, Shanxi Normal University, Linfen, China

Abstract

Nanoparticles (NPs) are becoming increasingly widespread in the environment. Free cadmium ions released from commonly used NPs under ultraviolet-B (UV-B) radiation are potentially toxic to living organisms. With increasing levels of UV-B radiation at the Earth's surface due to the depletion of the ozone layer, the potential additive effect of NPs and UV-B radiation on plants is of concern. In this study, we investigated the synergistic effect of CdTe quantum dots (CdTe-QDs), a common form of NP, and UV-B radiation on wheat seedlings. Graded doses of CdTe-QDs and UV-B radiation were tested, either alone or in combination, based on physical characteristics of 5-day-old seedlings. Treatments of wheat seedlings with either CdTe-QDs (200 mg/L) or UV-B radiation (10 KJ/m^2/d) induced the activation of wheat antioxidant enzymes. CdTe-QDs accumulation in plant root cells resulted in programmed cell death as detected by DNA laddering. CdTe-QDs and UV-B radiation inhibited root and shoot growth, respectively. Additive inhibitory effects were observed in the combined treatment group. This research described the effects of UV-B and CdTe-QDs on plant growth. Furthermore, the finding that CdTe-QDs accumulate during the life cycle of plants highlights the need for sustained assessments of these interactions.

Editor: Jin-Song Zhang, Institute of Genetics and Developmental Biology, Chinese Academy of Sciences, China

Funding: R.H. received funding from the National Natural Science Foundation of China No. 30671061 (http://www.nsfc.gov.cn/). The funders had no role in study design, data collection and analysis, decision to publish, or preparation of the manuscript.

Competing Interests: The authors have declared that no competing interests exist.

* Email: hhwrsl@163.com

♥ These authors contributed equally to this work.

Introduction

Quantum dots are semiconductor nanoparticles (NPs) that are increasingly used in industrial and biological applications [1–5]. The considerable amount of QDs released into the ecosystem raised concern due to their possible toxic effects in living organisms, particularly crop plants [6–8].

Cadmium telluride QDs (CdTe-QDs) are the most frequently used QDs [9–13]. However, the toxicity of CdTe-QDs is of increasing concern due to their small size (2–10 nm) and heavy metal formulation [14]. The cytotoxicity of CdTe-QDs was shown to be associated with their concentration and the duration of exposure [15]. CdTe-QDs release free cadmium ions (Cd^{2+}) under UV-B irradiation, leading to the disruption of DNA replication, the displacement of Zn^{2+} in Zn finger structures, and the generation of reactive oxygen species (ROS) [16,17]. Furthermore, the toxicity of CdTe-QDs has been shown to be much higher than that of Cd^{2+} [12] and that the cytotoxicity was not solely due to free Cd^{2+} [13]. Cellular toxicity was found to be closely related to the size of quantum dots, with smaller diameter QDs (2 nm) demonstrating more toxic effects than larger QDs (5 nm) [18]. These studies demonstrated the potential toxic effects of QDs on inducing antioxidative enzyme activity and DNA replication processes in plants.

Owing to their unique sedentary lifestyle, plants are predicted to be most affected by the accumulation of CdTe-QDs in the environment. This effect is likely to be further exacerbated by the increasing levels of solar UV-B radiation (290–320 nm) at the Earth's surface due to the depletion of stratospheric ozone. Current levels of UV-B during the cropping season are somewhere between 2–12 kJ/m^2/d on the Earth's surface, which exhibits an increase of 6–14% of UV-B radiation over the pre-1980 levels [19]. Based on the UV-B dose per 10 min, the computative calculation of the erythermal UV-B doses range from 2 kJ/m^2 to 9 kJ/m^2 on a clear-sky day on 31 August 2014 (themis.nl). It has been estimated that a 1% decrease in ozone will lead to an increase in UV-B for about 2%. GISS (Goddard Institute for Space Studies) modeling showed a springtime enhancement of erythemal UV doses of up to 14% in the Northern hemisphere and 40% in the Southern hemisphere [20]. Plants are continuously exposed to solar UV-B radiation as a result of their sessile lifestyle [21]. UV-B has the highest energy of any part of the daylight spectrum and has the potential to damage macromolecules including DNA [22], generate ROS, and impair cellular processes

Table 1. Light/dark cycles and treatment regimens with UV-B and CdTe-QDs.

Group	Treatment	Light	UV-B irradiation	Dark	CdTe-QDs treatment
CK	Control	8 h/d	-	16 h/d	dH_2O
B	UV-B	8 h/d	8 h/d	16 h/d	dH_2O
C	CdTe-QDs	8 h/d	-	16 h/d	CdTe-QDs solution
B+C	UV-B+CdTe-QDs	8 h/d	8 h/d	16 h/d	CdTe-QDs solution

[23–25]. Recently, UV-B radiation was found to act as a key environmental signal that initiated diverse responses in plants affecting metabolism, development and viability [26].

The impact of the environmental accumulation of CdTe-QDs in the presence of enhanced UV-B radiation on living organisms in particular plants is largely unknown. Further studies are needed to understand the interactions of plants, QDs, and enhanced UV-B radiation at the biochemical and molecular levels.

In this study, we investigated the individual and combined effects of QDs and UV-B radiation on an important crop, wheat. We assessed the effect of QDs and UV-B radiation, either alone or in combination, on seedling growth, the activation of antioxidant enzymes, the distribution of CdTe-QDs in root cells, and the induction of programmed cell death.

Materials and Methods

Synthesis of CdTe-QDs

Water-soluble CdTe-QDs were prepared as described [27]. Millipore water (100 mL) was degassed with argon for approximately 1 hour. $Cd(ClO_4)_2.6H_2O$ and 3-mercaptopropanoic acid were added and the pH adjusted to 11.3 with 1 M NaOH. The solution was gassed under argon for additional 30 min. H_2Te gas, generated from Al_2Te_3 by adding 0.5 M H_2SO_4 dropwise, was bubbled under a slow argon flow for approximately 10 min. The solution was subsequently refluxed for 2 hours. Samples were vacuum dried and stored in darkness before use.

QDs characterization

QDs were characterized using a JEM-2100 (JEOL, Japan) transmission electron microscope (TEM), and a D8 Advance (Bruker, Germany) X-ray diffractometer (Cu K_α). Samples were prepared by drying sample droplets from water dispersion onto a 100-mesh Cu grid coated with a lacey carbon film, which was dried prior to imaging by TEM. Absorption and emission spectra were collected using fluorescence spectroscopy (JASCO FP-8000).

Seedling cultivation and chlorophyll content

Wheat seeds (*Triticum aestivum* L. *cv.* JIN-8) were sterilised for 10 min with 1% NaClO and then washed for 10 min in running distilled water. Thirty seeds were cultured per Petri dish in a growth chamber at 25°C with 60% relative humidity and were watered daily. Four treatments, each with three replicates, were applied, starting on the day of seed germination as outlined in Table 1. Briefly, seeds were soaked with either QDs [8] or water for the control group all day long. UV-B radiation was applied 8 h/day during the light cycle. Five graded doses of UV-B radiation intensities were tested: (B1) 2.5 $KJ/m^2/d$; (B2) 5 $KJ/m^2/d$; (B3) 7.5 $KJ/m^2/d$; (B4) 10 $KJ/m^2/d$ and (B5) 12.5 $J/m^2/d$. CdTe-QDs were applied at five different concentrations: (C1) 25 mg/L; (C2) 50 mg/L; (C3) 100 mg/L; (C4) 200 mg/L and (C5) 400 mg/L. The effect of UV-B and QDs doses on seedling growth were determined by assessing height, root length, and the concentrations of malondialdehyde (MDA), soluble sugar and soluble protein. Five-day-old wheat seedlings were subsequently treated with the doses showing the most inhibitory effect and more detailed analyse including plant height, root length, biomass, and chlorophyll content were performed. After 5 d of growth, the plant height and root length were measured with a ruler. Twenty seedlings per replicate per treatment were randomly selected for analysis. A total of 90 seedlings were assessed for height, fresh weight (FW), and dry weight (DW). Fresh and dry weights were

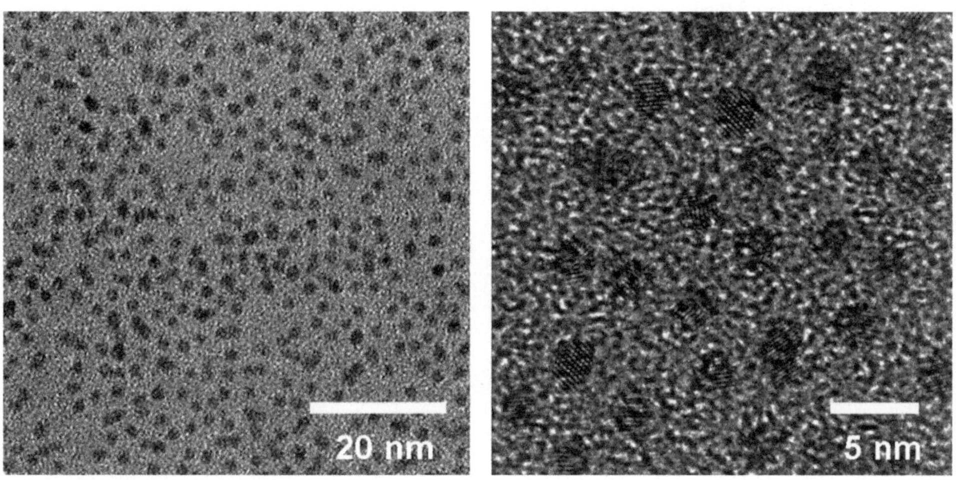

Figure 1. TEM image of synthesized CdTe-QDs.

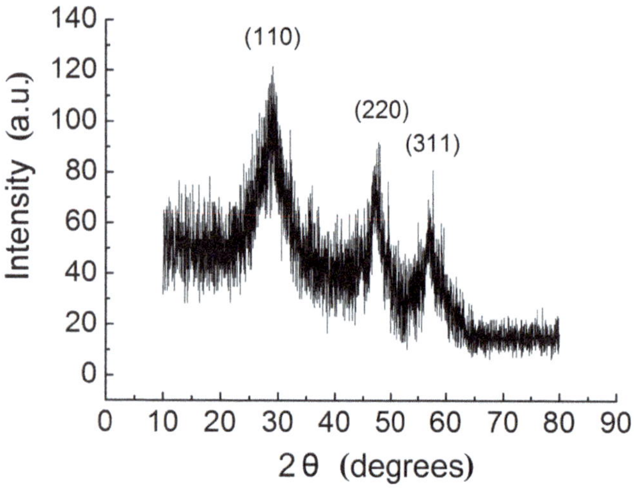

Figure 2. X-ray diffraction spectrum of CdTe-QDs.

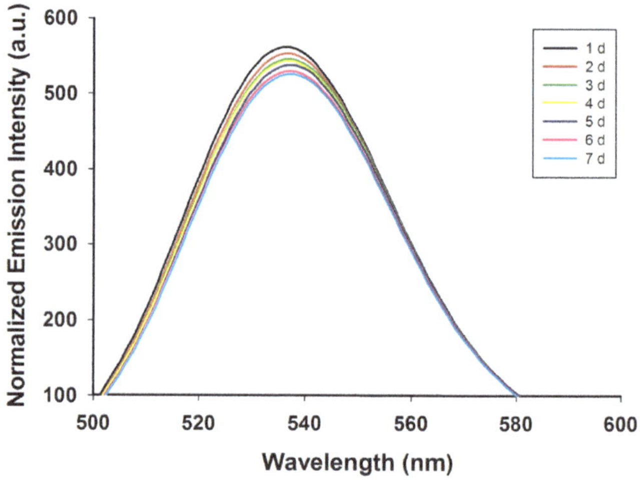

Figure 3. Emission spectra of QDs in H_2O over 7 days.

detected with an analytical balance. Chlorophyll content was measured as described [28].

Measurement of malondialdehyde (MDA), soluble sugar, and soluble protein concentrations

MDA concentration was determined using the trichloroacetic acid (TCA) method. Fresh tissues (1.0 g) were ground with SiO_2 in 2 mL 10% TCA. After centrifugation at 4000 rpm for 10 min, supernatants were removed, added to 2 mL 0.6% (w/v) thiobarbituric acid, and incubated in a 100°C water bath for 15 min. Reactions were stopped on ice. After centrifugation at 4000 rpm for 15 min, supernatants were assayed at 532 and 450 nm.

Total sugar concentration was determined using anthrone colorimetry. Dry plant tissues (50 mg) were triturated with 4 mL 80% ethyl alcohol. Supernatants were collected after 40 min continuous stirring in a water bath at 80°C. Activated carbon (10 mg) was used to decolorize the solution for 30 min, after which 5 mL anthrone were added and samples were incubated in a water bath at 100°C for 10 min. Samples were then cooled for 5 min before spectrophotometric absorbance assessment at 625 nm. Concentrations were determined using standard curves.

Soluble protein was extracted according to Zhao [29], using bovine serum albumin as a calibration standard. Fresh tissues (1.0 g) were ground in 6 mL distilled water and then centrifuged at 4000 rpm for 15 min. After centrifugation, supernatants were added to 5 mL Coomassie brilliant blue G-250 and incubated at room temperature for 15 min before spectroscopic assessment at 595 nm.

Cd^{2+} Accumulation

The concentration of Cd^{2+} within shoots and roots was analyzed using Inductively Coupled Plasma Mass Spectrometry (ICP-MS) [30]. After exposure to CdTe-QDs for 5 days, 4–7 plants were recovered and roots were rinsed thoroughly with deionized water to remove material that was neither adsorbed nor

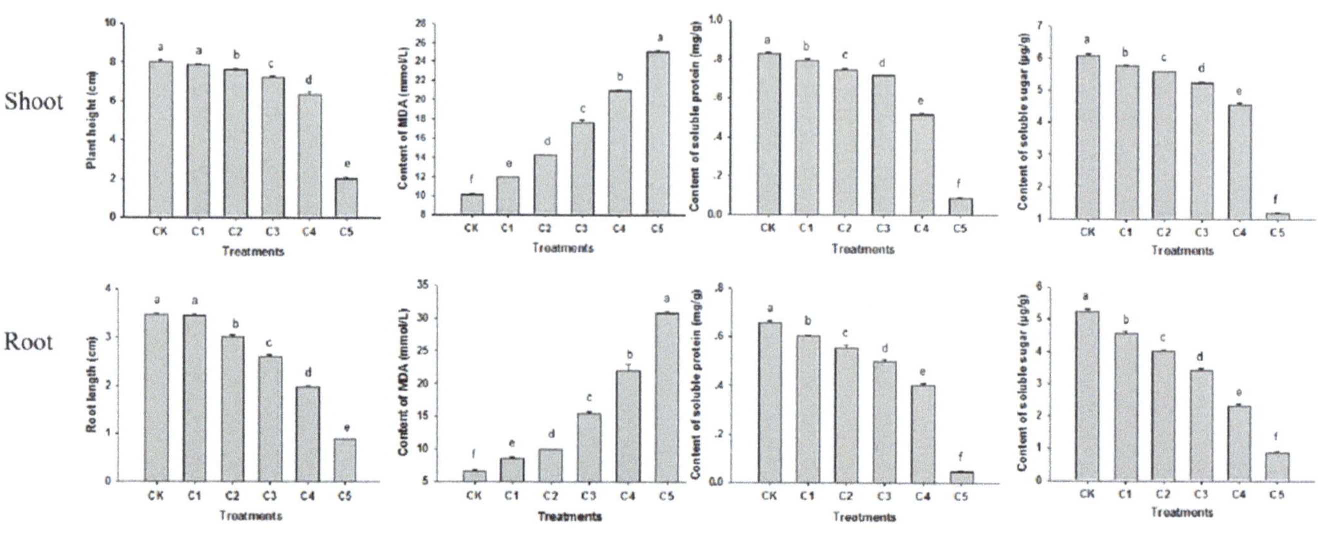

Figure 4. Effects of different CdTe-QDs concentrations on 5-day-old wheat seedlings. CK, control group; C, treatment groups with different concentrations of CdTe-QDs (C1, 25 mg/L; C2, 50 mg/L; C3, 100 mg/L; C4, 200 mg/L; C5, 400 mg/L). Data are means±SD (n = 3). Means with the same letter are not significantly different at Tukey's test ($p \leq 0.05$).

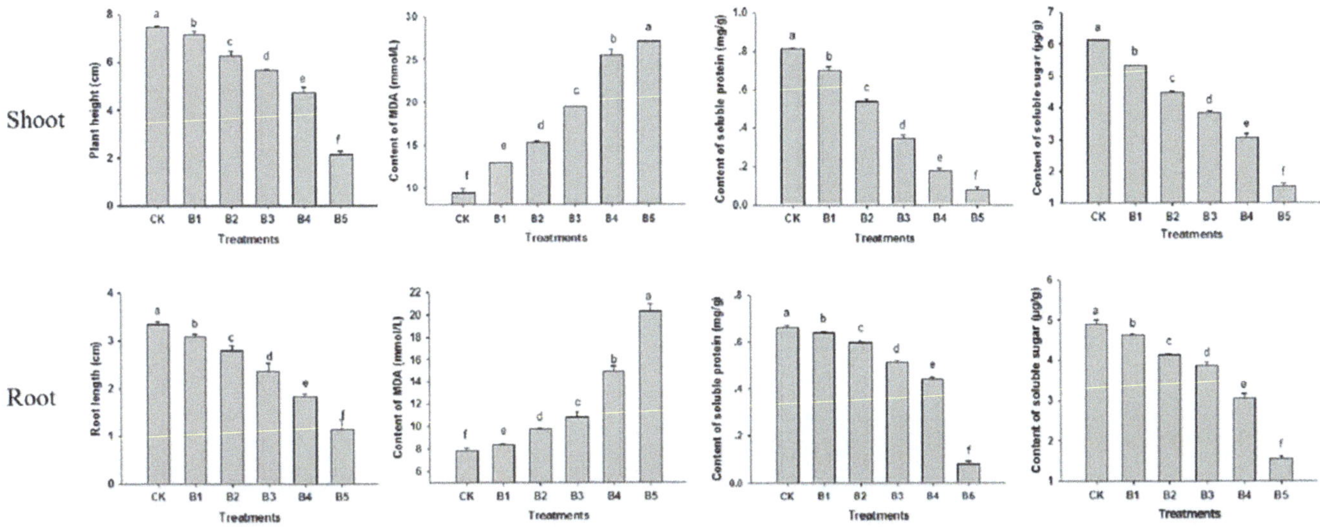

Figure 5. Effects of different enhanced UV-B radiation doses on 5-day-old wheat seedlings. CK, control group; B, treatment groups with graded doses of UV-B radiation (B1, 2.5 $KJ/m^2/d$; B2, 5 $KJ/m^2/d$; B3, 7.5 $KJ/m^2/d$; B4, 10 $KJ/m^2/d$; B5, 12.5 $KJ/m^2/d$). Data are means±SD (n = 3). Means with the same letter are not significantly different at Tukey's test ($p \leq 0.05$).

integrated into the plant tissues. Roots and leaves were separated, oven dried (70°C, 24 h) and weighed. Dried tissues were digested in 4:1 concentrated HNO_3:30% H_2O_2 for at least 2 h on a hotplate at 60°C. Samples were transferred to polypropylene tubes and centrifuged at 1000 g for 10 min. Supernatants were recovered and brought to 10 mL using deionized water. Cd analysis was performed using an X-Series ICP-MS instrument with an ICP-MS elements solution set used as standards.

Antioxidant Assays

Roots and shoots were separately homogenized, centrifuged and extracts assayed for catalase (CAT), guaiacol peroxidase (GPOX), superoxide dismutase (SOD), ascorbate peroxidase (APOX), dehydroascorbate reductase (DHAR) and glutathione reductase (GR) activities as previously described [31–35].

Measurement of ROS Production

Determination of H_2O_2 content in plant extracts. Frozen plant tissues (0.1 g) were homogenized in a 1:9 (w/v) phosphate buffer (50 mM, pH 6.0) at 4°C. The content of H_2O_2 was analyzed with a hydrogen peroxide assay kit (Beyotime, China) according to the manufacturer instruction. Briefly, test tubes containing 50 μL test solutions were placed at room temperature for 30 min and measured immediately with a spectrometer at a wavelength of 560 nm. Absorbance values were calibrated to a standard curve generated with known concentrations of H_2O_2 [36]. O_2^- levels were monitored by staining for 20 min in a solution of 2 mM NBT in 20 mM phosphate buffer (pH 6.1). The reaction was stopped by transferring the seedlings into distilled water. O_2^- content was quantified using the method of Ramel et al. [37]. The NBT-stained plantlets were ground in liquid nitrogen; the obtained powder was solubilized in 2 M KOH-dimethyl sulfoxide (1:1.16, v/v) and then centrifuged for 10 min at 12000 g. The A630 was immediately measured and compared with a standard curve obtained from known amounts of NBT in the KOH-dimethyl sulfoxide mixture [38].

Confocal Fluorescence Imaging Analysis

Cellular fluorescence was monitored using a FV-1000 Confocal system (Olympus, Co.) with a 1.4 NA, 60×oil immersion objective lens. Cell observations were performed on at least three replicate samples. Quantum dots were excited at 488 nm. Nuclei were stained with 4′,6-diamidino-2-phenylindole (DAPI) (Sigma, USA) and excited at 405 nm.

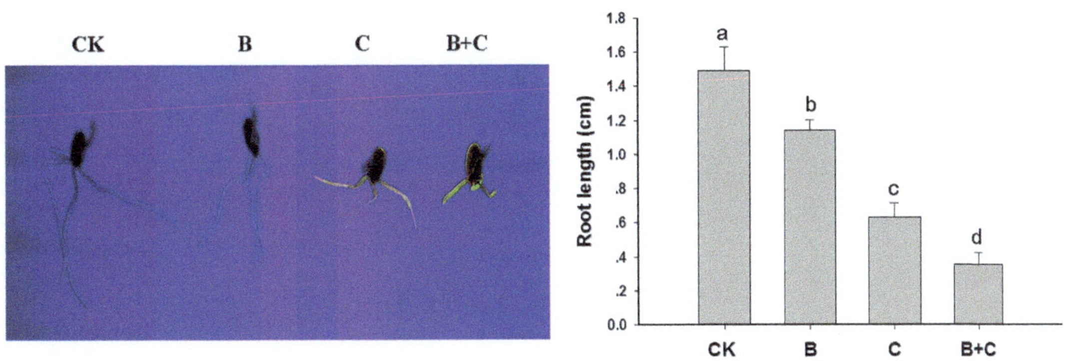

Figure 6. Two-day old seedlings visualized with UV light. Control, CK; UV-B treatment alone, B; CdTe-QDs treatment alone, C; combined CdTe-QDs and UV-B treatment, C+B.

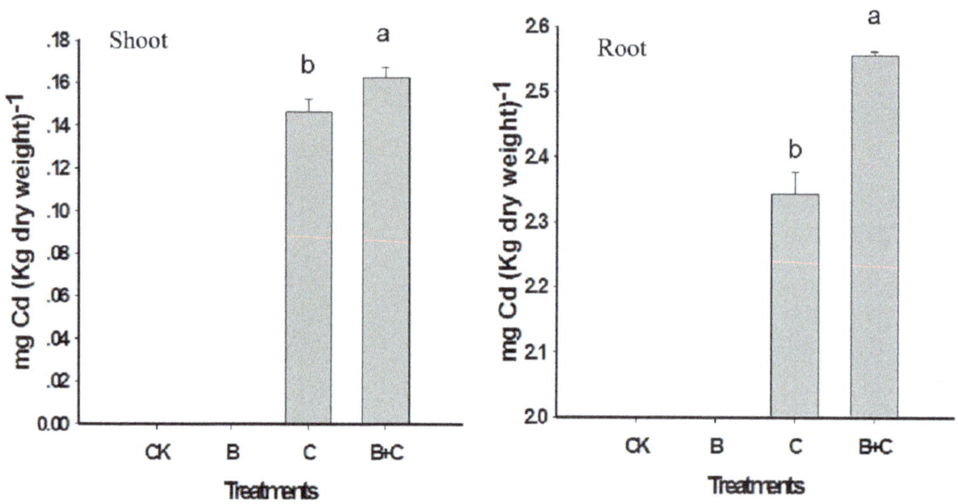

Figure 7. Cd concentration in tissues of 5-day-old wheat seedlings under different treatments. Control, CK; UV-B treatment alone, B; CdTe-QDs treatment alone, C; combined CdTe-QDs and UV-B treatment, C+B.

DNA laddering

Genomic DNA was extracted using a standard CTAB protocol [39]. DNA concentration was determined by spectrophotometry. DNA (5 μg) was analyzed using 1.5% agarose gel electrophoresis in the presence of ethidium bromide (5 μg/mL) and gels were visualized on a UV transilluminator.

Statistical Analysis

Data results are expressed as means ± standard deviations (SD). Statistical significance was assessed using one-way analysis of variance (ANOVA) tests using General Linear Model followed by Tukey test was performed using the SPSS 21.0 and Sigma-plot 10.0.

Results and Discussion

QDs characteristics

CdTe-QDs were synthesized and characterized using TEM. CdTe-QDs had an average diameter of 2.3±0.2 nm (Figure 1). X-ray diffraction (XRD) spectra scanned over the 2 theta (θ) range of 10–80° showed a cubic XRD structure of CdTe-QDs with diffractive peaks of 28°, 46°, and 57°, indicating well-crystallized QDs (Figure 2). The QDs exhibited narrow fluorescence emission spectra (λ max = 531 nm). Emission spectra collected over a period of 7 days are shown in Figure 3. QDs emission intensity decreased daily, with a total decrease of approximately 6.5% over 7 days (normalized intensity change from 561.8488 to 525.1871). These results indicated that almost 93.5% of QDs persisted during the experimental exposure period.

The effect of CdTe-QDs and UV-B irradiation on the growth of wheat seedlings

Wheat seedlings were treated with single agents of either CdTe-QDs or UV-B irradiation. Treatments of seedlings with five different concentrations of CdTe-QDs resulted in reductions in shoot height and root length, increased lipid oxidation as measured by the levels of the oxidation product MDA, as well as reductions in soluble sugar and soluble protein concentrations, in a dose dependent manner when compared with the control (CK) group (Figure 4). Of five CdTe-QDs concentrations tested,

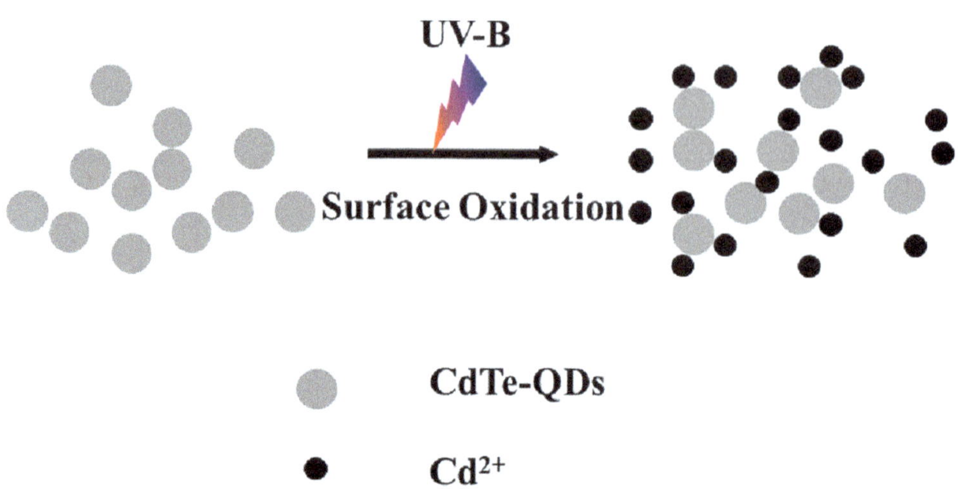

Figure 8. Schematic showing Cd^{2+} release from QDs as a result of UV-B oxidation of CdTe-QDs surfaces [41,42].

Table 2. Effects of QDs and UV-B radiation on growth of 5-day-old wheat seedlings.

Group	Treatments	Plant height (mm)	Root length (mm)	Fresh weight (g)	Dry weight (g)
CK	Control	75.63±1.379a	26.87±0.493a	0.1983±0.003a	0.0307±0.002a
B	UV-B	45.53±1.193c	16.26±0.404b	0.1217±0.011c	0.0143±0.003c
C	CdTe-QDs	51.23±1.101b	12.13±0.493c	0.1393±0.007b	0.0187±0.003b
B+C	UV-B+CdTe-QDs	31.47±1.026d	7.96±0.776d	0.0980±0.003d	0.0080±0.001d

Data are means±SD (n = 3). Means with the same letter are not significantly different at Tukey's test ($p \leq 0.05$).

the concentration of 200 mg/L (C4) was selected for further evaluations when taking into account of all parameters assessed (almost 75% shoot/root length reduction and 15 times MDA increase performed in the C5 treatment) (Figure 4). Treatments of seedlings with graded doses of UV-B irradiation led to similar growth inhibition as assessed by the above parameters (Figure 5). The UV-B dose of 10.0 KJ/m^2/d was selected for subsequent experiments since the dose higher than 10.0 KJ/m^2/d strongly inhibited wheat growth (Figure 5). This dose of UV-B irradiation should decrease the density of O_3 by 20% [40].

CdTe-QD accumulation in plant tissues

Wheat seedlings were randomly assigned to four treatment groups: control (CK), UV-B at 10 KJ/m^2/d (B), CdTe-QDs at 200 mg/L (C), and UV-B (10 KJ/m^2/d) plus CdTe-QDs (200 mg/L) (B+C). The uptake of CdTe-QDs was evident in two-day-old treated seedlings in the group treated by CdTe-QDs alone (C) and that treated by UV-B (10 KJ/m^2/d) plus CdTe-QDs (200 mg/L) (B+C) (Figure 6). Root length was reduced in all three treatment groups compared to the control group (Figure 6). CdTe-QDs fluorescence was observed in roots of seedlings treated with CdTe-QDs alone (C) and in seedlings received combined CdTe-QDs and UV-B (B+C), indicating CdTe-QDs nanoparticles uptake by roots.

Treatment of seedlings with CdTe-QDs resulted in a higher Cd accumulation in roots (2.344 mg/Kg DW) than in shoots (0.1461 mg/Kg DW) (Figure 7). This is likely due to the direct contact exposure of roots to the CdTe-QDs suspension. There was also a marked increase in Cd concentration in both shoots and roots of seedlings treated with CdTe-QDs in combination with UV-B irradiation (Figure 7). This increase in the levels of Cd^{2+} in the presence of UV-B radiation was reported previously as a consequence of QDs surface oxidation and Cd^{2+} release (Figure 8) [41,42]. Our results suggest a potential additive effect of CdTe-QDs and UV-B on plants.

Plant growth and chlorophyll content

The growth and chlorophyll contents of wheat seedlings exposed to different treatments were determined (Tables 2 and 3). UV-B radiation treatment resulted in a significantly reduction in plant height with an average height 30.1 mm shorter than that of controls (Table 2). Similarly, chlorophyll content, both chlorophyll *a* and *b*, were significantly lower than in controls (Table 3). The phytotoxic effects of CdTe-QDs treatment on plant height and chlorophyll content were less severe than those of UV-B radiation. However, root length was more severely affected in the CdTe-QDs group compared to the UV-B treated group (Table 2). Average root length in CdTe-QDs-treated plants was 14.74 mm shorter than in untreated controls. Total biomass was similar between the UV-B and QDs treated groups, but was significantly reduced compared to control untreated seedlings. The combined treatment group showed a significant reduction in plant height, root length, and total biomass when compared to the control group and the single treatment groups (Table 2). The enhanced toxicity in the combined treatment group may have been due to the release of free Cd^{2+} from CdTe-QDs in the presence of UV-B radiation.

Antioxidant defense in wheat seedlings

The cytotoxicity of QDs and UV-B radiaton have shown to be associated with the production of ROS [6,43,44]. H_2O_2 and O_2^- are the most common ROS produced during normal cellular metabolic processes. Their imbalance can be highly cytotoxic [35,45]. To determine the involvement of H_2O_2 and O_2^- during QDs and UV-B exposure, 5-day-old seedlings were examined for the presence of ROS following treatments with CdTe-QDs, UV-B, and combined CdTe-QDs and UV-B. The results showed that there was a marked increase in the concentrations of H_2O_2 in roots treated with CdTe-QDs, either alone or in combination with UV-B, compared to the UV-B treated and untreated groups. Furthermore, there was 11.5% increase in the levels of H_2O_2, from 215.37 nmol/g FW in the CdTe-QD treated group to 247.89 nmol/g FW in the combined CdTe-QD and UV-B treated

Table 3. Chlorophyll content in tissues of 5-day-old wheat seedlings.

Group	Treatments	Chlorophyll *a* (mg/g FW)	Chlorophyll *b* (mg/g FW)	Total chlorophyll (mg/g FW)	Chlorophyll *a*/*b* ratio
CK	Control	0.5943±0.0068a	0.2723±0.0042a	0.8667±0.0059a	2.1828±0.0510a
B	UV-B	0.4030±0.0076c	0.1970±0.0056c	0.6000±0.0026c	2.0474±0.0949b
C	CdTe-QDs	0.4990±0.0061b	0.2473±0.0057b	0.7463±0.0012b	2.0186±0.0714b
B+C	UV-B+CdTe-QDs	0.3443±0.0047d	0.1763±0.0084d	0.5207±0.0038d	1.9564±0.1165b

Data are means±SD (n = 3). Means with the same letter are not significantly different at Tukey's test ($p \leq 0.05$).

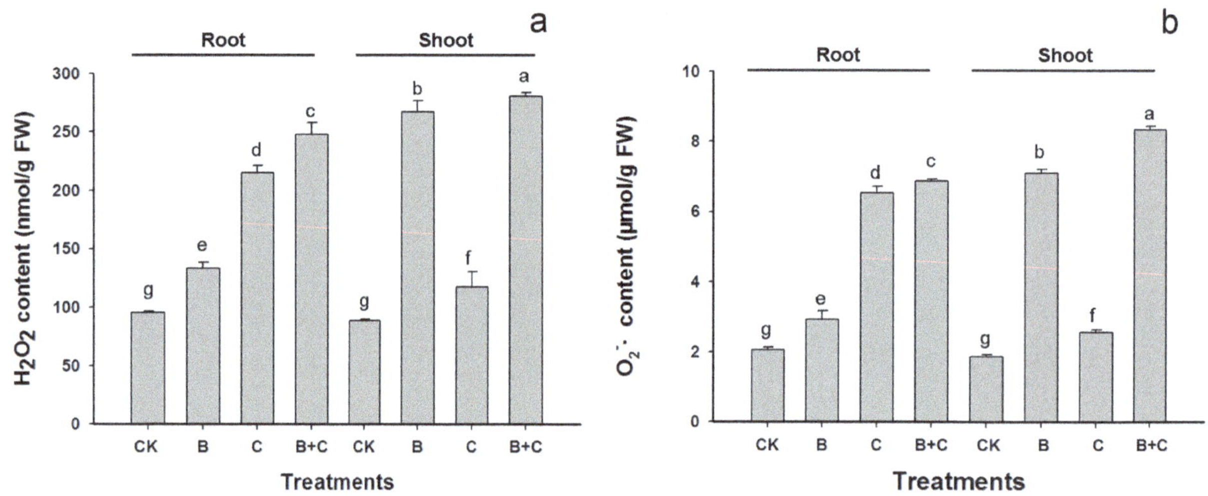

Figure 9. Contents of H_2O_2 and O_2^- in the root and shoot of 5 day old seedlings following different treatments. Data are means±SD (n = 3). Means with the same letter are not significantly different at Tukey's test ($p \leq 0.05$).

group (Figure 9). Similar results were observed for the O_2^- level. These results are consistent with the greater toxicity from higher levels of Cd following UV-B irradiation predicted for roots. We noted that such additive effect by UV-B was moderate in shoots where the production of H_2O_2 and O_2^- is found to be primarily associated with UV-B radiation. The relatively moderate effect in the Cd-associated ROS production in shoots is likely due to the low levels of Cd accumulation in shoots (Figure 7).

Plants use a highly complex antioxidant defense system to protect against oxidative stress that includes antioxidant enzymes such as SOD, CAT, GPOX, APOX, DHAR, and GR (Figure 10) [35,46]. The O_2^- generated from mitochondria, chloroplasts, endoplasmic reticulum, and the cell wall is reduced to H_2O_2 by SOD. H_2O is then formed from H_2O_2 through the activities of CAT, and APOX. DHAR and glutathione reductase (GR) participate in the transformation of glutathion (GSH) to oxidized form of glutathione (GSSG) and generation of nicotinamide adenine dinucleotide phosphate (NADPH). GSH can bind to free Cd^{2+}, forming $Cd(GSH)_2$ and leading to the reduction of the Cd toxicity. In addition, phytochelatins (PCs) may bind Cd to form PC-Cd-S, which can then be stored in the vacuole [47]. Finally,

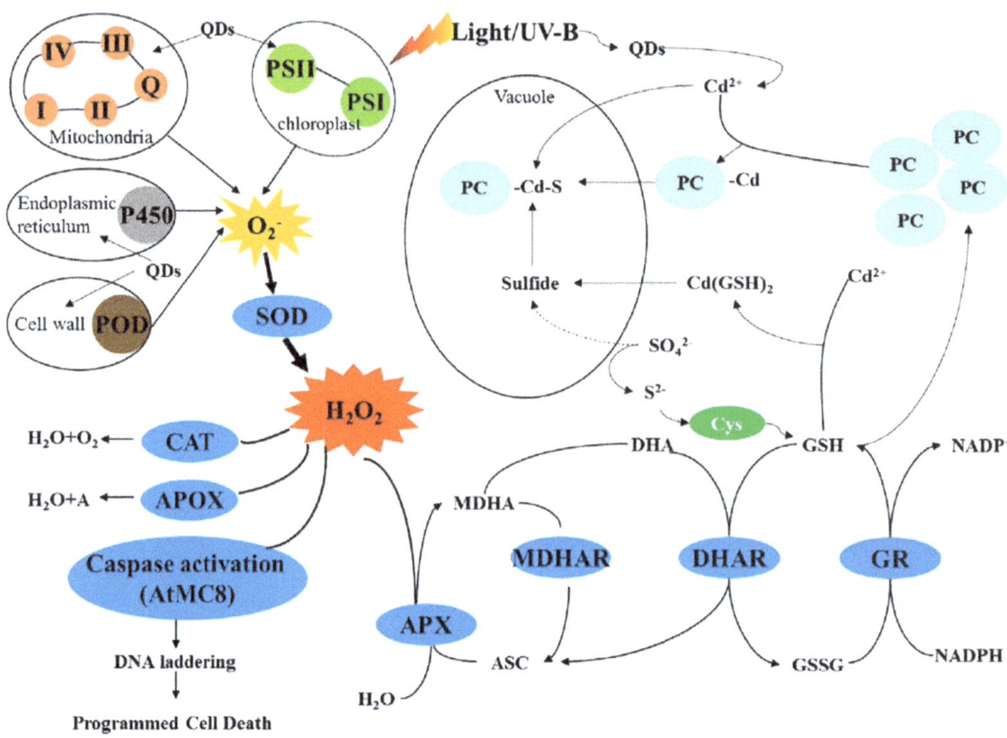

Figure 10. Schematic of the antioxidative defense system and Cd degradation mechanism in plants.

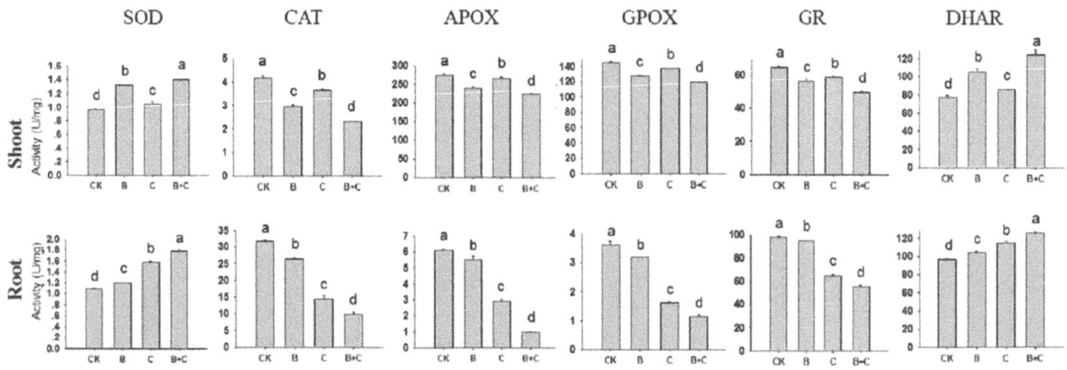

Figure 11. Antioxidant enzyme activities in shoots and roots of 5-day-old wheat seedlings. Data are means±SD (n = 3). Means with the same letter are not significantly different at Tukey's test ($p \leq 0.05$).

H_2O_2 activation of caspase can lead to programmed cell death [44,48].

Antioxidative defense enzymes activities were further assessed in the control, UV-B-only treated, CdTe-QDs treated, and UV-B plus CdTe-QDs treated seedlings (Figure 11). The inhibitory effect of CdTe-QDs upon antioxidant enzymes activities was milder than that of UV-B radiation treatment in shoots. Photosynthesis, which is susceptible to changes in light, mainly occurs in leaves [49]. Under ambient UV-B, the UV resistance locus 8 (UVR8) protein mediates antioxidant enzyme activities and regulates plant growth. However, it is currently unknown whether UV-B irradiation used in this study stimulates antioxidant enzyme activities in the same pathway.

In shoots, SOD and DHAR activities significantly increased in seedlings exposed to UV-B radiation when compared to controls. However, the other enzymes were significantly inhibited in the UV-B treated group compared to the control group. There were also significant difference in enzyme activities between the CdTe-QDs treated group and the control group, but the effects were not as marked as with the UV-B treated group (Figure 11 upper part). We inferred that CdTe-QDs affected metabolism in shoots through the release and transport of Cd^{2+} ions from QDs in root cells.

The cytotoxicity of Cd^{2+} has been known to be associated with oxidation stress in mammalian cells. We have demonstrated the additive effect of UV-B radiation on the Cd^{2+} toxicity in roots, manifested by the decreased levels of antioxidant enzymes such as CAT, APOX, GPOX and GR (Figure 11), and increased DNA fragmentation (Figure 12). The more pronounced effects in roots is presumably due to the release of additional Cd^{2+} as a result of surface oxidation of QDs under UV-B radiation [41,42]. We inferred that cytoplasmic diffusion of free Cd^{2+} occurred more rapidly than the diffusion of NPs. The greater root toxicity observed with the combined treatment compared to the single CdTe-QDs treatment may be the consequence of the increased accumulation of intracellular Cd^{2+}.

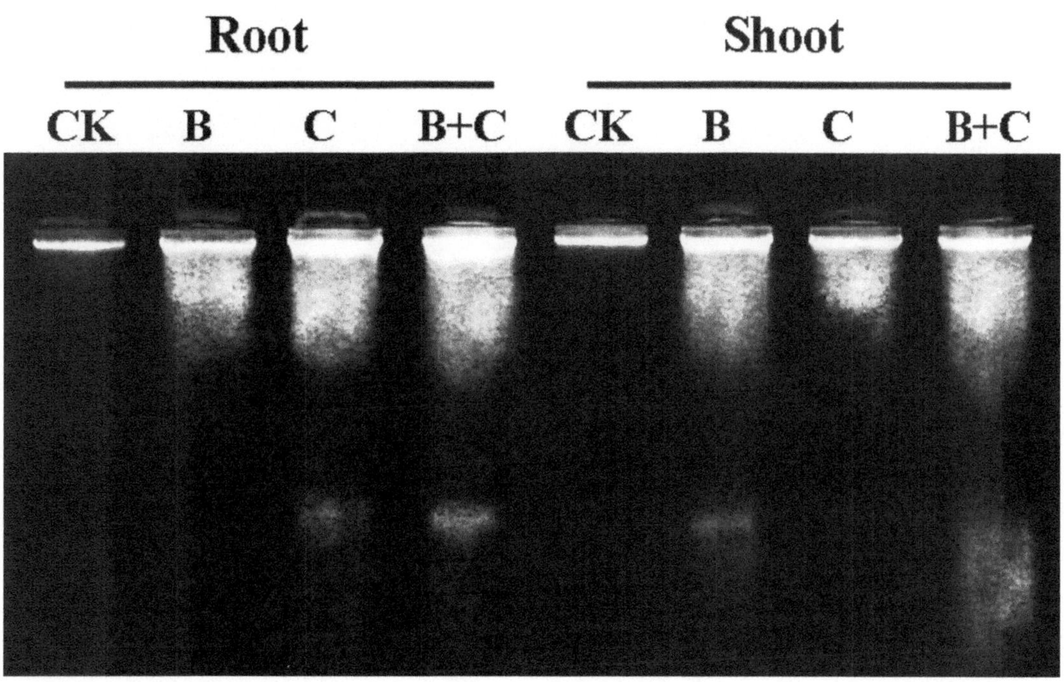

Figure 12. DNA laddering in wheat tissues. Electrophoresis gels of DNA extracted from wheat with different treatments. CK: control group; B: enhanced UV-B radiation; C: CdTe-QDs treatment; B+C: combined UV-B and CdTe-QDs treatment.

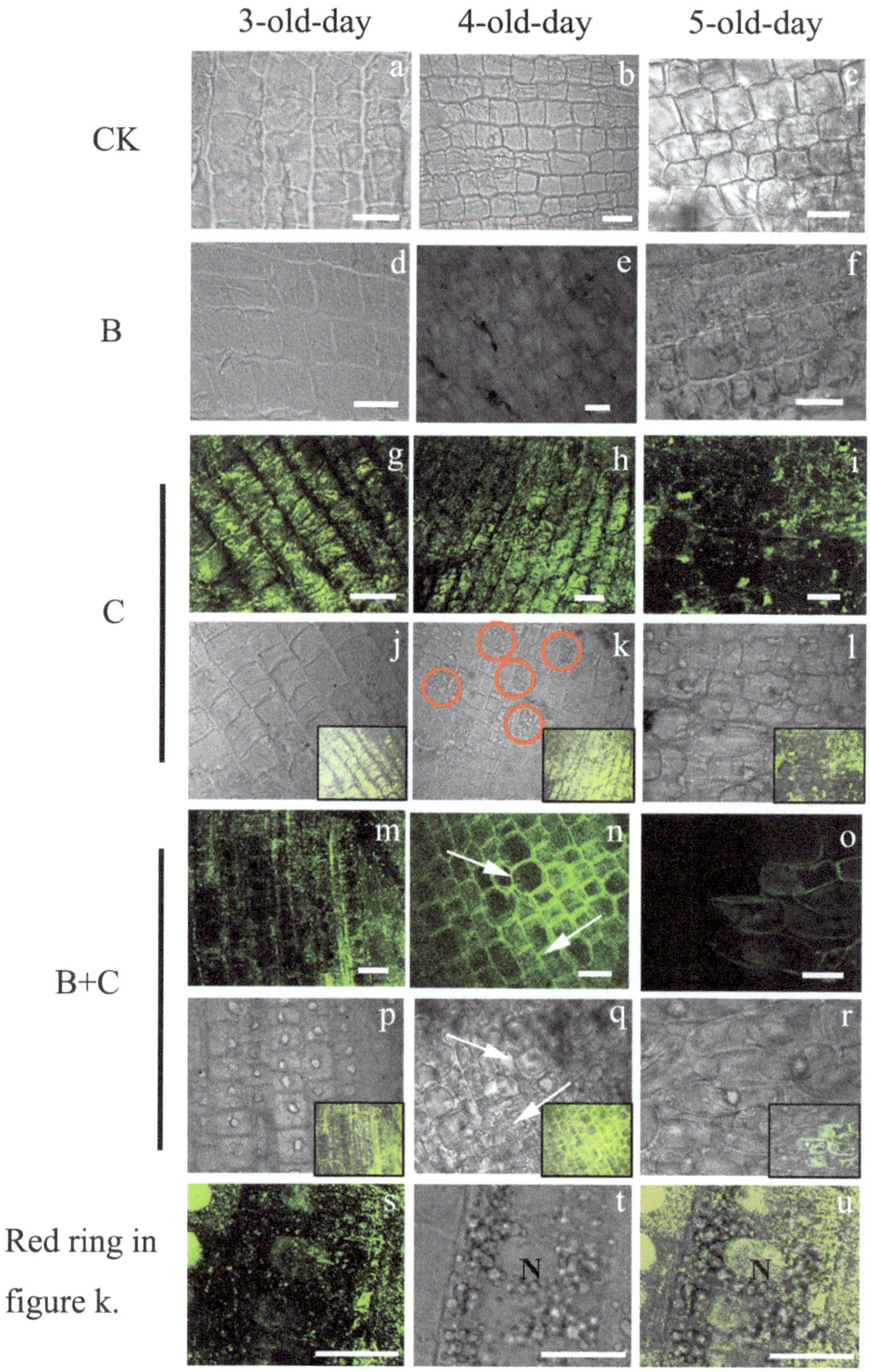

Figure 13. Confocal imaging of CdTe-QDs distribution in wheat root cells. CdTe-QDs are indicated by fluorescence imaging (g, h, i, m, n, o, s), DIC images demonstrate cell integrity (a, b, c, d, e, f, j, k, l, p, q, r, t). Regions highlighted in red in panel e are shown in panels s, t, and u. Pictures of the bottom right corner in figure j, k, l, p, q, r is the merge figures of g and j, h and k, i and l, m and p, n and q, o and r separately. Scale bars = 20 μm.

Distribution of CdTe-QDs in root cells

It has been demonstrated that CdSe-QDs, through the apoplastic pathway with the aid of silwet L-77, could be transferred into roots and were stable [50]. We therefore examined root cells of wheat seedling for the presence of CdSe-QDs using confocal microscopy (Figure 13). No fluorescence signal was detected in the untreated control (a, b, c) and in the UV-B treated cells (d, e, f). As shown in the DIC images, cells in the control group exhibited a clear outline of cell wall and regular shape ove three day period. In contrast, cells in the UV-B treated

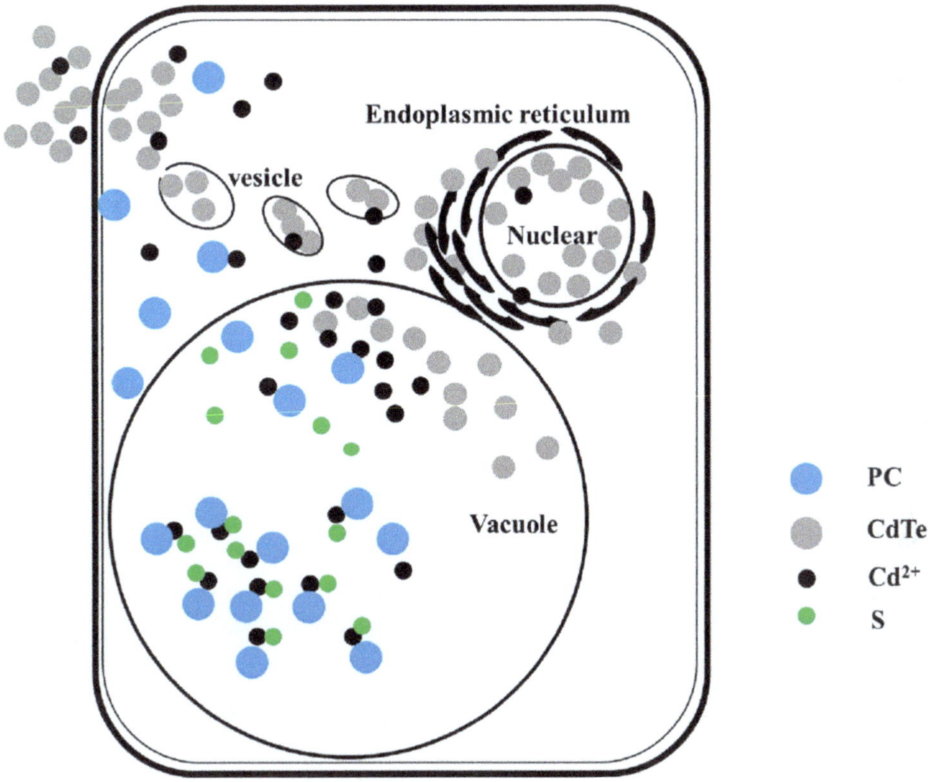

Figure 14. Schematic of a putative mechanism through which CdTe-QDs are processed by plant cells.

group showed the presence of intracellular vesicles on the fifth day and apparent abnormal rectangular shapes (f).

The presence of intracellular CdTe-QDs (g) was clearly visible after 3 days of CdTe-QDs treatment (g, j). Cell morphology appeared normal, with smooth surface and intact cell walls. After 4 days of treatment (h, k), vacuolation of cytoplasm appeared in some cells as shown ringed in red (k) and at higher magnification (s, t, u). Many small vesicles appeared around the nucleus (t), and CdTe-QDs were located inside the nucleus (u). This indicated the CdTe-QDs may enter the nucleus through the nuclear pore. CdTe-QDs were not uniformly distributed, but were found preferentially in and around the nucleus (u). A putative vesicle transport mechanism is shown in Figure 14. We propose that following the uptake of QDs by plant cells, PCs bind to Cd and form stable metal chelate complexes that are stored in the vacuole. Unbound excess QDs could spread into the nucleus leading to DNA damage. We inferred that NPs could produce a nanostructure that lead to an increase in the concentration of Cd around the nucleus, resulting in an increased cytotoxicity [12]. After 5 days of treatment (i, l), cell walls appeared looser and cells more variable in size (l), and the CdTe-QDs fluorescence intensity was reduced (i). The morphological characters resemble the features of programmed cell death.

In the combined UV-B and CdTe-QDs treated group, the situation is similar to the CdTe-QDs treated group. Cells exhibited normal morphology (p) on the third day and filled with CdTe-QDs (m). After 4 day treatment, cells are varied in size (n, q, white arrows). While following 5 day treatment, cells lost their square shape (r) and the fluorescence of CdTe-QDs was reduced dramatically (o).

Fifty cells were subsequently selected and the change in the fluorescence intensity over the treatment period was measured (Figure 15). CdTe-QD-derived fluorescence decreased by ~62.5% from the third to the fifth days of CdTe-QDs treatment and the decrease in intensity up to ~74.0% was observed in the combined treatment group. The results showed the UV-B radiation accelerated the CdTe-QDs-derived fluorescence decay, that is, ~11.5% fluorescence intensity decreased following the addition of UV-B radiation compared to single CdTe-QDs treatment. This was an order of magnitude higher than the inherent decrease in QDs emission intensity over 7 days (6.5%; Figure 3). The result indicated that CdTe-QDs were degraded by plant cells.

We detected the presence of QDs in shoots of seedlings but no fluorescence signal was observed in shoots. This suggested two possibilities. First, QDs may not have been transported to the top of the seedlings because QDs were only taken up by root epidermal cells and were not taken up by vascular tissues. Second, QDs that were taken up by root cells may have been degraded by PCs and stored in vacuoles, substantially decreasing the number of QDs were not available for transport into the shoot. Root cell took up the QDs and degraded them into Cd^{2+} partly. Cd^{2+} would bind to PCs or GSH, this is the cytotoxicity. While the other QDs changed ROS level and activated caspase, this would result in programmed cell death.

Programmed cell death in seedling cells

Apoptotic bodies visualized by DAPI staining were found in root and stem cells of all three treatment groups while absence in the control group (Figure 16), suggesting an induction of programmed cell death following the treatments.

DNA laddering or fragmentation is a feature of apoptosis [51]. Programmed cell death was investigated by using DNA ladder analysis (Figure 12). DNA fragmentation visualized as DNA

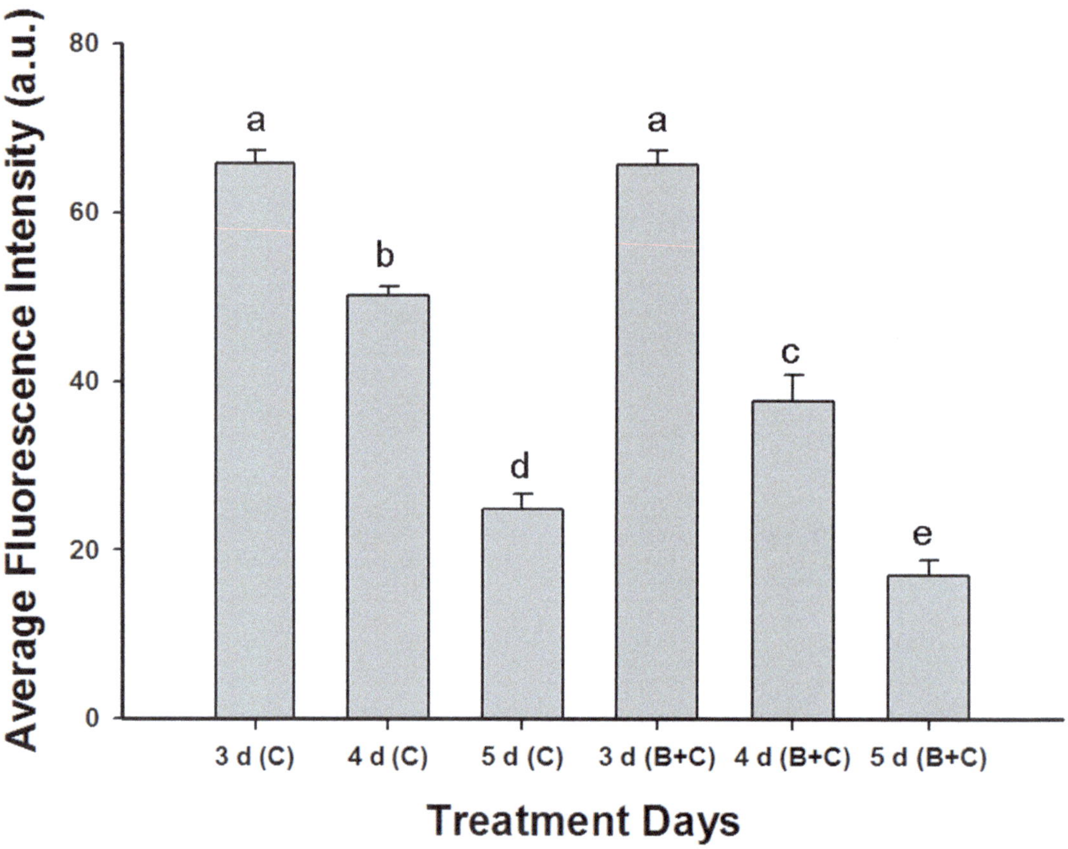

Figure 15. Average fluorescence intensity of QDs in 50 cells after different treatment durations. C, CdTe-QDs treatment alone; B+C, combined enhanced UV-B and CdTe-QDs treatment. Values are means±SD (n = 3). Means with the same letter are not significantly different at Tukey's test ($p \leq 0.05$).

laddering was evident in root cells of treatment groups involving CdTe-QDs. This may be due to nucleic acid damage caused by CdTe-QDs presence in the nuclei. Free Cd^{2+} would lead to further DNA damage [52]. As oxidation of aromatic DNA bases is the main source of DNA damage, an increase in intracellular ROS levels induced by CdTe-QDs would further induce DNA damage, leading to enhanced cytotoxicity. At the cellular level, Cd^{2+} alone could inhibit the biosynthesis of DNA, RNA and protein, and induce DNA strand breakage and lipid peroxidation [52]. In the combined treatment, Cd^{2+} release would be enhanced by UV-B radiation, which would further enhance the cytotoxicity.

DNA fragementation in shoots was detected in the treatment groups involcing UV-B radiation, with the combined treatments producing the most DNA laddering. UV-B radiation causes DNA damage [53] through various mechanisms including formation of cyclobutyl pyrimidine dimer (CPD) adducts, (6-4) photoproducts

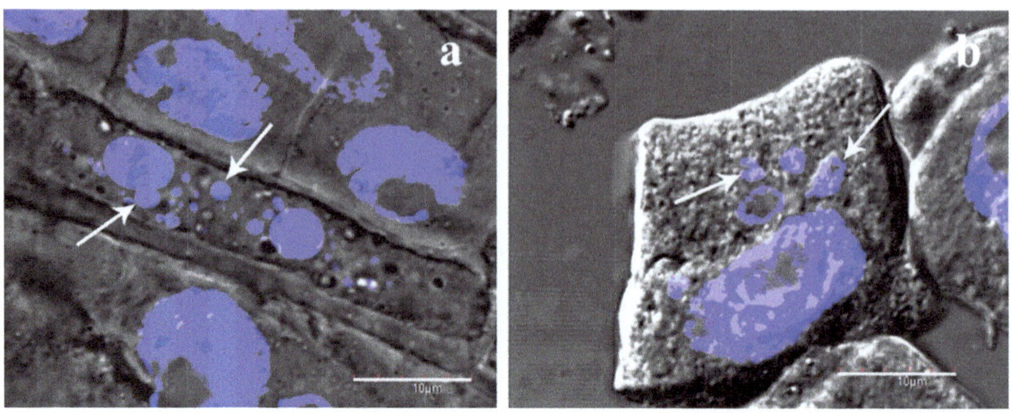

Figure 16. Cells undergoing programmed cell death. DAPI-stained DNA is shown in blue. Apoptotic bodies are indicated with white arrows. Shoot cells (a) and root cells (b) are shown. Scale bars = 10 μm.

((6-4) PPs), inter/intra cross links (ICLs), 8-oxoguanine, and DNA double-strand breaks (DSBs) [48]. We inferred that the damage caused by enhanced UV-B exposure in shoots masked the Cd^{2+} damage effects.

In summary, QDs are widely used and are becoming increasingly dispersed in the environment. Our study revealed the cytotoxicity of CdTe-QDs and UV-B radiation on wheat seedlings. Our results demonstrated that CdTe-QDs nanoparticles and UV-B radiation affected the antioxidant defense system and programmed cell death in the roots and shoots of wheat seedlings. At the doses used (200 mg CdTe-QDs/L and 10.0 $KJ/m^2/d$ enhanced UV-B), treatments induced negative effects on antioxidant enzyme activities and resulted in the formation of apoptotic bodies and DNA fragmentation. Combined treatments with both CdTe-QDs and enhanced UV-B exhibited the added deleterious effects. It is currently impossible to distinguish between Cd released from industrial waste and those released from QDs. The precise concentrations of Cd emissions from QDs are thus unknown. Therefore, in vitro studies such as ours are necessary for generating mechanistic insights into the specific effect of CdTe-QDs on agricultural plants.

Acknowledgments

We thank Dr Huiling Xu (Peter MacCallum Cancer Centre, Australia) for critical reading and editing the manuscript.

Author Contributions

Conceived and designed the experiments: RH. Performed the experiments: YG. Analyzed the data: HC. Contributed reagents/materials/analysis tools: YG. Contributed to the writing of the manuscript: HC.

References

1. Jones R (2009) Feynman's unfinished business. Nature nanotechnology 4: 785–785.
2. Djikanovic D, Kalauzi A, Jeremic M, Xu J, Micic M, et al. (2012) Interaction of the CdSe quantum dots with plant cell walls. Colloids Surf B Biointerfaces 91: 41–47.
3. Yang A, Zheng Y, Long C, Chen H, Liu B, et al. (2014) Fluorescent immunosorbent assay for the detection of alpha lactalbumin in dairy products with monoclonal antibody bioconjugated with CdSe/ZnS quantum dots. Food Chem 150: 73–79.
4. Sozer N, Kokini JL (2014) Use of quantum nanodot crystals as imaging probes for cereal proteins. Food Research International 57: 142–151.
5. Valizadeh A, Mikaeili H, Samiei M, Farkhani S, Zarghami N, et al. (2012) Quantum dots: synthesis, bioapplications, and toxicity. Nanoscale Research Letters 7: 1–14.
6. Rico CM, Hong J, Morales MI, Zhao LJ, Barrios AC, et al. (2013) Effect of Cerium Oxide Nanoparticles on Rice: A Study Involving the Antioxidant Defense System and In Vivo Fluorescence Imaging. Environmental Science & Technology 47: 5635–5642.
7. Miralles P, Church TL, Harris AT (2012) Toxicity, uptake, and translocation of engineered nanomaterials in vascular plants. Environmental science & technology 46: 9224–9239.
8. Nair R, Poulose AC, Nagaoka Y, Yoshida Y, Maekawa T, et al. (2011) Uptake of FITC labeled silica nanoparticles and quantum dots by rice seedlings: effects on seed germination and their potential as biolabels for plants. J Fluoresc 21: 2057–2068.
9. Nguyen KC, Willmore WG, Tayabali AF (2013) Cadmium telluride quantum dots cause oxidative stress leading to extrinsic and intrinsic apoptosis in hepatocellular carcinoma HepG2 cells. Toxicology 306: 114–123.
10. Liu Y, Wang P, Wang Y, Zhu Z, Lao F, et al. (2013) The influence on cell cycle and cell division by various cadmium-containing quantum dots. Small 9: 2440–2451.
11. Chen N, He Y, Su Y, Li X, Huang Q, et al. (2012) The cytotoxicity of cadmium-based quantum dots. Biomaterials 33: 1238–1244.
12. Su Y, Hu M, Fan C, He Y, Li Q, et al. (2010) The cytotoxicity of CdTe quantum dots and the relative contributions from released cadmium ions and nanoparticle properties. Biomaterials 31: 4829–4834.
13. Cho SJ, Maysinger D, Jain M, Röder B, Hackbarth S, et al. (2007) Long-term exposure to CdTe quantum dots causes functional impairments in live cells. Langmuir 23: 1974–1980.
14. Li X, Chen N, Su Y, He Y, Fan C, et al. (2013) Cytotoxicity of cadmium-based quantum dots. Chinese Science Bulletin (Chinese Version) 58: 1393.
15. Yu G, Tan Y, He X, Qin Y, Liang J (2014) CLAVATA3 Dodecapeptide Modified CdTe Nanoparticles: A Biocompatible Quantum Dot Probe for In Vivo Labeling of Plant Stem Cells. PloS one 9: e89241.
16. Gobe G, Crane D (2010) Mitochondria, reactive oxygen species and cadmium toxicity in the kidney. Toxicology letters 198: 49–55.
17. Cuypers A, Plusquin M, Remans T, Jozefczak M, Keunen E, et al. (2010) Cadmium stress: an oxidative challenge. Biometals 23: 927–940.
18. Lovrić J, Bazzi HS, Cuie Y, Fortin GR, Winnik FM, et al. (2005) Differences in subcellular distribution and toxicity of green and red emitting CdTe quantum dots. Journal of Molecular Medicine 83: 377–385.
19. Forster P, Thompson D, Baldwin M, Chipperfield M, Dameris M, et al. (2011) Scientific Assessment of Ozone Depletion: 2010. Global Ozone Research and Monitoring Project-Report 52: 516.
20. Taalas P, Amanatidis GT, Heikkilä A (2000) European conference on atmospheric UV radiation: overview. Journal of Geophysical Research: Atmospheres (1984–2012) 105: 4777–4785.
21. Choudhary KK, Agrawal SB (2014) Ultraviolet-B induced changes in morphological, physiological and biochemical parameters of two cultivars of pea (Pisum sativum L.). Ecotoxicol Environ Saf 100: 178–187.
22. Han R WX, Yue M (2002) Influence of He-Ne laser irradiation on the excision repair of cyclobutyl pyrimidine dimers in the wheat DNA. Chinese Science Bulletin 47: 818–821.
23. Hideg E, Jansen MA, Strid A (2013) UV-B exposure, ROS, and stress: inseparable companions or loosely linked associates? Trends Plant Sci 18: 107–115.
24. Wang G, Deng S, Li C, Liu Y, Chen L, et al. (2012) Damage to DNA caused by UV-B radiation in the desert cyanobacterium Scytonema javanicum and the effects of exogenous chemicals on the process. Chemosphere 88: 413–417.
25. Chen HZ ZJ, Du MT, Han R (2011) Influence of Enhanced UV-B Radiation on F-actin in Wheat Division Cells. Plant Diversity And Resources 33: 306–310.
26. Jenkins GI (2009) Signal Transduction in Responses to UV-B Radiation. Annu Rev Plant Biol: 407–431.
27. Gaponik N, Talapin DV, Rogach AL, Hoppe K, Shevchenko EV, et al. (2002) Thiol-capping of CdTe nanocrystals: an alternative to organometallic synthetic routes. The Journal of Physical Chemistry B 106: 7177–7185.
28. Schlemmer MR, Francis DD, Shanahan J, Schepers JS (2005) Remotely measuring chlorophyll content in corn leaves with differing nitrogen levels and relative water content. Agronomy Journal 97: 106–112.
29. Zhao S, Huang Q, Yang P, Zhang J, Jia H, et al. (2012) Effects of Ion Beams Pretreatment on Damage of UV-B Radiation on Seedlings of Winter Wheat (Triticum aestivum L.). Appl Biochem Biotechnol 168: 2123–2135.
30. Navarro DA, Bisson MA, Aga DS (2012) Investigating uptake of water-dispersible CdSe/ZnS quantum dot nanoparticles by Arabidopsis thaliana plants. J Hazard Mater 211–212: 427–435.
31. Lee YP, Kim SH, Bang JW, Lee HS, Kwak SS, et al. (2007) Enhanced tolerance to oxidative stress in transgenic tobacco plants expressing three antioxidant enzymes in chloroplasts. Plant Cell Rep 26: 591–598.
32. Verma S, Dubey RS (2003) Lead toxicity induces lipid peroxidation and alters the activities of antioxidant enzymes in growing rice plants. Plant Science 164: 645–655.
33. Shah K, Kumar RG, Verma S, Dubey R (2001) Effect of cadmium on lipid peroxidation, superoxide anion generation and activities of antioxidant enzymes in growing rice seedlings. Plant Science 161: 1135–1144.
34. Rao MV, Paliyath G, Ormrod DP (1996) Ultraviolet-B-and ozone-induced biochemical changes in antioxidant enzymes of Arabidopsis thaliana. Plant physiology 110: 125–136.
35. Hossain MA, Silva JATd, Fujita M (2011) Glyoxalase System and Reactive Oxygen Species Detoxification System in Plant Abiotic Stress Response and Tolerance: An Intimate Relationship.
36. Wang J, Wang Q, Li J, Shen Q, Wang F, et al. (2012) Cadmium induces hydrogen peroxide production and initiates hydrogen peroxide-dependent apoptosis in the gill of freshwater crab, Sinopotamon henanense. Comp Biochem Physiol C Toxicol Pharmacol 156: 195–201.
37. Ramel F, Sulmon C, Bogard M, Couée I, Gouesbet G (2009) Differential patterns of reactive oxygen species and antioxidative mechanisms during atrazine injury and sucrose-induced tolerance in Arabidopsis thaliana plantlets. BMC Plant Biology 9: 28.
38. Xu J, Yin H, Li Y, Liu X (2010) Nitric oxide is associated with long-term zinc tolerance in Solanum nigrum. Plant Physiol 154: 1319–1334.
39. Rogers SO, Bendich AJ (1985) Extraction of DNA from milligram amounts of fresh, herbarium and mummified plant tissues. Plant molecular biology 5: 69–76.
40. Zhi Qi MY, Rong Han, Xun-Ling Wang (2002) The damage repair role of He-Ne laser on plants exposed to different intensities of ultraviolet-B radiation. Photochemistry and Photobiology 75: 680–686.
41. Derfus AM, Chan WC, Bhatia SN (2004) Probing the cytotoxicity of semiconductor quantum dots. Nano letters 4: 11–18.

42. Kirchner C, Liedl T, Kudera S, Pellegrino T, Muñoz Javier A, et al. (2005) Cytotoxicity of colloidal CdSe and CdSe/ZnS nanoparticles. Nano Letters 5: 331–338.
43. Hideg É, Jansen MA, Strid Å (2013) UV-B exposure, ROS, and stress: inseparable companions or loosely linked associates? Trends in plant science 18: 107–115.
44. Huang W, Yang X, Yao S, LwinOo T, He H, et al. (2014) Reactive oxygen species burst induced by aluminum stress triggers mitochondria-dependent programmed cell death in peanut root tip cells. Plant Physiol Biochem 82C: 76–84.
45. Couee I, Sulmon C, Gouesbet G, El Amrani A (2006) Involvement of soluble sugars in reactive oxygen species balance and responses to oxidative stress in plants. J Exp Bot 57: 449–459.
46. Silva ECd, Albuquerque MBd, Neto ADdA, Junior CDdS (2013) Drought and Its Consequences to Plants – From Individual to Ecosystem.
47. Heiss S, Wachter A, Bogs J, Cobbett C, Rausch T (2003) Phytochelatin synthase (PCS) protein is induced in Brassica juncea leaves after prolonged Cd exposure. Journal of Experimental Botany 54: 1833–1839.
48. Nawkar GM, Maibam P, Park JH, Sahi VP, Lee SY, et al. (2013) UV-Induced Cell Death in Plants. Int J Mol Sci 14: 1608–1628.
49. Shen X, Zhou Y, Duan L, Li Z, Eneji AE, et al. (2010) Silicon effects on photosynthesis and antioxidant parameters of soybean seedlings under drought and ultraviolet-B radiation. J Plant Physiol 167: 1248–1252.
50. Hu Y, Li J, Ma L, Peng Q, Feng W, et al. (2010) High efficiency transport of quantum dots into plant roots with the aid of silwet L-77. Plant Physiol Biochem 48: 703–709.
51. Farage-Barhom S, Burd S, Sonego L, Mett A, Belausov E, et al. (2011) Localization of the Arabidopsis senescence- and cell death-associated BFN1 nuclease: from the ER to fragmented nuclei. Mol Plant 4: 1062–1073.
52. Fernández EL, Gustafson A-L, Andersson M, Hellman B, Dencker L (2003) Cadmium-induced changes in apoptotic gene expression levels and DNA damage in mouse embryos are blocked by zinc. Toxicological sciences 76: 162–170.
53. Rousseaux MC, Ballare CL, Giordano CV, Scopel AL, Zima AM, et al. (1999) Ozone depletion and UVB radiation: Impact on plant DNA damage in southern South America. Proceedings of the National Academy of Sciences of the United States of America 96: 15310–15315.

6

Exploring the Genes of Yerba Mate (*Ilex paraguariensis* A. St.-Hil.) by NGS and *De Novo* Transcriptome Assembly

Humberto J. Debat[1]◑, Mauro Grabiele[2,4]◑, Patricia M. Aguilera[2,4], Rosana E. Bubillo[3], Mónica B. Otegui[4], Daniel A. Ducasse[1], Pedro D. Zapata[4], Dardo A. Marti[2,4]*

1 Instituto de Patología Vegetal, Centro de Investigaciones Agropecuarias, Instituto Nacional de Tecnología Agropecuaria (IPAVE-CIAP-INTA), Córdoba, Argentina, **2** Instituto de Biología Subtropical, Universidad Nacional de Misiones (IBS-UNaM-CONICET), Posadas, Misiones, Argentina, **3** Estación Experimental Cerro Azul, Instituto Nacional de Tecnología Agropecuaria (EEA Cerro Azul-INTA), Misiones, Argentina, **4** Instituto de Biotecnología de Misiones, Facultad de Ciencias Exactas Químicas y Naturales, Universidad Nacional de Misiones (INBIOMIS-FCEQyN-UNaM), Misiones, Argentina

Abstract

Yerba mate (*Ilex paraguariensis* A. St.-Hil.) is an important subtropical tree crop cultivated on 326,000 ha in Argentina, Brazil and Paraguay, with a total yield production of more than 1,000,000 t. Yerba mate presents a strong limitation regarding sequence information. The NCBI GenBank lacks an EST database of yerba mate and depicts only 80 DNA sequences, mostly uncharacterized. In this scenario, in order to elucidate the yerba mate gene landscape by means of NGS, we explored and discovered a vast collection of *I. paraguariensis* transcripts. Total RNA from *I. paraguariensis* was sequenced by Illumina HiSeq-2000 obtaining 72,031,388 pair-end 100 bp sequences. High quality reads were *de novo* assembled into 44,907 transcripts encompassing 40 million bases with an estimated coverage of 180X. Multiple sequence analysis allowed us to predict that yerba mate contains ~32,355 genes and 12,551 gene variants or isoforms. We identified and categorized members of more than 100 metabolic pathways. Overall, we have identified ~1,000 putative transcription factors, genes involved in heat and oxidative stress, pathogen response, as well as disease resistance and hormone response. We have also identified, based in sequence homology searches, novel transcripts related to osmotic, drought, salinity and cold stress, senescence and early flowering. We have also pinpointed several members of the gene silencing pathway, and characterized the silencing effector Argonaute1. We predicted a diverse supply of putative microRNA precursors involved in developmental processes. We present here the first draft of the transcribed genomes of the yerba mate chloroplast and mitochondrion. The putative sequence and predicted structure of the caffeine synthase of yerba mate is presented. Moreover, we provide a collection of over 10,800 SSR accessible to the scientific community interested in yerba mate genetic improvement. This contribution broadly expands the limited knowledge of yerba mate genes, and is presented as the first genomic resource of this important crop.

Editor: Jin-Song Zhang, Institute of Genetics and Developmental Biology, Chinese Academy of Sciences, China

Funding: The authors have no support or funding to report.

Competing Interests: The authors have declared that no competing interests exist.

* Email: darmarti@gmail.com

◑ These authors contributed equally to this work.

Introduction

Ilex paraguariensis (Aquifoliaceae) is a dioecious crop tree native to the subtropical rainforest of Northeastern Argentina, Southwestern Brazil and Eastern Paraguay, where it is widely cultivated [1]. This evergreen holly is colloquially known as "yerba mate" or "erva mate" as it is mainly consumed as a nutritional and stimulant beverage named "mate", a type of hot infusion made from dried milled leaves and twigs of *I. paraguariensis*. Yerba mate is also extensively used to prepare infusions, concoctions and quenchers with similar purposes and, more recently as admixture in ice creams, candies and energy drinks [2], as well as in dyes, cosmetics and spa ingredients [3]. Antioxidant, anti-inflammatory, antimutagenic and lipid-lowering properties have been reported in *I. paraguariensis* [2], leading to an increasing interest in this tree. Yerba mate is an economically important crop cultivated and produced on a total area of more than 326,000 ha [4,5]. Argentina is the main producer with a total yield of over 880,000 t, representing ~85% of world-wide yerba mate production [6]. About 15% of total yerba mate production is exported to South American, European and Asian markets [7]. Besides the agricultural and economic importance of yerba mate, it is worth noting its profound and omnipresent influence in Latin American socio-cultural dynamics. Yerba mate widespread consumption embraces and extends ubiquitously, pervasively reaching every economic and cultural niche in South America [8–11]. To emphasize the relevance of yerba mate in South American tradition and its introduction and dissemination particularly in Argentinean culture, in 2009 a 5,000 people survey projected that while 81% of the Argentinean population consumes coffee, a striking 98% of the population consumes yerba mate [12].

Genetic improvement of *I. paraguariensis* has been limited by several factors. Agronomic evaluation and selection programs have been performed essentially to improve yield in this crop [13,1]. However, apart from this trait, very little is known about agronomically important loci on the limited available germplasm of this species. Moreover, yerba mate plants cannot be recognized as male or female prior to their first blooming, which takes 3 to 10 years post seed germination [14], delaying the selection of parentals for breeding purposes. Likewise, knowledge of sequences of interesting genes is needed to achieve genetic improvement based on molecular tools, a valuable information that is lacking in yerba mate. Currently, merely ~80 sequences originated from *I. paraguariensis* are available in the National Center for Biotechnology Information (NCBI) database, most of them corresponding to microsatellites. In addition, genetic information in the genus is scarce, annotated sequences are virtually inexistent and no expressed sequence tags (ESTs) libraries have been generated so far.

Massively parallel sequencing of RNA (RNA-Seq) is an efficient way to characterize the transcriptional landscape of a species and reveal its complexity [15]. It allows to investigate the transcriptome composition and expression and, in this direction, to explore and reveal the expressed profile of a defined organism [16]. This next-generation sequencing technology (NGS) is a simple and fast tool to analyze the transcriptome since it requires neither cloning library of the cDNAs nor any *a priori* knowledge of the species. Instead of this, RNA-Seq technology generates millions of short direct cDNA reads which are subsequently assembled to construct transcripts [15]. *De novo* transcriptome assembly is suitable in order to reconstruct full length transcripts from these short reads in organisms without a sequenced genome as reference. The most advanced algorithms to achieve this strategy consists in efficiently constructing and analyzing sets of *de Bruijn* graphs to construct and assemble transcripts and requires a great amount of parallel sequence short reads provided by high throughput sequencing technology [17].

This study presents the first analysis of the *I. paraguariensis* transcriptome. We employed the Illumina total RNA-Seq sequencing method to generate 72,031,388 pair-end 100 bp sequence reads. The obtained high quality reads were *de novo* assembled into 44,907 primary transcripts encompassing 40 million bases with an estimated coverage of 180X. Multiple sequence analysis allowed us to predict that yerba mate contains about 32,355 genes and 12,551 gene variants or isoforms. An initial analysis of these genes allowed us to identify and categorize members of more than 100 metabolic pathways. The transcriptome characterization of *I. paraguariensis* generated from our study is a very useful tool derived from a convenient and exhaustive approach of annotation and discovery of genes of several major metabolic pathways. The vast amount of information obtained would encourage and serve as reliable source in the path to the discovery of biologically and agronomically important traits, as well as for molecular markers development, gene mapping, analysis of genetic diversity and selective breeding in yerba mate.

Results and Discussion

RNA sequencing analysis and transcriptome *de novo* assembly

In order to shed light on the transcriptional landscape of yerba mate, total RNA was extracted from pooled leaves of *I. paraguariensis* breeding line Pg538 from INTA EEA-Cerro Azul, Misiones, Argentina. After initial quality controls the isolated RNA was sequenced by the Illumina HiSeq-2000 platform. A total of 72,031,388 100 bp pair-end reads were obtained (**Table 1**). An analysis of the sequencing run indicated an absence of cycle-wise multiplied calls of the same nt, an average high quality of Q = 36.3, a lack of positional biases in the call frequency for each base and a typical unimodal distribution of quality average and Kmer enrichment frequency (**Figure S1 a-e**).

Recently, towards the identification of phosphate starvation-responsive genes in wheat (*Triticum aestivum*), a similar NGS approach was employed based in *de novo* assembly of 73,8 million reads from RNA-seq libraries [18]. The extension of this sequencing process, similar to that of our study, was effective and sufficient to generate a comprehensive transcriptome in wheat in the absence of reference genome information.

After quality filtering, the sequencing reads were assembled with the Trinity software [17] and a transcriptome of 44,907 assembled sequences was obtained. A quality analysis of the assembly suggested a typical distribution and coverage of GC content, a lack of positional kmer enrichment, a high percentage coverage consistent with the sequence length distribution and a regular positional nucleotide contribution in the assembled transcripts (**Figure S2 a-f**). The transcriptome covers 39,969,375 bp, with a mean contig length of 890 bp, N50 of 1,430 bp and 8,353 sequences with a length of over the N50. Our *de novo* assembly utterly indicates that yerba mate presents an estimate of 32,356 genes and 12,551 gene variants or isoforms (**Table 2**).

Evaluating the yerba mate transcriptome by DEG analysis

It has been proposed that a comprehensive catalog of essential genes may constitute a minimal genome, forming a set of functional modules, which play key roles in eukaryotic metabolism [19]. In that direction, a catalog of over 356 genes has been assigned as essential in the cruciferous plant *Arabidopsis thaliana* [20]. To assess and estimate the "completeness" of the assembled yerba mate transcriptome, a DEG (Database of Essential Genes) analysis was performed (**Data S1**). Exploring the genes of yerba mate we observed that the orthologs of 97.2% of the *A. thaliana* essential genes were present in our assembled transcriptome. In a recent study [21] a highly representative *Nicotiana benthamiana* transcriptome was evaluated under a similar platform depicted as "Core eukaryotic genes dataset" (CEGMA) [22], which includes a widely conserved set of 248 eukaryotic proteins. In this CEGMA analysis, 95% of the core proteins were identified in the *N. benthamiana* transcriptome. In this scenario we consider that the yerba mate DEG score is indicative of an overall representative status of the transcript library produced.

Characterization and functional annotation of yerba mate transcripts

The assembled transcriptome of *I. paraguariensis* was sequentially subjected to homology searches using the BLASTX platform against the UniProt *viridiplantae* database (**Table S3**) and TAIR. BLASTX hits E-value distribution of assembled transcripts to TAIR *A. thaliana* proteins is presented in **Figure 1**. Using a cut-off value of 10E-05, over 77% of transcripts (31,787) attained a blast hit based on identity conservation (**Table S4**).

The obtained BLASTX hits were subsequently imported into the Blast2GO software, the KAAS server, and the agriGO platform, where gene ontology annotation, metabolic pathway profiling and GO categorization were performed. Over 217,655 GO term tags were identified in the yerba mate transcriptome, of which 4,341 were associated to KEGG ids corresponding to 315 KEGG maps (**Table S2**). An initial sorting of the yerba mate sequences based on GO terms is presented in **Figure 2** using the

Table 1. Yerba mate Illumina HiSeq-2000 sequencing run statistics.

Sequencing stats	*Ilex paraguariensis* RNA-seq
Total Bases	7,275,170,188
Read Count	72,031,388
GC (%)	45,38
N (%)	0,027
Q20 (%)	98,21
Q30 (%)	94,99
Average Q	36,3
read length	100 nt×2

A. thaliana transcriptome as background. The assembled transcripts are categorized by cell component (**Figure 2a**) where an enrichment in organelle and cell structural components was observed, molecular function (**Figure 2b**) that showed an elevated percentage of catalytic and binding representatives, and biological process (**Figure 2c**) where the distribution of sequences followed the typical frequency observed in *Arabidopsis*. A closer analysis of the GO associated yerba mate sequences (**Table S5**) by semantic similarity-based scatterplots representations and tree-maps (**Figure S3** and **Figure S4**) highlighted several terms based on p-values (circle size, rectangle size) associated to the GO enriched categories, such as growth, methylation and reproductive structure development on biological process (**a**), chloroplast, membrane enclosed lumen and ubiquitin ligase complex on cellular component (**b**) and chlorophyll binding, methyltransferase activity and sequence specific DNA binding on molecular function (**c**). An exhaustive analysis of the GO terms is presented as AgriGO generated plots of GO enrichment, significance and relationships in yerba mate based in biological process (**Figure S5**), cellular component (**Figure S6**) and molecular function (**Figure S7**). A 166 catalog of KEGG drawn maps representing the gene members of the yerba mate transcriptome extensively associated with numerous metabolic pathways is presented in **Data S2**. In order to explore the yerba mate genes, we approached a categorization of transcripts based in BLASTX. Overall, we have identified over 1,000 putative transcription factors of yerba mate (**Table S1**), 50 transcripts involved in heat-stress, more than 200 oxidative stress responsive putative genes, 30 transcripts associated with pathogen response, a significant number of transcripts associated with ribosome constituents, ribosome processing, trafficking, rRNA maturation, and ribosome assembly (**Figure S8**), as well as 60 assembled transcripts involved in disease resistance and 150 transcripts probably engaged in hormone response (**Figure S9**). We have also identified nearly 100 transcripts related to osmotic, drought, salinity and cold stresses, senescence, early flowering, and biosynthesis of sugars, flavonoids, carotenoids and chlorophyll (**Table S3**).

When compared with other plant species reported by previous studies that used the Illumina system and the Trinity software, the quality of our transcriptome sequence shows to be significantly high, which is evidenced in several aspects. First, the average length of the unigenes we observed is 890 bp which is comparable to that of chili pepper (*Capsicum frutescens*, 712 bp) [23], ramie (*Boehmeria nivea*, 824 bp) [24], *Salvia splendens* (779 bp and 812 bp for two different strains) [25] and peanut (*Arachis hypogaea*, 751 bp) [26] transcriptome sequences. Second, approximately 77% of the genes discovered in this study were successfully annotated for their putative functions. Previous reports of

Table 2. Yerba mate Trinity *de novo* assembled transcriptome statistics.

Assembly	*Ilex paraguariensis*
method	Trinity k25
assembled seq.	44,907
unigenes	44,906
gene families	32,355
gene variants	12,551
n: 100	44,907
n: N50	8,353
min	201 bp
median	544 pb
mean	890 bp
N50	1,430 bp
max	15,716 bp
sum	39,969,375 bp

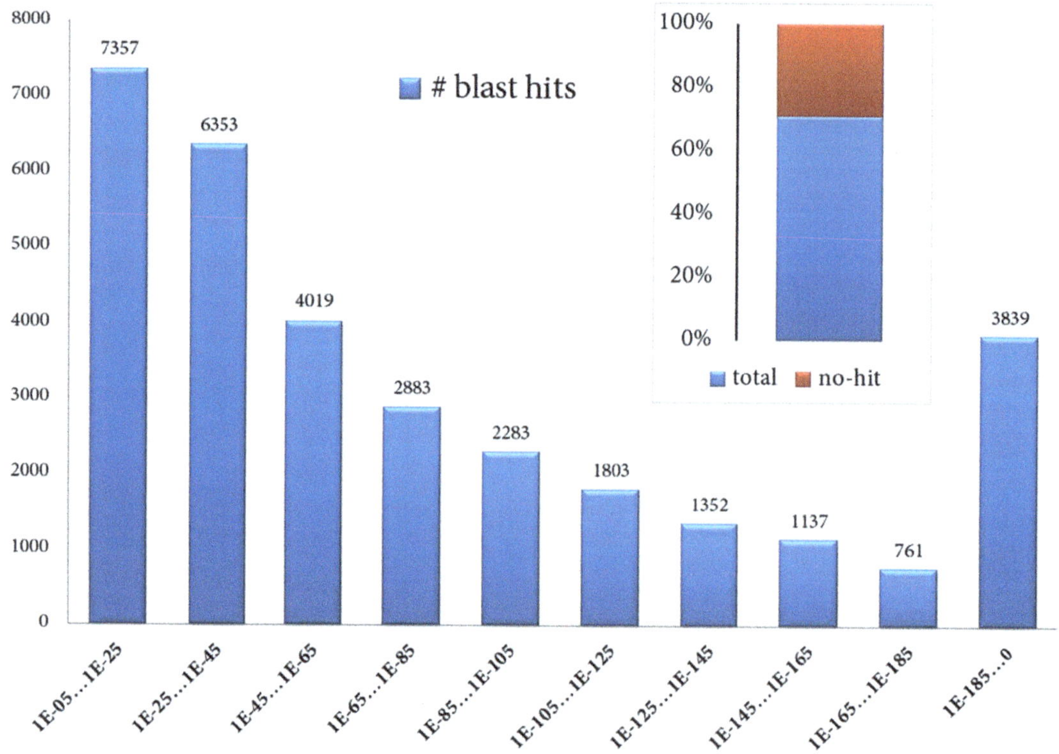

Figure 1. BLASTX hits E-value distribution of assembled transcripts to TAIR *Arabidopsis thaliana* proteins. Using a cut-off value of 10E-05, over 77% of transcripts (31,787 contigs) attained a blast hit based in identity conservation.

annotated genes in species such as *C. frutescens* (72.33%) [23], *Boehmeria nivea* (77.70%) [24], barnyardgrass (*Echinochloa crus-galli*, 57.45%) [27] and sugarcane varieties (*Saccharum officinarum* x *S. spontaneum*, 49.06%) [28] further support the notion of our assembly dataset being a fair representation of the yerba mate transcriptome.

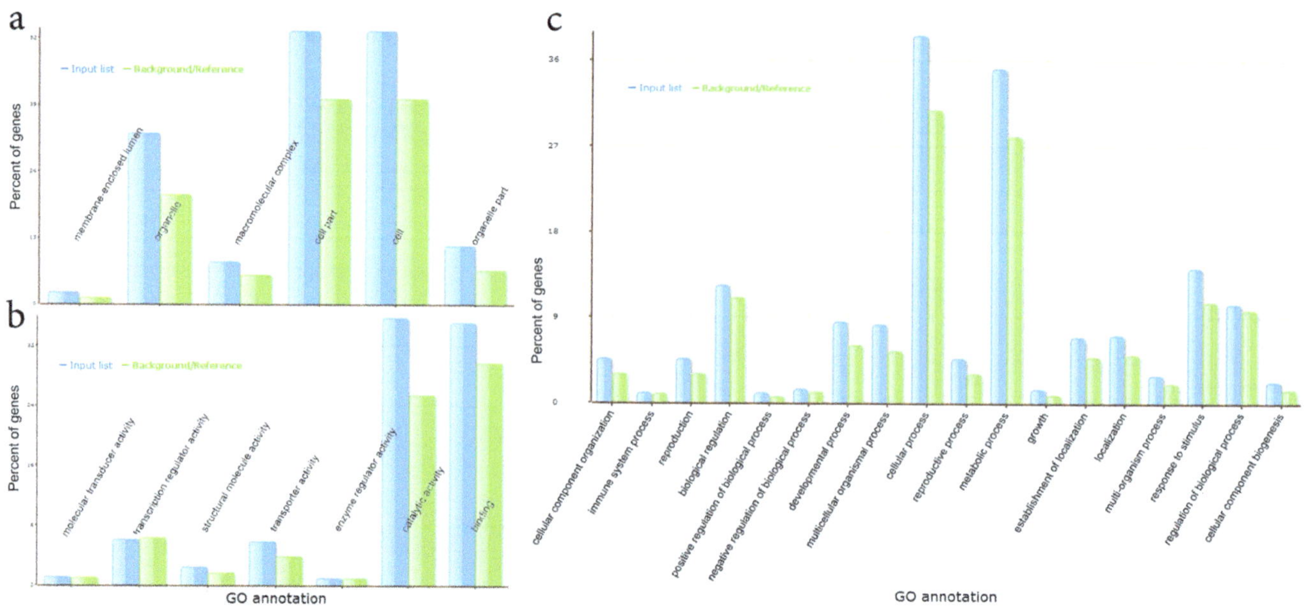

Figure 2. GO annotations obtained for the yerba mate transcriptome. Categorization by cell component (a), molecular function (b), and biological process (c). *Ilex paraguariensis* GO percentages are based on 31,787 BLASTX hits (blue), and the *Arabidopsis thaliana* transcriptome was employed as background (green).

Prediction of yerba mate SSRs

Simple sequence repeat (SSR) markers are well-known and widely used as valuable tools for assessing genetic diversity. SSRs are useful in the development of genetic maps, comparative genomics and marker-assisted selection breeding [28]. Thus, in parallel, the yerba mate transcripts library was comprehensively analyzed in search of SSRs. A total of 10,813 SSRs were identified in 8,449 sequences along the transcriptome. We analyzed our data and *in silico* predicted SSRs using 6,4,3,3,3 motifs repeats criteria for di-, tri-, tetra-, penta-, and hexa-nucleotides SSRs. In this context, the 2 nt motif repeats represented 40.9% of total SSRs found, while 3 nt motif repeats constituted roughly a 35.8% of total SSRs (**Figure 3a**). The most represented SSR corresponded to 2 nt motif ct/ag-tc/ga (**Figure 3b**) which encompassed over 84% of the 4,429 SSRs of 2 bp motif (**Table S11**). Among the tri-nucleotide motif repeats, with over 26% of the hits, aag/ctt-tct/aga-ttc/gaa are the most common SSR found in *I. paraguariensis* (**Figure 3c**). In most plant transcriptome studies, tri-nucleotide are the most frequent SSRs. However, the repeat motif abundance in plant transcriptomes is affected by the *in silico* determination of SSRs prediction criteria. For instance, several studies consider di-, tri-, tetra- penta- and hexa-nucleotides when diverse motif repeats are present, i.e. 6,5,4,4,4 in *Salvia splendens* [25], 6,5,5,5,5 in *Saccharum* spp [28], 6,5,5,4,4 in *Capsicum frutescens* [23], 6,4,3,3,3 in *Curcuma longa* [29], 4,4,4,4,4 in *Ipomoea batatas* [30]. In order to be consistent with the literature, we have *in silico* predicted SSRs using 6,4,3,3,3 motifs repeats criteria. In this background, di-nucleotides were the most representative SSR species, followed by tri-nucleotides. This non-standard distribution has also been described for *Salvia splendes* [25] with 39.9%/29.3% di- and tri-nucleotide frequencies, respectively, sweet potato with 43.3%/42.4% [30], rubber tree with 38%/34% [31] and several other plants such as cucumber [32], sesame [33], kiwi [34] and coffee [35] where di-nucleotides are also the most represented SSR species.

Transposable elements discovery

Several transposable elements (TE) were identified by sequence homology in yerba mate. In the literature, only a few sequences have been recently reported corresponding to yerba mate TE fragments obtained by DNA based methods [14,36], hence this is the first report describing actively yerba mate TEs. As expected for a transcriptional library, most of the sequences corresponded to Retro-Transposon elements, mainly Group Antigen polyprotein (GAG-Pol), reverse-transcriptase and RNAse H domain hits of Gypsy-like and Copia-like retro-elements. Also, a handful of Non-LTR retro-elements were identified and a few CACTA, En/Spm sub-class Transposons. The predicted repetitive elements were explored in detail, and representative results obtained by the NCBI conserved domain search web-service are presented as graphical summaries depicting typical TE domains such as Reverse Transcriptase domain in Gypsy-like elements and Transposase domain in En/Spm Transposons, of several yerba mate putative TEs (**Table S12**).

Organelles draft genome *de novo* assembly and analysis

Recent NGS based studies have emphasized in the abundance and wide extension of chloroplast and mitochondrial transcripts, postulating that most of the organelles genomic DNA is actively transcribed in plants [37–40]. In this scenario we surveyed the yerba mate transcriptome in order to generate a draft genome of both organelles based in RNA transcripts and sequence similarity to reference organelle sequences of slightly related plant species. Illumina reads were relaxedly mapped to a *Lactuca sativa* chloroplast sequence and a total of 10,798,227 reads comprised and sustained a high coverage library that was assembled into a consensus sequence draft of the yerba mate chloroplast (**Table 3**). The assembled *I. paraguariensis* chloroplast is predicted to be ~152,872 bp long, consisting in 51.6% of coding sequences, representing 83 protein coding genes, 37 transfer RNA genes and 7 ribosome RNA genes (**Figure 4**). The 83 protein coding genes included several ribosomal proteins, constituents of photosystem I & II, NADH dehydrogenases and ATP synthases among others (**Table 4** and **Table S6**). A sequence alignment of the *L. sativa* chloroplast complete sequence and yerba mate draft chloroplast shows extensive identity, in some regions exceeding 90%, particularly at gene transcripts such as 16 s rRNA, 23 s rRNA and several transfer RNA genes (**Figure S10**). Mapping of *I. paraguariensis* assembled transcripts to the chloroplast sequence draft shows an extensive and pervasive coverage (**Figure S11**). The assembled draft was subjected to microsatellite discovery (**Table S8**) and a total of 94 SSRs were identified, consisting mainly of 6 bp motif (57.5% of total SSRs).

A similar approach was employed in order to envisage a mitochondrial genome draft of yerba mate. A total of 1,265,566 Illumina reads were mapped to the *Helianthus annuus* mitochondrial genome sequence. In this case, most mapped reads corresponded mainly to the gene coding regions, and the assembled draft extended at about a third of the total predicted genome (**Table 3**). A sequence alignment of sunflower mitochondrial complete sequence and yerba mate mitochondrial sequence consensus (**Figure S12**) presented high identity at most of the 43 coding sequences that corresponded to protein coding genes such as Complex I NADH dehydrogenases, Complex V ATP synthases and ribosomal proteins (SSU and LSU), transfer RNA genes and ribosome RNA genes (**Table 5, Table S7**).

A *Mauve* alignment, which is preferred for rearranged genome sequences [41], was performed with the yerba mate and sunflower mitochondrial sequences. The higher identity, mostly confined to gene encoding regions, is represented hierarchically from white to red. The consensus *I. paraguariensis* sequence conserved most of the *Helianthus* gene annotations. As an example, the consensus sequence of *I. paraguariensis* at 76,000 bp coordinates presented high identity to the 230,000 bp coordinates of sunflower, corresponding to the ccmFn coding sequence (**Figure S13**). The assembled yerba mate consensus sequence was subjected to microsatellite discovery (**Table S9**) and a total of 69 SSRs were identified, consisting mainly of 6 bp motif (69.8% of total predicted SSRs).

The yerba mate RNA silencing and degradation pathway

A particular limited set of seventy six yerba mate transcripts identified yielded considerable similarity with several members of the RNA silencing and degradation pathway (**Table S3**). RNA interference is a post-transcriptional sequence-specific process of gene silencing that mediates resistance to both endogenous parasitic and exogenous pathogenic nucleic acids, and regulates the expression of protein-coding genes in eukaryotic organisms [42]. Among several enzymatic components of RNA interference such as Dicer-Like proteins, RNA dependent RNA polymerases, exosome members and dsRNA binding domain proteins, the family of Argonaute effectors (AGO) was also pinpointed in the yerba mate transcriptome. AGO and AGO-like proteins are the main RNA silencing effectors across kingdoms, and they mediate the cleavage of target RNAs using small RNAs of 20–24 nt as guides [43]. Argonaute 1 (AGO1) is responsible of two important circuits in plants: gene silencing of endogenous transcripts by microRNAs and virus resistance based in viral derived small

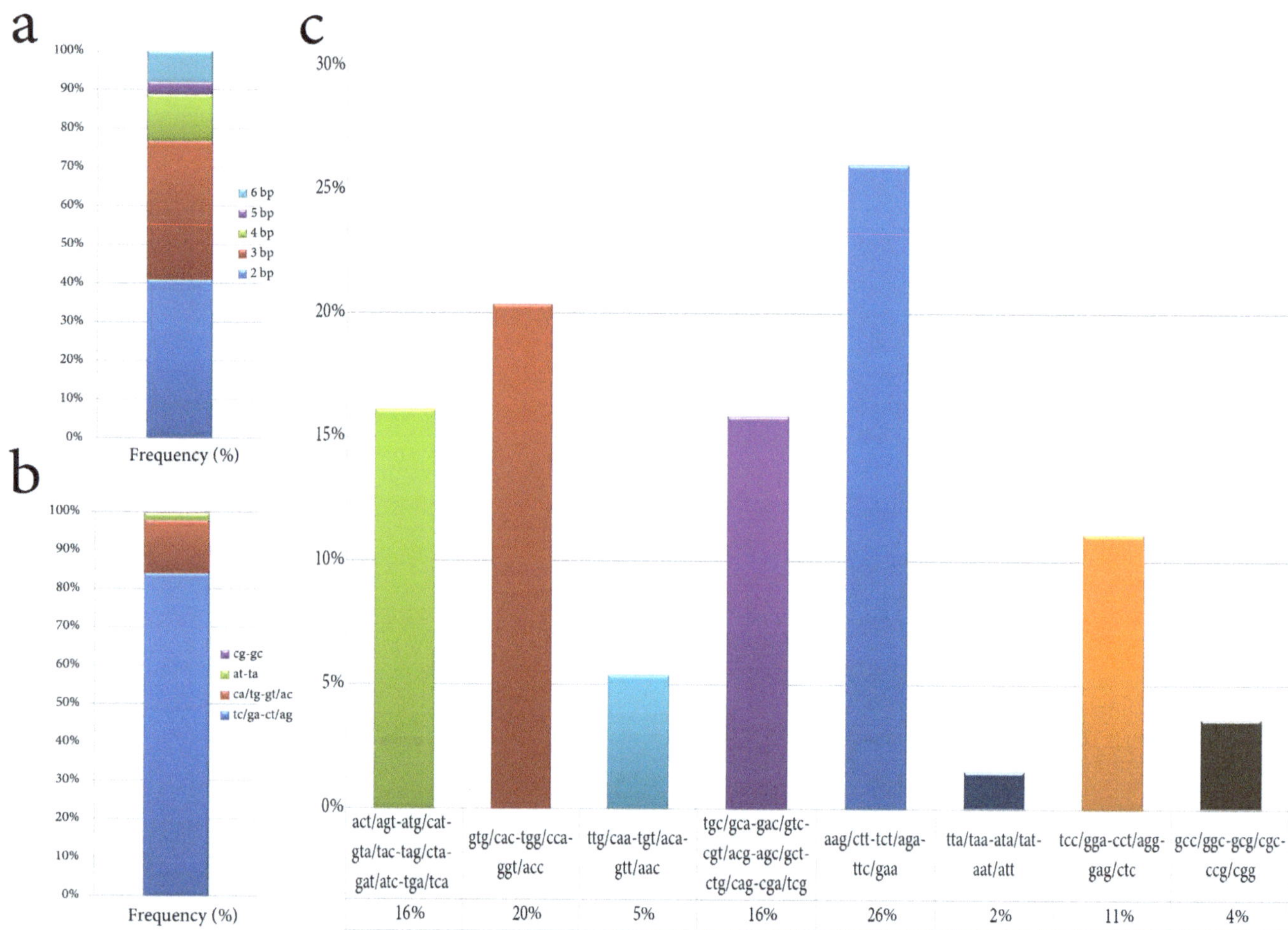

Figure 3. Proportion and frequencies of predicted SSRs in *Ilex paraguariensis* transcriptome. (a) Proportion of SSR predicted in yerba mate transcriptome categorized by k-mer length. (b) ct/ag-tc/ga account for 84% of di-nucleotide SSRs found in yerba mate. (c) Frecuency of tri-nucleotide SSRs predicted in yerba mate. With over 26% of the hits, aag/ctt-tct/aga-ttc/gaa are the most common SSR found in *Ilex paraguariensis*.

RNAS [44,45]. The yerba mate AGO1 predicted protein is estimated to be 1,062 aa in length and presented the typical AGO1 glycine rich domain (E-value = 1,8e-25), the PAZ domain (E-value = 4,3e-38) which is predicted to interact with single stranded small RNAs and the PIWI domain (E-value = 3,5e-112), responsible of the RNA-guided hydrolysis of single stranded-RNA (**Figure 5a**). Multiple protein alignment and secondary structure prediction of yerba mate, *N. benthamiana*, carrot and tomato AGO1 showed an important conservation in gene structure and domains. A phylogenetic tree based in Jukes-Cantor, neighbor-joining and 1000 bootstraps indicated that AGO1 from yerba mate is more related with carrot than *Solanaceae* AGO1 despite the basic genetic distance among them (**Figure 5 b–d, Figure S14, Figure S15**).

MicroRNAs (miRNAs) are small non-coding RNAs that modulate plant gene expression by means of gene silencing through sequence-specific inhibition of target mRNAs. MiRNAs derive from precise processing of precursor transcripts with stem-loop secondary-structure features that are recognized by a Dicer-like complex. Mature miRNAs are loaded predominantly onto AGO1 and target endogenous RNAs for their degradation or translational arrest [46]. By combining two *in silico* based approaches we engaged in an attempted characterization of putative miRNA precursors in yerba mate. A yerba mate miRNA sequence prediction report based in UEA small RNA workbench platform and canonical relaxed mapping of conserved precursor miRNAs to the yerba mate transcriptome, indicated the presence of at least 59 pre-miRNAs corresponding to 41 of both young and ancient miRNA families (**Table S10**). The miR156 gene family has been involved in the regulation of developmental timing, vegetative phase change, flowering and sex identity in plants [47–49]. In yerba mate several mature miRNAs were predicted based in sequence homology to miRBase [50] (**Figure 6a**). In the particular case of miR156, nine isoform variants were predicted with high sequence homology and minor mismatches. An insertion of a "A" at position 10 in miR156b and c forms, slightly affected the precursors secondary structure at the miRNA/miRNA* coordinates that can be observed as a bulge in **Figure 6b**. While the homology at the mature miR156 was high, the diversity among precursors of the miRNA gene family was extensive (**Figure 6d**). A library generated of predicted Squamosa Promoter Binding Protein-Like (SPL) mRNAs of yerba mate was evaluated as a target of Ipa-miR156. A strong interaction with a high expectation score was *in silico* predicted for SPL9, SPL6 and SPL4 with Ipa-miR156 (**Figure 6c**), which are typically conserved and validated targets of miR156 in plants [51]. These SPL genes significantly differed in their nucleotide sequence, however a strong conservation of the specific miR156 target could be

Table 3. General features of Yerba mate draft assembled organelles.

Feature	**Chloroplast**	**Mitochondrion**
Genome size (bp) estimated	~150,872	~301,093
contig coverage (bp)	118,064	90,151
total mapped 100 bp reads	10,798,227	1,265,566
GC content (%)	56.06	42.06
Coding sequences	127	43
Gene content (%)	51.6	9.03
No. of protein-coding gene	83	26
No. of introns	17	0
No. of tRNA genes	37	14
No. of rRNA operons	7	3
Sequence repeat	94	69

observed in the 3 genes (**Figure 6e**). The identification of transcripts related to sex identity, such as miR156 and SPL gene families, is of special interest in yerba mate. In this diclino-dioecious crop, plants cannot be recognized as male or female prior to their first blooming, which occurs between 3 and 10 years post seedling emergence [14], delaying considerably the selection of parentals for breeding purposes. So, a cost-effective early sex determination system would be promising for the yerba mate breeding programs. It is tempting to postulate, that perhaps the determination of expression levels of these particular genes during yerba mate plant development, may be employed as a gender predictor at early stages.

The yerba mate caffeine synthase

One of the most important constituents of yerba mate extracts is caffeine [52,53]. Caffeine is responsible for the stimulant effect of mate [54], and perhaps the underlying rationale of its profound influx in Latin American culture based on its effect on the body and mind and its properties that aid in staying awake and improving mental alertness after fatigue among others [55]. Caffeine or 1,3,7-trimethylxanthine is a crystalline xanthine alkaloid. Caffeine biosynthesis involves a series of reactions that direct the conversion of xanthosine to 7-methylxanthosine, to 7-methylxanthine to theobromine which is converted into caffeine [56]. The enzyme, assigned the name caffeine synthase (EC 2.1.1.160), catalyses the last two steps of caffeine biosynthesis, the conversion of 7-methylxanthine to caffeine via theobromine [57]. The gene encoding caffeine synthase (CS) was originally cloned from young tea leaves by Kato et al. [58]. Using the sequence of *Camellia sinensis* CS we probed our library and identified a sequence corresponding to the full length of an assembled transcript of the yerba mate transcriptome. The putative yerba mate assembled CS was identified by similarity to the *C. sinensis* CS complete mRNA (E-value = 8e-168). The yerba mate 1,491 nt CS transcript was assembled based in 16,851 100 bp reads with an average coverage of 1,113X. Exploring this transcript, a single ORF was predicted encompassing 1,098 nt between coordinates 113 to 1,210, encoding a 366 aa protein with a 61% identity (E-value = 2e-157) to the corresponding 369 aa *C. sinensis* CS protein, sharing the presence of the Methyltransferase_7 SAM dependent carboxyl methyltransferase domain involved in caffeine synthesis. With the predicted CS ORF, we performed multiple MUSCLE protein alignment and secondary structure prediction. By comparing *Coffea arabica*, *Theobroma cacao* and *C. sinensis* CS to *I. paraguariensis* predicted CS transcript an important conservation in gene structure and domains was observed (**Figure S16**). Since caffeine content is a desirable and important character in breeding programs of this crop, the preliminary and putative nature of the yerba mate predicted CS assembled transcript encourage further experimental validation, heterologous expression experiments and biochemical characterization of the full length CS coding sequence by traditional methods. After sequence annotation, we exploited the SWISS-MODEL algorithm to generate a yerba mate CS 3D prediction using *C. arabica* CS as a template (**Figure S17**). The 3D structure of the predicted *I. paraguariensis* caffeine synthase is presented in **Figure 7a** based in the X-ray crystallography solved structure of *C. arabica* CS (**Figure 7b**). A ribbon model of yerba mate CS (**Figure 7c**) and coffee CS (**Figure 7e**) suggested high conservation of secondary structure when superimposed (**Figure 7d**). A reconstruction of a mesh model of yerba mate CS is presented (**Figure 7f**) and compared with the coffee EM (**Figure 7h**), showing extensive quaternary structure similarity (**Figure 7g**).

Chlorogenic acid in yerba mate

Chlorogenic acid (CGA, caffeoyl quinate, KEGG compound C00852) is the major phenolic compound found in yerba mate [59]. CGA acts as an antioxidant in plants, is involved in resistance to insects and defense against fungal pathogens. Human CGA consumption and its antioxidant role have been associated to protection against degenerative age-related diseases [59]. The enzyme 4-coumaroyl-CoA: quinnate O-(hydroxycinnamoyl)transferase (HQC, E.C.: 2.3.1.133) and p-coumaroyl quinate 3′-hydroxylase (C3′H, E.C.: 1.14.13.36), corresponding to the flavonoid and stilbenoid metabolic pathways, respectively (**Data S2**, pages 50 and 145), are responsible and limiting for CGA biosynthesis in tobacco, coffee and switchgrass [60–62]. We examined our transcriptome and, interestingly we found four putative HQC and two C3′H gene versions sharing over 75% identity with the corresponding coffee enzymes, presenting the typical condensation and Cytochrome P450 domains, respectively. We noticed that these transcripts were amongst the 1–4% most abundantly expressed, ranging from 40 to 160 FPKM. Although further studies are required, we suggest that the presence of redundant and highly expressed Chlorogenic acid related genes might be directly associated with the important production of

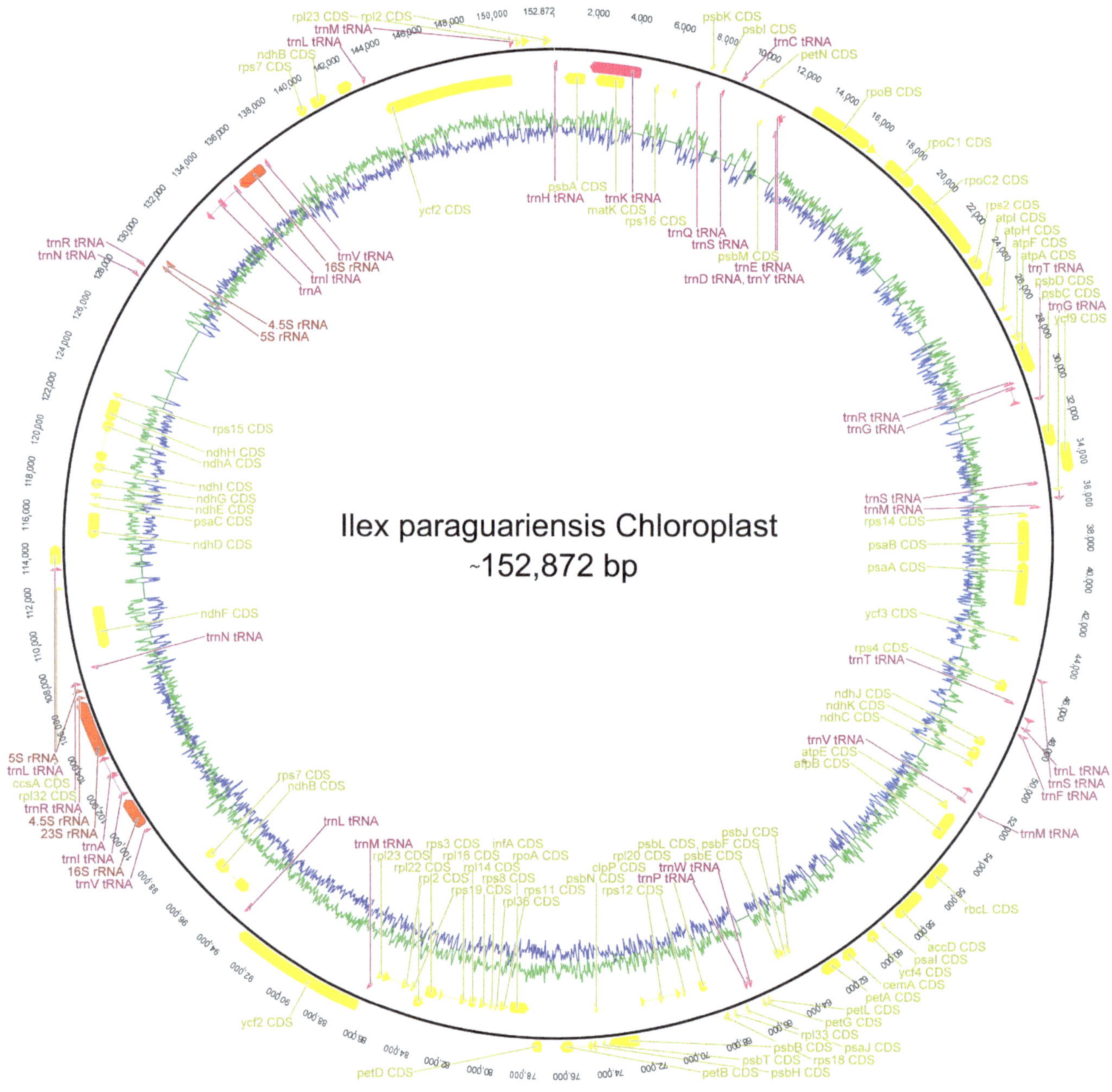

Figure 4. ***Ilex paraguariensis*** **chloroplast is predicted to be ~152,872 bp long, consisting in 51.6% of coding sequences, representing 83 protein coding genes (yellow), 37 transfer RNA genes (pink) and 7 ribosome RNA genes (red).** The 83 protein coding genes include several ribosomal proteins, constituents of photosystem I & II, NADH dehydrogenases and ATP synthases among others.

CGA in yerba mate and its significant effects on oxidative stress reduction when consumed [63].

Ilex paraguariensis PCR detection assay based in 5.8S and ITS2 assembled rRNA regions

In Argentina, *I. paraguariensis* is the only species authorized to be used to manufacture yerba mate products (Argentina law, Act 18.284, articles 1192–1193). It is interesting to note that a molecular based, standardized and cost effective method for *I. paraguariensis* detection is lacking in the literature. Moreover, *I. dumosa* is frequently used in mixtures with *I. paraguariensis*, and amounts above 1% are considered adulterants [64]. Based on our sequencing data we were able to assemble the *I. paraguariensis* rDNA corresponding regions of ITS1, 5.8S and ITS2. Multiple alignment of *I. paraguariensis* assembled 45S consensus sequence transcripts and *I. dumosa* homologous regions was performed (**Data S4**). This analysis showed that while ITS1 is highly conserved in both species, 5.8S and ITS2 are distinctive enough to differentiate them by means of PCR with species-specific primers. PCR amplifications were performed and agarose gel electrophoresis revealed that the designed primers from *I. paraguariensis* are species-specific and able to detect unequivocally genomic *I. paraguariensis* DNA. Moreover, by using the *I. dumosa* species-specific primers we were able to detect *I. dumosa* in DNA

Table 4. Yerba mate chloroplast encoded genes by category.

Chloroplast gene category	**Gene name**
Photosystem I	psaA, psaB, psaC, psaI, psaJ
Photosystem II	q
Cytochrome b/f complex	petA, petB, petD, petG, petL, petN
ATP synthase	atpA, atpB, atpE, atpF, atpH, atpI
NADH dehydrogenase	ndhA, ndhB, ndhC, ndhD, ndhE, ndhF, ndhG, ndhH, ndhI, ndhJ, ndhK
RubisCO large subunit	rbcL
RNA polymerase	rpoA, rpoB, rpoC1, rpoC2
Ribosomal proteins (SSU)	rps2, rps3, rps4, rps7(2), rps8, rps11, rps12, rps14, rps15, rps16, rps18, rps19
Ribosomal proteins (LSU)	rpl2(2), rpl14, rpl16, rpl20, rpl22, rpl23(2), rpl32, rpl33, rpl36
Other genes	clpP, matK, accD, ccs1, ccsA, infA, cemA
hypothetical	ycf2(2), ycf3, ycf4, ycf9
Transfer RNAs	trnA-UGC(2), trnC-GCA, trnD-GUC, trnE-UUC, trnF-GAA, trnG-UCC(2), trnH-GUG, trnI-CAU, trnI-GAU, trnK-UUU, trnL-CAA, trnL-GAG, trnL-UAA, trnL-UAG, trnM-CAU(4), trnN-GUU(2), trnP-UGG, trnQ-UUG, trnR-ACG(3), trnS-GCU(2), trnS-UGA, trnT-GGU(2), trnV-GAC(2), trnV-UAC, trnW-CCA, trnY-GUA
Ribosomal RNAs	rRNA 4.5 s(2), rRNA 5 s(2), rRNA 16 s(2), rRNA 23 s

preparations obtained of mix leaf tissue of fractions above 1% of *I. dumosa* respective to *I. paraguariensis* as previously reported [64] (**Data S4**). PCR amplifications with both species-specific primers is a versatile, simple and cost effective method to detect *I. paraguariensis* and the typical yerba mate products adulterant.

Conclusions

This is the first publicly available *I. paraguariensis* NGS study performed to investigate the entire yerba mate transcriptome, and our data provides the unique comprehensive transcriptome resource currently existing for yerba mate. In sum, through a systematic and exhaustive process of gene analysis and annotation, we have identified ~1,000 putative transcription factors, genes involved in heat and oxidative stress, pathogen response, as well as disease resistance and hormone response. We have also identified transcripts related to osmotic, drought, salinity and cold stress, senescence and early flowering. We have also pinpointed several members of the gene silencing pathway and characterized the silencing effector Argonaute1. We predicted a diverse supply of putative microRNA precursors involved in developmental processes. We developed a draft of the transcribed genomes of the yerba mate chloroplast and mitochondrion. The putative sequence and predicted structure of the caffeine synthase of yerba mate is presented. Finally, we provide here a collection of over 10,800 SSR accessible to the community interested in yerba mate genetic improvement.

The transcriptome characterization of *I. paraguariensis* generated from our study is a very useful tool derived from a convenient and exhaustive approach of annotation and discovery of genes of several major metabolic pathways in this important crop. The vast amount of information obtained would encourage and serve as reliable source in the path to the discovery of biological and agronomic interesting traits, as well as for molecular markers development, gene mapping, analysis of genetic diversity and selection breeding in yerba mate.

Table 5. Yerba mate mitochondrial encoded genes by category.

Mitochondrial gene category	**Gene name**
Complex I (NADH dehydrogenase)	nad3, nad4, nad5, nad6, nad8, nad9
Complex III (cytochrome c reductase)	cob
Complex IV (cytochrome c oxidase)	coxI, coxIII
Complex V (ATP synthase)	atp1, atp4, atp6, atp8, atp9
Cytochrome c biogenesis	ccmB, ccmC, ccmFc, ccmFn(2)
Ribosomal proteins (SSU)	rps4, rps12, rps13
Ribosomal proteins (LSU)	rpl5, rpl10, rpl16
Maturases	matR
Other genes	orf873
Transfer RNAs	trnC-GCA, trnD-GUC, trnE-UUC, trnF-GAA, trnG-GCC, trnH-GUG, trnK-UUU, trnL-UAA, trnM-CAU(2), trnN-GUU, trnP-UGG, trnQ-UUG, trnS-GCU, trnW-CCA, trnY-GUA
Ribosomal RNAs	rrn5, rrn18, rrn26

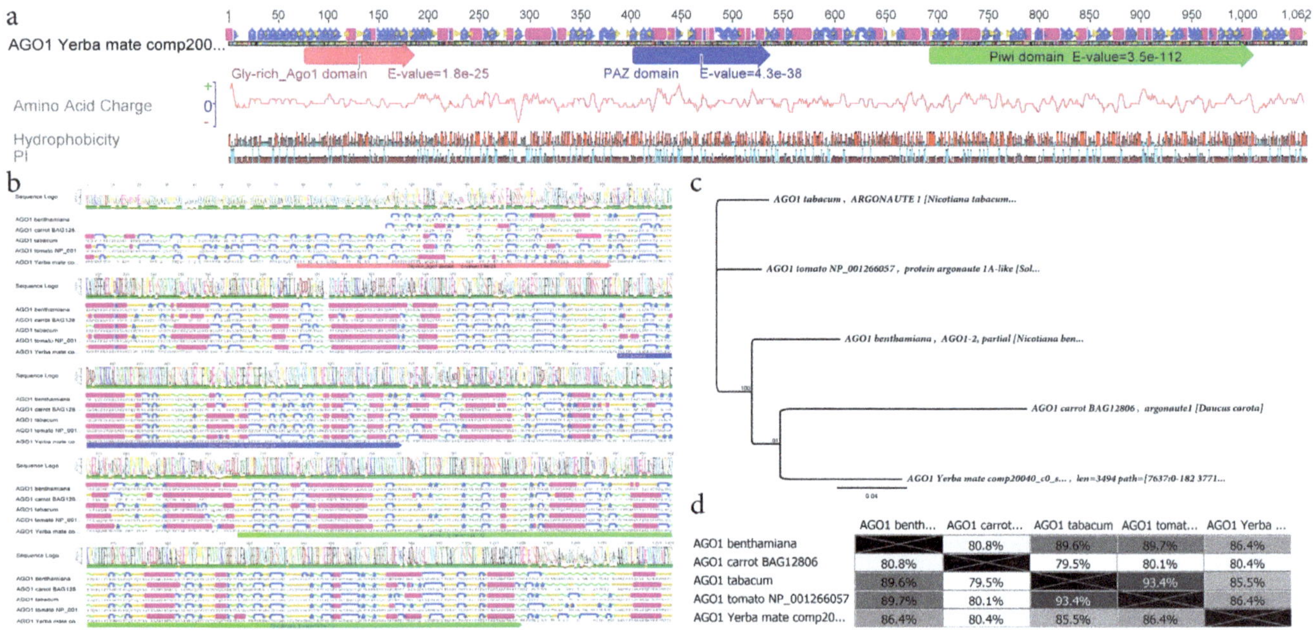

Figure 5. Yerba mate *Argonaute 1* (AGO1): characterization of the catalytic component of the miRNA pathway. (a) The predicted *Ilex paraguariensis* AGO1 protein is 1,062 aa in length and presents the typical AGO1 glycine rich domain, the PAZ domain which is predicted to interact with single stranded small RNAs and the PIWI domain, responsible of the RNA-guided hydrolysis of single stranded-RNA. (b) Multiple protein alignment and secondary structure prediction of yerba mate, *Nicotiana benthamiana*, carrot and tomato AGO1 showing an important conservation in gen structure and domains. (c) A phylogenetic tree based in Jukes-Cantor, neighbor-joining and 1000 bootstraps indicates that AGO1 from yerba mate is more related with carrot than *Solanaceae* AGO1 despite the basic genetic distance among them (d).

Materials and Methods

Plant materials and RNA extraction

Leaf samples at emerging, young, fully expanded, and early and late senescent stages from *I. paraguariensis* breeding line Pg538 from INTA EEA-Cerro Azul, Misiones, Argentina, were collected and immediately frozen in liquid Nitrogen. Total RNA was isolated from pooled leaf tissue with the RNeasy Plant Mini Kit (Qiagen Inc.) and supplemented with RNase-free DNase (Qiagen Inc.). To increase the depth of depletion of ribosomal RNA, a process with the RiboMinus Plant Kit (Life Sciences Inc.) was performed with the isolated RNA. The resulting RNA was evaluated in concentration and purity using Nanodrop 1000 (Thermo Inc.), and subsequently subjected to an integrity analysis by agar electrophoresis and by Bioanalizer 2100 (Agilent Inc.) to determine quality parameters by QC and RING.

RNA-seq library construction for Illumina sequencing

The resulting high quality RNA was employed for the generation of a cDNA library through TruSeq RNA Sample Preparation Kit (Illumina Inc.). The purified cDNA library was used for cluster generation on the Illumina Cluster Station and then sequenced on Illumina HiSeq 2000 following vendor instruction. A paired-end sequencing run with 100 nt read length for each read was performed for RNA-Seq. Raw sequencing intensities were then extracted and the bases were called using Illumina RTA software, followed by sequence quality filtering. The extracted sequencing reads were saved as a pair of fastq files for the first and second read, respectively.

Sequence data analysis and assembly

Quality reports and filtering of the sequencing run and assembly were generated using CLC Genomics Workbench v7.0.4 (http://www.clcbio.com/) and the RobiNA v1.2.4 software (http://mapman.gabipd.org/web/guest/robin). All raw reads generated from the sequencer were quality filtered and *de novo* assembled into contigs using the Trinity program [17] with optimal parameters of 25 kmer word and group pairs distance of 500. The abundance of assembled contigs/isoforms was estimated using RSEM (http://deweylab.biostat.wisc.edu/rsem/) following the Trinity protocol. The raw sequencing data with quality scores have been deposited in the NCBI Short Read Archive database under accession number SRP043293.

Sequence annotation

The obtained contigs were bulk analyzed in homology searches by BLASTX (http://blast.ncbi.nlm.nih.gov/Blast.cgi) to protein databases nr, Swiss-Prot (http://www.uniprot.org/), KEGG (http://www.genome.jp/kegg/) and COG (http://www.ncbi.nlm.nih.gov/COG/). The yerba mate transcripts were also investigated and analyzed with the Plant Ontology database (http://www.plantontology.org/), the database of essential genes DEG (http://www.essentialgene.org), Plaza 2.5 (http://bioinformatics.psb.ugent.be/plaza/) and alternatively with TAIR (http://www.arabidopsis.org/) the batch blast tool of the Rosaceae Genome Database (http://www.rosaceae.org/tools/batch_blast) and the PlantGDB BLAST (http://www.plantgdb.org/cgi-bin/blast/PlantGDBblast) using E≤1e-5 as threshold, retrieving best hits and functional annotations. The Blast2GO program www.blast2go.com/ (E≤1e-5) was also used to obtain GO annotation of genes. The GO annotations retrieved were subjected to enrichment analysis with the WEGO software (wego.genomics.org.cn/), Revigo web server (revigo.irb.hr/) and the AgriGO platform (bioinfo.cau.edu.cn/agriGO/).

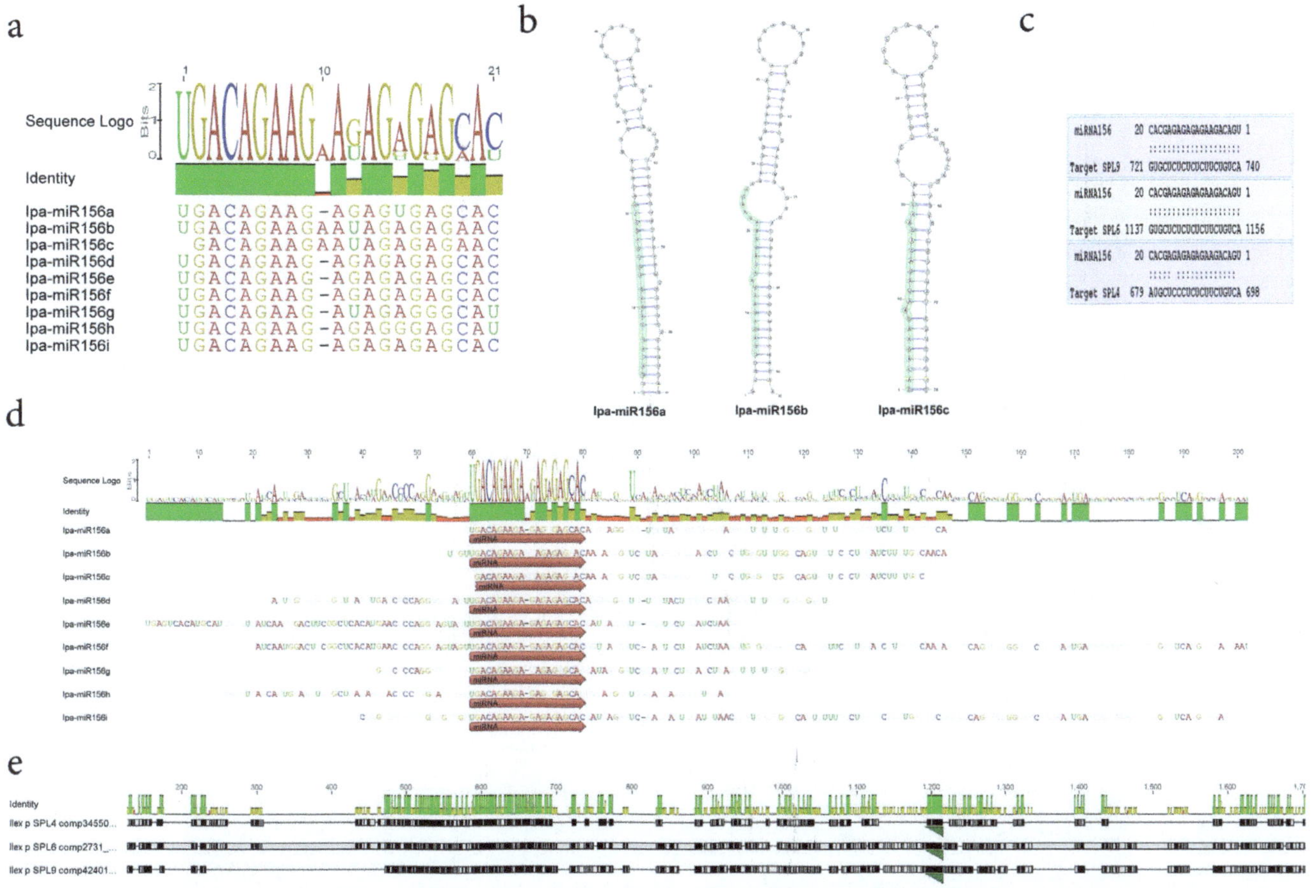

Figure 6. *MiR156* gene family in yerba mate. (a) Several mature miRNAs were predicted in yerba mate based in sequence homology to Mirbase. In the particular case of miR156, nine isoform variants were predicted with high sequence homology and minor mismatches. An insertion of a "A" at position 10 in miR156b and c forms, slightly affected the precursors secondary structure at the miRNA/miRNA* coordinates that can be observed as a bulge in (b). While the homology at the mature miR156 is high, the diversity among precursors of the miRNA gene family is extensive (d). A library generated of predicted SPL mRNAs of yerba mate was evaluated as a target of Ipa-miR156. A strong interaction with a high expectation score was *in silico* predicted for SPL9, SPL6 and SPL4 with Ipa-miR156 (c). These SPL genes significantly differ in their nucleotide sequence, however a strong conservation of the miR156 target can be observed in the 3 genes (green triangle, e).

Organelle genome assembly and annotation

To generate a draft of yerba mate transcribed chloroplast, Illumina reads were relaxedly mapped to a *Lactuca sativa* chloroplast sequence (Accession no. AP007232.1) employing the map to reference utility of the Geneious 7.0 software (Geneious assembler, medium sensitivity, iterations up to 5 times). A consensus sequence was generated with the mapped reads and aligned to the *Lactuca sativa* chloroplast complete sequence using a Geneious global aligment with free end gaps (93%, gap open penalty 12, gap extension penalty 3). The yerba mate chloroplast draft was annotated integrating *Lactuca sativa* chloroplast gene predictions and *in silico* based estimations obtained with the Dual Organellar GenoMe Annotator (DOGMA, http://dogma.ccbb.utexas.edu/). A similar process was employed to generate a yerba mate mitochondrial draft. In this latter case to compensate for significant sequence gaps a Mauve genome alignment of yerba mate and *Helianthus annuus* mitochondrial complete sequence (Accession no. KF815390.1) was generated. The annotation of the yerba mate transcribed mitochondrion draft was generated based in sequence homology to the respective sunflower predicted genes.

MicroRNA prediction

We analyzed both the assembled transcripts and the raw reads by two *in silico* based approaches in an attempted characterization of putative miRNA precursors in yerba mate. The UEA small RNA workbench platform (srna-workbench.cmp.uea.ac.uk/) with plant standard parameters and a cut-off P-value of 0.05 was employed using the yerba mate assembled transcriptome as reference. In parallel, a canonical relaxed mapping (similarity: 0.8 to 0.75, length fraction: 0.8, mismatch cost: 2, insertion cost: 2, deletion cost: 3) of conserved precursor miRNAs from miRBASE Release 20 (ftp://mirbase.org/pub/mirbase/CURRENT/miRNA.dat.gz) to the yerba mate transcriptome on the CLC Genomics Workbench v7.0.4 environment was generated and mapped reads were analyzed by eye and evaluated by proper secondary folding. The secondary structures of the predicted stem-loops secondary structures were solved using the mfold web server (http://mfold.rna.albany.edu/) using version 2.3 and adjusting folding temperature to 24°C, and Vienna type RNA structures were predicted with Context Fold (http://www.cs.bgu.ac.il/~negevcb/contextfold/).

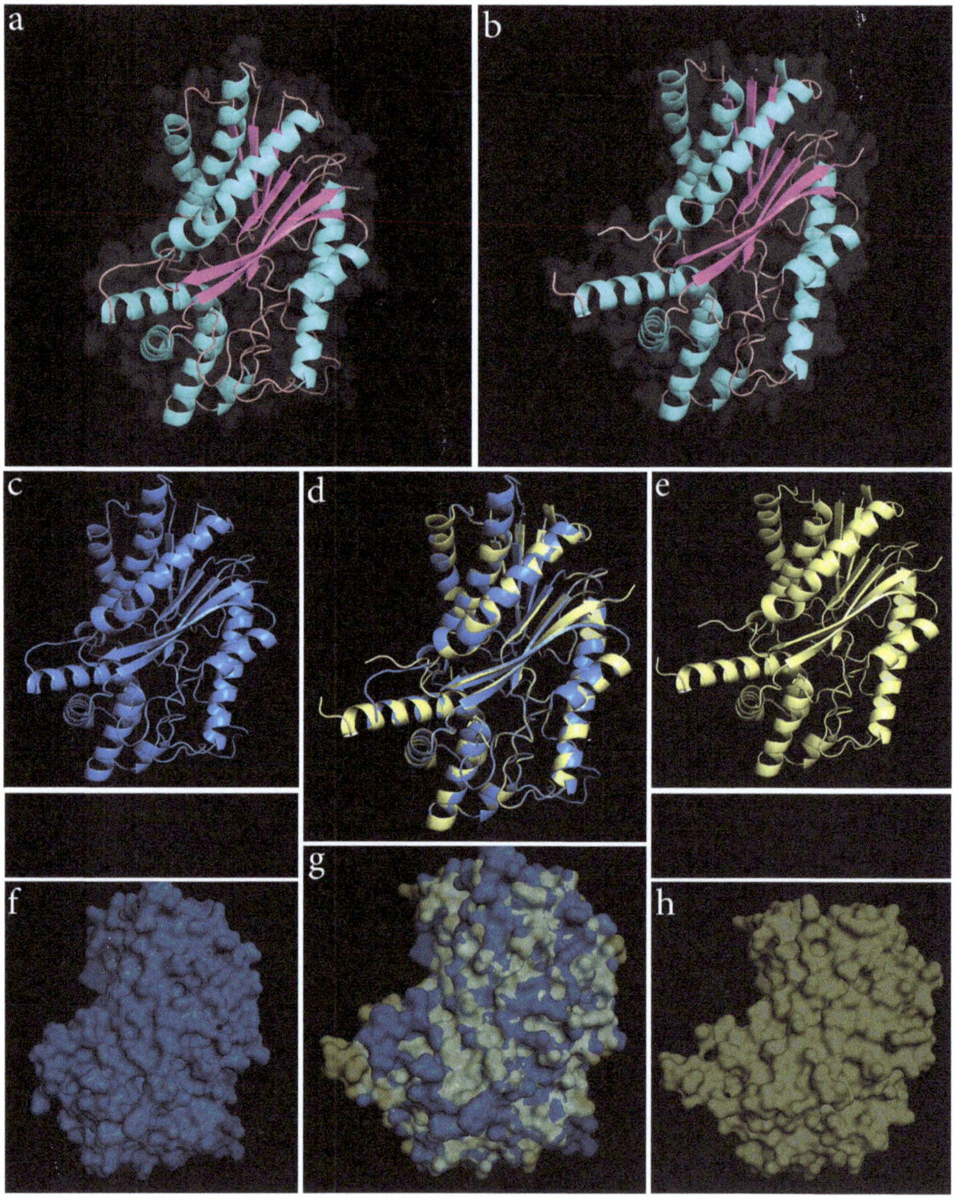

Figure 7. 3D structure of *Ilex paraguariensis* caffeine synthase. (CS). Employing the X-ray crystallography solved structure of *Coffea arabica* CS as a template (b), the 3D structure of yerba mate CS was predicted by the swiss-model algorithm (a). A ribbon model of yerba CS (c) and coffee CS (e) suggest high conservation of secondary structure when superimposed (d). A reconstruction of a mesh model of yerba CS is presented (f) and compared to the coffee EM (h), showing extensive quaternary structure similarity (g).

Caffeine synthase structure prediction

Coffea arabica (Acc. no: BAC43759.1), *Theobroma cacao* (Acc. no: BAE79730.1), *Camellia sinensis* (Acc. no: ABP98983.1) and *Ilex paraguariensis* predicted caffeine synthase coding sequences were subjected to multiple MUSCLE protein alignments (8 iterations, distance methods kmer6-6 and pctd-kimura, and clustering by UPGMB) and secondary structure prediction with the GOR method (Garnier-Osguthorpe-Robson) with Genious v7.0.

Yerba mate CS cuaternary structure was predicted with the automated protein structure homology-modelling server (swissmodel.expasy.org/; [65]) using as a template the solved structure of *Coffea arabica* CS (2efjA (2.00 A)). The pdb obtained structure was analyzed and rendered using the PyMol software version 1.7 (www.pymol.org/).

SSR detection

The GDR SSR Server [66,67] was employed to identify single sequence repeats in the transcriptome of yerba mate. 5 types of SSRs (di-nucleotide, tri-nucleotide, tetra-nucleotide, penta-nucleotide, hexa-nucleotide) were screened with an specific minimum number of 6, 4, 3, 3, 3 repeats respectively.

Transposable elements analysis

Transposable elements were first identified based in homology searches by NCBI-BLASTN and BLASTX. Specifically, Non-LTR Long Interspersed Elements (LINE) were pinpointed by the retrotransposase domain; LTR elements (Copia-like and Gypsy) were scanned with the LTR-FINDER software [68] and DNA transposons by domain based search. Every candidate was then explored and re-evaluated with NCBI conserved domain search (http://www.ncbi.nlm.nih.gov/Structure/cdd/wrpsb.cgi?).

Supporting Information

Figure S1 Illumina RNA sequencing analysis. (a) A lack of peaks reaching up to 100% at individual cycles in the homopolymer graph indicates an absence of a common technical artifact of cycle-wise multiplied calls of the same nt. (b) The Sequence quality plots allow an overview of the base call qualities assigned to each base by the base caller module of the sequencing pipeline. The plot shows the median (solid blue line), the 25th percentile and the 75th percentile (lower and upper bound of the light blue area) of the qualities at each position (cycle). (c) The base call frequency plot indicates a lack of positional biases in the call frequency for each base. (d) The plot shows the distribution of qualities averaged across the reads. In the yerba mate data, the average quality score is 36.3 (indicated by the red line). (e) The Kmer frequency check identifies short sequences that occur more often than expected. 10 Kmers that occur 3 times more often than expected are indicated.

Figure S2 Trinity *de novo* assembly report. (a) Distribution of GC-contents: The GC-content of a sequence is calculated as the number of GC-bases normalized to the total number of sequences. (b) Combined coverage of G and C bases: number of G and C bases observed at current position normalized to the total number of bases observed at that position. (c) The five most-overrepresented 5mers. The over-representation of a 5mer is calculated as the ratio of the observed and expected 5mer frequency. The expected frequency is calculated as product of the empirical nucleotide probabilities that make up the 5mer. (d) The number of sequences that support (cover) the individual base positions normalized to the total number of sequences. (e) Distribution of sequence lengths: x: sequence length in base-pairs y: number of sequences featuring a particular length normalized to the total number of sequences. (f) Coverages for the four DNA nucleotides.

Figure S3 Semantic similarity-based scatterplots of gene ontology categories representations of the yerba mate transcriptome based in biological process (a), cellular component (b) or molecular function (c). Circle size is estimated based in p-values associated to the GO categories. Color legend is represented as an inset in (a). The Revigo web server was used to generate the plots: http://revigo.irb.hr/.

Figure S4 Tree-map visualization of enriched GO categories of yerba mate transcriptome based in biological process (a), cellular component (b) or molecular function (c). Rectangle size is estimated based in p-values associated to the GO categories. The Revigo web server was used to generate the maps: http://revigo.irb.hr/.

Figure S5 AgriGO generated plot of GO enrichment in yerba mate based in biological process. Significance color levels and arrow types associated with GO relationships are represented as an inset at the superior left corner.

Figure S6 AgriGO generated plot of GO enrichment in yerba mate based in cellular component. Significance color levels and arrow types associated with GO relationships are represented as an inset at the superior left corner.

Figure S7 AgriGO generated plot of GO enrichment in yerba mate based in molecular function. Significance color levels and arrow types associated with GO relationships are represented as an inset at the superior left corner.

Figure S8 Profile of ribosomal proteins and ribosome biogenesis related transcripts obtained from the yerba mate transcriptome. Plant Ribosomes are constituted by 4 rRNAs and ~80 ribosomal proteins. In this study every major structural ribosome constituents **(a)**, and roughly every enzyme (**b**) responsible for ribosome processing, trafficking, rRNA maturation, and ribosome assembly were identified (green). Images credits: Kanehisa Laboratories, Japan.

Figure S9 Profile of plant hormone signal transduction related transcripts obtained from the yerba mate transcriptome. Plant development is regulated by endogenous signaling molecules including plant hormones. Perception of biological cues and signal transduction involves several hormone sensing and effector pathways. The major yerba mate enzymes involved in plant growth, cell division, stem growth, seed dormancy, senescence and cell elongation were identified in the assembled transcriptome (green). Images credits: Kanehisa Laboratories, Japan.

Figure S10 Genome alignment of *Lactuca sativa* chloroplast complete sequence (Accession no. AP007232.1) and yerba mate chloroplast. Identity is obtained based in 1 nt sliding window size and represented by color and bar height from 0% (red) to 100% (green). Annotations are depicted as protein coding genes (yellow), transfer RNA genes (pink) and ribosome RNA genes (red).

Figure S11 Mapping of *Ilex paraguariensis* assembled transcripts to the chloroplast sequence draft.

Figure S12 Genome alignment of sunflower mitochondrial complete sequence and yerba mate mitochondrial sequence consensus. Identity is obtained based in 1 nt sliding window size and represented by color and bar height from 0% (red) to 100% (green). Annotations are depicted as protein coding genes (yellow), transfer RNA genes (pink) and ribosome RNA genes (red).

Figure S13 Mauve genome alignment of yerba mate and *Helianthus annuus* mitochondrial complete sequence (Accession no. KF815390.1). Identity is represented hierarchically from white to red. The consensus *Ilex paraguariensis* sequence conserves most of the *Helianthus* gene annotations (rectangles). As an example, the consensus sequence of Ilex p. at 76,000 bp coordinates presents high identity to the 230,000 bp coordinates of sunflower (transparent bars), corresponding to the ccmFn coding sequence.

Figure S14 Bayesian phylogenic tree of the *Argonaute 1* (AGO1) genes of 35 plant species and yerba mate determined by the Geneious 7.0 platform. Values at the nodes indicate bootstrap support percentage obtained for 1,000 replicates.

Figure S15 Multiple gene alignment of *Argonaute 1* (AGO1) genes of 35 plant species and yerba mate. Identity is obtained based in 1 nt sliding window size and represented by color and bar height from 0% (red) to 100% (green).

Figure S16 Multiple MUSCLE protein alignment and secondary structure prediction of *Coffea arabica*, *Theobroma cacao*, *Camellia sinensis* and *Ilex paraguariensis* caffeine synthase showing an important conservation in gene structure and domains.

Figure S17 SWISS-MODEL report of yerba mate caffeine synthase 3D prediction using *Coffea arabica* CS as a template.

Table S1 Transcription factors predicted in yerba mate based in BLASTX hits descriptions employing the *viridiplantae* UniProt, or *Arabidopsis* TAIR protein database as reference.

Table S2 Gene Ontology terms associated to the yerba mate transcriptome. The GO terms were obtained employing the KAAS (KEGG Automatic Annotation Server, http://www.genome.jp/kegg/kaas/) for ortholog assignment and pathway mapping.

Table S3 Categorized yerba mate complete transcriptome best BLASTX hits, against the *viridiplantae* UniProt database.

Table S4 Yerba mate complete transcriptome best BLASTX hits, against the TAIR protein database.

Table S5 List and description of Gene Ontology, KO terms and mapsassociated to the yerba mate transcriptome obtained by the KAAS and deduced from BLASTX hits against *Arabidopsis* TAIR protein database and *viridiplantae* UniProt.

Table S6 Complete gene catalog list of yerba mate draft chloroplast and mitochondrion.

Table S7 *Helianthus annuus* mitochondrial gene coordinates and the analogous yerba mate mitochondrial draft ortholog gene sequence coordinate list.

Table S8 Predicted SSR summary, frequencies report and designed alternative primers for the yerba mate chloroplast genome.

Table S9 Predicted SSR summary, frequencies report and designed alternative primers for the yerba mate mitochondrial genome draft.

Table S10 Yerba mate miRNA sequence prediction report based in UEA small RNA workbench platform and canonical relaxed mapping of conserved precursor miRNAs to the yerba transcriptome on the CLC Genomics environment.

Table S11 Predicted SSR summary and frequencies report for the yerba mate complete transcriptome.

Table S12 Transposable element-like sequences present in the yerba mate transcriptome.

Data S1 DEG analysis of the yerba mate transcriptome.

Data S2 166 KEGG pathways models graphs representing most of yerba mate annotated genes (in green). Images credits: Kanehisa Laboratories, Japan.

Data S3 Alternative language abstract in Spanish.

Data S4 PCR detection assay of *I. paraguariensis* vs. *I. dumosa* based in 5.8S and ITS2 distinctive regions. Designed primer features, sample preparation and PCR conditions are presented as tables. Multiple alignment of *I. paraguariensis* assembled 45S consensus sequence vs. *I. dumosa* Acc. No. AJ492657 reveals that ITS1 is highly conserved in both species; however 5.8S and ITS2 are distinctive enough to differentiate them by means of PCR with species-specific primers. Primer bind locations are annotated in blue in both species. Agarose gel electrophoresis showing differential amplification according to primers specificity and reliability of the method is presented. M1: Promega 200 bp ladder molecular DNA ruler; M2: in-house control Marker; NTC: Non Template Control; 100%, 50%, 10%, 1%, 0.1%: percentage of *I. dumosa* (I.du) fraction respective to *I. paraguariensis* (I.pa) of leaf tissue employed for DNA isolation and subjected to specific PCR.

Acknowledgments

M.G. and D.A.M. are career members and P.M.A. is a postdoctoral research fellow of the Consejo Nacional de Investigaciones Científicas y Técnicas (CONICET-Argentina). H.J.D., R.E.B. and D.A.D. are researchers at the Instituto Nacional de Tecnología Agropecuaria (INTA-Argentina). We specially acknowledge Ing. Agr. Sergio D. Prat Kricun for his 40 years of field collection and selection of yerba mate plants from South America now available at the EEA Cerro Azul-INTA, Misiones. Authors also thank Universidad Nacional de Misiones for continuous support. Finally, we would like to thank Presidencia de la Nación Argentina, Ministerio de Ciencia, Tecnología e Innovación Productiva and Ministerio de Agricultura, Ganadería y Pesca for their permanent and compelling support of science, innovation, education and inclusion in Argentina.

Alternative language abstract

An Alternative language abstract in Spanish is available as **Data S3**.

Author Contributions

Conceived and designed the experiments: MG HJD DAM. Analyzed the data: HJD MG. Contributed reagents/materials/analysis tools: MG HJD DAM PMA REB DAD MBO PDZ. Wrote the paper: HJD MG PMA DAM. Reviewed and approved the final manuscript: HJD MG PMA REB MBO DAD PDZ DAM.

References

1. Giberti GC (1999) Recursos fitogenéticos relacionados con el cultivo y explotación de la Yerba Mate (*Ilex paraguariensis* St. Hil., Aquifoliáceas) en el cono sur de América. Acta Horticulturae 500: 137–144.
2. Bracesco N, Sanchez AG, Contreras V, Menini T, Gugliucci A (2011) Recent advances on *Ilex paraguariensis* research: Minireview. Journal of Ethnopharmacology 136: 378–384.
3. Gauer L, Cavalli-Molina S (2000) Genetic variation in natural populations of maté (*Ilex paraguariensis* A. St.-Hil., Aquifoliaceae) using RAPD markers. Heredity 84: 647–656.
4. Instituto Nacional de la Yerba Mate, Argentina. INYM website. Available: http://www.inym.org.ar/inym/imagenes/Estadisticas/sup%20cultivada%20depa.pdf. Accessed 2014 Jun 4.
5. Jerke G, Horianski MA, Salvatierra KA (2009) Evaluación de géneros micotoxigénicos en yerba mate elaborada. Revista de Ciencia y Tecnología (UNaM) 12: 41–45.
6. Sistema Integrado de Información Agropecuaria. Ministerio de Agricultura, Ganadería y Pesca, Presidencia de la Nación, Argentina. SIIA website. Available: http://www.siia.gob.ar/. Accessed 2014 Jun 4.
7. Canitrot L, Grosso MJ, Méndez A (2011) Complejo Yerbatero. Serie "Producción regional por complejos productivos". Ministerio de Economía y Finanzas Públicas, Argentina. Mecon website. Available: http://www.mecon.gov.ar/peconomica/docs/Complejo_Yerbatero.pdf. Accessed 2014 Jun 4.
8. Burtnik OJ (2006) Yerba Mate: Manual de Producción. INTA, AER Santo Tomé, Corrientes, Argentina. 52p.
9. Instituto Nacional de Tecnología Agropecuaria EEA Cerro Azul, Cambio Rural (1998) Yerba mate. Biblioteca para el Productor. 60p.
10. Sosa DA, Bárbaro S, Alvarenga FA, De Coll ODR, Ohashi DV, et al. (2011) Yerba mate. Manual de campo. 51p.
11. Ricca J (2012) El mate. Random House Mondadori Press. 288p.
12. Yerba mate Argentina website. Available: http://yerbamateargentina.org.ar/zona_archivo.php?archivo=descargas_02_Consumo_Cualitativo_YM_Congreso_Sud.ppt&titulo=Investigacion%20Cualitativa. Accessed 2014 Jun 4.
13. Belingheri LD, Prat Kricun SD (1997) Programa de mejoramiento genético de la Yerba Mate en el INTA. I Congresso Sul-Americano de Erva-Mate. II Reuniao Técnica do Cone Sul Sobre a Cultura da Erva-Mate: pp 267–277.
14. Gottlieb AM, Poggio L (2010) Genomic screening in dioecious "yerba mate" tree (*Ilex paraguariensis* A. St. Hill., Aquifoliaceae) through representational difference analysis. Genetica 138: 567–578.
15. Fan H, Xiao Y, Yang Y, Xia W, Mason AS, et al. (2013) RNA-Seq analysis of *Cocos nucifera*: Transcriptome sequencing and subsequent functional genomics approaches. PLoS ONE 8(3): e59997.
16. Shi CY, Yang H, Wei CL, Yu O, Zhang ZZ, et al. (2011) Deep sequencing of the *Camellia sinensis* transcriptome revealed candidate genes for major metabolic pathways of tea-specific compounds. BMC Genomics 12: 131.
17. Grabherr MG, Haas BJ, Yassour M, Levin JZ, Thompson DA, et al. (2011) Full-length transcriptome assembly from RNA-Seq data without a reference genome. Nature Biotechnology 29: 644–652.
18. Oono Y, Kobayashi F, Kawahara Y, Yazawa T, Handa H, et al. (2013) Characterisation of the wheat (*Triticum aestivum* L.) transcriptome by de novo assembly for the discovery of phosphate starvation-responsive genes: gene expression in Pi-stressed wheat. BMC Genomics 14: 77.
19. Luo H, Lin Y, Gao F, Zhang C-T, Zhang R (2013) DEG 10, an update of the Database of Essential Genes that includes both protein-coding genes and non-coding genomic elements. Nucleic Acids Research 42: D574-D580.
20. Meinke D, Muralla R, Sweeney C, Dickerman A (2008) Identifying essential genes in *Arabidopsis thaliana*. Trends in Plant Science 13: 483–491.
21. Nakasugi K, Crowhurst RN, Bally J, Wood CC, Hellens RP, et al. (2013) De novo transcriptome sequence assembly and analysis of RNA silencing genes of *Nicotiana benthamiana*. PLoS ONE 8: e59534.
22. Parra G, Bradnam K, Ning Z, Keane T, Korf I (2009) Assessing the gene space in draft genomes. Nucleic Acids Research 37: 289–297.
23. Liu S, Li W, Wu Y, Chen C, Lei J (2013) De novo transcriptome assembly in chili pepper (*Capsicum frutescens*) to identify genes involved in the biosynthesis of capsaicinoids. PLoS ONE 8(1): e48156.
24. Touming L, Siyuan Z, Qingming T, Ping C, Yongting Y, et al. (2013) De novo assembly and characterization of transcriptome using Illumina paired-end sequencing and identification of CesA gene in ramie (*Boehmeria nivea* L. Gaud). BMC Genomics 14: 125.
25. Ge X, Chen H, Wang H, Shi A, Liu K (2014) De novo assembly and annotation of *Salvia splendens* transcriptome using the Illumina platform. PLoS ONE 9(3): e87693.
26. Yin D, Wang Y, Zhang X, Li H, Lu X, et al. (2013) De novo assembly of the peanut (*Arachis hypogaea* L.) seed transcriptome revealed candidate unigenes for oil accumulation pathways. PLoS ONE 8(9): e73767.
27. Yang X, Yu X-Y, Li Y-F (2013) De novo assembly and characterization of the barnyardgrass (*Echinochloa crus-galli*) transcriptome using next-generation pyrosequencing. PLoS ONE 8(7): e69168.
28. Cardoso-Silva CB, Costa EA, Mancini MC, Balsalobre TWA, Canesin LEC, et al. (2014) De novo assembly and transcriptome analysis of contrasting sugarcane varieties. PLoS ONE 9(2): e88462.
29. Annadurai RS, Neethiraj R, Jayakumar V, Damodaran AC, Rao SN, et al. (2013) *De Novo* Transcriptome Assembly (NGS) of *Curcuma longa* L. Rhizome Reveals Novel Transcripts Related to Anticancer and Antimalarial Terpenoids. PLoS ONE 8(2): e56217.
30. Wang Z, Fang B, Chen J, Zhang X, Luo Z, et al. (2010) De novo assembly and characterization of root transcriptome using Illumina paired-end sequencing and development of cSSR markers in sweetpotato (*Ipomoea batatas*). BMC Genomics 11: 726.
31. Mantello CC, Cardoso-Silva CB, da Silva CC, de Souza LM, Scaloppi Junior EJ, et al. (2014) De Novo Assembly and Transcriptome Analysis of the Rubber Tree (*Hevea brasiliensis*) and SNP Markers Development for Rubber Biosynthesis Pathways. PLoS ONE 9(7): e102665.
32. Cavagnaro PF, Senalik DA, Yang L, Simon PW, Harkins TT, et al. (2010) Genome-wide characterization of simple sequence repeats in cucumber (*Cucumis sativus* L.). BMC genomics 11(1): 569.
33. Wei W, Qi X, Wang L, Zhang Y, Hua W, et al. (2011) Characterization of the sesame (*Sesamum indicum* L.) global transcriptome using Illumina paired-end sequencing and development of EST-SSR markers. BMC Genomics 12: 451.
34. Fraser LG, Harvey CF, Crowhurst RN, De Silva HN (2004) EST-derived microsatellites from *Actinidia* species and their potential for mapping. Theoretical and Applied Genetics 108: 1010–1016.
35. Aggarwal RK, Hendre PS, Varshney RK, Bhat PR, Krishnakumar V, et al. (2007) Identification, characterization and utilization of EST-derived genic microsatellite markers for genome analyses of coffee and related species. Theoretical and Applied Genetics 114: 359–372.
36. Gottlieb AM, Poggio L (2014) Quantitative and qualitative genomic characterization of cultivated *Ilex* L. species. Plant Genetic Resources: Characterization and Utilization 1–11. doi:10.1017/S1479262114000756
37. Hotto AM, Schmitz RJ, Fei Z, Ecker JR, Stern DB (2011) Unexpected diversity of chloroplast noncoding RNAs as revealed by deep sequencing of the *Arabidopsis* transcriptome. G3: Genes, Genomes, Genetics 1: 559–570.
38. Hotto AM, Germain A, Stern DB (2012) Plastid non-coding RNAs: emerging candidates for gene regulation. Trends in Plant Science 17: 737–744.
39. Zhelyazkova P, Sharma CM, Förstner KU, Liere K, Vogel J, et al. (2012) The primary transcriptome of barley chloroplasts: numerous noncoding RNAs and the dominating role of the plastid-encoded RNA polymerase. The Plant Cell Online 24: 123–136.
40. Small ID, Rackham O, Filipovska A (2013) Organelle transcriptomes: products of a deconstructed genome. Current Opinion in Microbiology 16: 652–658.
41. Darling AC, Mau B, Blattner FR, Perna NT (2004) Mauve: multiple alignment of conserved genomic sequence with rearrangements. Genome Research 14: 1394–1403.
42. Hannon GJ (2002) RNA interference. Nature 418: 244–251.
43. Bologna NG, Voinnet O (2014) Diversity, Biogenesis, and Activities of Endogenous Silencing Small RNAs In Arabidopsis. Annual Review of Plant Biology. doi: 10.1146/annurev-arplant-050213-035728
44. Vaucheret H, Vazquez F, Crété P, Bartel DP (2004) The action of ARGONAUTE1 in the miRNA pathway and its regulation by the miRNA pathway are crucial for plant development. Genes & Development 18: 1187–1197.
45. Baumberger N, Baulcombe DC (2005) Arabidopsis ARGONAUTE1 is an RNA Slicer that selectively recruits microRNAs and short interfering RNAs. Proceedings of the National Academy of Sciences of the United States of America 102: 11928–11933.
46. Debat HJ, Ducasse DA (2014) Plant microRNAs: Recent Advances and Future Challenges. Plant Molecular Biology Reporter. doi:10.1007/s11105-014-0727-z
47. Wu G, Poethig RS (2006) Temporal regulation of shoot development in *Arabidopsis thaliana* by miR156 and its target SPL3. Development 133: 3539–3547.
48. Wang JW, Czech B, Weigel D (2009) miR156-Regulated SPL transcription factors define an endogenous flowering pathway in *Arabidopsis thaliana*. Cell 138: 738–749.
49. Xing S, Salinas M, Höhmann S, Berndtgen R, Huijser P (2010) miR156-targeted and nontargeted SBP-box transcription factors act in concert to secure male fertility in *Arabidopsis*. The Plant Cell Online 22: 3935–3950.
50. Kozomara A, Griffiths-Jones S (2014) miRBase: annotating high confidence microRNAs using deep sequencing data. Nucleic Acids Research 42(D1): D68–D73.
51. Yang L, Conway SR, Poethig RS (2011) Vegetative phase change is mediated by a leaf-derived signal that represses the transcription of miR156. Development 138: 245–249.
52. Bastos DHM, Oliveira DM, Matsumoto RLT, Carvalho PO, Ribeiro ML (2007) Yerba mate: pharmacological properties, research and biotechnology. Med Aromat Plant Sci Biotechnol 1: 37–46.
53. Murakami ANN, Amboni RD, Prudêncio ES, Amante ER, Fritzen-Freire CB, et al. (2013) Concentration of biologically active compounds extracted from *Ilex paraguariensis* St. Hil. by nanofiltration. Food Chemistry 141: 60–65.
54. Silva RD, Bueno ALS, Gallon CW, Gomes LF, Kaiser S, et al. (2011) The effect of aqueous extract of gross and commercial yerba mate (*Ilex paraguariensis*) on intra-abdominal and epididymal fat and glucose levels in male wistar rats. Fitoterapia 82: 818–826.

55. Heckman MA, Weil J, Mejia D, Gonzalez E (2010) Caffeine (1, 3, 7-trimethylxanthine) in foods: a comprehensive review on consumption, functionality, safety, and regulatory matters. Journal of Food Science 75: R77–R87.
56. Ashihara H, Sano H, Crozier A (2008) Caffeine and related purine alkaloids: biosynthesis, catabolism, function and genetic engineering. Phytochemistry 69(4): 841–856.
57. Ashihara H, Monteiro AM, Gillies FM, Crozier A (1996) Biosynthesis of caffeine in leaves of coffee. Plant Physiology 111: 747–753.
58. Kato M, Mizuno K, Crozier A, Fujimura T, Ashihara A (2000) A gene encoding caffeine synthase from tea leaves. Nature 406: 956–957.
59. Deladino L, Teixeira A, Reta M, Molina García AD, Navarro AS, et al. (2013) Major phenolics in yerba mate extracts (*Ilex paraguariensis*) and their contribution to the total antioxidant capacity. Food and Nutrition 4: 154–162.
60. Niggeweg R, Michael AJ, Martin C (2004) Engineering plants with increased levels of the antioxidant chlorogenic acid. Nature Biotechnology 22(6): 746–754.
61. Lepelley M, Cheminade G, Tremillon N, Simkin A, Caillet V, et al. (2007) Chlorogenic acid synthesis in coffee: An analysis of CGA content and real-time RT-PCR expression of HCT, HQT, C3H1, and CCoAOMT1 genes during grain development in *C. canephora*. Plant Science 172(5): 978–996.
62. Escamilla-Treviño LL, Shen H, Hernandez T, Yin Y, Xu Y, et al. (2014) Early lignin pathway enzymes and routes to chlorogenic acid in switchgrass (*Panicum virgatum* L.). Plant Molecular Biology 84(4–5): 565–576.
63. Zenaro LC, Andrade LB, Santos P, Locatelli C (2014) Effects of aqueous extract of Yerba Mate (*Ilex Paraguariensis*) on the oxidative stress in rats fed a cafeteria diet. International Journal of Natural Sciences Research 2(3): 30–43.
64. Barchuk ML, Tiscornia MM, Giorgio EM, Fonseca MI, Zapata PD (2013) Diseño de un método molecular para la detección de *Ilex dumosa* en yerba mate elaborada utilizando una secuencia específca ubicada en la región ITS2 del DNA ribosómico. Revista de Ciencia y Tecnología (UNaM) 19: 28–34.
65. Kiefer F, Arnold K, Künzli M, Bordoli L, Schwede T (2009) The SWISS-MODEL Repository and associated resources. Nucleic Acids Research 37: D387–D392.
66. Jung S, Ficklin SP, Lee T, Cheng CH, Blenda A, et al. (2014) The Genome Database for Rosaceae (GDR): year 10 update. Nucleic Acids Research 42(D1): D1237–D1244.
67. Jung S, Staton M, Lee T, Blenda A, Svancara R, et al. (2008) GDR (Genome Database for Rosaceae): integrated web-database for Rosaceae genomics and genetics data. Nucleic Acids Research 36: D1034–D1040.
68. Zhao X, Wang H (2007) LTR_FINDER: an efficient tool for the prediction of full-length LTR retrotransposons. Nucleic Acids Research 35: W265–W268.

7

Development of a Heat-Shock Inducible Gene Expression System in the Red Alga *Cyanidioschyzon merolae*

Nobuko Sumiya[1,2]*, Takayuki Fujiwara[1], Yusuke Kobayashi[3¤], Osami Misumi[2,3], Shin-ya Miyagishima[1,2,4]*

1 Center for Frontier Research, National Institute of Genetics, Mishima, Shizuoka, Japan, **2** Japan Science and Technology Agency, CREST, Kawaguchi, Saitama, Japan, **3** Department of Biological Science and Chemistry, Faculty of Science, Yamaguchi University, Yamaguchi, Yamaguchi, Japan, **4** Department of Genetics, Graduate University for Advanced Studies (SOKENDAI), Mishima, Shizuoka, Japan

Abstract

The cell of the unicellular red alga *Cyanidioschyzon merolae* contains a single chloroplast and mitochondrion, the division of which is tightly synchronized by a light/dark cycle. The genome content is extremely simple, with a low level of genetic redundancy, in photosynthetic eukaryotes. In addition, transient transformation and stable transformation by homologous recombination have been reported. However, for molecular genetic analyses of phenomena that are essential for cellular growth and survival, inducible gene expression/suppression systems are needed. Here, we report the development of a heat-shock inducible gene expression system in *C. merolae*. *CMJ101C*, encoding a small heat shock protein, is transcribed only when cells are exposed to an elevated temperature. Using a superfolder GFP as a reporter protein, the 200-bp upstream region of *CMJ101C orf* was determined to be the optimal promoter for heat-shock induction. The optimal temperature to induce expression is 50°C, at which *C. merolae* cells are able to proliferate. At least a 30-min heat shock is required for the expression of a protein of interest and a 60-min heat shock yields the maximum level of protein expression. After the heat shock, the mRNA level decreases rapidly. As an example of the system, the expression of a dominant negative form of chloroplast division DRP5B protein, which has a mutation in the GTPase domain, was induced. Expression of the dominant negative DRP5B resulted in the appearance of aberrant-shaped cells in which two daughter chloroplasts and the cells are still connected by a small DRP5B positive tube-like structure. This result suggests that the dominant negative DRP5B inhibited the final scission of the chloroplast division site, but not the earlier stages of division site constriction. It is also suggested that cell cycle progression is not arrested by the impairment of chloroplast division at the final stage.

Editor: Stephan Neil Witt, Louisiana State University Health Sciences Center, United States of America

Funding: Funding provided by a Grant-in-Aid for Scientific Research from Japan Society for the Promotion of Science (no. 25251039 to S.M. http://www.jsps.go.jp/english/e-grants/), Core Research for Evolutional Science and Technology (CREST) of Program of Japan Science and Technology Agency (JST) (to S.M. http://www.jst.go.jp/kisoken/crest/en/). The funders had no role in study design, data collection and analysis, decision to publish, or preparation of the manuscript.

Competing Interests: The authors have declared that no competing interests exist.

* Email: nsumiya@nig.ac.jp (NS); smiyagis@nig.ac.jp (SM)

¤ Current address: Department of Botany, Graduate School of Science, Kyoto University, Sakyo-ku, Kyoto, Japan

Introduction

The unicellular red algae Cyanidiales are defined as thermo-acidophiles because they grow at an extremely low pH (0.05–5) and relatively high temperature (35–56°C) [1]. Among the Cyanidiales, the unicellular red alga, *Cyanidioschyzon merolae*, contains just one nucleus, mitochondrion and chloroplast per cell [2]. Cell division and organelle division are tightly synchronized by the light/dark cycle [3]. These features make this alga a model organism for studying the mechanisms of chloroplast and mitochondrial division [4]. In addition, the complete determination of the nuclear, mitochondrion and chloroplast genomes [5,6,7,8], as well as the establishment of genetic manipulation methods, such as transient plasmid expression [9] and gene targeting by homologous recombination [10,11,12], have contributed to the growing interest in *C. merolae* as a model organism. The *C. merolae* nuclear genome size is 16.5 Mbp with 4,774 protein-coding genes. The genome content is extremely simple, with a low level of redundancy in photosynthetic eukaryotes. In addition, most of the genes lack introns [7,8,13]. Thus, in addition to studies on chloroplast and mitochondrial division, *C. merolae* has become a promising organism for the study of cell biology and metabolisms in photosynthetic eukaryotes, such as vacuolar inheritance [14,15], the dynamics of the endoplasmic reticulum [16] and Golgi apparatus [17], nitrogen assimilation [18,19], chromosome organization [20], and circadian rhythms [21].

The simple nature of the cellular architecture and genome content of *C. merolae* has facilitated the characterization of fundamental intracellular phenomena. However, since such phenomena are essential to cellular growth and survival, inactivation or modification of the mechanisms by gene manipulation would likely be lethal. To overcome this critical issue, analyses of primary defects using conditional mutations, an approach which is not effective under permissive environmental conditions but can be effective under certain tightly restricted conditions, is greatly desired.

To this end, we have developed an inducible gene expression system using a heat-shock promoter in *C. merolae*, which we report in this study. As an example of an application of the system, expression of a dominant negative form of the DRP5B/CmDnm2 protein was induced by heat shock, after which the effect on chloroplast division and cell cycle progression was examined.

Materials and Methods

Algal culture

Cyanidioschyzon merolae 10D or stable transformants were grown in 50 mL of Allen's medium in a 100 mL test tube under continuous light (100 μE m^{-2} s^{-1}) with aeration (500 mL ambient air min^{-1}) [22].

Semi-quantitative RT-PCR

Cells were harvested by centrifugation at 800×*g*, frozen in liquid nitrogen and stored at −80°C until use. Total RNA was extracted with the RNeasy Mini Kit and RNase-Free DNase set (Qiagen, Venlo, Netherlands). cDNA was synthesized from the total RNA using 6 nucleotide random primers with ThermoScript RT (Life Technologies, Carlsbad, CA). The primers used for semi-quantitative RT-PCR are listed in Table S1.

DNA construction for the superfolder GFP (sfGFP) expressing strain

According to the amino acid sequence of original sfGFP [23], the oligonucleotides encoding sfGFP for the expression in *C. merolae* were artificially synthesized so as to introduce the mutations S30R, Y39N, Q80R, F99S, N105T, Y145F, M153T, V163A, I171V, A206V and L231H (accession no. AB971579) into the nucleotide sequence encoding the enhanced GFP (EGFP; [24]). For further construction, *Bam*HI and *Not*I sites were fused to the 5′- and 3′- terminals of the sequence, respectively.

To produce a strain which constitutively expresses sfGFP, a plasmid containing *APCC* promoter-*sfGFP-URA* was constructed by eliminating the *E2F orf* from the plasmid *APCC* promoter-*sfGFP-pmE2F-URA* [21]. The regions other than *pmE2F* were amplified from the original plasmid by the primers sfGFP720taaRbt and btUTR(+1), then circularized using an In-Fusion cloning kit (Takara Bio, Shiga, Japan).

To prepare a URA (the URA5.3 gene; CMK046C) expressing strain as a control, *APCC* promoter-*EGFP* was eliminated from the plasmid containing *APCC* promoter-*EGFP*-URA (pD184-O250-EGFP-URA_{Cm-Cm}; [11]). The regions other than *APCC* promoter-*EGFP* were amplified from the original plasmid using the primers d184(+25)Rura d184(+25)R and CmUra(−897)F and the fragment was circularized using an In-Fusion cloning kit. The primer sequences for the construction are listed in Table S1. The plasmid pD184-O250-EGFP-URA_{Cm-Cm} [11] was used to produce the strain constitutively expressing EGFP.

Construction of the plasmids for the transient transformation assays

To transiently express the sfGFP driven by the upstream region of *C. merolae CMJ101C* from a plasmid, we constructed plasmids in which a series of different lengths of the *CMJ101C* upstream region were fused to the *sfGFP orf* and the *Nos* terminator. *Bam*HI- and *Not*I-digested *sfGFP* was cloned between the *Bam*HI and *Not*I sites of the pI050P-GFP vector [25] to produce the pI050P-sfGFP plasmid. A series of different length of the upstream

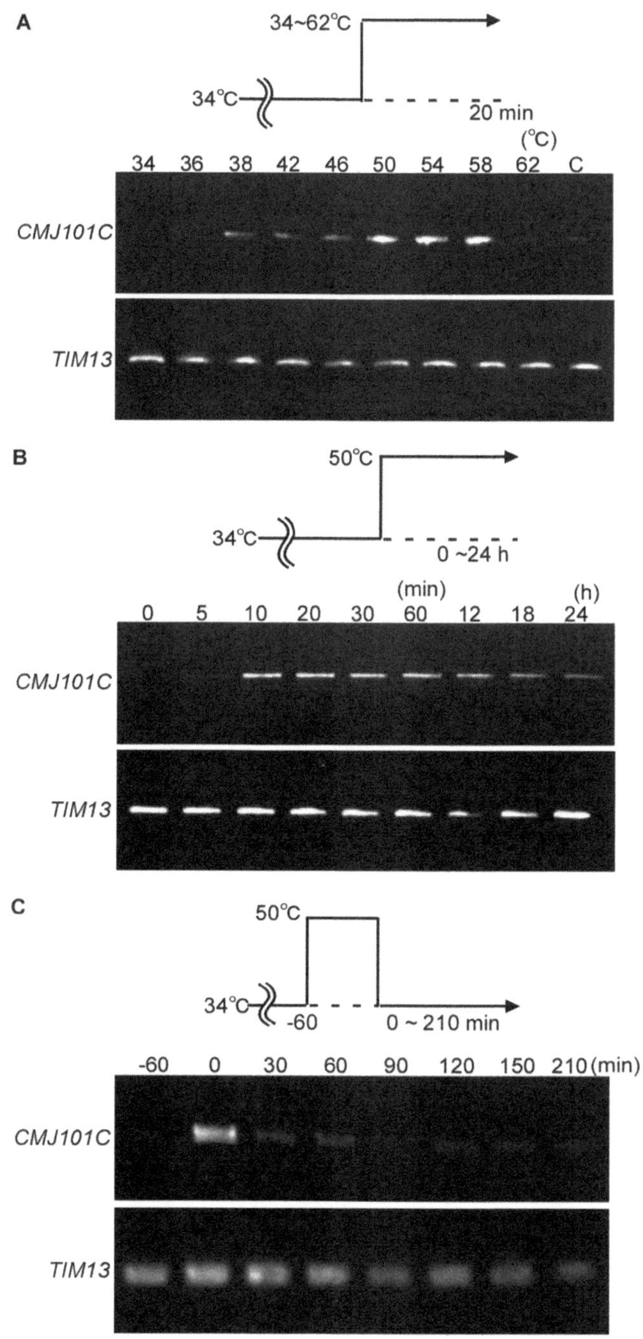

Figure 1. Patterns of *CMJ101C* mRNA induced expression by heat shock. (A) The expression pattern of *CMJ101C mRNA* after a 20 min heat shock at 34~62°C. (B) The expression pattern of *CMJ101C mRNA* after a 0~24 hr heat shock at 50°C. (C) *C. merolae* wild-type cells were cultured at 34°C before the heat shock. The expression pattern of *CMJ101C mRNA* before and 0~210 min after one-hour heat shock at 50°C. mRNA levels were examined by semi-quantitative RT-PCR. *TIM13* (mitochondrial intermembrane space complex subunit, *CMB148C*) was used as a quantitative control. C in (A) represents the mRNA level in the cells cultured at 42°C continuously.

region of *CMJ101C* (−1000 to +2, −750 to +2, −500 to +2, −345 to +2, −250 to +2, −200 to +2 and −100 to +2) were amplified by PCR. For the forward primers, 101_−1000_Xba, 101_−750_Xba, 101_−500_Xba, 101_−345_Xba, 101_−250_Xba, 101_−200_Xba and 101_−100_Xba were used. For the single reverse primer, 101_+2_*Bgl*II was used. The amplified fragments were digested with *Xba*I and *Bgl*II and cloned between the *Xba*I and *Bam*HI sites of pI050P-sfGFP. For convenience, the resultant plasmids were named P-1000, P-750, P-500, P-345, P-250, P-200, and P-100, respectively.

Construction of the plasmids for the stable transformants

To prepare the stable transformants which express sfGFP by 200, 250, or 345-bp upstream region of *CMJ101C orf*, upstream region comprising *CMJ101C*, *sfGFP* and the *nos* terminator was amplified with the primers CMD184D~+25-M13F_Inf_1 and M13 R. The amplified products were replaced with *APCC* promoter-*EGFP* in pD184-O250-EGFP-Ura_{Cm-Cm} [11] by ligation with a linearized vector that was amplified from the plasmid pD184-O250-EGFP-Ura_{Cm-Cm} by the primers CMD184C_+25R and M13R_CmUra-897~Inf_1 using In-Fusion cloning kit.

To produce the GFP-DRP5B or GFP-DRP5B K135A expressing the stable transformants, the stop codon of *sfGFP* in pI050P-sfGFP was eliminated. The promoter region of the resultant plasmid pI050P-sfGFP (-stop) was then substituted with a 200-bp upstream region of *CMJ101C*. The 200-bp upstream region of *CMJ101C* was amplified with the primers 101_−200_Xba and 101_+2_*Bgl*II. The amplified fragment was digested with *Xba*I and *Bgl*II and cloned between the *Xba*I and *Bam*HI sites of pI050P-sfGFP (-stop). The resultant plasmid was digested with *Not*I and *C. merolae DRP5B orf* (*CMN262C*), which was amplified with the primers GFP_Not_Dynamin_inf_1 and GFP_Not_Dynamin_inf_2, was inserted by the In-Fusion cloning kit. For the cloning of the dominant negative form of *DRP5B*, Lys-135 was replaced by Ala using the primers Dynamin_K44A_sense and Dynamin_K44A antisense. The primer sequences for the construction are listed in Table S1.

Transformation of *C. merolae*

Transient transformation was performed using a polyethylene glycol (PEG)-mediated protocol, as described previously [9]. To produce stable transformants, 6 μg of the PCR product obtained by the primers D184(1270)F and D184(+1448)R from the aforementioned plasmids were introduced via the PEG-mediated protocol into *C. merolae* M4, which has a mutation in the *URA* gene [10], and the transformants were selected as described previously [19].

Quantitative PCR Analyses

Quantitative-PCR analyses were performed using a StepOne-Plus Real-Time PCR System (Life Technologies) and a 20 μl reaction mixture containing 2 μl template DNA, 0.25 nM primers and 10 μl Power SYBR Green Master Mix (Life Technologies). To examine copy number of the introduced DNA fragment, standard curves were constructed using serially diluted solutions of DNA isolated from CmsfGFP-1 strain cells and the relevant sets of primers. The value of *GFP* (from 638 to 717 bp in the *orf*) in each strain was normalized with the data of a region (from 741 to 841 bp in *CMD184C orf*) located outside of the introduced DNA fragment and then the copy number was displayed relative to an EGFP strain, which contains ~1 copy of *EGFP* sequence [11]. To examine cell cycle progression of the heat-shocked culture, standard curves were constructed by using serially diluted solutions of cDNA that was prepared from S-200 cells cultured at 42°C under light with the relevant sets of primers. The value of *PCNA* (from 468 to 524 bp in the *orf*) and *CDC20* (from 419 to 473 bp in the *orf*) were normalized with the data of DRP3 (from 1981 to 2050 bp in the *orf*). The primer sequences for the quantitative-PCR analyses are shown in Table S1.

Immunoblot analyses

5×10^8 cells were harvested by centrifugation at 800×*g* for 5 min and stored at −80°C. Cells were suspended in SDS-PAGE sample buffer (50 mM Tris, pH 6.8, 6% 2-mercaptoethanol, 2% sodium dodecyl sulfate, 10% glycerol) and sonicated on ice for 6 cycles (10 s on and 30 s off at 310 W; Bioruptor UCW-310, Cosmobio, Tokyo, Japan). After centrifugation at 15,000×*g* for 10 min at 4°C, the supernatant fraction was separated by SDS-PAGE and analyzed by immunoblotting with a mouse monoclonal anti-GFP antibody (Living Colors A.v. Monoclonal Antibody (JL-8), Clontech, Mountain View, CA) as the primary antibody at a dilution of 1:1,000 and goat anti-mouse IgG-HRP antibody (Life Technologies) as the secondary antibody at a dilution of 1:2,000.

Fluorescence microscopy

The cells were examined using an epifluorescence microscope (BX51, Olympus, Tokyo, Japan) equipped with a CCD camera (DP70, Olympus). The GFP fluorescence and autofluorescence of chlorophyll were captured using an Olympus WIB cube.

DAPI (4′, 6-diamidino-2-phenylindole) staining

Cells were harvested by centrifugation at 2,000×*g* for 3 min and fixed in 1% glutaraldehyde in TAN (20 mM Tris-HCl, pH 7.6, 0.5 mM EDTA, 1.2 mM spermidine, 0.6 mM spermin, 7 mM 2-mercaptoethanol, 0.4 mM phenylmethyl- sufonyl fluoride). The cells were stained with 1.0 μg ml^{-1} DAPI by pressing them between a slide glass and a cover slip.

Results and Discussion

Expression pattern of *CMJ101C* mRNA

To develop an inducible gene expression system in *C. merolae* using a heat-shock promoter, we searched for *C. merolae HSP* genes that are transcribed only when cells are exposed to elevated temperature (i.e. with no basal level expression at a lower temperature).

In the *C. merolae* genome database (http://merolae.biol.s.u-tokyo.ac.jp/), 14 genes are predicted to encode heat shock proteins. Among these 14 genes, the transcripts of 13 were found in the previous expressed sequence tag (EST) data that was obtained from the culture at the optimal temperature (42°C), while the transcript of one gene, *CMJ101C* (A BLASTP search in the NCBI database showed that *CMJ101C* encodes a member of the Hsp20/alpha crystallin family of proteins) that was not found in the EST data. To examine whether *CMJ101C* is transcribed only at higher temperatures, *C. merolae* cells cultured at 34°C for one day were shifted to 34~62°C for 20 min. By RT-PCR analyses, *CMJ101C* mRNA was detected at only 38°C or higher (Figure 1A) and the mRNA level exhibited a maximum at 50°C.

To determine the optimum duration of the heat shock for inducing *CMJ101C* transcription, the cells were shifted from 34°C to 50°C. The mRNA level increased for 10 min after the temperature shift and the level was maintained for 24 hr (Figure 1B). After the 60-min heat shock, the *CMJ101C* mRNA level rapidly decreased over 30 min (Figure 1C), while a marginal transcript level was detected for at least 210 min, suggesting that *CMJ101C* transcription ceased immediately after the end of the heat shock. These results led us to explore the usage of the *CMJ101C* promoter for an inducible gene expression system in *C. merolae*.

sfGFP functions as a reporter protein in *C. merolae*

To examine the transcriptional activity of the *CMJ101C* promoter, we searched for GFP variants that are appropriate for

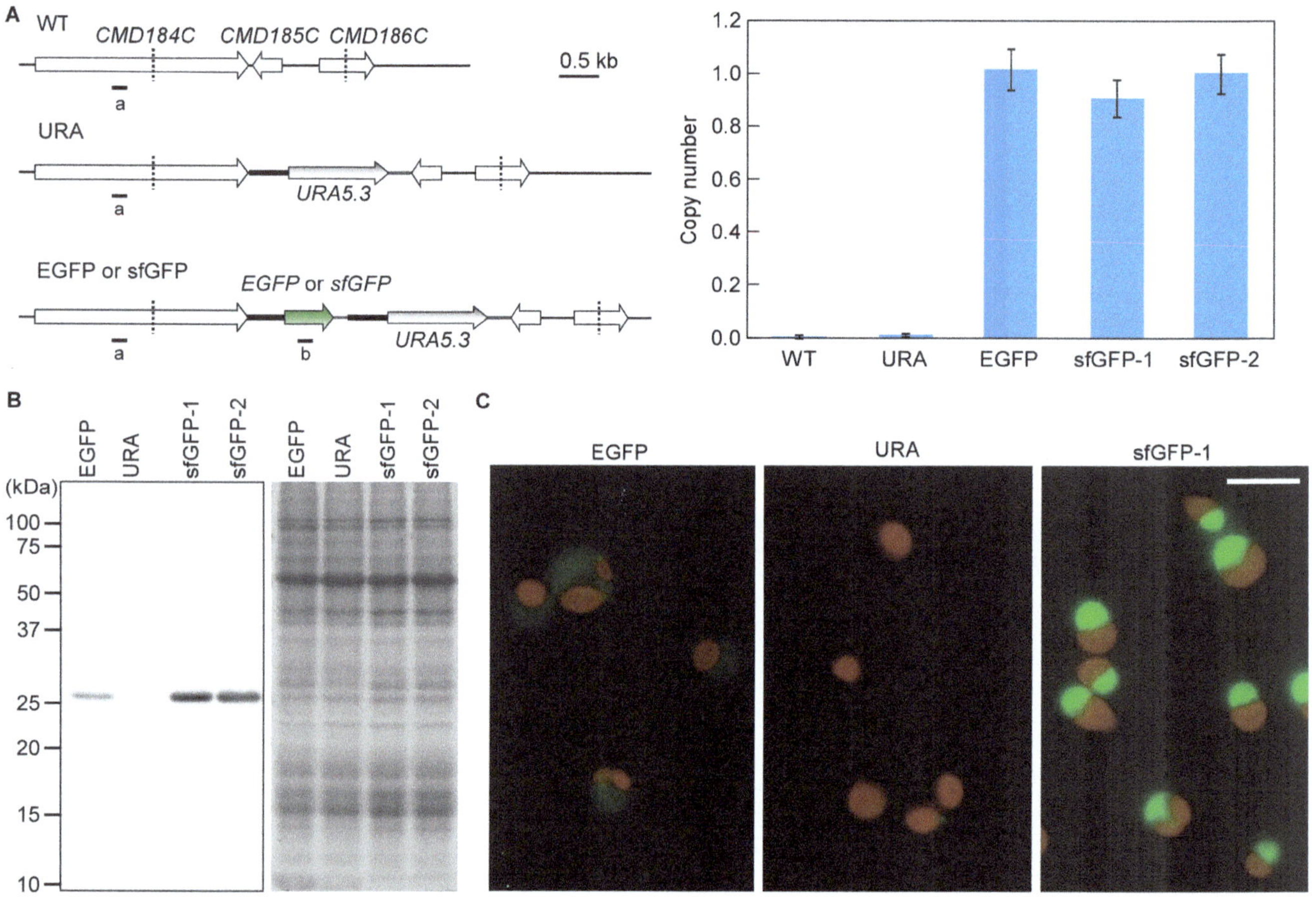

Figure 2. Comparison of the protein level and fluorescent intensity between EGFP and sfGFP expressed from the *C. merolae* genome. A single copy of *EGFP* or *sfGFP* was constitutively expressed by the *APCC* promoter from a *C. merolae* genomic-neutral locus. (A) Quantitative PCR analyses showing that a single copy of *EGFP* or *sfGFP* was integrated into the nuclear genome of the respective transformants. The CMD184C-mid region, which locates outside of the introduced DNA fragment (a) and GFP sequence (b), was investigated by quantitative PCR. The value of (b) in each strain was normalized against that of (a) to estimate the copy number. The copy number is shown relative to an *EGFP* strain, which contains ~1 copy of the *EGFP* sequence [11]. The bars indicate the SD ($n = 3$). (B) EGFP and sfGFP protein levels were compared by immunoblotting. Two sfGFP strains were isolated independently (sfGFP-1, sfGFP-2). The URA strain was used as a negative control, in which the *URA* gene (*URA5.3, CMK046C*) was introduced into the genomic-neutral locus of *C. merolae* M4 that has a mutation in the *URA5.3* gene. Coomassie Brilliant Blue (CBB) staining of the gel is shown as a loading control. (C) Fluorescence of EGFP and sfGFP were observed by fluorescence microscopy. The fluorescence from GFP and autofluorescence of chlorophyll were captured simultaneously. The images were captured using the same exposure time (1 sec). The scale bar = 5 μm.

C. merolae culture at higher temperatures (34~50°C). The folding of enhanced GFP (EGFP) is optimized at 37°C [24]. In contrast, superfolder GFP (sfGFP; [23]) exhibits bright fluorescence in extreme thermophiles, even at 70°C [26]. In order to compare the efficiency of EGFP and sfGFP as reporter proteins in *C. merolae*, we prepared strains that constitutively express EGFP or sfGFP (one copy of the *orf* was integrated into the genome, Figure 2A) under the *C. merolae* constitutive *APCC* promoter from a genomic-neutral locus [11]. By quantitative-PCR analyses, we confirmed that a single GFP sequence was integrated into the genome in the transformants (Figure 2A). Immunoblot analysis with an using the anti-GFP antibody showed that the level of sfGFP was higher than that of EGFP in *C. merolae* cells cultured at 42°C, a temperature which is optimal for *C. merolae* growth (Figure 2B). In addition, fluorescence microscopy revealed that sfGFP-expressing cells exhibited brighter fluorescence than EGFP-expressing cells (Figure 2C). Because the nucleotide sequences encoding the EGFP and sfGFP that we expressed in *C. merolae* do not contain any rare codons of *C. merolae*, the above results suggest that sfGFP is more stable than EGFP in *C. merolae* when the cells were cultured at 42°C. Thus, we used sfGFP as the reporter protein for further investigation.

Determination of the optimal *CMJ101C* promoter region for the heat-shock-inducible gene expression system

Upstream of the *CMJ101C orf*, the *CMJ100C orf* locates in the opposite orientation and the interval between the two *orfs* is 345 bp (Figure 3A). It is well known that there are conserved *cis*-regulatory promoter elements called heat shock elements in the upstream region of heat shock genes (reviewed in [27]), but there are no typical heat shock elements (at least of the three nGAAn elements) in the 345-bp region. In order to define the region that is optimal for transcription specifically in response to heat shock, we fused a series of *CMJ101C* upstream sequences of different length to *sfGFP orf* and the constructs (plasmids; P-100 to P-1,000) were transiently transformed into *C. merolae* wild-type cells (Figure 3A). The transient transformation of *C. merolae* cells resulted at most in 2~10% of transformants in a population [9]. After transforma-

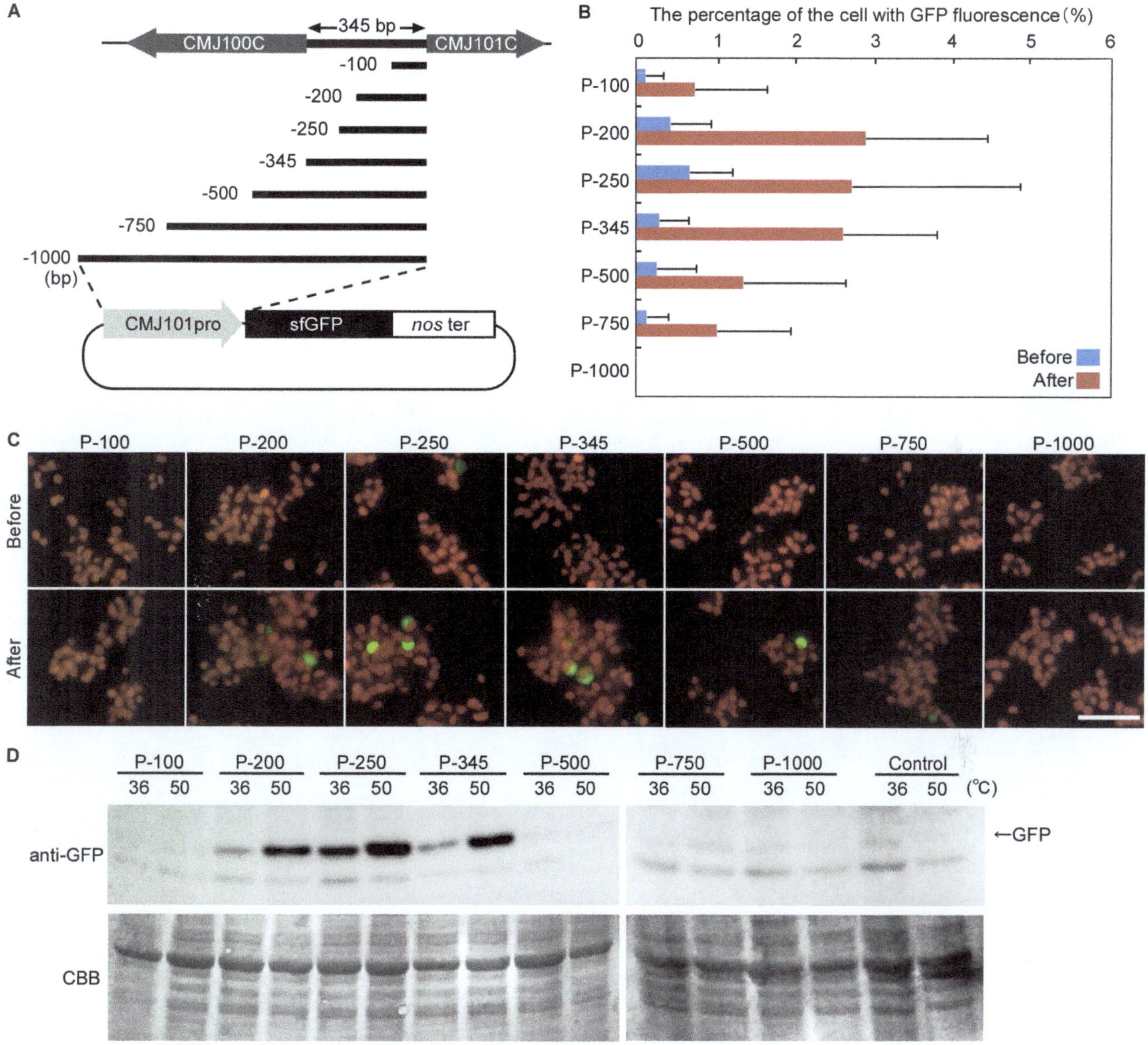

Figure 3. Transcriptional activities of the upstream region of *CMJ101C orf* before and after heat shock treatment in transiently transformed cells. Several different lengths of the upstream region of *CMJ101C orf* were respectively fused with *sfGFP orf*. The plasmids were transiently transformed into *C. merolae* wild-type cells. After transformation, cells were cultured at 36°C for 1 day and then shifted to 50°C for 1 h. Expression of the GFP protein was examined by fluorescence microscopy and immunoblotting. (A) Schematic diagram of the plasmids used. (B and C) The percentage of the cells with GFP fluorescence was compared before and after the 1-h 50°C heat shock treatment. The red is the autofluorescence of chlorophyll. The error bars represent the standard deviation calculated from three individual experiments ($n = 3$). 100 cells were counted for each experiment. The scale bar is 10 μm. (D) Immunoblotting with the anti-GFP antibody before (36°C) and after (50°C) the heat shock treatment. CBB staining of the PVDF membrane is shown as a loading control.

tion, cells were cultured at 36°C for 24 h and then the temperature was shifted to 50°C.

After 1-h heat shock at 50°C, GFP fluorescence-positive cells were detected, especially in the cultures transformed with P-200, P-250 and P-345. However, the fluorescence-positive cells were also observed before the heat shock treatment in the culture transformed with P-250 (Figures 3B, 3C). Consistent with this observation, the immunoblot analysis showed that GFP was expressed in the cultures transformed with P-200, P-250 and P-345 after 1-h heat shock at 50°C, and that GFP was expressed before the heat shock in the culture transformed with P-250 (Figure 3D).

The above results obtained by the transient expression of GFP from the plasmids suggest that a 200-bp or 345-bp upstream sequence of *CMJ101C orf* is suitable for the heat shock inducible gene expression system, but the plasmid copy number of the transformants varies with the method, so it is difficult to exactly compare the transcriptional activity of the upstream sequences before and after the heat shock. Thus, we prepared the stable transformants (S-200, S-250 and S-345) in which a single promoter (the 200, 250 or 345-bp upstream region of *CMJ101C*

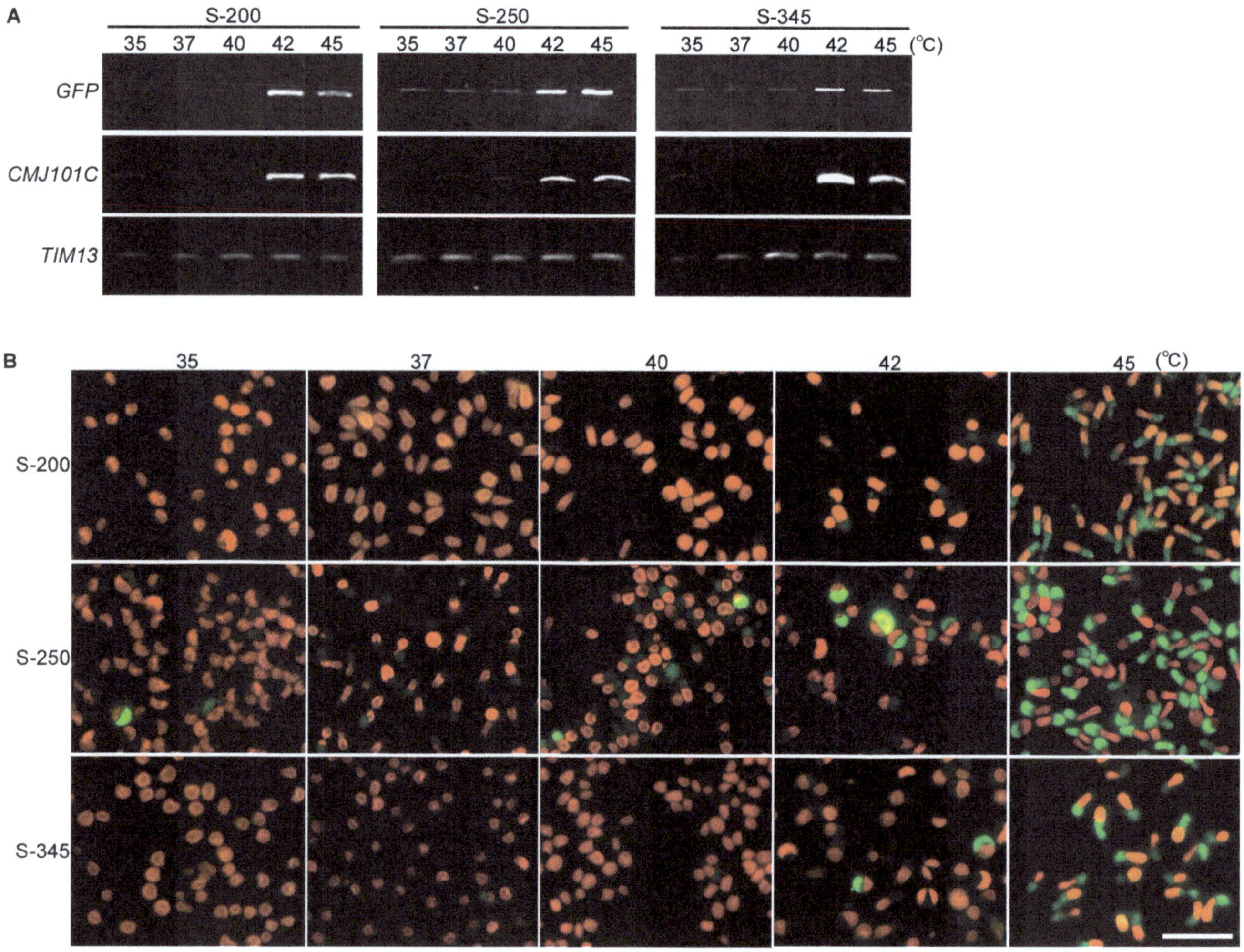

Figure 4. Basal transcriptional activities of the upstream region of *CMJ101C orf* at lower temperatures in the stable transformants. A single copy of *sfGFP orf* fused with the 200 (S-200), 250 (S-250) or 345 (S-345)-bp upstream region of *CMJ101 orf* (Figure 3A) was integrated into the *C. merolae* genomic-neutral locus [11]. The stable transformants were cultured at 35~45°C for 1 day. (A) Semi-quantitative RT-PCR showing the level of the *sfGFP* mRNA. *CMJ101C* and *TIM13* (*CMB148C*) were used as a positive and quantitative control, respectively. (B) Micrographs showing the GFP fluorescence and autofluorescence of chlorophyll (red). The scale bar is 10 μm.

orf) and *GFP orf* fusion was inserted into the same genomic-neutral site for further evaluation.

To evaluate the basal level of GFP expression before the induction by heat shock, the S-200, S-250 and S-345 stable transformants were cultured at 35, 37, 40, 42, and 45°C for one day. It was shown by semi-quantitative RT-PCR that *GFP* mRNA was detected in the S-200 cells at only 42°C and 45°C. In contrast, in the S-250 and S-345 cells, the mRNA was detected at all of the temperatures tested, even though a higher level of mRNA was detected at 42°C and 45°C (Figure 4A). Consistent with the results of RT-PCR, GFP fluorescence was detected in the S-200 cells at only 42°C and 45°C. In contrast, some of the S-250 and S-345 cells exhibited GFP fluorescence at lower temperatures (S-250 at 35, 37 and 40°C, and S-345 at 37 and 40°C)(Figure 4B). The strongest fluorescence was observed at 45°C in all of the transformants. These results indicate that the 200-bp upstream region of *CMJ101C orf* is the most suitable for the heat shock inducible system in the sense of there being little leakage of expression at lower temperatures.

Determination of the optimum temperature and duration for heat-shock inducible gene expression in the stable transformants

To determine the optimum temperature at which heat shock induces gene expression in the stable transformants, the S-200 cells were shifted from 34°C to 34–62°C and cultured for one hour. The semi-quantitative RT-PCR showed that the *GFP* mRNA level peaked at 50–58°C (Figure 5A). Immunoblotting and fluorescence microscopy showed that the GFP protein and fluorescent level peaked at 50–54°C (Figure 5B, 5C). Based on these results, we fixed the heat shock temperature at 50°C for further analyses.

To determine the optimum duration of the 50°C heat shock to induce gene expression in the stable transformants, the S-200 cells were shifted from 34°C to 50°C and cultured for 120 min. *GFP* mRNA was detected in within 5 min of the temperature shift and reached the maximum level at 30 min (Figure 6A). The GFP protein was detected at 30 min and reached the maximum level at 60 min (Figure 6B). GFP fluorescence-positive cells were detected at 30 min and the strongest fluorescence was detected at 60 min

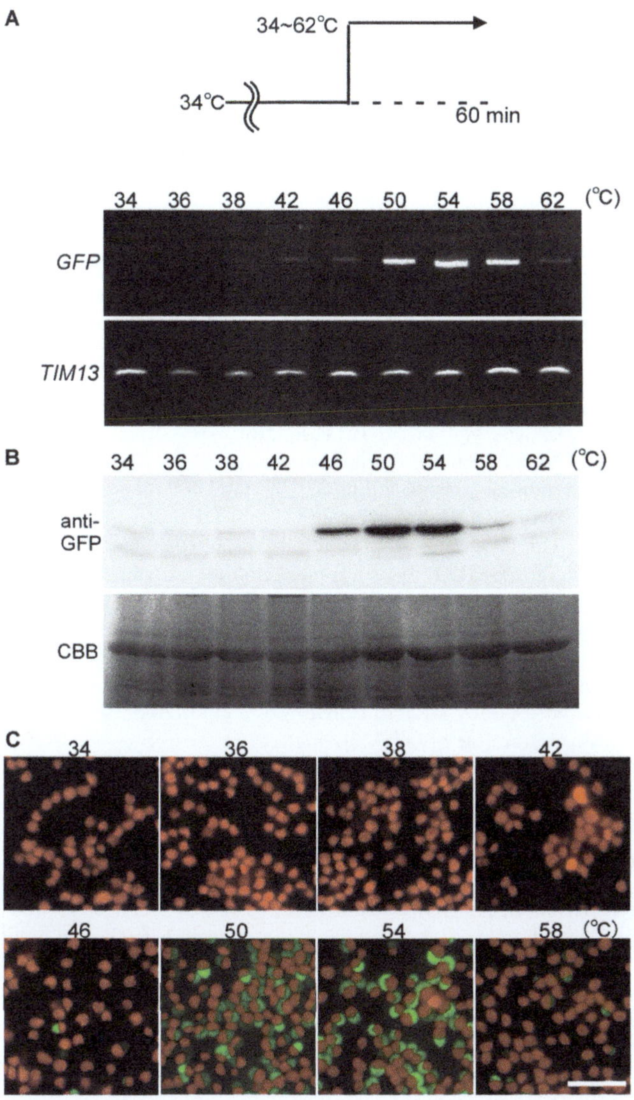

Figure 5. Optimum temperature for heat shock inducible expression in the stable transformants. The stable S-200 transformant (Figure 4) cultured at 34°C was shifted to 34~62°C and cultured for 60 min. (A) Semi-quantitative RT-PCR showing the *GFP* mRNA level. *TIM13* (*CMB148C*) was used as a quantitative control. (B) Immunoblotting with the anti-GFP antibody showing the GFP protein level. CBB staining of the PVDF membrane is shown as a loading control. (C) Micrographs showing the GFP fluorescence and autofluorescence of chlorophyll (red). The scale bar is 10 μm.

(Figure 6C). Thus, for the expression of GFP, at least 30-min heat shock is required and the protein expression level increased for a further 60 min after the temperature shift. Because *GFP* mRNA was detected for at least 24 hours in cells cultured at 50°C (Figure 6D), the duration of translational induction is likely to be able to be extended for particular purposes.

After the one hour heat shock at 50°C, the *GFP* mRNA level decreased rapidly (Figure 7A). In contrast, the GFP protein and fluorescence was detected even one day after the heat shock (Figures 7B, 7C). This difference between the mRNA and protein level is probably because of the high stability of GFP [28]. In conclusion, for the expression of a target gene by the 200-bp upstream region of *CMJ101C orf*, a heat shock at 50°C for at least

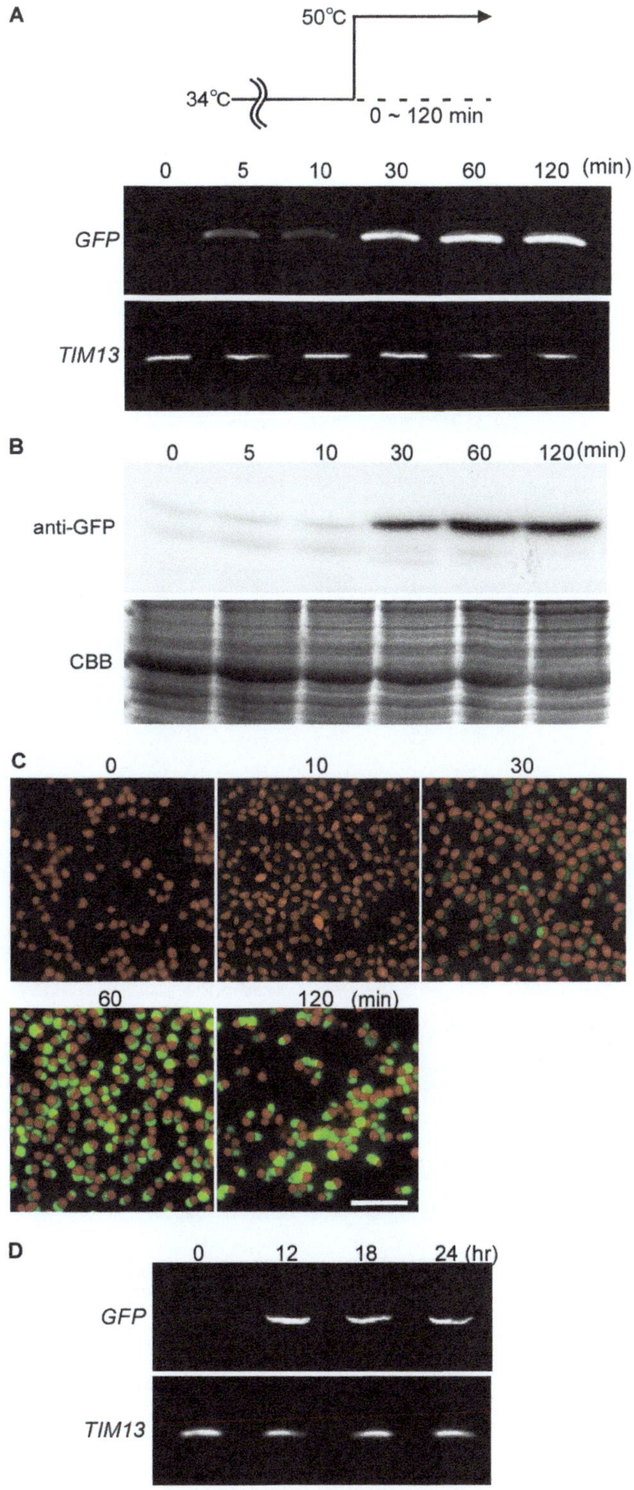

Figure 6. Duration-dependent effect of 50°C heat shock on the mRNA and protein levels in the stable transformants. The stable S-200 transformant (Figure 4) cultured at 34°C was shifted to 50°C and cultured for 120 min or 24 h. (A) Semi-quantitative RT-PCR showing the *GFP* mRNA level. *TIM13* (*CMB148C*) was used as a quantitative control. (B) Immunoblotting with the anti-GFP antibody showing the GFP protein level. CBB staining of the PVDF membrane is shown as a loading control. (C) Micrographs showing the GFP fluorescence and autofluorescence of chlorophyll (red). The scale bar is 10 μm. (D) Semi-quantitative RT-PCR showing the *GFP* mRNA level up to 24 h at 50°C. *TIM13* (*CMB148C*) was used as a quantitative control.

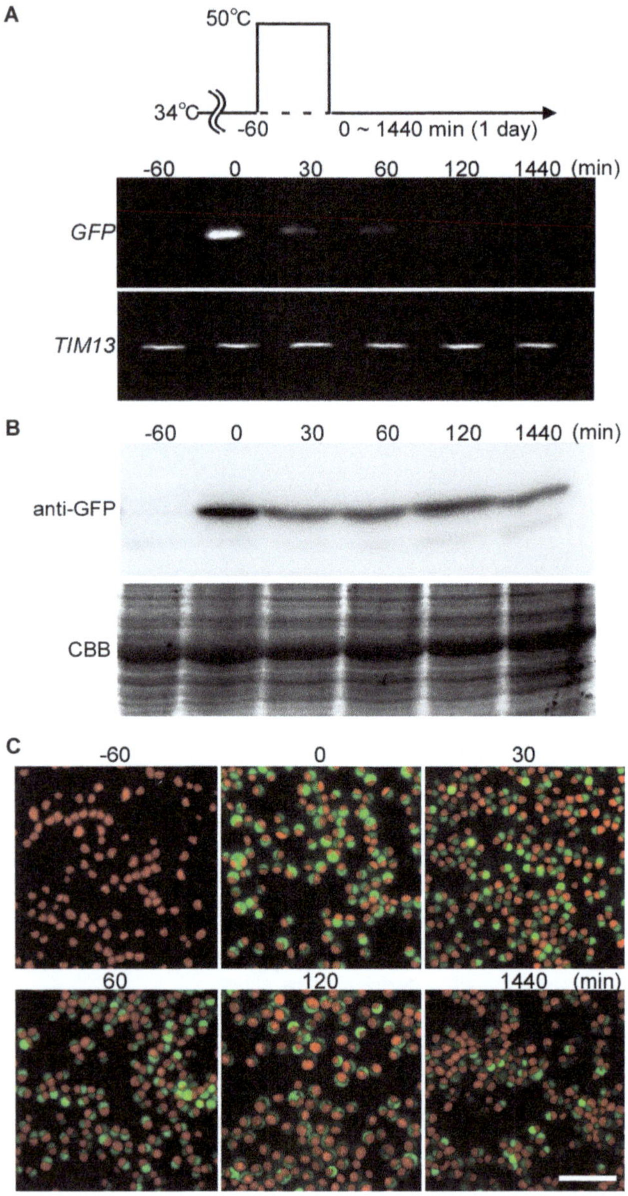

Figure 7. Change in the levels of *GFP* mRNA and GFP protein after 1-h 50°C heat shock in the stable transformants. The stable S-200 transformant (Figure 4) cultured at 34°C (−60 min) was shifted to 50°C for 1 h (0 min) and then returned to 34°C. (A) Semi-quantitative RT-PCR showing the *GFP* mRNA level. *TIM13* (*CMB148C*) was used as a quantitative control. (B) Immunoblotting with the anti-GFP antibody showing the GFP protein level. CBB staining of the PVDF membrane is shown as a loading control. (C) Micrographs showing the GFP fluorescence and autofluorescence of chlorophyll (red). The scale bar is 10 μm.

30 min is the optimum condition for achieving inducible protein expression in *C. merolae*.

Effect of DRP5B mutation on chloroplast division and cell cycle progression: An example of the application of the *C. merolae* heat-shock gene expression system

To apply the inducible gene expression system to an investigation of chloroplast division, we expressed a dominant negative form of DRP5B/CmDmm2 which has a mutation in the GTPase domain. Because the *C. merolae* cell, as in the case of many other algal species, contains only a single chloroplast, the blockage of chloroplast division is probably lethal and thus conditional mutation, such as the inducible expression system employed in this study, is required to characterize the effect of the mutation on chloroplast division.

DRP5B is a member of a eukaryotic dynamin family of self-assembling GTPases and localizes on the cytosolic side of the chloroplast division site, where it is involved in the division process [29,30]. Conventional dynamin, which is involved in endocytosis, consists of five domains: an N-terminal GTPase domain, a middle domain, a pleckstrin homology domain, a GTPase effecter domain and a proline-rich domain (Reviewed in [31], Figure 8A). GTPase activity is essential for membrane fission in endocytosis and the K44A mutation in human dynamin 1 is known to abolish the activity of dynamin binding GTP, and the expression of the K44A mutant protein exhibits dominant-negative effects [32]. The GTP binding domain is well conserved in the DRP5B of algae and plants (Figure 8A). It has been shown that DRP5B is required for normal chloroplast division based on the impairment of chloroplast division in DRP5B knockout or null mutants in *Arabidopsis thaliana* [30] and *Physcomitrella patens* [33]. In addition, DRP5B is recruited to the chloroplast division site before the onset of division site constriction and persists there throughout the division process [29]. However, it is not known how GTP-binding and/or GTP hydrolysis by DRP5B is involved in chloroplast division. To investigate the role of GTP-binding and/or hydrolysis by DRP5B in chloroplast division in *C. merolae*, we induced the expression of GFP-DRP5B or GFP-DRP5B K135A, the mutation of which corresponds to K44A of the human dynamin1 mutation, by the heat shock system in the stable transformants (Figure 8A).

In exponentially proliferating culture of both of the transformants, the GFP signal was not detected at 40°C. The cells were transferred to a dark condition to stop cell growth and new entrance of G1-phase cells into the S phase and then were heat-shocked at 50°C for 1 h (Figure 8B). After the heat shock, the GFP signal was detected at the chloroplast division site (Figures 8C, 8D. indicated with single, double and triple arrowheads), although additional fluorescent speckles were observed in the cytosol of the GFP-DRP5B K135A expressing cells (Figure 8D). Cells were heat-shocked again for 1 h (4–5 h in Figure 8B). After 24-h incubation under dark at 42°C, GFP-DRP5B cells had completed chloroplast division and subsequent cytokinesis, and thus GFP fluorescence was not detected. In contrast, GFP fluorescence was still observed in some GFP-DRP5B K135A cells (Figure 8D). After 1-h heat shock, GFP fluorescence was observed at the chloroplasts division site before and during the earlier and later stages of the division site constriction in the both transgenic lines. In contrast, after 24-h incubation under dark, GFP-DRP5B K135A cells with GFP fluorescence were at the final stage of chloroplast division and displayed aberrant shapes in which two daughter chloroplasts and the cells were still connected by a small DRP5B K135A positive tube-like structure (Figures 8D, E, indicated by the arrowheads in D). DNA-specific fluorescent dye (DAPI) -staining showed that these aberrant shaped cells contain two daughter nuclei on opposite sides of the tube-like structure (Figure 8F). These results suggest that the expression of DRP5B K135A inhibited the final scission of the division site, but not the earlier stage of division site constriction in the course of chloroplast division. It is also suggested that cell cycle progression is not arrested by the impairment of chloroplast division at the final stage, although the completion of cytokinesis is blocked, probably induced by obstruction brought about by the tubular bridge connecting the two daughter chloroplasts.

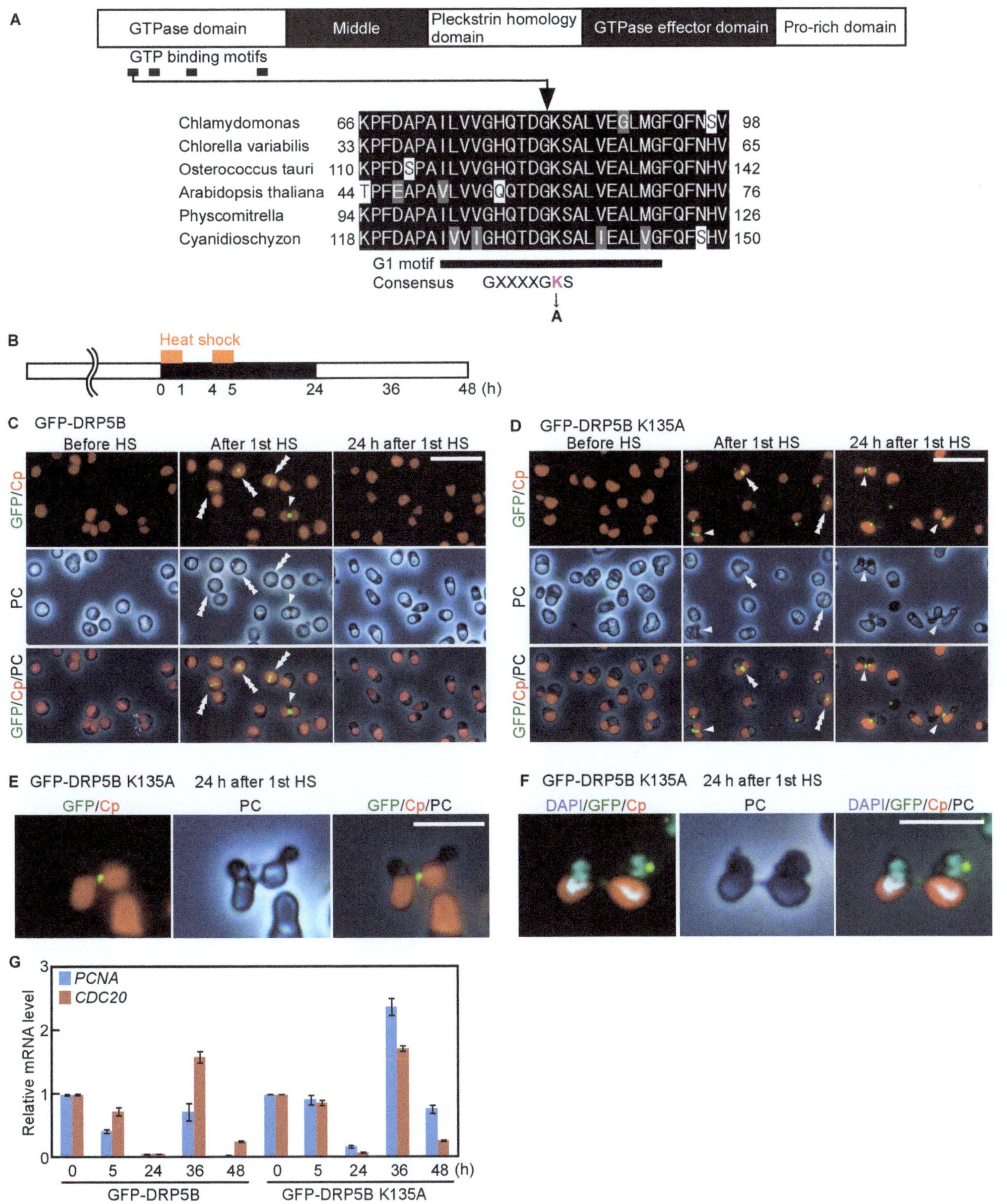

Figure 8. Effect of the DRP5B K135A dominant negative mutation on chloroplast division and cell cycle progression in *C. merolae*. (A) Schematic diagram of the domain composition of the conventional dynamin protein which is involved in endocytosis. In the GTPase domain, four GTP binding motifs (G1, G2, G3 and G4) contribute to the GTP binding. Partial alignment of the DRP5B proteins showed the conservation of the G1 motif. The DRP5B amino acid sequences (*Chlamydomonas reinhardtii*, GI: 30349146; *Chlorella variabilis*, GI: 552813628; *Ostreococcus tauri*, GI: 308803420; *Arabidopsis thaliana*, GI: 30349146; *Physcomitrella patens*, GI: 224434564; *Cyanidioschyzon merolae*, GI: 544214467) were aligned using ClustalW [35,36]. Alanine substitution for the conserved lysine residue within G1 of human dynamin 1 is known to result in a dominant negative effect

by abolishing GTP-binding activity [32]. (B to F) GFP-DRP5B or GFP-DRP5B K135A cells cultured at 42°C under light were transferred to dark condition to stop cell growth and new entrance into S phase. Then cells were heat-shocked at 50°C for 1 h x two times. After cultivation under dark for 12 h, cells were transferred to light condition to resume cell growth (B). GFP fluorescence (along with the red the autofluorescence of chlorophyll) phase-contrast (PC) images are shown (C, GFP-DRP5B, D and E, GFP-DRP5B K135A). The triple arrowheads indicate the chloroplast before division site constriction. The double arrowheads indicates the chloroplast in the earlier stage of division site constriction. The arrowhead indicates the chloroplast in the final stage of division. Images by DAPI staining (blue fluorescence) for GFP-DRP5B K135A 24 h after the heat shock are also shown in (F). The scale bars represent 10 µm in (C, D) and 5 µm in (E, F). (G) Quantitative RT-PCR analyses showing the change in the transcript levels of an S-phase marker *PCNA* (*CMS101C*) and an M-phase marker *CDC20* (*CMA138C*) in the GFP-DRP5B and GFP-DRP5B K135A expressing cells. *DRP3* was used as the internal control. The expression levels at 0 h were defined as 1.0. The bars indicate the standard deviation ($n = 3$).

When GFP-DRP5B and GFP-DRP5B K135A expressing cells were transferred to light condition after 24-h dark period, the transcript levels of both S-phase marker *PCNA* (CMS101C) and M-phase marker *CDC20* (CMA138C) [22] increased for 12 hour (24 h to 36 h in Figure 8G) and then decreased (36 h to 48 h in Figure 8G). These results indicate that both GFP-DRP5B K135A expressing cells entered and finished the next cell cycle even though final stage of chloroplast division and cytokinesis had been blocked.

In conclusion, we developed a heat-shock inducible gene expression system in *C. merolae*. By using the *CMJ101C* promoter, the heat shock at 50°C for 1 h resulted in the induction of maximum level of mRNA and protein expression. In addition, the results show that the level of expression is adjustable by changing the duration or temperature of the administered heat shock.

As an example of an application of the system, here we have examined the function of DRP5B in chloroplast division by expressing the dominant negative form. Although it has not been tested in this study, this system is probably also applicable to the knockdown expression of genes of interest by heat shock induction of antisense RNA expression. The *C. merolae* genome does not encode RNAi machinery components such as Dicer [34]. However, gene expression knockdown by antisense RNA has been achieved by the transient introduction of plasmids in *C. merolae* [25].

The use of the authentic *URA* (*CMK046C*) selection marker guarantees single-copy insertion of constructs into the genome [11,20]. In contrast, a recent study showed that the use of an artificial *URA* gene results in genomic insertion of multiple copies (5~50 copies) of a construct into genomic-neutral loci [11]. By combining the heat shock inducible expression system and the artificial *URA* marker, it is likely to be possible to further increase the induction level of mRNA or antisense RNA.

Acknowledgments

We thank Dr. M. Ohnuma (College of Science, Rikkyo University) for providing the pI050P-GFP vector and Ms. A. Yamashita (Center for Frontier Research, National Institute of Genetics) for technical assistance.

Author Contributions

Conceived and designed the experiments: NS SM. Performed the experiments: NS TF YK OM SM. Analyzed the data: NS TF YK OM SM. Contributed reagents/materials/analysis tools: NS TF YK OM SM. Wrote the paper: NS SM.

References

1. Reeb V, Bhattacharya D (2010) The Thermo-Acidophilic Cyanidiophyceae (Cyanidiales). In: Seckbach J, Chapman DJ, editors.Red Algae in the Genomic Age: Springer Netherlands. pp. 409–426.
2. Kuroiwa T, Kuroiwa H, Sakai A, Takahashi H, Toda K, et al. (1998) The division apparatus of plastids and mitochondria. Int Rev Cytol 181: 1–41.
3. Suzuki K, Ehara T, Osafune T, Kuroiwa H, Kawano S, et al. (1994) Behavior of mitochondria, chloroplasts and their nuclei during the mitotic cycle in the ultramicroalga *Cyanidioschyzon merolae*. Eur J Cell Biol 63: 280–288.
4. Kuroiwa T (2010) Mechanisms of organelle division and inheritance and their implications regarding the origin of eukaryotic cells. Proc Jpn Acad Ser B Phys Biol Sci 86: 455–471.
5. Ohta N, Sato N, Kuroiwa T (1998) Structure and organization of the mitochondrial genome of the unicellular red alga *Cyanidioschyzon merolae* deduced from the complete nucleotide sequence. Nucleic Acids Res 26: 5190–5198.
6. Ohta N, Matsuzaki M, Misumi O, Miyagishima SY, Nozaki H, et al. (2003) Complete sequence and analysis of the plastid genome of the unicellular red alga *Cyanidioschyzon merolae*. DNA Res 10: 67–77.
7. Matsuzaki M, Misumi O, Shin IT, Maruyama S, Takahara M, et al. (2004) Genome sequence of the ultrasmall unicellular red alga *Cyanidioschyzon merolae* 10D. Nature 428: 653–657.
8. Nozaki H, Takano H, Misumi O, Terasawa K, Matsuzaki M, et al. (2007) A 100%-complete sequence reveals unusually simple genomic features in the hot-spring red alga *Cyanidioschyzon merolae*. BMC Biol 5: 1–8.
9. Ohnuma M, Yokoyama T, Inouye T, Sekine Y, Tanaka K (2008) Polyethylene Glycol (PEG)-Mediated Transient Gene Expression in a Red Alga, *Cyanidioschyzon merolae* 10D. Plant Cell Physiol 49: 117–120.
10. Minoda A, Sakagami R, Yagisawa F, Kuroiwa T, Tanaka K (2004) Improvement of Culture Conditions and Evidence for Nuclear Transformation by Homologous Recombination in a Red Alga, *Cyanidioschyzon merolae* 10D. Plant Cell Physiol 45: 667–671.
11. Fujiwara T, Ohnuma M, Yoshida M, Kuroiwa T, Hirano T (2013) Gene targeting in the red alga *Cyanidioschyzon merolae*: single- and multi-copy insertion using authentic and chimeric selection markers. PLoS ONE 8: e73608.
12. Watanabe S, Sato J, Imamura S, Ohnuma M, Ohoba Y, et al. (2014) Stable expression of a GFP-reporter gene in the red alga *Cyanidioschyzon merolae*. Biosci Biotech Biochem 78: 175–177.
13. Misumi O, Matsuzaki M, Nozaki H, Miyagishima SY, Mori T, et al. (2005) *Cyanidioschyzon merolae* Genome. A Tool for Facilitating Comparable Studies on Organelle Biogenesis in Photosynthetic Eukaryotes. Plant Physiol 137: 567–585.
14. Yagisawa F, Nishida K, Yoshida M, Ohnuma M, Shimada T, et al. (2009) Identification of novel proteins in isolated polyphosphate vacuoles in the primitive red alga *Cyanidioschyzon merolae*. Plant J 60: 882–893.
15. Fujiwara T, Kuroiwa H, Yagisawa F, Ohnuma M, Yoshida Y, et al. (2010) The coiled-coil protein VIG1 is essential for tethering vacuoles to mitochondria during vacuole inheritance of *Cyanidioschyzon merolae*. Plant Cell 22: 772–781.
16. Yagisawa F, Fujiwara T, Kuroiwa H, Nishida K, Imoto Y, et al. (2011) Mitotic inheritance of endoplasmic reticulum in the primitive red alga *Cyanidioschyzon merolae*. Protoplasma. 249: 1129–1135.
17. Yagisawa F, Fujiwara T, Ohnuma M, Kuroiwa H, Nishida K, et al. (2012) Golgi inheritance in the primitive red alga, *Cyanidioschyzon merolae*. Protoplasma. 250: 943–948.
18. Imamura S, Kanesaki Y, Ohnuma M, Inouye T, Sekine Y, et al. (2009) R2R3-type MYB transcription factor, CmMYB1, is a central nitrogen assimilation regulator in *Cyanidioschyzon merolae*. Proc Natl Acad Sci U S A 106: 12548–12553.
19. Imamura S, Terashita M, Ohnuma M, Maruyama S, Minoda A, et al. (2010) Nitrate Assimilatory Genes and Their Transcriptional Regulation in a Unicellular Red Alga *Cyanidioschyzon merolae*: Genetic Evidence for Nitrite Reduction by a Sulfite Reductase-Like Enzyme. Plant Cell Physiol 51: 707–717.
20. Fujiwara T, Tanaka K, Kuroiwa T, Hirano T (2013) Spatiotemporal dynamics of condensins I and II: evolutionary insights from the primitive red alga *Cyanidioschyzon merolae*. Mol Biol Cell 24: 2515–2527.

21. Miyagishima SY, Fujiwara T, Sumiya N, Hirooka S, Nakano A, et al. (2014) Translation-independent circadian control of the cell cycle in a unicellular photosynthetic eukaryote. Nat Commun 5: 3807.
22. Miyagishima SY, Suzuki K, Okazaki K, Kabeya Y (2012) Expression of the nucleus-encoded chloroplast division genes and proteins regulated by the algal cell cycle. Mol Biol Evol.29: 2957–2970.
23. Pedelacq J-D, Cabantous S, Tran T, Terwilliger TC, Waldo GS (2006) Engineering and characterization of a superfolder green fluorescent protein. Nat Biotech 24: 79–88.
24. Cormack BP, Valdivia RH, Falkow S (1996) FACS-optimized mutants of the green fluorescent protein (GFP). Gene 173: 33–38.
25. Ohnuma M, Misumi O, Fujiwara T, Watanabe S, Tanaka K, et al. (2009) Transient gene suppression in a red alga, *Cyanidioschyzon merolae* 10D. Protoplasma 236: 107–112.
26. Cava F, De Pedro MA, Blas-Galindo E, Waldo GS, Westblade LF, et al. (2008) Expression and use of superfolder green fluorescent protein at high temperatures in vivo: a tool to study extreme thermophile biology. Environ Microbiol 10: 605–613.
27. Schöffl F, Prändl R, Reindl A (1998) Regulation of the Heat-Shock Response. Plant Physiol 117: 1135–1141.
28. Cubitt AB, Heim R, Adams SR, Boyd AE, Gross LA, et al. (1995) Understanding, improving and using green fluorescent proteins. Trends Biochem Sci 20: 448–455.
29. Miyagishima SY, Nishida K, Mori T, Matsuzaki M, Higashiyama T, et al. (2003) A plant-specific dynamin-related protein forms a ring at the chloroplast division site. Plant Cell 15: 655–665.
30. Gao H, Kadirjan-Kalbach D, Froehlich JE, Osteryoung KW (2003) ARC5, a cytosolic dynamin-like protein from plants, is part of the chloroplast division machinery. Proc Natl Acad Sci U S A 100: 4328–4333.
31. Ferguson SM, De Camilli P (2012) Dynamin, a membrane-remodelling GTPase. Nat Rev Mol Cell Biol 13: 75–88.
32. van der Bliek AM, Redelmeier TE, Damke H, Tisdale EJ, Meyerowitz EM, et al. (1993) Mutations in human dynamin block an intermediate stage in coated vesicle formation. J Cell Biol 122: 553–563.
33. Sakaguchi E, Takechi K, Sato H, Yamada T, Takio S, et al. (2011) Three dynamin-related protein 5B genes are related to plastid division in *Physcomitrella patens*. Plant Sci 180: 789–795.
34. Casas-Mollano JA, Rohr J, Kim E-J, Balassa E, van Dijk K, et al. (2008) Diversification of the Core RNA Interference Machinery in *Chlamydomonas reinhardtii* and the Role of DCL1 in Transposon Silencing. Genetics 179: 69–81.
35. Thompson JD, Higgins DG, Gibson TJ (1994) CLUSTAL W: improving the sensitivity of progressive multiple sequence alignment through sequence weighting, position-specific gap penalties and weight matrix choice. Nuc Acids Res 22: 4673–4680.
36. Larkin MA, Blackshields G, Brown NP, Chenna R, McGettigan PA, et al. (2007) Clustal W and Clustal X version 2.0. Bioinformatics 23: 2947–2948.

8

AtRD22 and AtUSPL1, Members of the Plant-Specific BURP Domain Family Involved in *Arabidopsis thaliana* Drought Tolerance

Vokkaliga Thammegowda Harshavardhan[1], Le Van Son[2,3], Christiane Seiler[1], Astrid Junker[4], Kathleen Weigelt-Fischer[4], Christian Klukas[5], Thomas Altmann[4], Nese Sreenivasulu[1,6], Helmut Bäumlein[2], Markus Kuhlmann[1]*

1 Research Group Abiotic Stress Genomics, Interdisciplinary Center for Crop Plant Research (IZN), Halle (Saale), Germany, and Leibniz Institute of Plant Genetics and Crop Plant Research (IPK) Gatersleben, Stadt Seeland, OT Gatersleben, Germany, **2** Research Group Gene Regulation, Leibniz Institute of Plant Genetics and Crop Plant Research (IPK) Gatersleben, Stadt Seeland, OT Gatersleben, Germany, **3** National Key Laboratory of Gene Technology, Institute of Biotechnology Vietnam, Academy of Science and Technology, Hanoi, Vietnam, **4** Research Group Heterosis, Leibniz Institute of Plant Genetics and Crop Plant Research (IPK) Gatersleben, Stadt Seeland, OT Gatersleben, Germany, **5** Research Group Image Analysis, Leibniz Institute of Plant Genetics and Crop Plant Research (IPK) Gatersleben, Stadt Seeland, OT Gatersleben, Germany, **6** Grain Quality and Nutrition Center, International Rice Research Institute (IRRI), Metro Manila, Philippines

Abstract

Crop plants are regularly challenged by a range of environmental stresses which typically retard their growth and ultimately compromise economic yield. The stress response involves the reprogramming of approximately 4% of the transcriptome. Here, the behavior of *AtRD22* and *AtUSPL1*, both members of the *Arabidopsis thaliana* BURP (BNM2, USP, RD22 and polygalacturonase isozyme) domain-containing gene family, has been characterized. Both genes are up-regulated as part of the abscisic acid (ABA) mediated moisture stress response. While *AtRD22* transcript was largely restricted to the leaf, that of *AtUSPL1* was more prevalent in the root. As the loss of function of either gene increased the plant's moisture stress tolerance, the implication was that their products act to suppress the drought stress response. In addition to the known involvement of AtUSPL1 in seed development, a further role in stress tolerance was demonstrated. Based on transcriptomic data and phenotype we concluded that the enhanced moisture stress tolerance of the two loss-of-function mutants is a consequence of an enhanced basal defense response.

Editor: Jin-Song Zhang, Institute of Genetics and Developmental Biology, Chinese Academy of Sciences, China

Funding: The work was funded by Vietnamese government. The junior Research Group Abiotic Stress Genomics (MK and NS) is funded by IZN (Interdisciplinary Center for Crop Plant Research, Halle (Saale), Germany. VTH and CS are funded by BMBF (GABI-GRAIN: Integrative genomics approach for exploring seed quality and yield under terminal drought; FKZ 0315041A and FKZ 0315041C) and IB-BMBF (Ind09/526). The funders had no role in study design, data collection and analysis, decision to publish, or preparation of the manuscript.

Competing Interests: The authors have declared that no competing interests exist.

* Email: kuhlmann@ipk-gatersleben.de

Introduction

Abiotic stress factors such as moisture stress, salinity, extreme temperature and variable light intensity can disturb plant metabolism and growth. It has been estimated that crop yield losses due to such stresses lie in the order of 50% [1], so increasing the resilience of crop plants will be an important contributor to yield stability. Abiotic stress affects both photosynthesis and photorespiration, as well as having an impact on the energy and redox status of the plant cell. One of the most damaging products of stress is the group of compounds referred to as reactive oxygen species (ROS). The plant response also includes the induced synthesis of certain enzymes and low molecular weight compounds associated with antioxidant activity, redox regulators, chaperones such as heat shock proteins and late embryogenesis abundant proteins, water and ion transporters, the production of compatible osmolytes to maintain cellular water content and the fine tuning of proteolysis involved in programmed cell death. Numerous attempts, with varying levels of success, have been made to genetically engineer the production of some of these components with a view to enhancing abiotic stress tolerance [2].

The phytohormone abscisic acid (ABA) is involved in the regulation of expression of many stress-responsive genes, although other stress responsive genes are known to be regulated in an ABA-independent manner [3–5]. The regulation of several stress-inducible genes is mediated by the interaction of bZIP transcription factors (ABFs) with ABA-response elements (ABRE) in target gene promoters [6].

The *Arabidopsis thaliana* gene *RESPONSIVE TO DEHYDRATION22* (*AtRD22*), originally identified as a gene which responded to dehydration [7,8], encodes a member of the BURP protein family, members of which share a highly conserved BURP domain at their C terminus [9]), sometimes also referred to as a U domain [10,11]. BURP proteins appear to be plant-specific; some examples are the *Vicia faba* unknown seed protein (USP) [10], the

Brassica napus microsporogenesis-specific protein BNM2 [12–14], the *Panicum maximum* apomixis-specific protein [15] and the wheat pollen protein RAFFTIN [16]. The soybean genome harbors 23 BURP protein encoding genes, of which 17 are responsive to stress [17–20]. The 18 *Populus trichocarpa* BURP family members fall into five recognizable sub-families [21]. There are 15 related genes in maize, one of which is specifically expressed in the root cortex parenchyma [22] while the sorghum genome harbors 11 homologs [23]. The ectopic expression of the *A. thaliana* gene *AtUSPL1* has been shown to distort seed development and to alter the morphology of seed lipid vesicles [24].

AtRD22 is up-regulated by moisture stress, salinity stress and exogenously supplied ABA [8] and its induction has been used as a marker for abiotic stress [25–29]. The heterologous expression in both *A. thaliana* and rice of the soybean gene *GmRD22* enhances salinity stress tolerance [30]. Members of the BURP family are up-regulated by stress in rice [31], soybean [30] and maize [23]. Members of the BURP family have been described in relation to stress conditions. In cotton, an RD22-like protein interacts with an α-expansin and the over expression of both proteins simultaneously promotes growth and fruit weight [32].Remarkably, the expression programmes active during stress response partially overlap with gene expression during early embryogenesis and seed desiccation [24]. Most likely ancient stress response genes have been recruited to protect seed tissue from dehydration stress in drying seeds [33,34]. Here, a combination of genetic, molecular and physiological approaches has been applied to isolate and characterize T-DNA insertion mutants of *AtRD22* and *AtUSPL1*.

Materials and Methods

Plant material and growth conditions

A. thaliana seeds (ecotype Col-0 and the three T-DNA insertion lines SALK_146066 (*rd22-1*), WiscDsLox481-484P12 (*rd22-2*) and SALK_022325 (*uspl1*)) obtained from the European Arabidopsis Stock Center (NASC) were stratified before imbibing them for three days at 4°C in the dark. The resulting germinated seedlings were grown in soil for four weeks under 60% relative humidity, a constant temperature of 22°C and under a 16 h photoperiod (light intensity of 120 μmol m^{-2} s^{-1}). Drought stress was applied by a. active dehydration under low humidity for 1–5 days and b. withholding water for 1–5 days; the relative soil water content was monitored using an HH2 moisture meter (delta-T devices, Cambridge, UK). Moisture stress was also mimicked in two week old seedlings by transferring them for three days on a MS basal medium containing one of 150 mM or 300 mM NaCl, 100 μM ABA, 4% w/v trehalose, 4% w/v sorbitol, 4% w/v glucose, 4% w/v fructose, 4% w/v sucrose, 300 mM mannitol, 15% w/v PEG 6000 or 4% w/v PEG 20000.

Detailed information is provided in Figure 1/Figure S4. Primers used for genotyping are listed in Table S1. Drought stress was applied by withdrawal of water and relative soil water content [%] was controlled with HH2 moisture meter (delta-T devices Ltd, Cambridge, UK).

Plasmid construction and plant transformation

Standard protocols [35] were used to prepare plasmids for Gateway Cloning Technology (Invitrogen GmbH, Karlsruhe, Germany). To achieve its ectopic expression, the *AtUSPL1* coding sequence was cloned into the pBENDER GATEWAY vector (a gift from B. Weisshaar, MPI, Cologne).

A 962 bp fragment harboring the *AtUSPL1* promoter was amplified, re-sequenced and inserted into the *Sph*I/*Hind*III

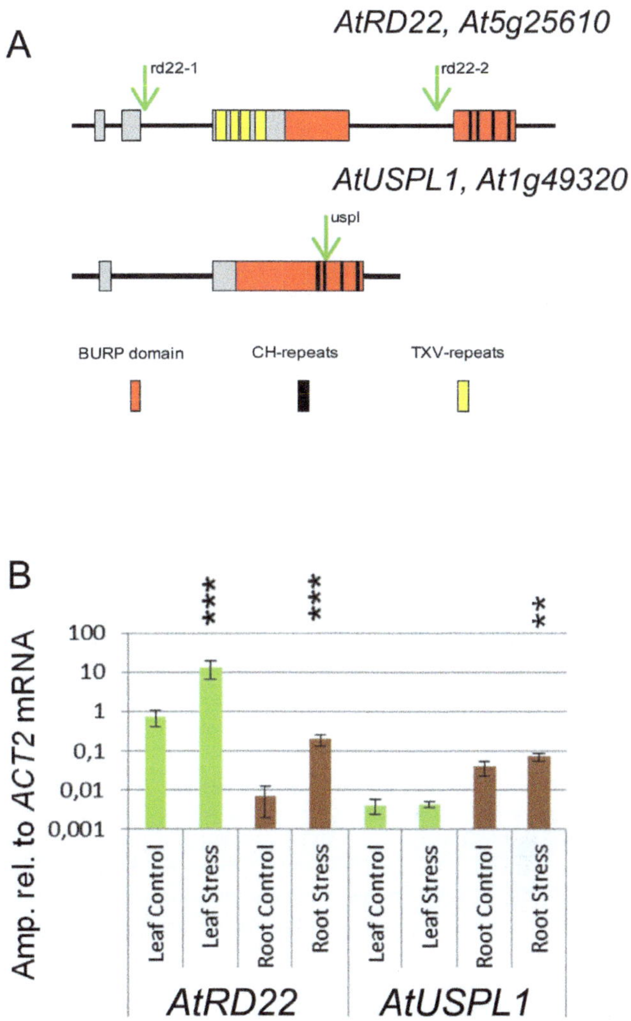

Figure 1. *AtRD22* and *AtUSPL1*, members of the BURP gene family. A. Scheme of BURP-domain containing proteins AtRD22 and AtUSPL1. BURP (named after BNM2, USP, RD22 and Polygalacturonase isozyme) proteins are identified by their C-terminal BURP-domain (red). The BURP domains contain 4 CH-repeats (black). In comparison to AtUSPL1, AtRD22 contains an additional motif, the TXV repeats (yellow), in AtUSPL1 no repetitive domain structure can be found. AGI ID is given in brackets. Position of T-DNA insertions for the used loss-of function mutants is indicated by a green arrow. B. Quantification of *AtRD22* and *AtUSPL1* mRNA in leaves and roots. The bar diagram indicates the amplification of *AtRD22* (left) and *AtUSPL1* mRNA relative to *ACT2* mRNA (amp. rel. to *ACT2* mRNA). Leaf (green) and root (brown) tissue of well watered and drought stressed (2% RWC) plants (N = 5) was analyzed. Asterisks indicate significant differences (T-test: *** p<0.01, ** = p< 0.05).

cloning site of pBIN101 (Clontech) in front of the *uidA GUS* reporter gene. Stable transformation of *A. thaliana* was performed following [36] and subsequent GUS staining to detect transgene expression following [37].

Automated plant phenotyping and image analysis

After 14 days of pre-cultivation pots (diameter 10 cm) were transferred to an automated phenotyping system composed of a conveyor belt transport system and three imaging chambers for plant imaging in the visible (390–750 nm) and NIR (1400–1510 nm with detection peak at 1452 nm) wavelength ranges as well as capturing fluorescence emission (520–750 nm) as well as a watering and weighing station (LemnaTec 3D Scanalyzer; LemnaTec, GmbH, Aachen, Germany) installed in a climate controlled plant cultivation chamber (environmental conditions: 8 h/16 h night/day, 20/22°C (night/day), relative air humidity of 60% and 240 µmol m^{-2} s^{-1} light intensity). Ten blocks (corresponding to ten biological replicates) with each 15 plants (2×5 genotypes under stress conditions, 1×5 under control conditions) were used. Half of the blocks were placed on opposite sides of the cultivation chamber and pots within one block were randomized in order to account for lane or block effects. For reasons of acclimation, plants were grown under control conditions for another week (till day 21 after sowing) before stress application was started. Automated watering was performed every 2nd day to reach defined target values of the pot weight (70% field capacity for control, no watering for soil drought stress). Relative soil water content [%] was controlled in parallel to pot weight with HH2 moisture meter (delta-T devices, Cambridge, UK). The growth rate was calculated as Relative Growth Rate (RGR) = (log(LA[tn+1]) - log(LA[tn]))/((tn+1) - tn) following [38]; Leaf area (LA), timepoint n (tn). Statistical significant growth differences were estimated by Welch t-test from values for plant area determined from top view (visual [39]). Detailed information on calculation of NIR intensities and Cameras used are provided in Table S3.

Estimation of chlorophyll content

Chlorophyll and pheophytin (a chlorophyll degradation product were extracted from the aerial tissue of 6–7 seedlings using a 1:1 acetone:DMSO mixture, and the absorbance of the extract was recorded at 663, 645 and 553 nm [40] using the UVIKON XS 60/99-90286 spectrophotometer. The experiments comprised three independent replicates. Since the chlorophyll content in the mutant plants grown under non-stressed conditions differed from that present in wild type plants, a relative value was calculated to derive the effect of the stress treatment (Figure S6C). Chlorophyll and pheophytin contents were estimated according to [41], and subjected to a one way ANOVA.

RNA isolation qRT-PCR and Northern analysis

RNA was isolated using an RNeasy kit (Qiagen, Hilden, Germany). For the purposes of Northern blotting, 10 µg RNA was loaded into each lane of a 1.2% w/v agarose, 15% v/v formaldehyde gel, electrophoresed, then transferred passively onto a Hybond N+ membrane (Amersham, GE Healthcare, Waukesha, USA) using 10× SSC as the transfer buffer. The RNA was cross-linked to the membrane by UV irradiation. An *AtUSPL1* probe was amplified from *A.thaliana* genomic DNA using primer pair USPa/b (sequences given in Table S1) and labeled with α-^{32}P dCTP via random priming (Ready Prime Labeling, Pharmacia, GE Healthcare, Waukesha, USA). Membrane/probe hybridizations were carried out in Church hybridization solution [42] at 65°C with pre-hybridization for 6 h and hybridization for 16 h. The membrane was washed for 15 min twice each in 2xSSC, 0.1% SDS, 0.5xSSC, 0.1% SDS and 0.1xSSC, 0.1% SDS at 65°C. Signals were detected and quantified with a Bio-Imaging analyser BAS2000 or X-ray film (Fuji Photo Film Co. Ltd., Tokyo, J). Quantitative real time PCRs (qRT-PCR) were run following [43], in order to assess transcript abundances in leaf and root tissue of six week old plants (five replicate RNA extractions per biological sample, and three technical replications per RNA extract). All primer sequences are given in Table S1.

RNA Isolation, target synthesis and microarray hybridization

Total RNA was isolated from the aerial tissue of two week old seedlings using the TRIzol reagent (Invitrogen) and RNAeasy columns (Qiagen) (see Figure S7 for detailed information). The integrity of the RNA was monitored by agarose gel electrophoresis, and its concentration estimated using a NanoDrop device (Peqlab, Erlangen, Germany), following the manufacturer's protocol. The integrity of the RNA was further confirmed using an Agilent 2100 Bioanalyzer in conjunction with the RNA 6000 Nano assay (Agilent Technologies, Böblingen, Germany). RNA was processed for use on an Affymetrix Arabidopsis ATH1 Genome Array (ATLAS Niolabs GmbH, Berlin, Germany).

Microarray data analysis

The ATH1 chip, which assays 22,392 unique genes, was used to contrast the transcriptomes of wild type Col-0, *rd22* and *uspl1* plants grown under both control and moisture stressed conditions. Probe set to target gene mappings were taken from the TAIR Web site: ftp://ftp.arabidopsis.org/home/tair/Microarrays/Affymetrix/affy_ATH1 _array_elements2010-12-20.txt. The microarray experiment and basic data interpretation was performed by ATLAS biolabs GmbH (Berlin, Germany) by GeneChip Operating System (GCOS) 1.4. To ensure reliability of the analyses, each GeneChip experiment was performed with two biological replicates. After logarithmic transformation of the data, the average expression for all experimental samples for this probe set was subtracted from each individual expression value, thus leading to a positive value in case of above-average expression levels and a negative value in case of below-average expression levels. GeneSpring GX software (Agilent) was used to gene-wise normalize the expression data. To identify potentially differentially expressed genes, the fold changes >2 and <2 of expression values were identified. This was done separately for the Col-0 and mutant series at different conditions. For the Col-0 and respective mutant sample, a simple moderated t-test was performed and P values were corrected using the Benjamini and Hochberg [44] false discovery rate control, applying standard limma procedures. Differentially expressed genes, between Col-0 and mutant samples were identified for both conditions using the limma nestedF procedure, applying a significance threshold of 0.5 in combination with Benjamini-Hochberg false-discovery rate control and a minimal log2-fold change value of 2. A functional categorization of the differentially transcribed genes was derived using Mapman software [45].

Results

The *BURP* gene family

The *A. thaliana* genome contains five *BURP* family genes; these include both *AtUSPL1 (At1g49320)* and *AtRD22 (At5g25610)* (Figure 1A), but also three genes encoding proteins sharing similarity with the tomato non-catalytic β-subunit of polygalacturonase. The latter have been proposed to be designated as *AtPG1* (*At1g60390*), *AtPG2* (*At1g70370*) and *AtPG3* (*At1g23760*). Based on an alignment of related sequences extracted from various species, the family can be subdivided into eight sub-families [23], according to which *AtRD22* belongs to the ATRD22-like subgroup, *AtUSPL1* to the *BNM2*-like sub-family, and *AtPG1-3* to the PG1β-like subfamily. The *A. thaliana* genome has no representative of either the V-VIII or the VfUSP like sub-groups. The N-terminal regions of *AtPG1-3* each include sequences encoding 21 FXXY–N_{9-11} repeats (of unknown function) [24]. AtRD22 contains four TXV-repeats, while AtUSPL1 has no repetitive features. Based on the domain structure and phenotype this study is restricted to the functional analysis of *AtRD22* and *AtUSPL1*.

Transcription of *AtRD22* and *AtUSPL1* during development and in response to stress

Some *Arabidopsis BURP* genes were described to be preferentially expressed in early embryogenesis [24], their involvement in stress response was less obvious. An exception of this is *AtRD22*, which was shown previously to be induced under drought treatments [8].

The transcription profiles of *AtRD22* and *AtUSPL1* were recovered from the Genevestigator database ([46] Figure S1A) and verified using qRT-PCR and Northern blotting. Archival microarray data suggested that while *AtRD22* transcript is abundant throughout plant development in the aerial part of the plant, that of *AtUSPL1* is low and is restricted to the root. In non-stressed plants, *AtRD22* transcription is highest in the leaf, and particularly so in the guard cells [46]. *AtUSPL1* transcript is most abundant in the root (Figure 1B, Figure S1B), but is also detectable in the aerial part of the plant early in development. In the root, *AtUSPL1* transcription is stimulated by exposure to either mannitol or NaCl, as is that of *AtRD22* in the leaf [47]. Here, moisture stress up-regulated *AtRD22* transcription was detected, particularly in the leaf. *AtUSPL1* transcript was detectable in the root, and its level was enhanced by the imposition of moisture stress (Figure 1B). *AtUSPL1* expression was also assayed by tracking GUS expression produced by the *pAtUSPL1::GUS* transgene. GUS activity was detected in the young leaf, in the hypocotyl and in the stem (Figure 2A). In the silique, its expression was only detectable in the mature seed funiculum (Figure 2B), while in the developing flower and stem, a low level of expression was observed (Figure 2C). GUS activity was especially strong in the root tip (Figure 2D). Compared to the strong *AtRD22* promoter activity in aerial parts of the plant [26], *ProAtUSPL1* shows strong transcriptional activity in the root tissue.

According to *in silico* data, treatment with ABA strongly induces *AtRD22* in the leaf, and *AtUSPL1* responds similarly in the root. *AtRD22* is also inducible by exposure to mannitol, glucose, nitrate, high levels of illumination, high temperature and salinity (Figure S2). The gene is down-regulated when the plant is treated with paclobutrazol, an inhibitor of gibberellin synthesis, as well as with cycloheximide, a general inhibitor of protein synthesis, and with syringolin, an inhibitor of cell proliferation.

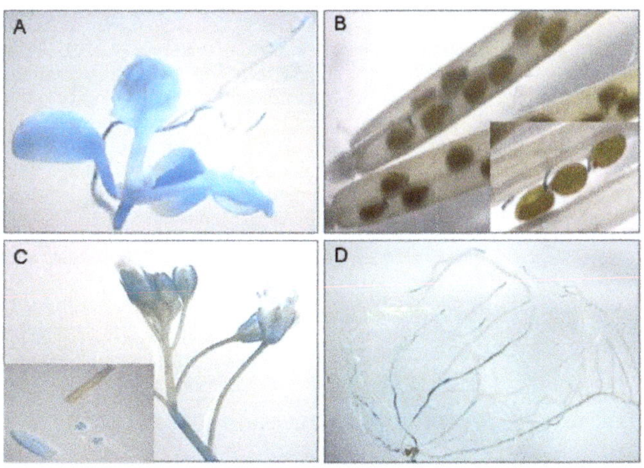

Figure 2. Histochemical *ProAtUSPL1::GUS* activity in transgenic Arabidopsis plants. The *AtUSPL1* promoter activity was determined by histochemical localisation of GUS activity derived from the transgenic *ProATUSPL1::GUS* reporter gene. Activity indicated by blue colour can be seen in A) seedling; B) in funiculus of mature seeds; C) in flowers and stems; and D) in roots.

The behavior of *AtRD22* and *AtUSPL1* T-DNA insertion mutants

In wild type plants, the expression of the two selected *BURP*-gene family members was confirmed by quantitative RT real-time PCR (Figure 1B) and microarray analysis of the aerial part and among them *AtRD22* transcript is abundant under stress treatments (Figure S3). The loss of function mutants achieved by T-DNA insertions were analysed for characterization of *AtRD22* and *AtUSPL1* genes in the functional context of drought stress tolerance (Figure S4A). For *AtRD22* two independent T-DNA insertion alleles were used: *rd22-1* and *rd22-2*. For *AtUSPL1* the mutant line *uspl1* was used. To analyse the functional loss of both members of the BURP-gene family the *rd22-1/uspl1* double mutant was analysed. In the two analysed T-DNA insertion lines *rd22-1* and *uspl1* no mRNA of the respective gene was detectable by semi-quantitative PCR and Northern analysis (Figure S4B), which is corroborated by microarray analysis (Figure S4C). Therefore we assume that the used T-DNA mutants represent a loss of function mutation of the respective gene.

Loss of AtRD22 and AtUSPL1 lead to enhanced drought tolerance

The response of the loss-of-function *AtRD22 and AtUSPL1* T-DNA insertion mutants to moisture stress is summarized in Figure 3 and Figure S4/S5. Two independent *AtRD22* (*rd22-1* and *rd22-2*) and one *AtUSPL1* (*uspl1*) mutants were analysed, along with the *rd22-1/uspl1* double mutant. In both *rd22-1* and *uspl1,* transcript of the mutated gene was not detected based on either semi-quantitative PCR or Northern blotting (Figure S4B), corroborating the prediction of microarray analysis (Figure S4C). When subjected to moisture stress for 2–3 days, wild type plants became discolored as a result of an accumulation of anthocyanin and their growth ceased, whereas both the single and double mutant plants remained green and showed no evidence of any growth retardation (Figure 3A). Exposure to a longer period of moisture stress discriminated between wild type and the mutants in a similar manner (Figure S5A).

Detection of plant growth via automated phenotyping

To quantify the drought response of the mutant plants an automated phenotyping platform (LemnaTec) was used. The growth of 30 plants of control (Col-0), *rd22-1*, *rd22-2*, *uspl1* and *rd22-1/uspl1* each under defined control and drought stress condition was analysed. Drought stress was applied from 21 DAS by complete withdrawal of water. To monitor the growth of the plants, leaf area was estimated from top view images ([39,48], Figure S5B, Table S2) and used for calculation of relative growth rates [38]. An elevated growth rate of the *rd22-1* single mutant plant as well as for the double mutant plants compared to wild type could be found in the early phase of the experiment (22 DAS, Figure 3B). The first significant drought related difference in growth rate was detectable after 7 days without watering (28 DAS, Figure S5B). A significant difference in the growth rate of the *rd22/uspl1* double mutant plants compared to the wild type plants was detectable almost throughout the entire experiment initially from day 24 till day 32 after sowing under stress conditions (Figure 3B, Figure S5B).

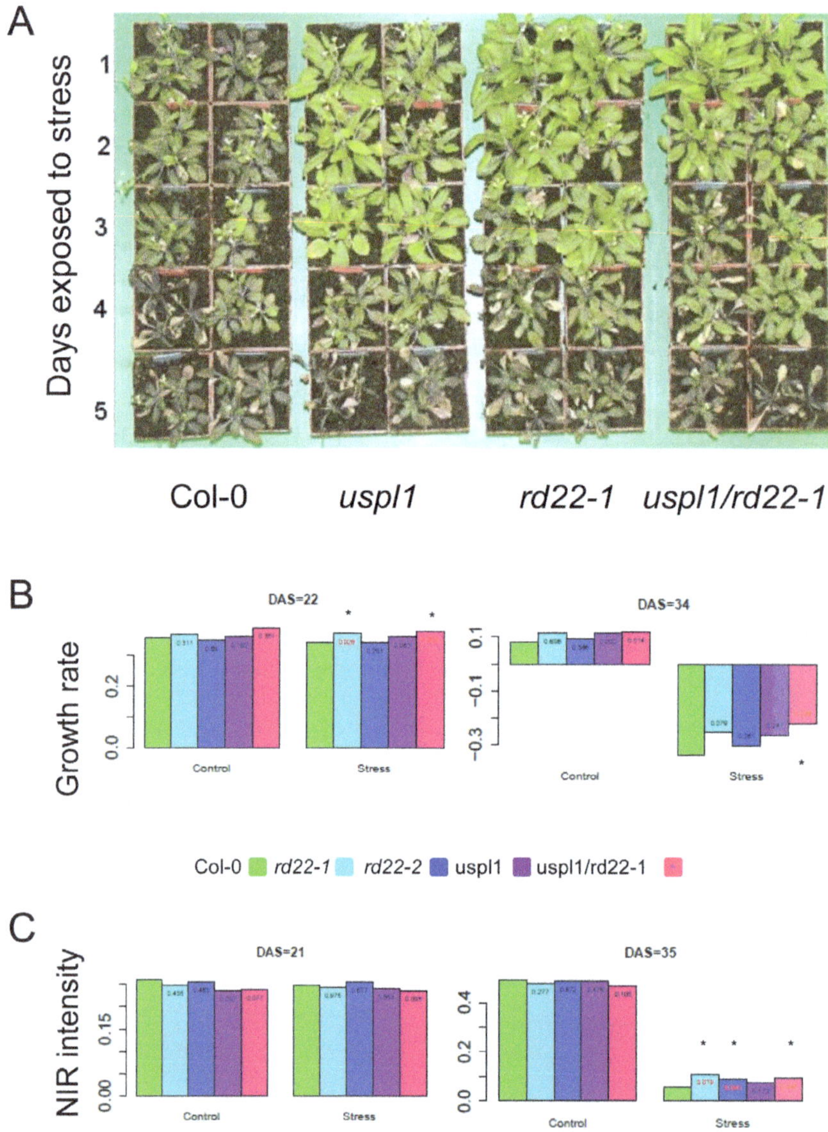

Figure 3. Influence of drought stress treatment on single and double loss of function mutants. A) Four weeks old wild type (Col-0), single and double mutant plants (*rd22-1* and *uspl1, rd22-1/uspl1*) were drought stressed for 1–5 days before they were returned to climate chamber conditions. Time of stress treatment in days is indicated left. B) Growth rates of plants under control and drought stress conditions. Bars indicate the growth rates at 22 days after sowing (DAS) for early phase of drought stress and 34 for the late phase of drought stress. For application of drought stress stop of watering started at 21 DAS. Wild type (Col-0): green bar; *rd22-1*: bright blue bar; *rd22-2*: dark blue bar; *uspl1*: purple bar; *rd22-1/uspl1*pink bar. Original data: Figure S5 B, Table S2, Asterisks indicate significant differences ($p<0.01$). C) NIR reflection as a water content-related parameter. Bars indicate the NIR intensity at 21 days after sowing (DAS) for start of experiment and at 35 DAS for the end of experiment. Wild type (Col-0): green bar; *rd22-1*: bright blue bar; *rd22-2*: dark blue bar; *uspl1*: purple bar; *rd22-1/uspl1*pink bar. $N_{control}=5$, $N_{stress}=10$ plants. Original data in Table S3.

Plant senescence as consequence of the drought stress was monitored by quantifying the ratio of yellow to green pixels. Leaves of the mutant plants *rd22-1*, *uspl1* and the double mutant showed a lower accumulation of yellow stained material, indicative for reduced senescence (33 DAS, Figure S5C). In order to analyse the relative water status of the plants the reflected near-infrared (NIR) radiation (1450 nm) from leaves was detected [49]. NIR intensity is calculated as 1 – NIR reflectance and illustrates the relation to the water content of the leaves. The drought stress related decrease of the NIR intensity was detectable at 33 DAS (Table S3). All genotypes analysed showed a similar level of detectable NIR reflectance in the beginning of the experiment and continuously under control conditions. At the end of the drought stress exposure *rd22* and *rd22/uspl1* double mutant plants showed a higher NIR intensity, indicating higher water content in the leaves (Figure 3C).

Plant growth response to salinity and osmotic stress and exposure to ABA

To further investigate the role of *AtRD22* and *AtUSPL1* during salinity stress and correlating responses on the transcriptional level, wild type and mutant plants were grown on plates with the respective treatments. Two-weeks-old seedlings were transferred for three days to MS basal medium containing one of the following stress inducing compounds: 150 mM and 300 mM NaCl, 100 μM ABA, 4% trehalose, 4% sorbitol, 4% glucose, 4% fructose, 4%

sucrose, 300 mM manitol, 15% PEG 6000 and 4% PEG 20000. Although these conditions are artificially mimicking drought, they were chosen to achieve a uniform plant response to the stimulus. When exposed to either 150 mM NaCl or 100 μM ABA, only minor signs of stress symptoms were apparent, but in the former case, the growth of both the wild type and single mutant plants was retarded (Figure S6A). Wild type and *rd22-1* plants exhibited the least extent of leaf bleaching, while the double mutant and particularly the *uspl1* single mutant, were more visibly affected. Exposure to the various sugars had only a mild effect on plant growth, but a general tendency was for the mutant plants to be more vigorous than the wild type ones. Therefore NaCl 150 mM conditions were chosen for further transcriptome and chlorophyll analysis.

Chlorophyll and pheophytin content

After a four day exposure to moisture stress, the quantity of chlorophyll a and b in the *rd22-1* mutant and the double mutant leaves was less than that in non-stressed plants (Figure S6C). Since the contents of chlorophyll a and chlorophyll b were strongly correlated with one another, subsequent measurements considered the total chlorophyll content. Pheophytin is a degradation product of chlorophyll, that accumulates during senescence, dark [50] and salt stress [51]. In plants grown in the presence of 300 mM NaCl, the content of chlorophyll (Figure 4A) fell sharply, while that of pheophytin rose (Figure 4B). However when challenged by a lesser level of stress (100 μM ABA or 150 mM NaCl), while the chlorophyll content was reduced, there was no measurable increase in pheophytin content (Figure 4C). Chlorophyll content in the *rd22-1* and *uspl1* mutants was more severely reduced in the presence of either 150 mM NaCl or 100 μM ABA. The *uspl1* mutant was the most compromised genotype with respect to chlorophyll content when the medium was supplemented with fructose, while the *rd22-1* and the double mutant plants out-performed the wild type and *uspl1* mutant plants when the stress was imparted by PEG. To clearly indicate this different chlorophyll degradation in the *rd22-1* single and double mutant plants the amount [%] of total clorophyll is indicated relative to the amount of clorophyll at control conditions (Figure 4C). Compared to the amount of chlorophyll under control conditions the *rd22-1* and *uspl1* mutant plants show the strongest reduction upon 150 mM NaCl and 100 μM ABA. In addition, the reduction of chlorophyll in the *uspl1* mutant with fructose show a major difference compared to the other genotypes. In the *rd22-1* and *rd22-1/uspl1* mutant plants the amount of chlorophyll is higher on fructose, 150 mM NaCl and PEG supplemented media compared to the wild type and *uspl1* mutant plants. Taken into account that the chlorophyll content in the mutant in the unstressed condition varies from the wild type a relative value is indicated to show the change due to the treatment (Figure S6B).

The conclusion was that AtRD22 acts to suppress chlorophyll degradation under moisture stress.

Transcriptome analysis

The transcriptional responses to exposure to 150 mM NaCl and to 4% w/v trehalose overlapped by about 10% (Figure S6A). The 150 mM NaCl treatment resulted in a changed transcript abundance for 913 of the 22,392 genes represented on the ATH1 chip. Salinity stress and both the trehalose and sorbitol treatments marginally up-regulated *AtRD22*, while the other *BURP* family genes were hardly affected (Figure S6D). The four genes showing the greatest change in transcript abundance between wild type and mutant plants grown under non-stressed conditions (Table 1) were *At4g33720* (encoding CAP, a PR protein under the control of the DREB2A tanscription factor, [52]: up-regulated by 12 fold in both *rd22-1* and *uspl1* compared to wild type); *At5g49700* (encoding a DNA-binding protein associated with moisture stress [53]: up-regulated by 56 fold in *rd22-1* and by 40 fold in *uspl1*); *At2g27550* (encoding ATC, a systemic inhibitor of floral initiation [54]: up-regulated by 23 fold in both *rd22-1* and *uspl1*); and *At2g44380* (encoding a DC1-domain containing protein, involved in the abiotic stress response, [55]: up-regulated by 22 fold in both *rd22-1* and *uspl1*). The full set of differentially transcribed genes detected in response to the various treatments is given in Table S4 (see also Figure S7). A total of 77 genes displayed differential transcription between wild type and *rd22-1* plants grown under non-stressed conditions; 31 out of these are associated with either biotic or abiotic regulatory pathways (Figure S7A) and included eight encoding a peroxidase putatively involved in H_2O_2 degradation. The absence of AtUSPL1 only re-programmed 18 genes, of which five (one encoding a peroxidase) are associated with either biotic or abiotic regulatory pathways (Figure S7B). Exposure to 150 mM NaCl resulted in the up-regulation of a number of genes associated with either moisture stress or pathogen defense (Table 2). In *rd22-1*, out of the 764 differentially transcribed genes, 231 fell into one of these two categories (Figure S7C), while in the *uspl1* mutant, only 18 genes were up-regulated, of which five were associated with either the biotic or the abiotic stress response (Figure S7D). In the latter mutant, *AtRD22* was strongly up-regulated in the presence of salinity stress. When the stress was applied by the addition of trehalose to the medium, 55 of the 171 differentially transcribed genes identified in the *rd22-1* mutant were associated with either the abiotic or the biotic stress response (Figure S7E). Over 50% (111 out of 212) of the genes responding to trehalose treatment reacted similarly to salinity stress (Figure S3A, Table S4). The set of common up- regulated genes included the transcription factor genes *AtMYB15* (*At3g23250*) and *WRKY33* (*At2g38470*), *LPT3* and *LPT4* (encoding lipid transporters), *At1g35910* (a putative trehalose-6-phosphate phosphatase), *At1g61120* (terpene synthase 04), *At1g16850* (unknown function) and *At1g78410* (VQ motif-containing protein). Although fewer genes were induced by the trehalose (212) than by the 150 mM NaCl (913) treatment, the conclusion was that the two stress agents must affect a similar class of gene. The response to both stress agents also included the down-regulation of several photosynthesis-related genes.

Discussion

RD22 and USPL1 have suppressor function during drought stress

AtRD22 and *AtUSPL1* were transcribed predominantly in, respectively, the leaf and the root. The induction of both genes by moisture stress generated a slight growth reduction. While the induction by moisture stress of *AtRD22* has been noted previously, this was not the case for *AtUSPL1*, most likely due to a concentration on the short-term stress response. Microarray-based transcriptomic analyses have shown that exposure to either NaCl or trehalose induces not only the up-regulation of salinity, trehalose and moisture stress responsive genes, but also the down-regulation of photosynthesis-related genes. The latter is implied by both the induction of leaf chlorosis and senescence and the measured fall in the chlorophyll content of the leaves of stressed plants. The present data confirmed that both salinity and trehalose activate the ABA-mediated moisture stress response. [56–58]. The sole *BURP* gene family to be up- regulated in the aerial part of the plant was *AtRD22,* consistent with observations based on the expression of the transgene *pAtRD22::GUS* [26].

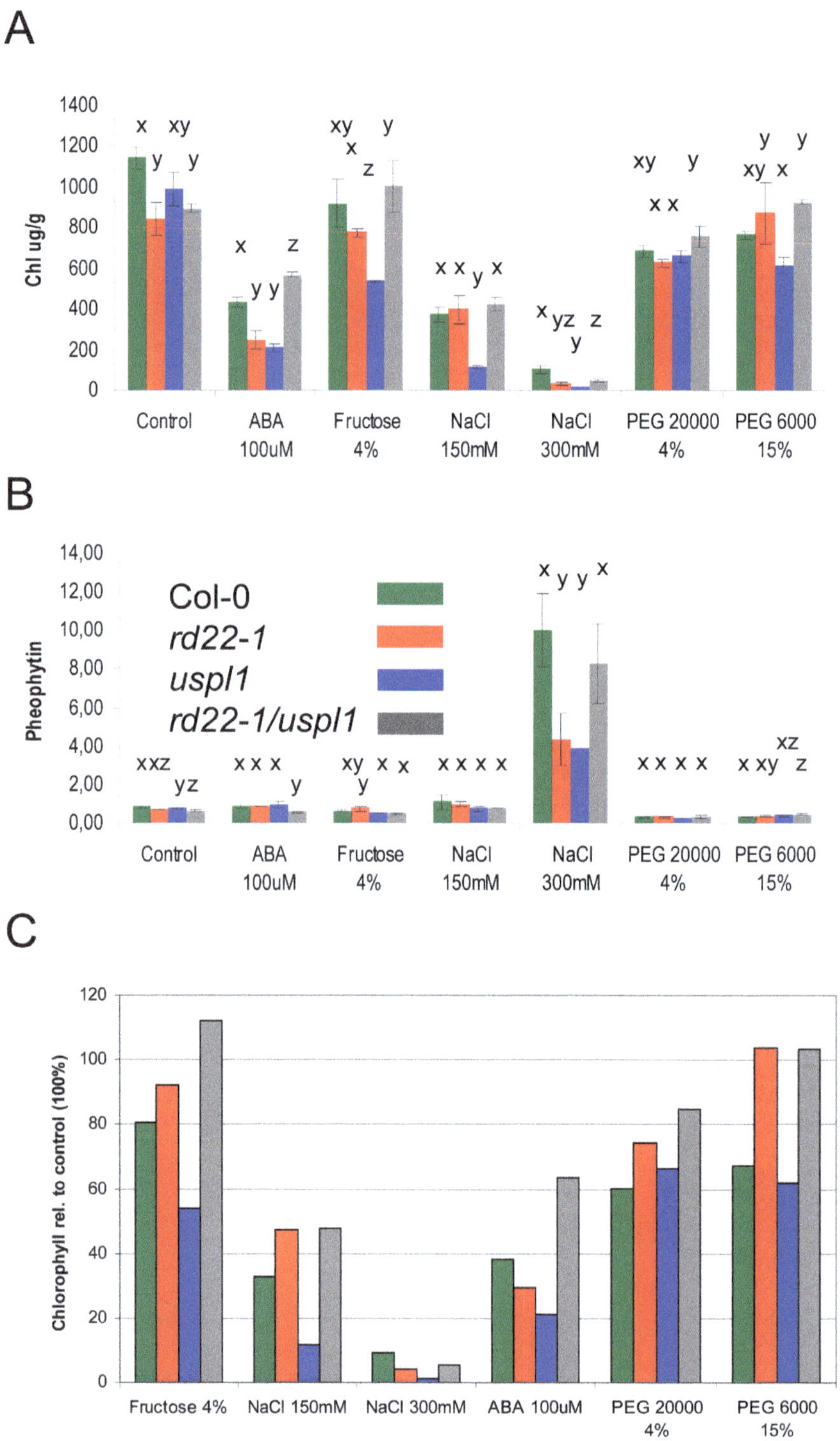

Figure 4. Chlorophyll and pheophytin content in single and double *BURP* mutants. The bars represent the total A) chlorophyll and B) pheophytin content in leaves from wild type (Col-0, green bar), single and double mutant plants (*rd22-1*, red bar, *uspl1*, blue bar and *rd22-1/uspl1*, grey bar). Chlorophyll a and b was determined separately (Figure S6A) and subsumed as total chlorophyll content. Error bar represents standard error. N = 5–6 plants in duplicate. Statistical analysis was performed by oneway ANOVA at alpa = 0.05 Tukey post hoc test: same letters indicate no difference, different letters indicate significant difference. C) The bars show the total chlorophyll content [%] relative to unstressed control plants.

The root specificity of *AtUSPL1* expression was confirmed both by the behavior of the *pAtUSPL1::GUS* transgene and by Northern hybridization experiments.

The T-DNA insertion mutants of both *AtRD22* and *AtUSPL1* produced no detectable relevant transcript, so each was taken as a genuine loss-of- function allele. In both cases, a reduced photosynthesis phenotype was exhibited, as reflected by a fall in

Table 1. Top differential expressed genes in A) *rd22-1* and B) *uspl1* at standard growth conditions in the aerial part of 2 week old seedlings.

A) *rd22*

Probe Set ID	fold change	Gene	AGI ID	put. Function	Probe Set ID	fold change	Gene	AGI ID	put. Function
257162_s_at	59		At3G24290	amm. transporter	265364_at	−45		At2G13330	Transposon
248564_at	56		At5G49700	DNA binding	246908_at	−40	RD22	At5G25610	Disease Defense
249052_at	47	PDF1.2	At5G44420	Disease Defense	251677_at	−38	ORG3	At3G56980	DNA bind
245276_at	41	ATHB-2	At4G16780		257453_at	−30		At1G65130	hydrolase
265102_at	39		At1G30870	Peroxidase	261684_at	−29		At1G47400	
259276_at	38		At3G01190	PER27 Peroxidase	265665_at	−28		At2G27420	
263096_at	38	AHB1	At2G16060		249645_at	−26	THI2.2	At5G36910	toxin receptor
264567_s_at	37		At1G05250	Peroxidase	266472_at	−26			
254044_at	35	XTR9	At4G25820	hydrolase	254170_at	−22		At4G24430	lyase
254200_at	34		At4G24110		247252_at	−22		At5G64770	

B) *uspl1*

Probe Set ID	fold change	Gene	AGI ID	put. Function	Probe Set ID	fold change	Gene	AGI ID	put. Function
247718_at	42	LTP4	At5G59310	Lipid transfer	252795_at	−38			
248564_at	40		At5G49700	DNA bind	246948_at	−23	CYP71B12	At5G25130	Cytochr. P450
261226_at	31	ATEXPA11	At1G20190	Expansin	250207_at	−22	CHS	At5G13930	
253667_at	30		At4G30170	peroxidase	256940_at	−19		At3G30720	
252374_at	24	ARR5	At3G48100	transc. Reg.	267168_at	−2		At2G37770	aldo/keto red.
254979_at	23		At4G10550	subtilase					
254832_at	23		At4G12490	protinase inhibitor					
266209_at	23	ATC	At2G27550	A.th centroradialis					
267385_at	22		At2G44380	DC domain					
247717_at	22	LTP3	At5G59320	Lipid transfer					

Complete list of differential expressed genes in Table S4.

Table 2. Differential expressed genes in A) *rd22-1* and B) *uspl1* on medium containing 150mM NaCl in the aerial part of 2 week old seedlings.

A) *rd22* on NaCl		
up regulated genes		
Probe Set ID	**Fold change**	**Gene Symbol**
247215_at	40	**PROPEP3**
266017_at	38	
252984_at	34	**ELI3-2**
252487_at	33	
256627_at	33	
246340_s_at	32	**FAMT**
264005_at	31	AGP2
257206_at	29	
267035_at	26	**AGT3**
259975_at	25	
253872_at	24	RD26
254101_at	24	**AMY1**
266267_at	23	**ATGSTU4**
256245_at	22	**HSP70**
255502_at	21	
247308_at	20	
256436_at	20	
266142_at	11	
256603_at	11	
258791_at	10	**PR4**
264415_at	6	**RAP2.6**
257517_at	4	
258277_at	4	**PAD3**
260919_at	3	
249481_at	2	
down regulated		
Probe Set ID	**Fold change**	**Gene Symbol**
246366_at	−27	
256772_at	−22	**BGAL1**
250366_at	−21	
261118_at	−20	
261684_at	−20	
267264_at	−17	**SCPL11**
258497_at	−11	**COL2**
261413_at	−11	**PLL5**
246908_at	−7	RD22
247450_at	−2	
261351_at	−2	
266363_at	−1	
B) *uspl1* on NaCl		
up regulated genes		
Probe Set ID	Fold change	Gene Symbol
261684_at	42	
258497_at	21	**COL2**
266363_at	20	
247308_at	18	

Table 2. Cont.

B) ***uspl1*** **on NaCl**		
247215_at	15	**PROPEP3**
257206_at	14	
249481_at	13	
261118_at	12	
246908_at	11	RD22
256245_at	11	**HSP70**
261351_at	11	
255502_at	11	
250366_at	11	
264415_at	10	**RAP2.6**
266267_at	10	**ATGSTU4**
256225_at	10	
256603_at	4	
258277_at	2	**PAD3**
266017_at	2	
256627_at	1	
267035_at	1	**AGT3**
down regulated		
Probe Set ID	**Fold change**	**Gene Symbol**
246366_at	−27	
266142_at	−24	
258791_at	−23	**PR4**
267264_at	−20	**SCPL11**
254101_at	−17	**AMY1**
259975_at	−15	
246340_s_at	−15	**FAMT**
256436_at	−15	
256772_at	−14	**BGAL1**
257517_at	−14	
260919_at	−13	
252487_at	−13	
252984_at	−11	**ELI3-2**
264005_at	−11	AGP2
247450_at	−11	
253872_at	−11	RD26
261413_at	−7	**PLL5**

both chlorophyll content under non-stressed growing conditions and photosynthesis-associated gene transcript abundance in stressed plants. Both under prolonged moisture stress mimicking conditions and actual moisture stress, stress tolerance was enhanced in all of the mutants. The mutants' performance (and particularly that of the double mutant), as measured by either the development of leaf area or by the rate of leaf tissue senescence, showed that they were more tolerant of moisture stress than were wild type plants. NIR intensity appeared to reliably reflect relative water content, which was higher in each of the mutants than in wild type.

Given that photosynthesis is clearly compromised by moisture stress, it was not surprising that the transcription of several photosynthesis-associated genes was altered when the plants were exposed to moisture stress. Although the mutant plants' transcriptomes were indistinguishable from that of the wild type with respect to photosynthesis-related genes, nevertheless their leaves contained less chlorophyll than the leaves of wild type plants raised under non-stressed conditions. The reduction in chlorophyll content in the mutants induced by exposure to either NaCl or PEG was less severe than that in wild type plants. Notably, under moisture stress conditions, the leaves of double mutant plants retained more chlorophyll than those of wild type ones. The retention of chlorophyll can be expected to support a higher rate of photosynthesis, so that less transpiration is required to generate a given quantity of assimilate. Since the plant's capacity to retain its water is improved, its water content under moisture stress conditions was greater (Figure 3C).

The observed up-regulation of peroxidase encoding genes in the mutants suggested a secondary effect of the loss-of-function mutations. The transcriptomic data implied that the oxidative state within the mutant plants differed from that within the wild type, with a knock-on effect on gene expression in both ABA-dependent and ABA-independent pathways [59]. Both the phenotype of the mutants as well as the transcriptional response of both genes suggested that under moisture stress conditions, their products exerted a suppressor function. Similar conclusions have been drawn regarding the effect of mutations of genes encoding SnRK2 [60] the effect of which is an almost complete abolition of the ABA response. In particular, *AtRD22* is not transcribed in these mutants when the growing medium is supplemented with ABA reflecting moisture stress condition. Under normal growing conditions, the *SnRK2* mutants exhibit reduced growth, which (along with plant survival) is improved by the addition of ABA to the growing medium. The inference is that the BURP domain containing proteins AtRD22 and AtUSPL1 act as suppressors of the ABA-mediated moisture stress response.

RD22 act as suppressor predominantly in the leaf, while USPL1 act in the root

Although both *AtRD22* and *AtUSPL1* exhibited organ-specific transcription, the loss of function of each gene induced a comparable improvement in the plant's tolerance to moisture stress, at least at the level of the vigour of the leaves. The sensitivity of the microarray platform rules out any trans-silencing effects of either T-DNA insertion on other members of the BURP gene family. Since this sort of silencing has been associated with several T-DNA insertion mutations [61], the assumption is that both *AtRD22* and *AtUSP1* are involved in a holistic moisture stress response. The microarray analysis demonstrated that the two loss-of-function mutations resulted in the induction of a partially overlapping set of genes in the aerial part of the plant, even though *AtUSPL1* was transcribed specifically in the root. Not only was the tolerance of moisture stress enhanced in both mutants, but also their chlorophyll content was reduced. As the lack of *AtRD22* and *AtUSP1* led to an enhanced tolerance to moisture stress, the proposed suppressor function of the BURP domain-containing proteins during an episode of moisture stress includes an organ-specific component. Both proteins are part of an ABA mediated moisture stress response pathway, with AtRD22 functioning mainly in the aerial part of the plant and AtUSPL1 in the root.

Increased drought resistance might be correlated to an increased defence gene response

The transcriptomic analysis of the two mutants revealed an increased transcript abundance compared to the wild type levels with respect to various genes associated with the response to biotic stress (for example, the gene *PDF1.2* was up-regulated by nearly 50 fold). This class of genes was differentially transcribed both under non-stressed and moisture stressed conditions. Several peroxidase encoding genes were also up-regulated. Peroxidases are known to represent an integral component of the plant's hypersensitive response [62]. H_2O_2 is also known as signal molecule during the stress response [63,64]. A loss of control over the production of H_2O_2 might result in the described reduction of chlorophyll content [65,66] and increased water content under drought stress conditions. Such regulation would subsequently lead to better performance of the mutant plants under drought stress.

The up-regulation of peroxidase encoding genes implies a level of linkage between the ABA-mediated moisture stress response and defense against pathogen invasion. Such a connection has been proposed in a suggested model for the function of OCP3, a homeodomain transcription factor. The loss of *OCP3* function results in an enhanced tolerance to moisture stress and at the same time an increased sensitivity to ABA. The abundance of *AtRD22* transcript in the *ocp3* mutant is not different to that in the wild type, and the plant's susceptibility to *Botrytis cinerea* infection is reduced [67]. The *ocp3* mutant was initially identified and named after the phenotype of constitutive over expression of a cationic peroxidase [68]. Here, *AtOCP3* was marginally down regulated when the plants were challenged by NaCl, but there was no transcriptional difference between wild type and either of the two mutants. The implication is that the BURP-containing proteins act to enhance the plant's moisture stress response via the up-regulation of peroxidase encoding genes. In soybean, a direct interaction between RD22 and cell wall-localized peroxidases has been described [69]. The up- regulation of H_2O_2 detoxifying enzymes enhances moisture stress tolerance [70] as well as explaining the link with the pathogen defense response.

A reduced chlorophyll content, in conjunction with an elevated level of transcription of defense response genes under non-stressed growing conditions, suggests that the mutant plants are primed to mount a stronger and/or more rapid set of measures to prevent the accumulation of ROS. A change in the accumulation of H_2O_2 would not only have an impact on the defense response [71], but also on the response to moisture stress [63,72]. Due to the overall elevated expression of H_2O_2-detoxifying enzymes such responses could be limited, leading to an enhanced drought resistance. Our study provides the first functional approach to investigate *BURP*-domain encoding gene function in addition to the previously published structural comparisons.

Supporting Information

Figure S1 Expression profile of the *Arabidopsis thaliana BURP* gene family. A) Expression profile of the *Arabidopsis thaliana BURP* gene family. Data obtained from Genevestigator database (Zimmerman *et al.*, 2004). Relative expression of *AtRD22* (red) and *AtUSPL1* (blue) is given for the different developmental stages of Arabidopsis life cycle (left to right: germinating seed, seedling, Young rosette, developed rosette, bolting, young flower, developed flower, flowers and siliques, mature siliques, senescence). B) Expression of *AtUSPL1* confirmed by Northern Blot analysis. Expression of *AtUSPL1* was determined from root, leaf, shoot, young silique and total flower tissue of *Arabidopsis thaliana* (*Col-0*) plants. For equal loading of the RNA samples probing of the membrane with specific probe against housekeeping mRNA of *AtROC1* (rotamase cyclophilin, renamed in *AtCYP1*) was performed. C) Expression of *AtRD22* obtained from Arabidopsis eFP Browser (Winter *et al.*, 2007). The expression of selected stimuli (Cold: 4°C, Osmotic: 300 mM Mannitol, Salt: 150 mM NaCl and Drought: air steam 15 min) is displayed for the aerial as well as the hypogeic part of the plant. *AtRD22* expression is induced in the aerial part of the plant after applying osmotic, salt stress and slightly increased after loss of water due to airstream treatment. D) Expression of *AtUSPL1* obtained from Arabidopsis eFP Browser (Winter *et al.*, 2007). The expression of selected stimuli (Cold: 4°C, Osmotic: 300 mM mannitol, Salt: 150 mM NaCl and Drought: air steam 15 min) is displayed for the aerial as well as the hypogeic part of the plant. *AtUSPL1* expression is induced in the hypogeic part of the plant after applying osmotic, salt stress and slightly increased after loss of water due to airstream treatment.

Figure S2 Expression of the BURP domain containing gene family in *Arabidopsis thaliana*. Expression analysis of *AtRD22* (*At5G25610*) and *AtUSPL1* (*At1G49320*) obtained from Genevestigator database (Zimmerman et al., 2004) displaying induced and reduced expression after different conditions and stresses. Displayed are only changes in expression upon stress/ treatment above threefold with a statistic significance $p<0.001$. Red indicates up-regulation; Green indicates down-regulation.

Figure S3 Comparison of gene expression in Arabidopsis wild type plants grown on 150 mM NaCl and 4% trehalose supplemented MS medium. BURP gene family mRNA in *Col-0* under selected stress conditions. Bars indicate the expression pattern obtained by microarray analysis using ATH1 chip: *AtRD22* (red): 246908_at; *AtUSPL1* (blue): 262388_at; *AtPG1* (dark green): 265131_at; *AtPG2* (green): 264277_at; *AtPG3* (bright green): 264315_at. Displayed is the rel. Abundance of mRNA.

Figure S4 Characterization of BURP mutant plants. A) Scheme of the *AtRD22* and *AtUSPL1* gene model. In blue the exon-intron structure within the coding region of the respective gene is given. The protein structure refers to Figure 1. In the encoded protein parts are given in aminoacids [aa]. And the size of the fulllenght protein is given below the gene description. Position of T-DNA insertions of used mutant plants are indicated by black lines. Mutant alleles for *rd22*: *rd22-1* (SALK_146066) and *rd22-2* (WiscDsLox481-484P12). Mutant alleles for *uspl1*: *uspl1*: (SALK_022325). The T-DNA insertion line SALK_146066 (*rd22-1*) is based on pROK2 conferring kanamycin resistance and the WiscDsLox481-484P12 (*rd22-2*) is based on pWiscDs-Lox conferring phosphinotricin (BASTA) resistance. *uspl1* T-DNA insertion lines SALK_022325 (referred to as *uspl1*, based on pROK2 conferring kanamycin resistance from Nottingham Arabidopsis Stock Centre) was analyzed. The position of the T-DNA insertion in *At1G49320 (AtUSPL1)* was determined by PCR and subsequent sequencing. The position of the T-DNA insertion are depicted and confirmed by PCR. Double mutant *rd22-1/uspl1* line was generated by crossing SALK_146066 and SALK_022325 and identified in the F_3 generation by PCR. Kanamycin and phosphinotricin resistance of the plants was tested on germination medium (1 MS salts; 10 g/l sucrose) plates with 40 mg/l kanamycin or 20 mg/l glufosinate-ammonium under long day conditions. B) Analysis of used T-DNA insertion mutants. The absence of *AtRD22* and *AtUSPL1* mRNA in homozygous *rd22* and *uspl1* mutant plants was determined by Northern Blot analysis (left) and semi quantitative RT-PCR. C) Analysis of BURP-gene family mRNA in *rd22-1* and *uspl1* mutants by microarray analysis on MS medium. Bars indicate the rel. expression signal obtained by microarray analysis using ATH1 chip from each single experiment: *AtRD22* (red): 246908_at; *AtUSPL1* (blue): 262388_at; *AtPG1* (dark green): 265131_at; *AtPG2* (green): 264277_at; *AtPG3* (bright green): 264315_at.

Figure S5 Increased drought stress resistance of the *rd22* and *uspl1* mutant plants. A) The plants were drought stressed by withdrawal of water. Top: day 0 (80% RWC in the soil); Bottom: appearance of plants after 8 days without watering. B) Top: Projected area of wild type and mutant plants under control conditions and drought stress (dotted line) obtained by lemnatec phenotyping; Drought stress was started 21 days after sawing (DAS). Middle: Growth rates calculated based on Poorter and Lewis 1986 for individual days. Wild type (Col-0): green line; *rd22-1*: bright blue line; *rd22-2*: dark blue line; *uspl1*: purple line; *rd22-1/uspl1*pink line (+/- s.e.m.). Bottom: Statistical analysis or growth rates at 28 DAS. Asterisks indicate significant differences ($p<0.05$) between control and stress. C) Estimation of senescence after drought stress. The graph indicates the ration of yellow to green pixels in the plant area of the analysed top view images from day 33. Wild type (Col-0): green bar; *rd22-1*: bright blue bar; *rd22-2*: dark blue bar; *uspl1*: purple bar; *rd22-1/uspl1*: pink bar. $N_{control} = 5$, $N_{stress} =$ 10 plants. Asterisks indicate significant differences ($p<0.05$).

Figure S6 A) Differentially expressed genes (Col-0) between 150 mM NaCl containing medium and standard growth conditions sorted by relation to pathway (Mapman). Top: Differentially expressed genes (Col-0) between 150 mM NaCl containing medium and standard growth conditions sorted by relation to pathway (Mapman). The different numbers indicate different categories/pathways and are described in the table. Bottom: Differentially expressed genes (Col-0) between standard growth conditions and 4% trehalose containing medium. The bar diagram indicates the percentage of common (red) and inverse (yellow) regulated genes upon salt (grey) and sugar (blue) treatment. Approximately half of the genes induced by 4% trehalose treatment are also reacting on 150 mM NaCl treatment. B) Influence of ABA and NaCl on single and double loss of function mutants. Growth phenotypes of wild type (Col-0), single and double mutant plants (*rd22-1* and *uspl1, rd22-1/uspl1*) on standard MS-medium and 150 mM NaCl, 300 mM NaCl and 100 μM ABA supplemented MS-Medium. Two week old seedlings were transferred for 3 days to the MS basal and supplemented medium. C) Chlorophyll and pheophytin content of wild type and *rd22-1*, *uspl1* and *rd22-1/uspl1* mutant plants on different supplemented media. Two week old seedlings were transferred to the MS basal + one of the following stress treatments for 4 days: 100 μM ABA, 4% fructose, 150 mM NaCl, 300 mM NaCl, 4% PEG 20000 and 15% PEG 6000. From Top to bottom: Chlorophyll a and b content. The error bar represents standard error. Content was estimated from 5-6 plants in duplicate. Statistical analysis was performed by oneway ANOVA at alpa = 0.05 Tukey post hoc test: same letters indicate no difference, different letters indicate significant difference. Chlorophyll a and b and pheophytin content [%] relative to unstressed control plants.

Figure S7 Differentially expressed genes in the *rd22-1* and *uspl1* plants grown on control, 150 mM NaCl and 4% trehalose containing medium. List of used categories is given in Figure 3 A. Top: Display of top regulated category (Bin 20, biotic and abiotic stress pathways) of differential regulated genes. Bottom: Overview of all differentially regulated genes mapped to categories (Bins) by MAPMAN. For RNA extraction tissue was grinded in liquid nitrogen and the homogenized powder was added to 1 ml TRIZOL and incubated at RT for 5 min. Samples were centrifuged at 10.000 rpm for 10 min and the supernatant was transferred to a new tube. 200 μl of chloroform were added and incubated at room temperature for 2–3 min. Samples were again centrifuged as described above and the aqueous supernatant was transferred to the QIAshredder column and centrifuged for 30 s at 10000 rpm. 350 μl of RLT buffer (plus β-mercaptoethanol) and 250 μl of absolute ethanol were added to the flow-through and passed through an RNAeasy spin column. All the following steps were performed as described in the manufacturer's protocol followed by in-column DNAse digestion. A) Schematic display of differentially expressed genes in *rd22-1* by

MAPMAN. 31 out of 77 differential regulated genes are mapping to biotic and abiotic stress pathways. B) Schematic display of differentially expressed genes in *uspl1* by MAPMAN. 5 out of 18 differential regulated genes are mapping to biotic and abiotic stress pathways. C) Schematic display of differentially expressed genes in *rd22-1* on 150 mM NaCl by MAPMAN. 231 out of 764 differentially regulated genes are mapping to biotic and abiotic stress pathways. D) Schematic display of differentially expressed genes in *uspl1* on 150 mM NaCl by MAPMAN. 7 out of 12 differentially regulated genes are mapping to biotic and abiotic stress pathways. E) Schematic display of differentially expressed genes in *rd22* on 150 mM NaCl by MAPMAN. 55 out of 171 differentially regulated genes are mapping to biotic and abiotic stress pathways.

Table S1 Primer information.

Table S2 Average projected plant area (mm2) from top images.

Table S3 Average near-infrared intensity as observed from top images, obtained using a Nir 300 camera from VDS Vosskühler (now Allied Vision Technologies). High values indicate relative high water content.

Table S4 List of differentially expressed genes.

Acknowledgments

We thank Elke Liemann and Jana Lorenz for excellent technical assistance, Dijun Chen for support with statistical analysis of lemnatec data, Ingo Mücke for his support using the automated phenotyping system.

Author Contributions

Conceived and designed the experiments: NS HB MK TA. Performed the experiments: VTH CS LVS KWF. Analyzed the data: AJ CK. Wrote the paper: MK AJ TA.

References

1. Bray EA, Bailey-Serres J, Weretilnyk E (2000) Responses to abiotic stresses. Biochemistry and Molecular Biology of Plants, Rockville, MD, USA: American Society of Plant Physiologists. pp.1158–1203.
2. Sreenivasulu N, Sopory SK, Kishor PBK (2007) Deciphering the regulatory mechanisms of abiotic stress tolerance in plants by genomic approaches. Gene 388: 1–13.
3. Agarwal PK, Agarwal P, Reddy MK, Sopory SK (2006) Role of DREB transcription factors in abiotic and biotic stress tolerance in plants. Plant Cell Rep 25: 1263–1274.
4. Hasegawa PM, Bressan RA, Zhu JK, Bohnert HJ (2000) Plant Cellular and Molecular Responses to High Salinity. Annu Rev Plant Physiol Plant Mol Biol 51: 463–499.
5. Yamaguchi-Shinozaki K, Shinozaki K (2006) Transcriptional regulatory networks in cellular responses and tolerance to dehydration and cold stresses. Annu Rev Plant Biol 57: 781–803.
6. Kasuga M, Liu Q, Miura S, Yamaguchi-Shinozaki K, Shinozaki K (1999) Improving plant drought, salt, and freezing tolerance by gene transfer of a single stress-inducible transcription factor. Nature Biotechnology 17: 287–291.
7. Yamaguchi-Shinozaki K, Shinozaki K (1992) A novel Arabidopsis DNA binding protein contains the conserved motif of HMG-box proteins. Nucleic Acids Res 20: 6737.
8. Yamaguchi-Shinozaki K, Shinozaki K (1993) The plant hormone abscisic acid mediates the drought-induced expression but not the seed-specific expression of rd22, a gene responsive to dehydration stress in Arabidopsis thaliana. Mol Gen Genet 238: 17–25.
9. Hattori J, Boutilier KA, van Lookeren Campagne MM, Miki BL (1998) A conserved BURP domain defines a novel group of plant proteins with unusual primary structures. Mol Gen Genet 259: 424–428.
10. Baumlein H, Boerjan W, Nagy I, Bassuner R, Van Montagu M, et al. (1991) A novel seed protein gene from Vicia faba is developmentally regulated in transgenic tobacco and Arabidopsis plants. Mol Gen Genet 225: 459–467.
11. Wohlfarth T, Braun H, Kirik V, Kolle K, Czihal A, et al. (1998) Regulation and evolution of seed globulin genes. Journal of Plant Physiology 152: 600–606.
12. Boutilier KA, Gines MJ, DeMoor JM, Huang B, Baszczynski CL, et al. (1994) Expression of the BnmNAP subfamily of napin genes coincides with the induction of Brassica microspore embryogenesis. Plant Mol Biol 26: 1711–1723.
13. Teerawanichpan P, Xia Q, Caldwell SJ, Datla R, Selvaraj G (2009) Protein storage vacuoles of Brassica napus zygotic embryos accumulate a BURP domain protein and perturbation of its production distorts the PSV. Plant Molecular Biology 71: 331–343.
14. Treacy BK, Hattori J, Prudhomme I, Barbour E, Boutilier K, et al. (1997) Bnm1, a Brassica pollen-specific gene. Plant Molecular Biology 34: 603–611.
15. Chen LZ, Guan LM, Seo M, Hoffmann F, Adachi T (2005) Developmental expression of ASG-1 during gametogenesis in apomictic guinea grass (Panicum maximum). Journal of Plant Physiology 162: 1141–1148.
16. Wang AM, Xia Q, Xie WS, Datla R, Selvaraj G (2003) The classical Ubisch bodies carry a sporophytically produced structural protein (RAFTIN) that is essential for pollen development. Proceedings of the National Academy of Sciences of the United States of America 100: 14487–14492.
17. Batchelor AK, Boutilier K, Miller SS, Hattori J, Bowman LA, et al. (2002) SCB1, a BURP-domain protein gene, from developing soybean seed coats. Planta 215: 523–532.
18. Datta N, Lafayette PR, Kroner PA, Nagao RT, Key JL (1993) Isolation and Characterization of 3 Families of Auxin down-Regulated Cdna Clones. Plant Molecular Biology 21: 859–869.
19. Granger C, Coryell V, Khanna A, Keim P, Vodkin L, et al. (2002) Identification, structure, and differential expression of members of a BURP domain containing protein family in soybean. Genome 45: 693–701.
20. Xu HL, Li YX, Yan YM, Wang K, Gao Y, et al. (2010) Genome-scale identification of Soybean BURP domain-containing genes and their expression under stress treatments. Bmc Plant Biology 10.
21. Shao YH, Wei G, Wang L, Dong Q, Zhao Y, et al. (2011) Genome-Wide Analysis of BURP Domain-Containing Genes in Populus trichocarpa. Journal of Integrative Plant Biology 53: 743–755.
22. Held BM, John I, Wang H, Moragoda L, Tirimanne TS, et al. (1997) Zrp2: a novel maize gene whose mRNA accumulates in the root cortex and mature stems. Plant Mol Biol 35: 367–375.
23. Gan D, Jiang H, Zhang J, Zhao Y, Zhu S, et al. (2011) Genome-wide analysis of BURP domain-containing genes in maize and sorghum. Mol Biol Rep 38: 4553–4563.
24. Van Son L, Tiedemann J, Rutten T, Hillmer S, Hinz G, et al. (2009) The BURP domain protein AtUSPL1 of Arabidopsis thaliana is destined to the protein storage vacuoles and overexpression of the cognate gene distorts seed development. Plant Molecular Biology 71: 319–329.
25. Abe H, Yamaguchi-Shinozaki K, Urao T, Iwasaki T, Hosokawa D, et al. (1997) Role of arabidopsis MYC and MYB homologs in drought- and abscisic acid-regulated gene expression. Plant Cell 9: 1859–1868.
26. Iwasaki T, Yamaguchishinozaki K, Shinozaki K (1995) Identification of a Cis-Regulatory Region of a Gene in Arabidopsis-Thaliana Whose Induction by Dehydration Is Mediated by Abscisic-Acid and Requires Protein-Synthesis. Molecular and General Genetics 247: 391–398.
27. Park MY, Chung MS, Koh HS, Lee DJ, Ahn SJ, et al. (2009) Isolation and functional characterization of the Arabidopsis salt-tolerance 32 (AtSAT32) gene associated with salt tolerance and ABA signaling. Physiol Plant 135: 426–435.
28. Sanchez JP, Chua NH (2001) Arabidopsis PLC1 is required for secondary responses to abscisic acid signals. Plant Cell 13: 1143–1154.
29. Song HM, Zhao RM, Fan PX, Wang XC, Chen XY, et al. (2009) Overexpression of AtHsp90.2, AtHsp90.5 and AtHsp90.7 in Arabidopsis thaliana enhances plant sensitivity to salt and drought stresses. Planta 229: 955–964.
30. Wang HM, Zhou L, Fu YP, Cheung MY, Wong FL, et al. (2012) Expression of an apoplast-localized BURP-domain protein from soybean (GmRD22) enhances tolerance towards abiotic stress. Plant Cell and Environment 35: 1932–1947.
31. Ding Z, Li S, An X, Liu X, Qin H, et al. (2009) Transgenic expression of MYB15 confers enhanced sensitivity to abscisic acid and improved drought tolerance in Arabidopsis thaliana. J Genet Genomics 36: 17–29.
32. Xu B, Gou JY, Li FG, Shangguan XX, Zhao B, et al. (2013) A Cotton BURP Domain Protein Interacts With -Expansin and Their Co-Expression Promotes Plant Growth and Fruit Production. Molecular Plant 6: 945–958.
33. Schallau A, Kakhovskaya I, Tewes A, Czihal A, Tiedemann J, et al. (2008) Phylogenetic footprints in fern spore- and seed-specific gene promoters. Plant Journal 53: 414–424.
34. Sreenivasulu N, Radchuk V, Strickert M, Miersch O, Weschke W, et al. (2006) Gene expression patterns reveal tissue-specific signalling networks controlling programmed cell death and ABA-regulated maturation in developing barley seeds (vol 47, pg 310, 2006). Plant Journal 47: 987–987.

35. Orkin S (1990) Molecular-Cloning - a Laboratory Manual, 2nd Edition - Sambrook, J, Fritsch, Ef, Maniatis, T. Nature 343: 604–605.
36. Bechthold N, Elli J, Pelletier G (1993) *In planta Agrobacterium*-mediated transfer by infiltration of adult *Arabidopsis thaliana* plants. CR Acad Sci Paris 316: 871–883.
37. Jefferson RA, Kavanagh TA, Bevan MW (1987) Gus Fusions - Beta-Glucuronidase as a Sensitive and Versatile Gene Fusion Marker in Higher-Plants. Embo Journal 6: 3901–3907.
38. Poorter H, Lewis C (1986) Testing Differences in Relative Growth-Rate - a Method Avoiding Curve Fitting and Pairing. Physiologia Plantarum 67: 223–226.
39. Klukas C, Chen D, Pape JM (2014) IAP: an open-source information system for high-throughput plant phenotyping. Plant Physiol.
40. Govind G, Harshavardhan VT, Patricia JK, Dhanalakshmi R, Senthil Kumar M, et al. (2009) Identification and functional validation of a unique set of drought induced genes preferentially expressed in response to gradual water stress in peanut. Mol Genet Genomics 281: 591–605.
41. Hiscox JD, Israelstam GF (1979) Method for the Extraction of Chlorophyll from Leaf Tissue without Maceration. Canadian Journal of Botany-Revue Canadienne De Botanique 57: 1332–1334.
42. Church GM, Gilbert W (1984) Genomic sequencing. Proceedings of the National Academy of Sciences of the United States of America 81: 1991–1995.
43. Seiler C, Harshavardhan VT, Reddy PS, Hensel G, Kumlehn J, et al. (2014) Abscisic acid flux alterations result in differential abscisic acid signaling responses and impact assimilation efficiency in barley under terminal drought stress. Plant Physiol 164: 1677–1696.
44. Benjamini Y, Hochberg Y (1995) Controlling the False Discovery Rate - a Practical and Powerful Approach to Multiple Testing. Journal of the Royal Statistical Society Series B-Methodological 57: 289–300.
45. Thimm O, Blasing O, Gibon Y, Nagel A, Meyer S, et al. (2004) MAPMAN: a user-driven tool to display genomics data sets onto diagrams of metabolic pathways and other biological processes. Plant Journal 37: 914–939.
46. Zimmermann P, Hirsch-Hoffmann M, Hennig L, Gruissem W (2004) GENEVESTIGATOR. Arabidopsis microarray database and analysis toolbox. Plant Physiology 136: 2621–2632.
47. Winter D, Vinegar B, Nahal H, Ammar R, Wilson GV, et al. (2007) An "Electronic Fluorescent Pictograph" Browser for Exploring and Analyzing Large-Scale Biological Data Sets. Plos One 2.
48. Klukas C, Pape JM, Entzian A (2012) Analysis of high-throughput plant image data with the information system IAP. J Integr Bioinform 9: 191.
49. Munns R, James RA, Sirault XR, Furbank RT, Jones HG (2010) New phenotyping methods for screening wheat and barley for beneficial responses to water deficit. J Exp Bot 61: 3499–3507.
50. Schelbert S, Aubry S, Burla B, Agne B, Kessler F, et al. (2009) Pheophytin Pheophorbide Hydrolase (Pheophytinase) Is Involved in Chlorophyll Breakdown during Leaf Senescence in Arabidopsis. Plant Cell 21: 767–785.
51. Santos CLV, Campos A, Azevedo H, Caldeira G (2001) In situ and in vitro senescence induced by KCl stress: nutritional imbalance, lipid peroxidation and antioxidant metabolism. Journal of Experimental Botany 52: 351–360.
52. Qin F, Sakuma Y, Tran LSP, Maruyama K, Kidokoro S, et al. (2008) Arabidopsis DREB2A-interacting proteins function as RING E3 ligases and negatively regulate plant drought stress-responsive gene expression. Plant Cell 20: 1693–1707.
53. Catala R, Ouyang J, Abreu IA, Hu Y, Seo H, et al. (2007) The Arabidopsis E3 SUMO ligase SIZ1 regulates plant growth and drought responses. Plant Cell 19: 2952–2966.
54. Huang NC, Jane WN, Chen J, Yu TS (2012) Arabidopsis thaliana CENTRORADIALIS homologue (ATC) acts systemically to inhibit floral initiation in Arabidopsis. Plant Journal 72: 175–184.
55. Li C, Lv J, Zhao X, Ai X, Zhu X, et al. (2010) TaCHP: a wheat zinc finger protein gene down-regulated by abscisic acid and salinity stress plays a positive role in stress tolerance. Plant Physiol 154: 211–221.
56. Barrero JM, Rodriguez PL, Quesada V, Piqueras P, Ponce MR, et al. (2006) Both abscisic acid (ABA)-dependent and ABA-independent pathways govern the induction of NCED3, AAO3 and ABA1 in response to salt stress. Plant Cell Environ 29: 2000–2008.
57. Hwang IW, Goodman HM (1995) An Arabidopsis-Thaliana Root-Specific Kinase Homolog Is Induced by Dehydration, Aba, and Nacl. Plant Journal 8: 37–43.
58. Rosado A, Schapire AL, Bressan RA, Harfouche AL, Hasegawa PM, et al. (2006) The Arabidopsis tetratricopeptide repeat-containing protein TTL1 is required for osmotic stress responses and abscisic acid sensitivity. Plant Physiology 142: 1113–1126.
59. Shinozaki K, Yamaguchi-Shinozaki K (1992) [Plant genes induced by drought stress and ABA]. Tanpakushitsu Kakusan Koso 37: 1190–1199.
60. Fujii H, Verslues PE, Zhu JK (2011) Arabidopsis decuple mutant reveals the importance of SnRK2 kinases in osmotic stress responses in vivo. Proc Natl Acad Sci U S A 108: 1717–1722.
61. Daxinger L, Hunter B, Sheik M, Jauvion V, Gasciolli V, et al. (2008) Unexpected silencing effects from T-DNA tags in Arabidopsis. Trends in Plant Science 13: 4–6.
62. Almagro L, Gomez Ros LV, Belchi-Navarro S, Bru R, Ros Barcelo A, et al. (2009) Class III peroxidases in plant defence reactions. J Exp Bot 60: 377–390.
63. Desikan R, Cheung MK, Bright J, Henson D, Hancock JT, et al. (2004) ABA, hydrogen peroxide and nitric oxide signalling in stomatal guard cells. Journal of Experimental Botany 55: 205–212.
64. Hancock JT, Desikan R, Neill SJ (2001) Role of reactive oxygen species in cell signalling pathways. Biochem Soc Trans 29: 345–350.
65. Kar RK, Choudhuri MA (1987) Possible Mechanisms of Light-Induced Chlorophyll Degradation in Senescing Leaves of Hydrilla-Verticillata. Physiologia Plantarum 70: 729–734.
66. Hynninen PH, Kaartinen V, Kolehmainen E (2010) Horseradish peroxidase-catalyzed oxidation of chlorophyll a with hydrogen peroxide Characterization of the products and mechanism of the reaction. Biochimica Et Biophysica Acta-Bioenergetics 1797: 531–542.
67. Ramirez V, Coego A, Lopez A, Agorio A, Flors V, et al. (2009) Drought tolerance in Arabidopsis is controlled by the OCP3 disease resistance regulator. Plant Journal 58: 578–591.
68. Coego A, Ramirez V, Gil MJ, Flors V, Mauch-Mani B, et al. (2005) An Arabidopsis homeodomain transcription factor, OVEREXPRESSOR OFCATIONIC PEROXIDASE 3, mediates resistance to infection by necrotrophic pathogens. Plant Cell 17: 2123–2137.
69. Wang YN, Liu C, Li KX, Sun FF, Hu HZ, et al. (2007) Arabidopsis EIN2 modulates stress response through abscisic acid response pathway. Plant Molecular Biology 64: 633–644.
70. Miao Y, Lv D, Wang P, Wang XC, Chen J, et al. (2006) An Arabidopsis glutathione peroxidase functions as both a redox transducer and a scavenger in abscisic acid and drought stress responses. Plant Cell 18: 2749–2766.
71. Mullineaux P, Ball L, Escobar C, Karpinska B, Creissen G, et al. (2000) Are diverse signalling pathways integrated in the regulation of arabidopsis antioxidant defence gene expression in response to excess excitation energy? Philos Trans R Soc Lond B Biol Sci 355: 1531–1540.
72. Neill S, Barros R, Bright J, Desikan R, Hancock J, et al. (2008) Nitric oxide, stomatal closure, and abiotic stress. J Exp Bot 59: 165–176.

9

Arabidopsis mTERF15 Is Required for Mitochondrial *nad2* Intron 3 Splicing and Functional Complex I Activity

Ya-Wen Hsu[1,2], Huei-Jing Wang[2], Ming-Hsiun Hsieh[2], Hsu-Liang Hsieh[1], Guang-Yuh Jauh[2,3]*

1 Institute of Plant Biology, National Taiwan University, Taipei, 116, Taiwan, ROC, **2** Institute of Plant and Microbial Biology, Academia Sinica, Nankang, Taipei, 11529, Taiwan, ROC, **3** Biotechnology Center, National Chung-Hsing University, Taichung, 402, Taiwan, ROC

Abstract

Mitochondria play a pivotal role in most eukaryotic cells, as they are responsible for the generation of energy and diverse metabolic intermediates for many cellular events. During endosymbiosis, approximately 99% of the genes encoded by the mitochondrial genome were transferred into the host nucleus, and mitochondria import more than 1000 nuclear-encoded proteins from the cytosol to maintain structural integrity and fundamental functions, including DNA replication, mRNA transcription and RNA metabolism of dozens of mitochondrial genes. In metazoans, a family of nuclear-encoded proteins called the mitochondrial transcription termination factors (mTERFs) regulates mitochondrial transcription, including transcriptional termination and initiation, via their DNA-binding activities, and the dysfunction of individual mTERF members causes severe developmental defects. *Arabidopsis thaliana* and *Oryza sativa* contain 35 and 48 mTERFs, respectively, but the biological functions of only a few of these proteins have been explored. Here, we investigated the biological role and molecular mechanism of Arabidopsis mTERF15 in plant organelle metabolism using molecular genetics, cytological and biochemical approaches. The null homozygous T-DNA mutant of *mTERF15*, *mterf15*, was found to result in substantial retardation of both vegetative and reproductive development, which was fully complemented by the wild-type genomic sequence. Surprisingly, mitochondria-localized mTERF15 lacks obvious DNA-binding activity but processes mitochondrial *nad2* intron 3 splicing through its RNA-binding ability. Impairment of this splicing event not only disrupted mitochondrial structure but also abolished the activity of mitochondrial respiratory chain complex I. These effects are in agreement with the severe phenotype of the *mterf15* homozygous mutant. Our study suggests that Arabidopsis mTERF15 functions as a splicing factor for *nad2* intron 3 splicing in mitochondria, which is essential for normal plant growth and development.

Editor: Emanuele Buratti, International Centre for Genetic Engineering and Biotechnology, Italy

Funding: This work was supported by research grants from Academia Sinica (Taiwan; http://ipmb.sinica.edu.tw/index.html/), the National Science and Technology Program for Agricultural Biotechnology (NSTP/AB, 098S0030055-AA, Taiwan) and the Ministry of Science and Technology (99-2321-B-001-036-MY3 and 102-2321-B-001-040-MY3; http://www.most.gov.tw/) to G.-Y. Jauh. The funders had no role in study design, data collection and analysis, decision to publish, or preparation of the manuscript.

Competing Interests: The authors have declared that no competing interests exist.

* Email: jauh@gate.sinica.edu.tw

Introduction

Mitochondria, which originated through the endosymbiosis of α-proteobacteria into ancestral host cells, are the cellular powerhouses and play vital roles in diverse eukaryotic cell processes through the production of ATP and various metabolic intermediates [1,2]. Recent studies also suggest that dysfunctional mitochondria are involved in many neurodegenerative diseases such as aging and cognitive decline in a wide range of metazoans, including humans [3]. Maintaining the structural and metabolic integrity of this semi-autonomous organelle is essential for the normal function of eukaryotic cells. Nevertheless, over the course of symbiotic evolution, the majority of mitochondrial genes migrated into the nuclear genome of the original host, leaving an incomplete set of essential genes in the mitochondrial genomes of most organisms, including plants [4–6]. Complicated and dynamic communication and coordination between the nucleus and mitochondria greatly impact many fundamental cellular processes in, and even the lives of, most eukaryotes [7]. Indeed, based on the complete sequence of the Arabidopsis mitochondrial genome, it has been reported that 57 mitochondrial genes encode the subunits of multiprotein complexes that are required for the respiratory chain, heme and cytochrome assembly, and mitochondrial ribosomes [4]. Additionally, plant mitochondria are more complex than those found in other kingdoms and exhibit unique RNA metabolism, including RNA transcription, splicing, editing, degradation, and translation [8–10]. The proteins involved in these processes are predominantly encoded by the nuclear genome and are imported into mitochondria after protein synthesis. For example, recent studies suggested that one protein family, called the mitochondrial transcription termination factors (mTERFs), plays important roles in regulating the organellar transcription machinery.

The mTERF proteins were first identified two decades ago as regulators of transcription termination in human mitochondria [11]. Phylogenetic analyses of mTERF homologs in metazoans and plants revealed the presence of 4 subfamilies, mTERF1 to mTERF4 [12]. These proteins share a common 30-amino-acid repeat module called the mTERF motif [13]. The proteins within this family possess diverse numbers and arrangements of these motifs, yet the folding patterns of these proteins are similar. Moreover, crystal structure studies of mTERF1, mTERF3 and mTERF4 suggest that the helical structure of the mTERF motifs may be essential for their nucleic acid-binding activities [14,15]. Human mTERF1 binds specific sites located at the 3′-end of the 16S rRNA and $tRNA^{Ler(UUR)}$ genes to terminate mitochondrial transcription [16]. Additionally, mTERF1 binds to the mitochondrial transcription initiation site to create a DNA loop that allows for the recycling of the transcriptional machinery [16]. This simultaneous link between mitochondrial transcriptional initiation and termination sites may explain the high rate of mitochondrial rRNA biogenesis. mTERF2 binds to mitochondrial DNA (mtDNA) in a non-specific manner and associates with nucleoids. *mTERF2* loss-of-function mice exhibit myopathies, memory deficits, and impaired respiratory function due to a reduction in the number of mitochondrial transcripts [17,18]. *mTERF3* is essential, and mTERF3-knockout mice die during the early stages of embryogenesis; mTERF3 non-specifically interacts with mtDNA promoter regions and mediates the repression of mitochondrial transcription initiation [19]. Mouse NSUN4 (NOL1/NOP2/Sun domain family, member 4) is a methyltransferase involved in the methylation of ribosomes. Mouse mTERF4 first binds 16S and 12S rRNAs, then forms a stoichiometric complex with NSUN4, which is essential for proper mitochondrial ribosomal assembly and translation [14,20,21].

The genome of Arabidopsis contains 35 identified mTERFs, which vary in the number and arrangement of their mTERF motifs [22]. Several recently characterized Arabidopsis mTERF proteins displayed diverse and complicated functions during mitochondrial and plastid transcription, such as *SINGLET OXYGEN-LINKED DEATH ACTIVATOR10* (*SOLDAT10*) [23], *BELAYA SMERT/RUGOSA2* (*BSM/RUG2*) [22,24], *MTERF DEFECTIVE IN ARABIDOPSIS 1* (*MDA1*) [25] and *SUPPRESSOR of hot1-4 1* (*SHOT1*) [26]. SOLDAT10, the first Arabidopsis mTERF to be characterized, is targeted to plastids and regulates plastid homeostasis. Arabidopsis *soldat10* mutants suppressed the $^1O_2$-mediated cell-death and growth-retardation phenotypes of the conditional *fluorescent* (*flu*) mutant, in which excess levels of the reactive oxygen species (ROS) $^1O_2$ were generated within chloroplasts after the dark-to-light shift [23]. BSM/RUG2 is essential for embryogenesis and postembryonic development in Arabidopsis and displays DNA-binding activity towards plastid DNA. Additionally, BSM/RUG2 localizes to both plastids and mitochondria, and the perturbation of BSM/RUG2 affects the processing and steady-state levels of plastid transcripts [22] and decreases the expression of mitochondrial transcripts [24]. Arabidopsis MDA1 is a plastid-localized protein involved in abiotic stress tolerance. *MDA1* mutation leads to significant defects in plant growth and chloroplast development but enhanced tolerance to salt and osmotic stress, perhaps via reduced sensitivity to abscisic acid [25]. Mutating *SHOT1* suppresses the heat-hypersensitive phenotype caused by the loss of heat shock protein 101 in Arabidopsis [26]. However, SHOT1 is a mitochondrial-resident mTERF protein, and the thermotolerance phenotype of *shot1* may be caused by reduced ROS-mediated oxidative damage during heat stress. Additionally, SHOT1 plays an important role in regulating the expression of mitochondrial-encoded genes and mitochondrial genome copy number. Although it is known that many mTERF mutations cause defective development and stress responses, studies of the precise molecular functions of these mTERF proteins are limited in Arabidopsis organelles.

In Arabidopsis, several protein families that participate in organellar RNA splicing, including those containing the chloroplast RNA splicing and ribosome maturation (CRM) domain [27–29], the pentatricopeptide repeat (PPR) [30–35], the plant organellar RNA recognition (PORR) domain [36,37] and the ACCUMULATION OF PHOTOSYSTEM ONE (APO) motif [38]. These domains are recognized as RNA binding domains and interact with specific elements within introns. Although several splicing factors have been identified in plastids and mitochondria, the detailed mechanism of RNA splicing in plant organelles is still being investigated. Recently, maize mTERF4 was identified as a splicing factor associated with several introns and splicing factors in chloroplasts [39].

In the present study, we investigated another Arabidopsis mTERF protein, mTERF15, and identified its role in post-transcriptional modification of mitochondrial RNA processing. Mutating *mTERF15* significantly disturbed normal plant vegetative and reproductive growth, which may be the result of defective mitochondrial development and/or function. Interestingly, mTERF15 is a mitochondria-localized protein that binds to RNA but not to double-stranded DNA (dsDNA). Moreover, an analysis of mitochondrial intron splicing in *mterf15* mutants revealed defective RNA splicing of *nad2* intron 3, which led to a deficiency in complex I activity. The possible molecular function of mTERF15 in regulating Arabidopsis mitochondrial RNA metabolism is discussed.

Materials and Methods

Plant material and growth conditions

The *mTERF15* T-DNA insertion mutant (SALK_134099) was obtained from the Arabidopsis Biological Resource Center (Ohio State University, USA). Seeds from *Arabidopsis thaliana* ecotype Columbia-0 and *mterf15*/+ plants were surface-sterilized by chloride gas after stratification for 3 d at 4°C and then either germinated on half-strength Murashige and Skoog medium (Duchefa Biocheme, The Netherlands) containing 1% sucrose and 0.7% phytoagar (Duchefa) or grown in soil inside a growth chamber on a 16-h light/8-h dark cycle at 22°C.

Complementation of the *mterf15* mutant

Approximately 3.3 kb of the genomic sequence corresponding to locus *At1g74120*, including a 2-kb putative promoter and the 1.3-kb coding region, was amplified using Phusion polymerase (Finnzymes, http://www.finnzymes.com) and then ligated into a modified *pPZP221* vector containing a C-terminal GFP sequence followed by the Nopaline synthase (NOS) terminator. The primers used for cloning are in Table S1. After verification of the DNA sequence, the construct was introduced into *Agrobacterium tumefaciens* strain GV3101. Arabidopsis plants heterozygous for *mTERF15* were transfected with *Agrobacterium* by the vacuum infiltration method [40]. Seeds corresponding to the T_1 generation were collected and grown on solid half-strength Murashige and Skoog medium (Duchefa) containing 1% sucrose, 0.7% phytoagar (Duchefa) and the antibiotic G418 at 100 µg/µl. Leaves from T_1 plants were harvested for genotyping. Plants with an *mterf15* homozygous background harboring *mTEFR15p::mTERF15-GFP* were collected for further studies.

Ultrastructural sample preparation and tetramethylrhodamine methyl ester perchlorate (TMRM) staining

For transmission electron microscopy, samples were frozen in a high-pressure freezer (Leica EMPACT2, 2000–2050 bar). Freeze

Table 1. The phenotype of progeny from SALK 134099 heterozygotes (*mterf15/+*) with abnormal morphology.

	Normal seedlings	Seedlings with retarded growth	Non-germinated seeds	Ratio (normal: abnormal)	*p*-value (χ^2 value, 3:1)
Wild-type	308 (99.68)	0	1 (0.32)	*NA	NA
SALK_134099 heterozygotes	246 (74.77)	27 (8.21)	56 (17.02)	2.96:1	0.924

Phenotypes were recorded after seeds were germinated on half-strength MS solid medium and then incubated for 7 d with a 16-h light/8-h dark cycle at 22°C.
*NA, Not available.
Data indicates total number of plants observed and numbers in brackets indicate their corresponding percentage.

substitution involved the use of anhydrous acetone (containing 1% OsO_4 and 0.1% UA) with a Leica EM AFS2. The samples were kept at −85°C for 3 d, at −60°C for 1 d, at −20°C for 1 d, and at 0°C for 1 d and then brought to room temperature. The samples were infiltrated and embedded in Spurr resin. Ultrathin sections (70–90 nm) were cut using a Reichert Ultracut S or Leica EM UC6 (Leica, Vienna, Austria) and collected on 100 mesh copper grids. After staining with 5% uranyl acetate in 50% methanol for 10 min and 0.4% lead citrate for 4 min, sections were observed under a Philips CM 100 Transmission Electron Microscope at 80 KV; images were captured using a Gatan Orius CCD camera.

TMRM staining (Molecular Probes, Invitrogen) was performed to examine the mitochondrial membrane potential in protoplasts. Isolated protoplasts were stained with 20 nM TMRM and immediately observed under a laser scanning confocal microscope (Zeiss, LSM510, Carl Zeiss, http://www.zeiss.com). TMRM fluorescence and chlorophyll autofluorescence were induced by excitation with a 543-nm HeNe laser and a 488-nm Argon laser. The emission signals from TMRM and from chlorophyll were recorded with using 565- to 615- and 650-nm band-pass filters, respectively. The fluorescence intensity and area were quantified using LSM510 software (Carl Zeiss, http://www.zeiss.com).

Northwestern blot analysis

Northwestern blot analysis was performed to test the RNA-binding ability of mTERF15 as described previously [41]. Briefly, *mTERF15* cDNA lacking the 34 N-terminal amino acids was ligated into the pET52b vector to produce the mTERF15-strep recombinant protein. The recombinant proteins were expressed in *E. coli* strain BL21 and incubated with StrepTactin-Sepharose resins (GE Healthcare) according to the manufacturer's instructions. The resins were washed extensively with incubation buffer before elution of the recombinant proteins. The mTERF15-strep recombinant protein and the MBP-strep control protein were transferred from 12% SDS-polyacrylamide gels onto a PVDF membrane and then renatured overnight in renaturation buffer (0.1 M Tris-HCl, pH 7.5 and 0.1% (v/v) NP-40) at 4°C. After 4 washes with renaturation buffer (15 min), blots were blocked with blocking buffer [(10 mM Tris-HCl, pH 7.5, 5 mM Mg acetate, 2 mM DTT, 5% (w/v) BSA and 0.01% (v/v) Triton X-100)] at room temperature for 5 min. Full length of *nad2* intron 3 was *in vitro* synthesized using SP6 polymerase (Promega). Total RNA and *in vitro* synthesized *nad2* intron3 were labeled with γ^{32}P ATP using T4 polynucleotide kinase (Fermentas). Hybridization was performed at 4°C in the presence of 5 μg labeled RNA. Blots were washed 4 times (5 min each) in a washing buffer (10 mM Tris-HCl, pH 7.5, 5 mM Mg acetate, 2 mM DTT) to remove unbound RNA. The radioisotope signals were detected via autoradiography on X-ray film (Kodak).

RNA isolation and qRT-PCR analysis

Total RNA was isolated using the Qiagen RNeasy Mini Kit (QIAGEN) and then treated with TURBO DNA-free DNase (Ambion) to remove contaminating genomic DNA. Reverse transcription of RNA was performed using the M-MLV transcriptase system (Invitrogen) and random hexamer primers.

Gene expression analysis via qRT-PCR was carried out using Power SYBR Green Supermix (Applied Biosystems) running on an ABI Prism 7000 sequence detection system (Applied Biosystems). Primer sets were designed as described previously [31] (Table S2) to monitor the expression of genes involved in the splicing of junctions within mitochondria.

Northern blot analysis

A total amount of 5 μg RNA was denatured at 65°C for 10 min and then separated on 2% formaldehyde-agarose gels in MOPS buffer and transferred to a nylon membrane (PerkinElmer). The *nad2* probes were labeled using the Rediprime II DNA labeling system (GE Healthcare). Probe primers are listed in Table S1. The prehybridization, hybridization and washing steps were performed following the manufacturer's instructions.

Crude mitochondria isolation and complex I Activity assay

Crude mitochondria were prepared as described previously [36,42]. Approximately 200 mg fresh tissue was extracted in 2 ml extraction buffer (75 mM MOPS-KOH, pH 7.6, 0.6 M sucrose, 4 mM EDTA, 0.2% polyvinylpyrrolidone 40, 8 mM cysteine, 0.2% bovine serum albumin) at 4°C. The lysate was centrifuged first at 1300×g for 5 min (twice), after which the supernatant was centrifuged at 22,000×g for 20 min at 4°C. The resulting pellet (which usually contains most of the thylakoid and mitochondrial membranes) was resuspended in buffer (10 mM MOPS-KOH, pH 7.2, 0.3 M sucrose) and stored at −80°C.

Clear native electrophoresis was performed as described previously [43]. Crude mitochondria were washed with distilled water, resuspended with buffer (50 mM NaCl, 50 mM imidazole/HCl, 2 mM 6-aminohexanoic acid, 1 mM EDTA, pH 7.0) and solubilized by adding DDM (10%) at a DDM-to-protein ratio of 2.5 (g/g). After centrifugation at 100,000×g for 15 min, 10% glycerol and 0.02% Ponceau S were added to the supernatant for 4–16% native PAGE with anode buffer (25 mM imidazole/HCl, pH 7.0) and cathode buffer (50 mM Tricine, 7.5 mM imidazole, pH 7.0, 0.01% DDM and 0.05% DOC). Clear native gels were washed 3 times with distilled water for 5 min and incubated with 1 mM nitroblue tetrazolium and 0.14 mM NADH in 0.1 M Tris, pH 7.4. The reaction was stopped with 40% methanol and 10% acetic acid after a dark blue signal appeared on the gel.

Results

Disruption of the Arabidopsis *mTERF15* gene results in significant growth and developmental retardation

A screen of Arabidopsis T-DNA insertional mutants from the SALK collection allowed us to identify several mutants that were defective in post-embryonic development and/or seed germination. We found one T-DNA mutant showing abnormal phenotype in progenies segregated from SALK_134099 heterozygotes. The progeny of SALK_134099 heterozygotes showed growth retardation at a ratio of 3:1 (normal to defective) (χ^2, $p = 0.924$; Table 1). Most abnormal progenies derived from SALK_134099 heterozygous plants were unable to germinate on solid medium (17.02%), although a few grew slowly (8.21%, Table 1). The SALK_134099 homozygous mutant containing a T-DNA insertion in the coding region of Arabidopsis *mTERF15* (At1g74120; Figure 1A) and showing significant growth and development retardation was named *mterf15* (Figure 1B). The full-length *mTERF15* transcript was not detected in the homozygous *mterf15* plant (Figure 1C), suggesting that all of the observed phenotypes resulted from the null mutation of the *mTERF15* allele (SALK_134099). Developmental defects were widespread in homozygous *mterf15* plants. Compared with wild-type seedlings, *mterf15* seedlings were small in stature, had small cotyledons and displayed reduced growth of primary roots (Figure 1D). The rosette leaves of *mterf15* were smaller, wrinkled and twisted whereas the wild type plant showed larger and smooth round-shaped leaves. Moreover, the extended petiole region of *mterf15* was shorter than the wild type plant. Compared with 1.5-month-old wild-type plants that were ready for blossoming, 3-month-old adult *mterf15* plants were dwarfed and just about ready for flowering (Figure 1E). The flowers of *mterf15* plants were abnormal, with misplaced petals, extremely defective anthers and a limited amount of pollen (Figure 1F, upper panel). Unlike wild-type siliques, mature *mterf15* siliques were much shorter (Figure 1G), and no viable seeds were found (Figure 1F, lower panel).

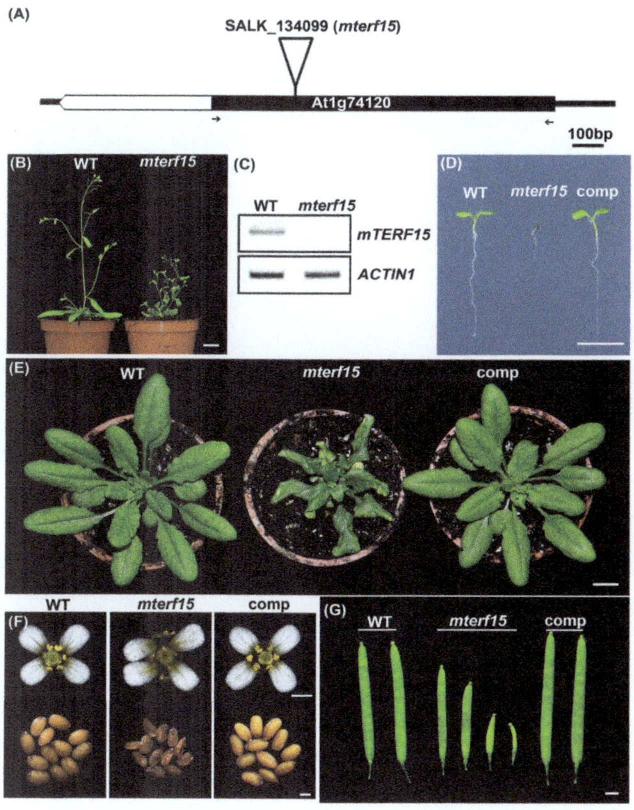

Figure 1. Phenotypes associated with *mterf15* mutants. **(A)** Schematic diagram of the Arabidopsis *mTERF15* gene (At1g74120) and the single T-DNA insertion within the coding region for *mterf15* (SALK_134099). Primers used to analyze *mTERF15* transcripts are indicated by black arrows. **(B)** Comparison of whole-plant morphology between a 1.5-month-old wild-type plant and a 3-month-old *mterf15* mutant. Scale bar = 1 cm. **(C)** RT-PCR analysis of *mTERF15* expression in 7-d-old wild-type (WT) and *mterf15* seedlings. The housekeeping gene *ACTIN1* was used as an internal control. **(D)** Morphology of 7-d-old seedlings for wild-type and *mterf15* lines and a complemented line transformed with *mTERF15p::mTERF15-GFP*. Scale bar = 1 cm. **(E)** Morphology and arrangement of leaves for 1-month-old wild-type, 2-month-old *mterf15* and 1-month-old *mterf15* complementation lines. Scale bars = 1 cm **(F)** Morphology of flowers and seeds from the wild-type, *mterf15* and complementation lines. Scale bars = 1 mm (flowers) and 200 μm (seeds). **(G)** Morphology of siliques from the wild-type, *mterf15* and complementation lines. Scale bars = 1 mm.

To further verify that these defects were indeed caused by the disruption of the *mTERF15* allele, *mterf15* heterozygotes were complemented with the *mTERF15* native promoter-driven *mTERF15* coding sequence tagged with *GFP* (*mTERF15p::mTERF15-GFP*). The T_2 progeny of homozygotes for the *mTERF15* mutation which harbors at least one copy of the transgene *mTERF15p::mTERF15-GFP* showed similar phenotypes as 7-d-old wild-type seedlings and adult plants (Figure 1 and Figure S1). All of the above results show that functional *mTERF15* is essential for normal plant growth and development.

mTERF15, a plant-specific and mitochondria-localized protein, is important for mitochondrial biogenesis and membrane potential

The Arabidopsis *mTERF15* locus contains a single exon and appears to encode a 51-kDa protein. Architecture analysis with SMART (http://smart.embl-heidelberg.de) revealed that mTERF15 contains 5 typical mTERF motifs (Figure S2A). A search of the NCBI database (http://www.ncbi.nlm.nih.gov) revealed mTERF15 homologs present in *Vitis vinifera*, *Populus trichocarpa*, *Brachypodium distachyon*, *Sorghum bicolor* and *Zea mays*. Comparing protein similarities and identities between flowering plant mTERF15 orthologs and metazoan mTERF1, Arabidopsis mTERF15 is evolutionally closer to plant mTERF15 orthologs than metazoan mTERF1. This result suggests that mTERF15 is a plant-specific mTERF protein (Table S3). The phylogenetic tree branched into 2 groups, monocots and dicots (Figure S2B). Amino acid sequence alignments of all 6 homologs revealed stronger conservation of the C terminus compared with the N terminus. All of these homologs appeared to contain 5 mTERF motifs with a conserved proline residue at position 8, except for the first motif (Figure S2A).

In silico analysis of the BIO-array (http://bar.utoronto.ca/efp_arabidopsis/cgi-bin/efpWeb.cgi) and Genevestigator (https://www.genevestigator.com/gv/plant.jsp) microarray databases suggested various expression levels of *mTERF15* in different tissues, with the highest and lowest expression levels in mature seeds and pollen, respectively. However, in the Genevestigator microarray database, *mTERF15* showed the highest expression in sperm cells and the lowest in leaf phloem protoplasts. Quantitative real-time PCR spatiotemporal analysis in Arabidopsis revealed *mTERF15* expression in all organs, with high levels of accumulation in seedlings, flowers and siliques (Figure S3).

Transient protoplast assays and stable transgenic studies suggested that mTERF15 is a mitochondrial protein [22]. To confirm the subcellular localization of mTERF15, we observed the root hairs of 7-d-old *mTERF15p::mTERF15- GFP* transgenic seedlings via confocal microscopy after staining with the mitochondrial indicator MitoTracker orange. The co-localization of the mTERF15-GFP and MitoTracker signals demonstrated that mTERF15 is a mitochondrial protein (Figure S4).

We further investigated the morphology of chloroplasts and mitochondria from the leaves of 1-month-old wild-type and 2-month-old *mterf15* plants by transmission electron microscopy. The sizes and features of *mterf15* chloroplasts were similar to those of wild-type chloroplasts (Figure 2A, D). Although the sizes of the mitochondria did not differ between *mterf15* and wild-type plants, the inner membrane systems were aberrant in *mterf15* mitochondria; wild-type mitochondria (Figure 2B–2C) showed normal cristae structures with clear inner spaces, whereas *mterf15* mitochondria (Figure 2E–2F) lacked any obvious cristae structure. Therefore, the integrity of the mitochondrial inner membrane system may have been compromised, potentially resulting in decreased mitochondrial membrane potential.

To assess mitochondrial membrane integrity, we treated protoplasts from leaves of the wild-type, complementation line and *mterf15* plants with tetramethylrhodamine methyl ester (TMRM), a lipophilic, cationic, red-orange fluorescent dye that accumulates within mitochondria in living cells. The intensity of fluorescence is associated with the mitochondrial membrane potential [44,45]. We observed large and strong fluorescent TMRM signals in protoplasts from the wild-type and complementation lines, which suggests that mitochondria were active in these lines (Figure 2G, I); however, TMRM signals were slightly weaker in *mterf15* protoplasts (Figure 2H). The mean fluorescence intensity for *mterf15* protoplasts was reduced by ~80% compared with wild-type protoplasts (Figure 2J). These results suggest that mTERF15 is crucial for maintaining mitochondrial biogenesis and membrane integrity.

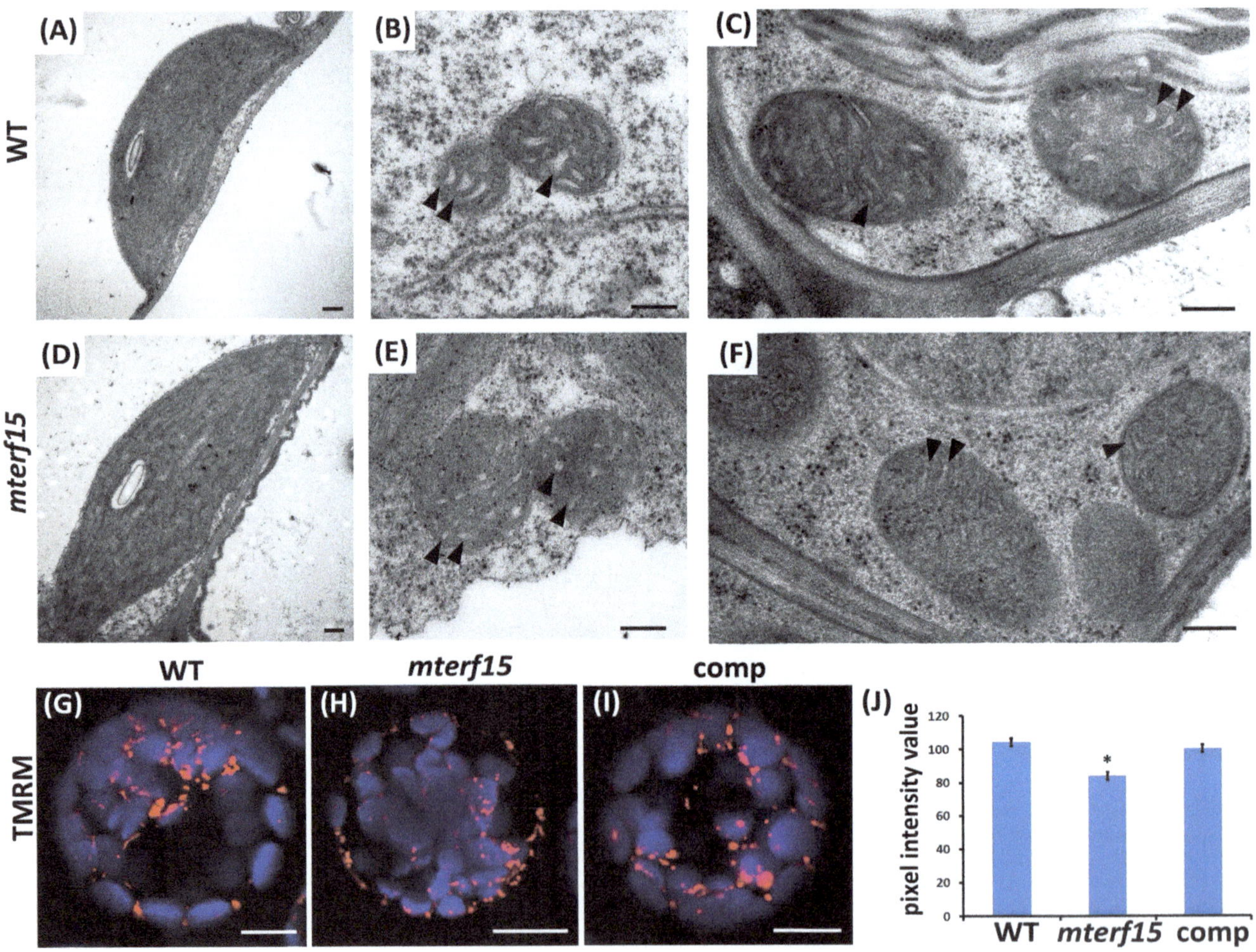

Figure 2. Morphological analysis of mitochondria in wild-type (WT) and *mterf15* plants. (**A** and **D**) Transmission electron microscopy of thin sections corresponding to chloroplasts from 1-month-old wild-type (**A**) and 2-month-old *mterf15* plants (**D**). Sections were processed by chemical fixation. Scale bar = 2 μm. (**B, C, E, F**) Transmission electron microscopy of thin sections corresponding to mitochondria from wild-type (**B, C**) and *mterf15* (**E, F**) plants. Images B and E correspond to chemically fixed leaves. Images C and F correspond to leaves fixed by high-pressure freezing. Scale bar = 1 μm. Arrowheads pointed the inner membrane spaces in mitochondria. (**G, H, I**) Overlay of signals from chlorophyll autofluorescence (blue) with signals from the mitochondrial membrane potential marker tetramethylrhodamine methyl ester perchlorate (TMRM; red) in protoplasts from the wild-type (**G**), *mterf15* (**H**) and complementation lines (**I**). Scale bar = 10 μm. (**J**) Quantification of the mean fluorescent intensity after TMRM staining. Data are mean ± SD of 15 individual protoplasts from the wild-type, *mterf15* and complementation lines. Asterisks represent significant differences (*P<0.5, **P<0.01, ***P<0.001; Student's *t* test) relative to wild type.

mTERF15 is an RNA-binding protein that is involved in mitochondrial *nad2* intron 3 splicing and normal complex I activity

Because most eukaryotic mTERFs function as nucleic acid-binding proteins, we performed *in vitro* binding studies with dsDNA-cellulose and northwestern blot analyses to determine the dsDNA- and RNA-binding abilities of mTERF15. After incubation with commercial dsDNA-cellulose resin, most of the mTERF15 was present in the unbound fraction (Figure S5A), which implies a lack of dsDNA-binding activity for mTERF15. Additionally, southwestern blot analysis was employed to validate the DNA-binding ability of mTERF15. As shown in Figure S5B, there is no DNA-binding signal for the mTERF15 protein in southwestern blots. The above results imply that mTERF15 may not be a DNA-binding protein. Therefore, we investigated the RNA-binding ability of mTERF15 by northwestern blot analysis (Figure 3). Strep-tagged mTERF15 lacking the first N-terminal 34 amino acids, which contain a putative targeting signal (MW = 48 kDa), was separated by SDS-PAGE, transferred to a PVDF membrane and probed with a specific antibody against strep-tagged or radiolabeled total RNA prepared from 7-d-old Arabidopsis seedlings. The presence of the mTERF15-strep and MBP-strep fusion protein were confirmed by immunoblot analysis using an antibody against the strep-tag, and 48-kDa bands corresponding to mTERF15-strep were detected in a dose-dependent manner in the RNA-binding assay but not in the control. These results suggest that Arabidopsis mTERF15 functions as an RNA-binding protein.

Previously, Babiychuk *et al.* (2011) showed that one mTERF protein, BSM, is required for embryogenesis and plays a role in chloroplast intron splicing. Therefore, we examined 23 intron-splicing events in mitochondria by quantitative RT-PCR (qRT-PCR). In mitochondria, 9 genes require intron splicing for the complete maturation of their respective transcripts, and 4 of these contain a single intron: *rps3*, *cox2*, *ccmFc* and *rpl2*. The others (*nad1*, *nad2*, *nad4*, *nad5* and *nad7*) contain multiple introns and are required for the normal function of mitochondrial complex I. Using the specific primer sets designed by Longevialle *et al.* (2007), we examined individual splicing events in mitochondria obtained from wild-type, *mterf15* and complemented plants. All splicing events showed similar results in both the wild-type and complemented plants; however, the RNA splicing of *nad2* intron 3 (*nad2-3*) was significantly reduced in *mterf15* plants (Figure 4A).

In Arabidopsis, the mature *nad2* mRNA is formed by merging exons from 2 different transcripts, *nad2a* and *nad2b*, through one *trans*-splicing and 3 *cis*-splicing events. The genomic fragment of *nad2a* carries exons 1 and 2 of *nad2* separated by *nad2* intron 1. The other 3 exons of *nad2*, exons 3 through 5, are from *nad2b* transcripts (Figure 4B). To validate the results of the qRT-PCR analysis, we used RT-PCR to examine the accumulation of un-spliced *nad2b* transcripts in *mterf15* plants; the amplification of *nad2* exon 5 by primers 3F and 3R was used for normalization (lower panel in Figure 4B). RT-PCR with primers 1F and 2R revealed no mature transcripts from the splicing of *nad2* intron 3 in *mterf15* plants (Figure 4B). Additionally, the transcript levels were greater in *mterf15* plants than in wild-type plants upon amplification of the exon–intron junction with the primer sets 1F&1R and 2F&2R. This *nad2* intron 3 splicing defect was further confirmed by northern blot analysis. The *nad2b* transcripts revealed by probes for *nad2* exon 3 and 5 showed no mature *nad2* mRNAs in *mterf15* plants (m in Figure 4C) and was accompanied by the overaccumulation of un-spliced transcripts (u in Figure 4C). A similar accumulation of un-spliced transcripts in *mterf15* was confirmed using probes specific to *nad2* intron 3 (u in Figure 4C). This splicing defect was fully rescued in complemented plants. Another *cis*-splicing transcript, *nad1* exon 3, was used as a positive control, with comparable results to those observed in qRT-PCR: the splicing was normal, but the accumulation of *nad1-3* was greater in *mterf15* plants than in wild-type and complemented plants (Figure 4C).

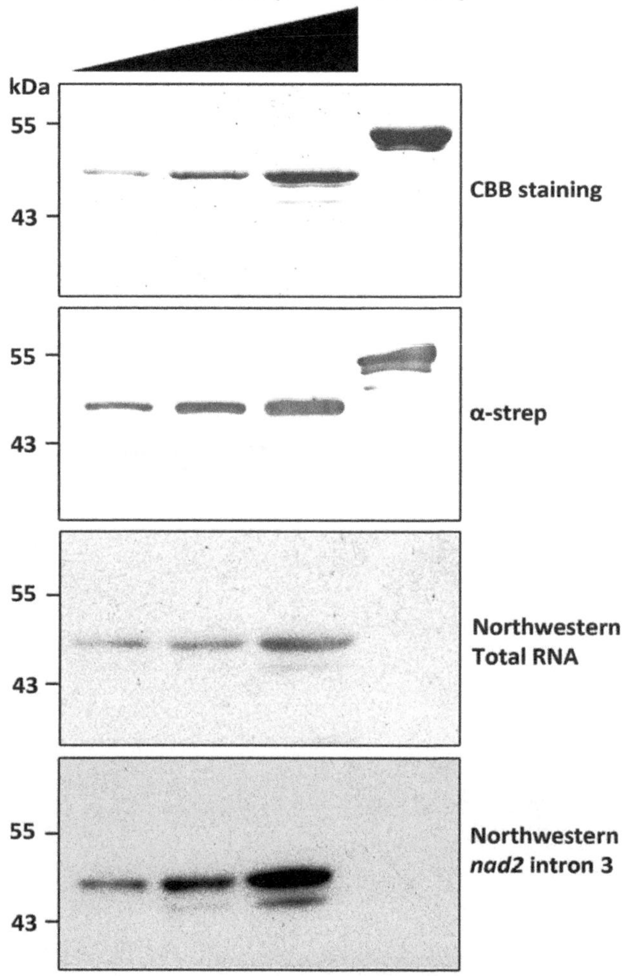

Figure 3. Northwestern blot assay of RNA-binding by mTERF15 recombinant protein. The first panel shows Coomassie brilliant blue (CBB) staining of MBP-strep and mTERF15-strep lacking the first 34 N-terminal amino acids. The second panel shows immunoblot analysis with anti-strep antibody to recognize the strep-tag. The third panel shows the northwestern blot analysis using radioisotope-labeled total RNA isolated from 7-d-old seedlings. The final panel shows the northwestern blot analysis using radioisotope-labeled *nad2* intron 3; signals are observed on the PVDF membrane corresponding to the RNAs bound to mTERF15.

Because mTERF15 functions as an RNA-binding protein and is involved in *nad2* intron 3 splicing, mTERF15 may directly interact with *nad2* intron 3. First, we confirmed the interaction between mTERF15 and full-length *nad2* intron 3 (~2.6 kb) by northwestern blot analysis (Figure 3, lowest panel). The interaction between mTERF15 and *nad2* intron 3 was stronger than the interaction with total RNA. We first examined the interaction between mTERF15 and full-length *nad2* intron 3 (~2.6 kb) by northwestern blot analysis (Figure 3, lowest panel). The interaction between mTERF15 and *nad2* intron 3 was stronger than the interaction with total RNA. Then RNA-EMSA (RNA-Electro-

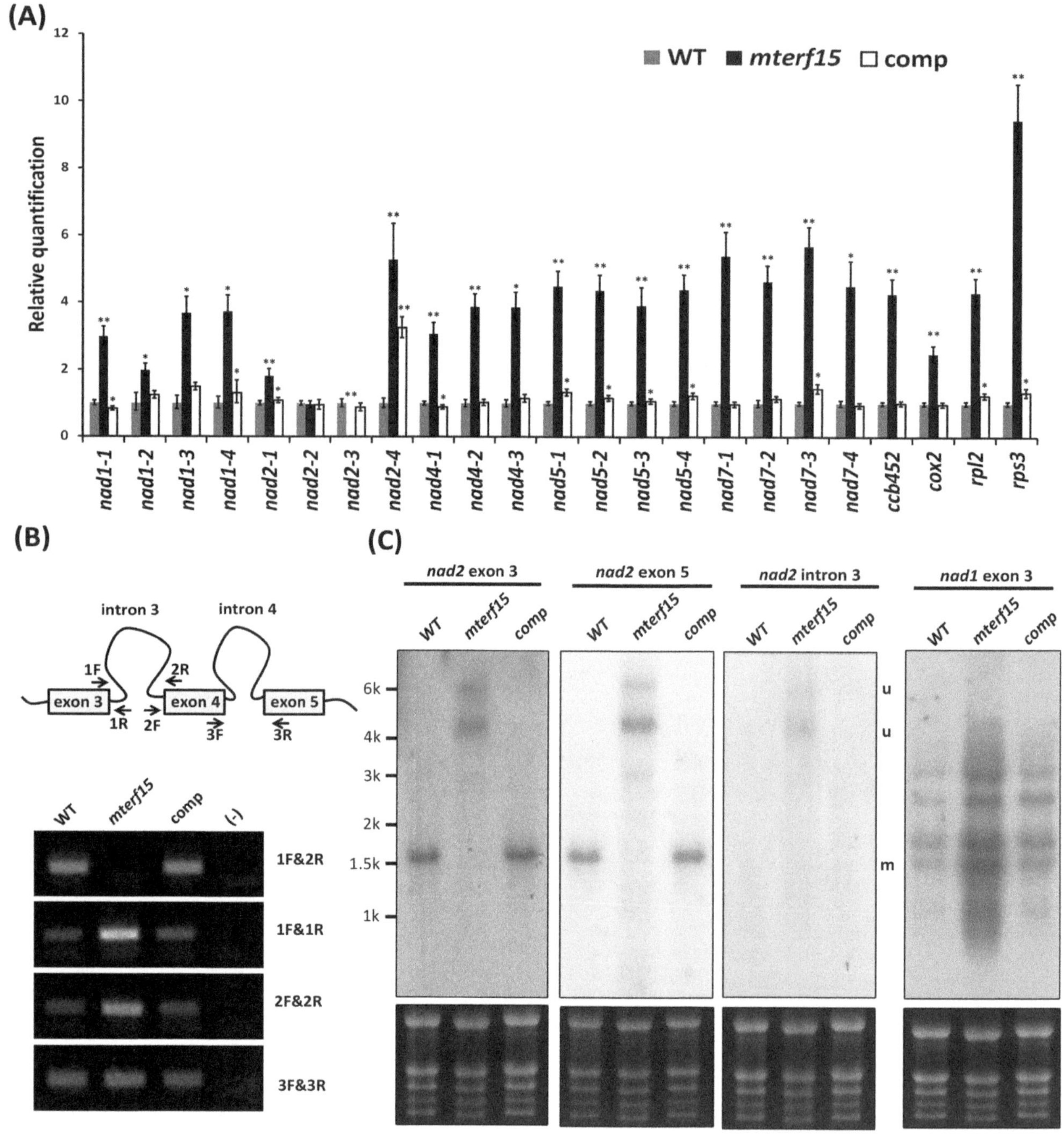

Figure 4. Mitochondrial splicing efficiency in *mterf15* mutants. (**A**) qRT-PCR analysis of mitochondrial RNA splicing in 7-d-old seedlings from the wild-type, *mterf15* and complementation lines. Data were normalized to the expression of 5 housekeeping genes (*YSL8, RPL5B, UBC, ACTIN2/ ACTIN8* and *TUB6*). Asterisks represent significant differences (*P<0.5, **P<0.01; Student's *t* test) relative to wild type. (**B**) Diagram of *nad2b* transcripts including exon 3 to 5 and RT-PCR analysis of transcript levels from the *nad2* exon–intron junction. Primers used for RT-PCR analysis are indicated with black arrows. (**C**) Northern blot analysis using probes recognizing *nad2b* transcripts (*nad2* exon 3 and *nad2* exon 5), *nad2* intron 3, and *nad1* exon 3. The radioisotope signal is shown in the top panel, and the rRNA signals shown in the lower panel were used as loading controls. Marker sizes are indicated. m, mature transcript; u, un-spliced transcript.

phoresis Mobility Shift Assay) was used to confirm the RNA-binding activity of mTERF15 (Figure S6). Additionally, two unrelated mitochondrial intron sequences, *ccmF*$_C$ and *rpl2*, were used as the control probes in RNA-EMSA. Since the full length of *nad2* intron 3 is too large to be used as the probe for RNA-EMSA, we divided it into five fragments (approximately 500 bp each), named Fragment 1-5 (Figure S6A). Fragment 1, part of domain 1, is the largest and substantially varies in sequence and secondary structure among characterized group II introns. Fragment 5 contains phylogenetically conserved and typical RNA structure found in all mitochondrial introns [51]. As shown in Figure S6B, only Fragment 1 could be recognized by mTERF15 in a dosage-dependent manner. Neither Fragments 2–5 nor *ccmFc* and *rpl2* could be recognized by mTERF15. Above results suggest that mTERF15 is an RNA-binding protein and is involved in the proper splicing of *nad2* intron 3.

In Arabidopsis, the *nad2* gene encodes a subunit of mitochondrial complex I (an NADH:ubiquinone oxidoreductase), which is required to translocate protons from the mitochondrial matrix into the intermembrane space. Complex I activity was therefore investigated in *mterf15*, in which the splicing of *nad2* intron 3 is defective. Isolated crude mitochondria were subjected to complex separation on clear native gels to visualize protein complexes by silver staining (Figure 5A) or to complex I activity assays (Figure 5B). The amount and activity of complex I were greatly reduced in *mterf15*, whereas the wild-type and complementation lines showed normal activity. Therefore, intron-splicing defects in *nad2* intron 3 led to reduced complex I formation and activity. Disturbing mitochondrial complex I impairs energy generation and leads to the extreme retardation of diverse aspects of growth and development in *mterf15* plants (Figure 1).

The steady-state levels of several mitochondrial proteins, including Nad9, Cox2, Cox3, AtpA, alternative oxidase (AOX) and the channel-forming protein Porin, were examined by immunoblot analysis of crude mitochondrial extracts from wild-type and *mterf15* plants (Figure 6). The steady-state levels of 3 proteins, Cox2, Cox3 and AtpA, were greater in *mterf15* plants than in wild-type plants. The level of alternative oxidase (AOX) was significantly increased in *mterf15* plants, which is consistent with a defect in NADH oxidation [46]. The level of another complex I subunit, Nad9, was decreased in *mterf15* plants, which may be due to an imbalance in the stoichiometry of multi-subunit complexes [47].

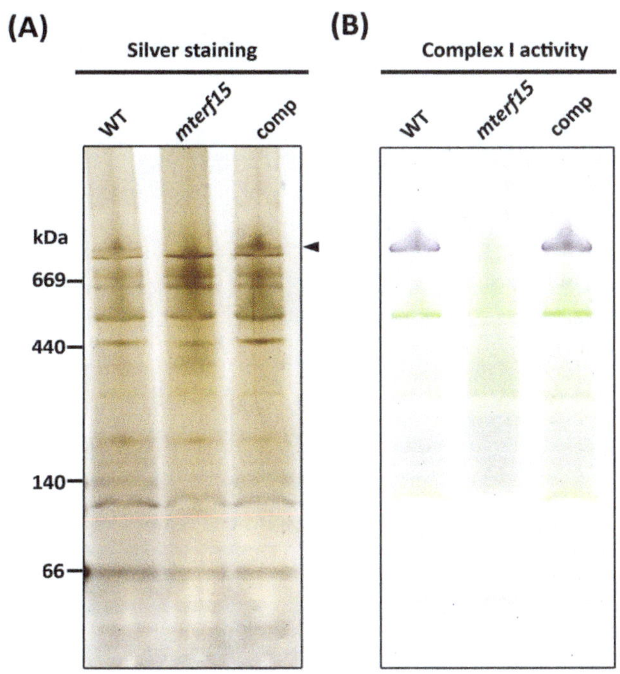

Figure 5. Clear native PAGE analysis of mitochondrial complex I activity. (**A**) Silver staining of the protein amount corresponding to crude mitochondria. (**B**) Complex I activity in crude mitochondria. Activity was visualized as a purple-blue color resulting from the utilization of the substrate (NADH) and the electron acceptor (nitroblue tetrazolium). Arrowhead indicated the band corresponding to Complex I.

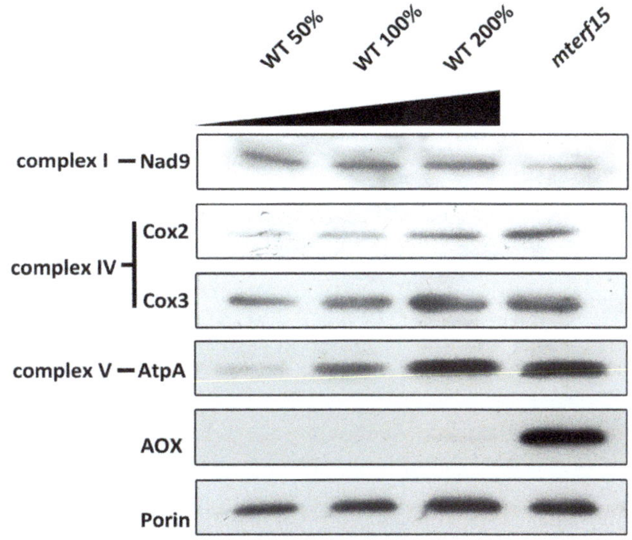

Figure 6. Relative accumulation of mitochondrial proteins in *mterf15* mutants. Crude mitochondria were isolated from the wild-type, *mterf15* and complementation lines. Following protein quantification based on the Bradford assay, equal amounts of protein were subjected to immunoblot analysis using various antibodies recognizing Nad9 [55], Cox2, Cox3, AtpA [56], and AOX [57], as well as Porin.

Discussion

Our genetic, molecular and biochemical results indicate that mTERF15 is a unique, mitochondria-localized protein (Figures S2 and S4) with RNA-binding activity (Fig. 3). This protein is critical for post-transcriptional modification in the RNA splicing of mitochondrial *nad2* intron 3 (Figure 4). Mutating *mTERF15* impaired the normal activity of mitochondrial respiratory chain complex I (Figure 5), thereby resulting in abnormal mitochondrial development (Figure 2) and the widespread retardation of plant growth and development (Figure 1).

mTERFs constitute a broad family of eukaryotic proteins that are essential for the initiation and termination of organellar transcription and the translation and replication of the organellar transcription machinery [12]; however, mounting evidence suggests an emerging role for these proteins in RNA splicing in plants [39]. The human genome contains only 4 genes encoding mTERF paralogs, whereas the Arabidopsis genome contains at least 35 genes for mTERF proteins with a diverse arrangement and number of mTERF motifs. Using transient assays and stable transgenic plants expressing GFP-tagged mTERF proteins, Babiychuk *et al.* (2011) identified 11 and 17 mTERF proteins that localize to chloroplasts and mitochondria, respectively. However, only a few mTERFs have been well studied and are essential for vegetative growth [24–26] and embryogenesis [22,23]. Whether Arabidopsis mTERF proteins share similar conserved molecular functions with their mammalian counterparts or have evolved additional regulatory mechanisms is unclear. In contrast to animal cells, plant cells harbor 2 types of nucleoid-containing organelles – chloroplasts and mitochondria. Moreover, plant mitochondrial genomes are much larger and more complex, requiring intron splicing for proper gene expression [48]. The biological functions of mTERF proteins may be complicated in plant cells because recent co-expression analyses of the 35 Arabidopsis mTERF members indicated the association of mTERF proteins with DNA and RNA metabolism [49]. Recently, it was found that Zm-mTERF4, an ortholog of Arabidopsis BSM/RUG2, is required for the splicing of several RNAs necessary for plastid translation in maize [39]. Therefore, an understanding of the mTERF family may provide new insights into the plant-specific functions of mTERF proteins in the transcriptional and post-transcriptional regulation of organellar nucleoids, as reported in this study.

In flowering plants, recombinogenic events such as the creation of intron-split genes greatly affect the dynamic nature of the mitochondrial genome [50]. The Arabidopsis mitochondrial genome contains 23 group II introns, and most are found dispersed in *nad* genes. The exons, including the flanking introns of these genes dispersed among the mtDNA, are transcribed and are the mRNAs generated by the splicing machinery. Until now, knowledge of the splicing machinery found in Arabidopsis mitochondria had been limited by the difficulty of organellar genome manipulation. However, a growing number of studies have revealed the involvement of a nuclear-encoded splicing factor in the excision of these introns [51]. Several proteins are involved in *nad2* intron splicing. For example, 2 proteins, ABA overly-sensitive 5 (ABO5) and RCC1/UVR8/GEF-like 3 (RUG3), were identified as splicing factors regulating the *cis*-splicing of *nad2* intron 3 [35,52]. *ABO5*, which encodes a PPR protein, was isolated from a mutational screen for ABA sensitivity and is required for *cis*-splicing of *nad2* intron 3 in mitochondria [35]. Another protein, *RUG3*, which encodes an RCC1/UVR8-like protein, is responsible for the efficient splicing of *nad2* introns 2 and 3 in mitochondria [52]. The mTERF15 protein, identified in this study, is required for *nad2* intron 3 RNA splicing, as demonstrated by RT-PCR and by northern blotting showing the accumulation of *nad2* intron 3 in *mterf15* plants (Figure 4B–C). Moreover, RNA-binding assays also demonstrated the direct interaction between mTERF15 and *nad2* intron 3. Our results suggest that mTERF15 is a new splicing factor in mitochondria. The next steps are to identify the specific elements of *nad2* intron 3 that are required for the binding of mTERF15 and to explore the relationships and/or potential interactions between ABO5, RUG3 and mTERF15 in *nad2* intron 3 splicing.

The *nad2* intron 3 splicing defect led to the loss of mitochondrial complex I activity in *mterf15*. Similar results were observed with the *RUG3* mutant [52]. Both mutants showed reduced root growth, small stature, and growth retardation. Another complex I mutant, *ndufs4* (NADH dehydrogenase [ubiquinone] fragment S subunit 4), also exhibited a similar phenotype [53]. Although approximately one-third of mitochondrial ATP production is associated with complex I, the presence of alternative NADH dehydrogenases allows the electron transport chain to bypass complex I when complex I efficiency is reduced or when plants encounter stress [54]. Interestingly, *abo5*, *rug3* and *ndufs4* plants produced viable seeds and propagated to the next generation, whereas *mterf15* showed a more severe phenotype. Homozygous *mterf15* plants showed a similar phenotype as these mutants in terms of complex I deficiency; however, seeds from homozygous *mterf15* plants could not germinate, even in solid medium over three months. It is possible that mTERF15 in maternal tissues of heterozygotic *mterf15* embryo may partially contribute to the developing homozygous *mterf15* seeds, which is completely absent in the developing seeds of homozygous *mterf15* plants. This suggests that the germination defect of *mterf15* is not only due to embryo lethality but also caused by aberrant reproductive organs, especially maternal tissues. The defective maternal tissues may reduce the flow of nutrient into the endosperm and eventually cause seed lethality [58]. Therefore, mTERF15 may have additional roles in mitochondria, as observed in other eukaryotes, during embryogenesis and seed development.

Here, we identify mTERF15 as an RNA-binding protein that is required for *nad2* intron 3 splicing in mitochondria. The disruption of this splicing event leads to decreased complex I activity as well as growth and developmental retardation in *mterf15* mutants. Further studies of the detailed mechanism of mTERF15 are required to reveal the molecular mechanisms involved in post-transcription regulation in Arabidopsis mitochondria.

Supporting Information

Figure S1 *mTERF15p::mTERF15-GFP* fully complements developmental defects found in *mterf15* homozygous mutant. 7-day-old seedlings from 6 independent complementation lines, homozygotes for *mTERF15* mutation and harbouring at least one copy of the transgenes *mTERF15p::mTERF15-GFP*. Seeds were surface-sterilized and germinated on half MS medium under 22°C with light period control (16-h light/8-h dark).

Figure S2 Protein alignment and phylogenetic tree of mTERF15 and homologs. (A) Protein sequence alignment of mTERF15 and its homologs in *Vitis vinifera* (Vv), *Populus trichocarpa* (Pt), *Brachypodium distachyon* (Bd), *Sorghum bicolor* (Sb) and *Zea mays* (Zm). The alignment was generated by use of CLUSTALW, which creates pairwise alignments to calculate the divergence between pairs of sequences. Five mTERF motifs present in all homologs are underlined in black. Asterisks indicate the location of highly conserved proline residues. (B) Phylogenetic tree of mTERF15 and its homologs. The phylogenetic tree was created with use of Molecular Evolutionary Genetics Analysis (MEGA) by the Unweighted Pair Group Method with Arithmatic Mean (UPGMA) method.

Figure S3 Spatial expression profile of *mTERF15*. qRT-PCR analysis of the expression of *mTERF15* in seedlings, roots, stems, rosette leaves, flowers, siliques and pollen. Primers used in this experiment are in Table S1.

Figure S4 Subcellular localization of mTERF15 protein. Confocal microscopy of the localization of the mTERF15 protein in root hairs of the *mterf15* complementation line harboring mTERF15p::mTERF15-GFP. (A) Signal corresponding to the mTERF15-GFP fusion protein; (B) signal corresponding to the MitoTracker marker; and (C) merged signals from GFP and the MitoTracker marker. Scale bar = 10 μm.

Figure S5 DNA-binding assay with mTERF15 recombinant protein. (A) *In vitro* binding studies of dsDNA-cellulose with mTERF15-GST and GST. mTERF15-GST and GST were incubated with dsDNA-cellulose for 2 h at room temperature. The resin was washed with washing buffer 3 times and after different salt concentrations to elute bound protein. The detailed experimental procedure was previously described (Wobbe and Nixon, 2013). (B) Southwestern blot assay of DNA-binding ability by mTERF15-strep. The upper panel is Coomassie brilliant blue (CBB) staining of MBP-strep and mTERF15-strep without the first N-terminal 34 amino acid. The lower panel is southwestern blot analysis with radioisotope-labeled HindIII-digested mtDNA. **Methods of Figure S5: DNA-binding assay with dsDNA-cellulose resin.** This experimental procedure has been described previously (Wobbe & Nixon, 2013). Briefly, 1.8 μM of the mTERF15-GST fusion protein and GST control were incubated with 60 mg dsDNA-cellulose resin (Sigma) in DCBB buffer [50 mM HEPES, pH 8.0, 50 mM NaCl, 1 mM EDTA, 1 mM beta-mercaptoethanol and protease inhibitor cocktail (Roche)] at room temperature for 2 hr and under gentle rotation. The resin was washed 3 times with DCBB buffer, then bound protein was eluted at different salt concentrations (100, 300, 600 and 1000 mM NaCl in DCBB buffer). The samples corresponding to each step were collected and analyzed on SDS-PAGE. **Southwestern blot assay.** The recombinant protein were separated on 12% SDS-polyacrylamide gels and transferred onto a PVDF membrane and then renatured overnight in renaturation buffer (0.1 M Tris-HCl, pH 7.5 and 0.1% (v/v) NP-40) at 4°C. After 4 washes with renaturation buffer (15 min), blots were blocked with blocking buffer [(10 mM Tris-HCl, pH 7.5, 2 mM DTT, 5% (w/v) BSA and 0.01% (v/v) Triton X-100)] at room temperature for 5 min. Mitochondrial DNA was isolated from crude mitochondria fraction, digested with HindIII and labeled with radioisotope using T4 polynucleotide kinase (Fermentas). Hybridization was performed at 4°C overnight in the presence of 5 μg labeled mtDNA. Blots were washed 4 times (5 min) in washing buffer (10 mM Tris-HCl, pH 7.5, 2 mM DTT) to remove unbound DNA. Radioisotope signals from PVDF membrane were detected via autoradiography on X-ray film (Kodak).

Figure S6 RNA-EMSA of mTERF15 recombinant protein with different mitochondrial transcribed intron fragments. (A) The diagram of *nad2* intron 3 and the regions and sizes of 5 fragments in *nad2* intron 3 used for assays. (B) Different concentration mTERF15 recombinant proteins (20, 10, 5 and 0 nM) were incubated with 100 nM radioisotope labeled indicated RNA probes. **Methods of Figure S6:** mTERF15 fusion protein were purified as mentioned in Northwestern blot analysis. For *in vitro* transcription, mitochondrial introns were amplified and cloned to pJET vector. Primer sets are nad2int3-1F "ACGCCAAGCTATTTAGGTGACACTATAGAATACGGGCGGCTGTAGGACGGAC" and nad2int3-1R "CTGTTCACCGTTGGATCTCGCC" for *nad2* intron 3 Fragment 1; nad2int3-5F "ACGCCAAGCTATTTAGGTGACACTATAGAATACGCGTGTTATCTGAAGGGAGCACG" and nad2int3-5R "GGGGGAGGGGGTTTTCTTCG" for *nad2* intron 3 Fragment 5; rpl2-F "ACGCCAAGCTATTTAGGTGACACTATAGAATACATGAGACCAGGGAGAGCAAGAGCAC" and rpl2-mid-intron-R "CGTTGCTAAGCCAAGGTCCC" for *rpl2* intron; ccmFc-F "ACGCCAAGCTATTTAGGTGACACTATAGAATACATGGTCCAACTACATAACTTTTTC" and ccmFc-mid-intron-R "GCTTTGCCAACACAACATTAGG" for $ccmF_C$ intron. Then, RNA probes were *in vitro* synthesized using SP6 polymerase (Promega) and labeled with γ^{32}P ATP using T4 polynucleotide kinase (Fermentas). Different concentration of mTERF15 recombinant proteins (20, 10, 5 and 0 nM) were incubated with 100 nM *in vitro* synthesized RNAs for 30 min at room temperature in binding buffer (20 mM Tris/HCl pH 7.5, 180 mM NaCl, 2 mM dithiothreitol, 17 μg/μl BSA, 0.5 mM EDTA, and 20 μg/ml heparin). The mixtures were separated by 5% polyacrylamide gel and gel was dried. Signals were detected via autoradiography on X-ray film (Kodak).

Table S1 Primer sequences used in mTERF125 cloning, gene expression, and probe preparation for northern blotting and RT-PCR for the *nad2* exon–intron junction.

Table S2 Primer sequences for determining RNA splicing efficiency in mitochondria.

Table S3 Protein identity and similarity of mTERF15 with its plant homologs and mTERF1 in metazoans.

Acknowledgments

We thank Dr. Géraldine Bonnard (Institut de Biologie Moléculaire des Plantes, Centre National de la Recherche Scientifique, Strasbourg, France) for generously sharing the anti-NAD9 antibody, Ms. Mei-Jane Fang for her assistance with qPCR (DNA Analysis Core Laboratory, Institute of Plant and Microbial Biology, Academia Sinica) and Dr. Wann-Neng Jane for assistance with transmission and scanning electron microscopy (Plant Cell Biology Core Laboratory, Institute of Plant and Microbial Biology, Academia Sinica).

Author Contributions

Conceived and designed the experiments: YWH HJW GYJ. Performed the experiments: YWH HJW. Analyzed the data: YWH MHH HLH GYJ. Contributed reagents/materials/analysis tools: YWH MHH HLH GYJ. Wrote the paper: YWH MHH GYJ.

References

1. Andersson SG, Zomorodipour A, Andersson JO, Sicheritz-Ponten T, Alsmark UC, et al. (1998) The genome sequence of *Rickettsia prowazekii* and the origin of mitochondria. Nature 396: 133–140.
2. Gray MW, Burger G, Lang BF (1999) Mitochondrial evolution. Science 283: 1476–1481.
3. Gkikas I, Petratou D, Tavernarakis N (2014) Longevity pathways and memory aging. Front Genet 5: 155.
4. Unseld M, Marienfeld JR, Brandt P, Brennicke A (1997) The mitochondrial genome of *Arabidopsis thaliana* contains 57 genes in 366,924 nucleotides. Nat Genet 15: 57–61.
5. Martin W, Herrmann RG (1998) Gene transfer from organelles to the nucleus: how much, what happens, and why? Plant Physiol 118: 9–17.
6. Martin W, Stoebe B, Goremykin V, Hapsmann S, Hasegawa M, et al. (1998) Gene transfer to the nucleus and the evolution of chloroplasts. Nature 393: 162–165.
7. Welchen E, Garcia L, Mansilla N, Gonzalez DH (2014) Coordination of plant mitochondrial biogenesis: keeping pace with cellular requirements. Front Plant Sci 4: 551.
8. Liere K, Weihe A, Borner T (2011) The transcription machineries of plant mitochondria and chloroplasts: composition, function, and regulation. J Plant Physiol 168: 1345–1360.
9. Binder S, Brennicke A (2003) Gene expression in plant mitochondria: transcriptional and post-transcriptional control. Philos Trans R Soc Lond B Biol Sci 358: 181–188; discussion 188–189.
10. Hammani K, Giege P (2014) RNA metabolism in plant mitochondria. Trends Plant Sci 19: 380–389.
11. Kruse B, Narasimhan N, Attardi G (1989) Termination of transcription in human mitochondria: identification and purification of a DNA binding protein factor that promotes termination. Cell 58: 391–397.
12. Linder T, Park CB, Asin-Cayuela J, Pellegrini M, Larsson NG, et al. (2005) A family of putative transcription termination factors shared amongst metazoans and plants. Curr Genet 48: 265–269.
13. Roberti M, Polosa PL, Bruni F, Manzari C, Deceglie S, et al. (2009) The MTERF family proteins: mitochondrial transcription regulators and beyond. Biochim Biophys Acta 1787: 303–311.
14. Yakubovskaya E, Guja KE, Mejia E, Castano S, Hambardjieva E, et al. (2012) Structure of the essential MTERF4:NSUN4 protein complex reveals how an MTERF protein collaborates to facilitate rRNA modification. Structure 20: 1940–1947.
15. Yakubovskaya E, Mejia E, Byrnes J, Hambardjieva E, Garcia-Diaz M (2010) Helix unwinding and base flipping enable human MTERF1 to terminate mitochondrial transcription. Cell 141: 982–993.
16. Martin M, Cho J, Cesare AJ, Griffith JD, Attardi G (2005) Termination factor-mediated DNA loop between termination and initiation sites drives mitochondrial rRNA synthesis. Cell 123: 1227–1240.
17. Pellegrini M, Asin-Cayuela J, Erdjument-Bromage H, Tempst P, Larsson NG, et al. (2009) MTERF2 is a nucleoid component in mammalian mitochondria. Biochim Biophys Acta 1787: 296–302.

18. Wenz T, Luca C, Torraco A, Moraes CT (2009) mTERF2 regulates oxidative phosphorylation by modulating mtDNA transcription. Cell Metab 9: 499–511.
19. Park CB, Asin-Cayuela J, Camara Y, Shi Y, Pellegrini M, et al. (2007) MTERF3 is a negative regulator of mammalian mtDNA transcription. Cell 130: 273–285.
20. Camara Y, Asin-Cayuela J, Park CB, Metodiev MD, Shi Y, et al. (2011) MTERF4 regulates translation by targeting the methyltransferase NSUN4 to the mammalian mitochondrial ribosome. Cell Metab 13: 527–539.
21. Spahr H, Habermann B, Gustafsson CM, Larsson NG, Hallberg BM (2012) Structure of the human MTERF4-NSUN4 protein complex that regulates mitochondrial ribosome biogenesis. Proc Natl Acad Sci U S A 109: 15253–15258.
22. Babiychuk E, Vandepoele K, Wissing J, Garcia-Diaz M, De Rycke R, et al. (2011) Plastid gene expression and plant development require a plastidic protein of the mitochondrial transcription termination factor family. Proc Natl Acad Sci U S A 108: 6674–6679.
23. Meskauskiene R, Wursch M, Laloi C, Vidi PA, Coll NS, et al. (2009) A mutation in the Arabidopsis mTERF-related plastid protein SOLDAT10 activates retrograde signaling and suppresses $^1O_2$-induced cell death. Plant J 60: 399–410.
24. Quesada V, Sarmiento-Manus R, Gonzalez-Bayon R, Hricova A, Perez-Marcos R, et al. (2011) Arabidopsis *RUGOSA2* encodes an mTERF family member required for mitochondrion, chloroplast and leaf development. Plant J 68: 738–753.
25. Robles P, Micol JL, Quesada V (2012) Arabidopsis MDA1, a nuclear-encoded protein, functions in chloroplast development and abiotic stress responses. PLoS One 7: e42924.
26. Kim M, Lee U, Small I, des Francs-Small CC, Vierling E (2012) Mutations in an Arabidopsis mitochondrial transcription termination factor-related protein enhance thermotolerance in the absence of the major molecular chaperone HSP101. Plant Cell 24: 3349–3365.
27. Asakura Y, Barkan A (2007) A CRM domain protein functions dually in group I and group II intron splicing in land plant chloroplasts. Plant Cell 19: 3864–3875.
28. Asakura Y, Bayraktar OA, Barkan A (2008) Two CRM protein subfamilies cooperate in the splicing of group IIB introns in chloroplasts. RNA 14: 2319–2332.
29. Zmudjak M, Colas des Francs-Small C, Keren I, Shaya F, Belausov E, et al. (2013) mCSF1, a nucleus-encoded CRM protein required for the processing of many mitochondrial introns, is involved in the biogenesis of respiratory complexes I and IV in Arabidopsis. New Phytol 199: 379–394.
30. de Longevialle AF, Hendrickson L, Taylor NL, Delannoy E, Lurin C, et al. (2008) The pentatricopeptide repeat gene *OTP51* with two LAGLIDADG motifs is required for the *cis*-splicing of plastid *ycf3* intron 2 in *Arabidopsis thaliana*. Plant J 56: 157–168.
31. de Longevialle AF, Meyer EH, Andres C, Taylor NL, Lurin C, et al. (2007) The pentatricopeptide repeat gene *OTP43* is required for *trans*-splicing of the mitochondrial *nad1* Intron 1 in *Arabidopsis thaliana*. Plant Cell 19: 3256–3265.
32. Koprivova A, des Francs-Small CC, Calder G, Mugford ST, Tanz S, et al. (2010) Identification of a pentatricopeptide repeat protein implicated in splicing of intron 1 of mitochondrial *nad7* transcripts. J Biol Chem 285: 32192–32199.
33. Chateigner-Boutin AL, des Francs-Small CC, Delannoy E, Kahlau S, Tanz SK, et al. (2011) OTP70 is a pentatricopeptide repeat protein of the E subgroup involved in splicing of the plastid transcript *rpoC1*. Plant J 65: 532–542.
34. Khrouchtchova A, Monde RA, Barkan A (2012) A short PPR protein required for the splicing of specific group II introns in angiosperm chloroplasts. RNA 18: 1197–1209.
35. Liu Y, He J, Chen Z, Ren X, Hong X, et al. (2010) *ABA overly-sensitive 5 (ABO5)*, encoding a pentatricopeptide repeat protein required for *cis*-splicing of mitochondrial *nad2* intron 3, is involved in the abscisic acid response in Arabidopsis. Plant J 63: 749–765.
36. Francs-Small CC, Kroeger T, Zmudjak M, Ostersetzer-Biran O, Rahimi N, et al. (2012) A PORR domain protein required for *rpl2* and *ccmF$_c$* intron splicing and for the biogenesis of *c*-type cytochromes in Arabidopsis mitochondria. Plant J 69: 996–1005.
37. Kroeger TS, Watkins KP, Friso G, van Wijk KJ, Barkan A (2009) A plant-specific RNA-binding domain revealed through analysis of chloroplast group II intron splicing. Proc Natl Acad Sci U S A 106: 4537–4542.
38. Watkins KP, Rojas M, Friso G, van Wijk KJ, Meurer J, et al. (2011) APO1 promotes the splicing of chloroplast group II introns and harbors a plant-specific zinc-dependent RNA binding domain. Plant Cell 23: 1082–1092.
39. Hammani K, Barkan A (2014) An mTERF domain protein functions in group II intron splicing in maize chloroplasts. Nucleic Acids Res doi: 10.1093/nar/gku112.
40. Bechtold N, Pelletier G (1998) In planta Agrobacterium-mediated transformation of adult *Arabidopsis thaliana* plants by vacuum infiltration. Methods Mol Biol 82: 259–266.
41. Thangasamy S, Chen PW, Lai MH, Chen J, Jauh GY (2012) Rice LGD1 containing RNA binding activity affects growth and development through alternative promoters. Plant J 71: 288–302.
42. Pineau B, Layoune O, Danon A, De Paepe R (2008) L-galactono-1,4-lactone dehydrogenase is required for the accumulation of plant respiratory complex I. J Biol Chem 283: 32500–32505.
43. Wittig I, Karas M, Schagger H (2007) High resolution clear native electrophoresis for in-gel functional assays and fluorescence studies of membrane protein complexes. Mol Cell Proteomics 6: 1215–1225.
44. Cupp JD, Nielsen BL (2012) *Arabidopsis thaliana* organellar DNA polymerase IB mutants exhibit reduced mtDNA levels with a decrease in mitochondrial area density. Physiol Plant 49: 91–103.
45. Udy DB, Belcher S, Williams-Carrier R, Gualberto JM, Barkan A (2012) Effects of reduced chloroplast gene copy number on chloroplast gene expression in maize. Plant Physiol 160: 1420–1431.
46. Rasmusson AG, Geisler DA, Moller IM (2008) The multiplicity of dehydrogenases in the electron transport chain of plant mitochondria. Mitochondrion 8: 47–60.
47. Adam Z, Ostersetzer O (2001) Degradation of unassembled and damaged thylakoid proteins. Biochem Soc Trans 29: 427–430.
48. Gray MW (1989) Origin and evolution of mitochondrial DNA. Annu Rev Cell Biol 5: 25–50.
49. Kleine T (2012) *Arabidopsis thaliana* mTERF proteins: evolution and functional classification. Front Plant Sci 3: 233.
50. Bonen L (2008) *Cis*- and *trans*-splicing of group II introns in plant mitochondria. Mitochondrion 8: 26–34.
51. de Longevialle AF, Small ID, Lurin C (2010) Nuclearly encoded splicing factors implicated in RNA splicing in higher plant organelles. Mol Plant 3: 691–705.
52. Kuhn K, Carrie C, Giraud E, Wang Y, Meyer EH, et al. (2011) The RCC1 family protein RUG3 is required for splicing of *nad2* and complex I biogenesis in mitochondria of *Arabidopsis thaliana*. Plant J 67: 1067–1080.
53. Meyer EH, Tomaz T, Carroll AJ, Estavillo G, Delannoy E, et al. (2009) Remodeled respiration in *ndufs4* with low phosphorylation efficiency suppresses Arabidopsis germination and growth and alters control of metabolism at night. Plant Physiol 151: 603–619.
54. Millar AH, Whelan J, Soole KL, Day DA (2011) Organization and regulation of mitochondrial respiration in plants. Annu Rev Plant Biol 62: 79–104.
55. Lamattina L, Gonzalez D, Gualberto J, Grienenberger JM (1993) Higher plant mitochondria encode an homologue of the nuclear-encoded 30-kDa subunit of bovine mitochondrial complex I. Eur J Biochem 217: 831–838.
56. Luethy MH, Horak A, Elthon TE (1993) Monoclonal Antibodies to the α- and β-Subunits of the Plant Mitochondrial F1-ATPase. Plant Physiol 101: 931–937.
57. Elthon TE, Nickels RL, McIntosh L (1989) Monoclonal antibodies to the alternative oxidase of higher plant mitochondria. Plant Physiol 89: 1311–1317.
58. Chaudhury AM, Berger F (2001) Maternal control of seed development. Semin Cell Dev Biol 12: 381–386.

10

Variability among the Most Rapidly Evolving Plastid Genomic Regions is Lineage-Specific: Implications of Pairwise Genome Comparisons in *Pyrus* (Rosaceae) and Other Angiosperms for Marker Choice

Nadja Korotkova[1,2,3☯], Lars Nauheimer[1,2☯], Hasmik Ter-Voskanyan[3,4], Martin Allgaier[5], Thomas Borsch[1,2,3*]

1 Institut für Biologie/Botanik, Systematische Botanik und Pflanzengeographie, Freie Universität Berlin, Berlin, Germany, **2** Dahlem Centre of Plant Sciences (DCPS), Berlin, Germany, **3** Botanischer Garten und Botanisches Museum Berlin-Dahlem, Berlin, Germany, **4** Institute of Botany, National Academy of Sciences of Republic Armenia, Yerevan, Armenia, **5** The Berlin Center for Genomics in Biodiversity Research (BeGenDiv), Berlin, Germany

Abstract

Plastid genomes exhibit different levels of variability in their sequences, depending on the respective kinds of genomic regions. Genes are usually more conserved while noncoding introns and spacers evolve at a faster pace. While a set of about thirty maximum variable noncoding genomic regions has been suggested to provide universally promising phylogenetic markers throughout angiosperms, applications often require several regions to be sequenced for many individuals. Our project aims to illuminate evolutionary relationships and species-limits in the genus *Pyrus* (Rosaceae)—a typical case with very low genetic distances between taxa. In this study, we have sequenced the plastid genome of *Pyrus spinosa* and aligned it to the already available *P. pyrifolia* sequence. The overall *p*-distance of the two *Pyrus* genomes was 0.00145. The intergenic spacers between *ndhC–trnV*, *trnR–atpA*, *ndhF–rpl32*, *psbM–trnD*, and *trnQ–rps16* were the most variable regions, also comprising the highest total numbers of substitutions, indels and inversions (potentially informative characters). Our comparative analysis of further plastid genome pairs with similar low *p*-distances from *Oenothera* (representing another rosid), *Olea* (asterids) and *Cymbidium* (monocots) showed in each case a different ranking of genomic regions in terms of variability and potentially informative characters. Only two intergenic spacers (*ndhF–rpl32* and *trnK–rps16*) were consistently found among the 30 top-ranked regions. We have mapped the occurrence of substitutions and microstructural mutations in the four genome pairs. High AT content in specific sequence elements seems to foster frequent mutations. We conclude that the variability among the fastest evolving plastid genomic regions is lineage-specific and thus cannot be precisely predicted across angiosperms. The often lineage-specific occurrence of stem-loop elements in the sequences of introns and spacers also governs lineage-specific mutations. Sequencing whole plastid genomes to find markers for evolutionary analyses is therefore particularly useful when overall genetic distances are low.

Editor: Damon P. Little, The New York Botanical Garden, United States of America

Funding: The study was carried out as part of the project "Developing tools for conserving the plant diversity of the Transcaucasus" funded by VolkswagenStiftung (http://www.volkswagenstiftung.de/nc/en.html), grant number AZ85021. The funders had no role in study design, data collection and analysis, decision to publish, or preparation of the manuscript.

Competing Interests: The authors have declared that no competing interests exist.

* Email: t.borsch@bgbm.org

☯ These authors contributed equally to this work.

Introduction

Clarifying species limits and reconstructing phylogenetic relationships in clades with recently diverged species is challenging. Levels of genetic divergence are often low while at the same time large numbers of samples need to be analysed. The same applies to analysing phylogeographic patterns, where many individuals from different populations need to be included. Due to the often complex modes of speciation in angiosperms, evidence from uniparentally inherited organellar genomes and the recombined nuclear genome is needed to unravel evolutionary histories [1–3]. This is also the case in the genus *Pyrus* where — like in many Rosaceae — polyploidy, hybridization, and reticulate evolution occur. Estimates of *Pyrus* diversity vary between 50 and 80 species [4,5] and 20 taxa alone have been described from the southern Caucasus [6,7]. Similarly, the numbers of accepted species differ between treatments as a consequence of poorly understood species limits. *Pyrus* is a typical case for evolutionary and taxonomic analyses of diverse species groups in flowering plants that require the inclusion of hundreds of individuals. Before entering into large-scale sampling, we were interested to find the genomic regions with the best information potential for generating haplotype networks and inferring phylogenetic relationships. In this study, we focus on the plastid genome.

Along the same line of argumentation, Shaw et al. [8,9] inspired to employ a broader spectrum of noncoding and rapidly evolving plastid markers in phylogenetic analyses of closely related species. Shaw et al. [8] sequenced a wide range of plastid markers for three species across angiosperms and later compared plastid genome pairs of three lineages of angiosperms (*Atropa* and *Nicotiana* for the asterids, *Lotus* and *Medicago* for the rosids, and *Oryza* and *Saccharum* for the monocots) [9]. Their studies resulted in a set of 32 regions that ranked highest in their number of potentially informative characters (defined as sum of substitutions, indels and inversions following [8] and abbreviated as "PICs"). This set was consequently suggested to generally contain the most variable and phylogenetically most informative genomic regions in angiosperm plastid genomes. However, the question remains how to best select four or five of the total top 32 regions, as many species-level evolutionary studies require.

Noncoding genomic regions such as introns and spacers often contain stem-loops and other specific structural elements that can be highly dynamic and are AT-rich. This results in a mosaic-like pattern of conserved and variable elements [10]. Considering that certain stem-loop elements within given introns and spacers are often unique to restricted lineages [11,12], lineage specificity in the overall variability of genomic regions is to be expected. In several recent comparative analyses of angiosperm plastid genomes [13,14] different genomic regions were depicted as the most variable. Nonetheless, these results need to be considered with care because some of the respective authors worked with pairs of hardly differentiated genomes while others had pairs of genomes with high *p*-distances. We expect that taxon-specific differences caused by certain sequence elements will be less prominent when more distant genomes are studied.

Next-generation sequencing techniques greatly facilitate the analysis of whole plastid genomes [15–17]. To date, phylogenomic studies of plastid genomes in land plants often just relied on concatenated sequences of the conserved genes, neglecting the information from the noncoding regions. In other cases, the authors included rather few taxa for which plastid genome sequences were automatically assembled from the respective 454 or Illumina runs, without completing parts of low coverage or areas with difficulties to obtain correct sequences. However, especially those might be informative at and below the species level (e.g., AT-rich stretches of DNA including microsatellites) [18–20]. On the other hand, there are recent studies which used completely annotated plastid genomes to detect infraspecific variability in species of *Olea* [21], *Colocasia* [22], or *Phalaenopsis* [23], or to find genomic regions with the highest number of potentially informative characters in more distant genome pairs of angiosperm genera [9,24–26].

We have sequenced the plastid genome of *Pyrus spinosa* using 454 pyrosequencing in order to compare it with the published plastid genome sequence of *P. pyrifolia* [27]. In our *Pyrus* genome pair, the proportion of sites at which the two sequences are different (*p*-distances) is almost 10-fold lower than in the genome pairs studied by Shaw et al. [9]. For further comparison, we selected three fully annotated plastid genome pairs using the criterion of low *p*-distances (≤0.005) similar to *Pyrus*. Here we wanted to represent another rosid pair (*Oenothera parviflora* and *O. argillicola*; Onagraceae), an asterid pair (*Olea europaea* and *O. woodiana*; Oleaceae) and a monocot pair (*Cymbidium tortisepalum* and *C. sinense* (Orchidaceae).

The goals of this study were (1) to find the most variable regions of the *Pyrus* plastid genome and to propose plastid markers for species-level evolutionary studies in *Pyrus*, (2) to assess the variability of plastid genome regions based on comparable genome-pairs with overall low *p*-distances (0.0005 to 0.005) in major lineages of angiosperms, (3) to clarify if there are universal or lineage-specific rankings of variability within the group of about 35 top variable genomic regions, and (4) to evaluate if there are lineage specific differences in molecular evolutionary patterns that could cause the variability of genomic regions.

Material and Methods

DNA extraction, 454 pyrosequencing, genome assembly and annotation

Pyrus spinosa was sampled from the living collection of the Botanical Garden Berlin-Dahlem (Acc. No. 248458110, IPEN-Nr. TR-0-B-2484581, origin: Turkey: Kastamonu, Pontic Mountains around Küre, leg.: Ern, Krone 7145, 9/1981, voucher at B). The leaf tissue was silica-dried and total genomic DNA was extracted using the NucleoSpin Plant II kit (Macherey Nagel) according to the manufacturer's instructions.

Shotgun sequencing from total genomic DNA was performed on a Roche 454 GS-FLX Titanium sequencer (Roche Applied Science, Indianapolis, Indiana, USA). The 454 run (1/4 plate) resulted in 120,255 reads with an average of 400 bp after removing the adaptor sequences.

An initial mapping assembly with MIRA 4 [31] using *Pyrus pyrifolia* as reference resulted in 4191 reads mapped to a single contig with an average coverage of 13.44. However, reads with larger indels, not occurring in the reference, were not incorporated into the contigs what lead to an incorrect genome sequence. To remove the bias of the reference sequence, the reads were *de novo* assembled to contigs using the Roche GS *De Novo* Assembler (Newbler) v.2.6 which resulted in 836 large contigs (N50 = 829), and with Mira 4 [28], which resulted in 1125 large contigs (N50 = 1072, N90 = 538, N95 = 519). All these contigs were mapped on the *Pyrus pyrifolia* plastid genome (GenBank acc. no. NC015996; Terakami et al. [27]) using Geneious 7 to produce a consensus sequence. The combined method of mapping *de novo* contigs recovered nine indels (maximum length 71 bp), which were not found with mapping alone. Finally the second inverted repeat was manually inserted into the consensus sequence.

The positions of protein coding genes, rRNAs, tRNAs and the inverted repeats were annotated with the help of DOGMA [29] and Geneious 7. All coordinates of exons, reading frames and the positions of tRNAs were manually checked by aligning the respective genes of *Nicotiana tabacum* L. (NC001879) to the *Pyrus spinosa* sequence in PhyDe [30] because DOGMA tends to incorrectly place the start and stop codons and often does not annotate small exons. In case of more deviating gene sequences (e.g. *matK* or *ycf1*), the *Pyrus* gene sequences were translated to amino acid sequences to correctly annotate the reading frame.

Verification by Sanger sequencing. Pyrosequencing is limited in that the exact number of nucleotides within longer homonucleotide stretches (polyAs or polyTs) cannot be reliably determined [16,31]. Our initial assembly contained several homonucleotide stretches and AT-rich sequence motifs. In our data, ambiguously called bases were frequent in homonucleotide stretches with more than six of the same nucleotides. To validate the sequence in such parts, we applied the Sanger method (electrophoresis was done at Macrogen Europe, Amsterdam, The Netherlands). Primers for amplification and sequencing were taken from the literature or designed in this study (see Table S1). Pherograms were checked by eye for peaks and corresponding quality scores to ensure that the polyA/T stretch was correctly read. All Sanger sequencing reads were unambiguous with no overlapping peaks after the polyA/T stretches. The respective

Table 1. GenBank accession numbers and references for the plastid genomes used in this study.

Species	GenBank accession number	Reference
Pyrus spinosa	HG737342	this study
Pyrus pyrifolia	NC015996	Terakami et al. [27]
Cymbidium tortisepalum	NC021431	Yang et al. [24]
Cymbidium sinense	NC021430	Yang et al. [24]
Oenothera parviflora	NC010362	Greiner et al. [66]
Oenothera argillicola	EU262887	Greiner et al. [67]
Olea woodiana	NC015608	Besnard et al. [68]
Olea europaea	NC015401	Besnard et al. [68]

reads were aligned with the previously assembled genome sequence in Geneious 7 and the consensus sequence was corrected accordingly. The *Pyrus spinosa* plastid genome sequence is available in EMBL under accession HG737342.

Pairwise genome comparisons and calculation of sequence divergence. In addition to *Pyrus*, we took three other plastid genome pairs from published sources to represent closely related species, a further rosid genus, an asterid and a monocot genus. Genome sequences had to be complete and fully annotated. The aligned genome pairs had to show an overall distance of $p<0.005$ (Table 1). All genome sequences were aligned in PhyDe using a motif alignment approach [32,33]. The pairwise alignments are provided as File S1, S2, S3, and S4.

Sequences of all introns and intergenic spacers larger than 100 bp were extracted from the alignments. The number of single nucleotide polymorphisms (SNPs) and indels for each sequence pair were counted with a script in R (v. 3.0.2). PICs were then determined in the sense of Shaw et al. [8] as the sum of all substitutions and indels. *P*-distances (proportion of differing nucleotide sites in the two sequences compared) of the regions were calculated by dividing the number of SNPs by the length of the regions without counting indel positions. The two parts of the *trnK* intron were analysed separately.

To assess the *p*-distances of the genome pairs used by Shaw et al. [8], we have aligned the genomes of *Lotus japonicus* (NC002694) and *Medicago truncatula* (AC093544); *Nicotiana tabacum* (NC001879) and *Atropa belladonna* (NC004561.1); *Saccharum* hybrid (NC005878) and *Oryza sativa* (NC008155) using MAFFT v. 7 [34], and calculated the *p*-distances of these genomes using PAUP* v. 4.0b10 [35].

To compare the whole genome variability apart from specific regions, a sliding window approach was performed counting the number of SNPs and indels and calculating the AT-content for 500 bp slots of the consensus sequences. The genome comparisons were visualized using Circos v. 0.64 [36].

Molecular evolution within genomic regions

In order to assess the role of the base composition in variable sequence parts, i.e., indels and nucleotides around SNPs, we calculated their AT contents and compared them with the overall AT content of the whole genomes (consensus of pairwise aligned genomes). Three groups of indels were distinguished: (1) length variable poly-n loci that consist of a single nucleotide that is repeated at least sevenfold, (2) simple sequence repeats (SSRs) that show one repetition of a motif of multiple nucleotides, inverted repeats, or inversions, and (3) indels that do not fall in the former categories.

Further, AT contents of nucleotides adjacent to SNPs were calculated in intervals of increasing size (1–10, 20, 50, and 100 bp in each direction). A script was written in R v.3.0.2, which distinguishes the indels and regions around SNPs, calculates the AT contents, and displays their distributions.

The lineage-specific occurrence of substitutions and microstructural mutations was examined in more detail on the example of group II introns (*atpF*, *rpl16*) that strongly deviated in variability among our four genome pairs. These introns possess a mosaic-like structure of conserved and variable sequence elements. The variable parts usually correspond to the structurally and functionally least constrained terminal stem-loops, which appear in the respective RNA secondary structure. We first annotated the domains of the *atpF* and *rpl16* introns by comparing our sequences with the consensus alignment of Michel et al. [37].The RNA secondary structures of individual domains were then predicted using RNAstructure 5.6 (available at http://rna.urmc.rochester.edu/RNAstructure.html) using the algorithm of Mathews et al. [38]. The "fold as RNA" option was implemented to allow for U–G pairings.

Selecting genomic regions as markers for evolutionary studies in *Pyrus*. Our aim was not only to find the most variable plastid regions in *Pyrus* but also to select several regions to be best used in evolutionary studies of *Pyrus*. Thus, efficient

Table 2. Sequence statistics for the four genome pairs compared.

Genome pair	p-distance	Aligned length [bp]	Length difference	SNPs	Indels
Pyrus spinosa/P. pyrifolia	0.00145	160607 bp	227 bp	230	173
Olea europaea/O. woodiana	0.00294	156091 bp	30 bp	458	112
Oenothera parviflora/O. argillicola	0.00122	165952 bp	1690 bp	199	173
Cymbidium tortisepalum/C. sinense	0.0008	155833 bp	79 bp	124	62

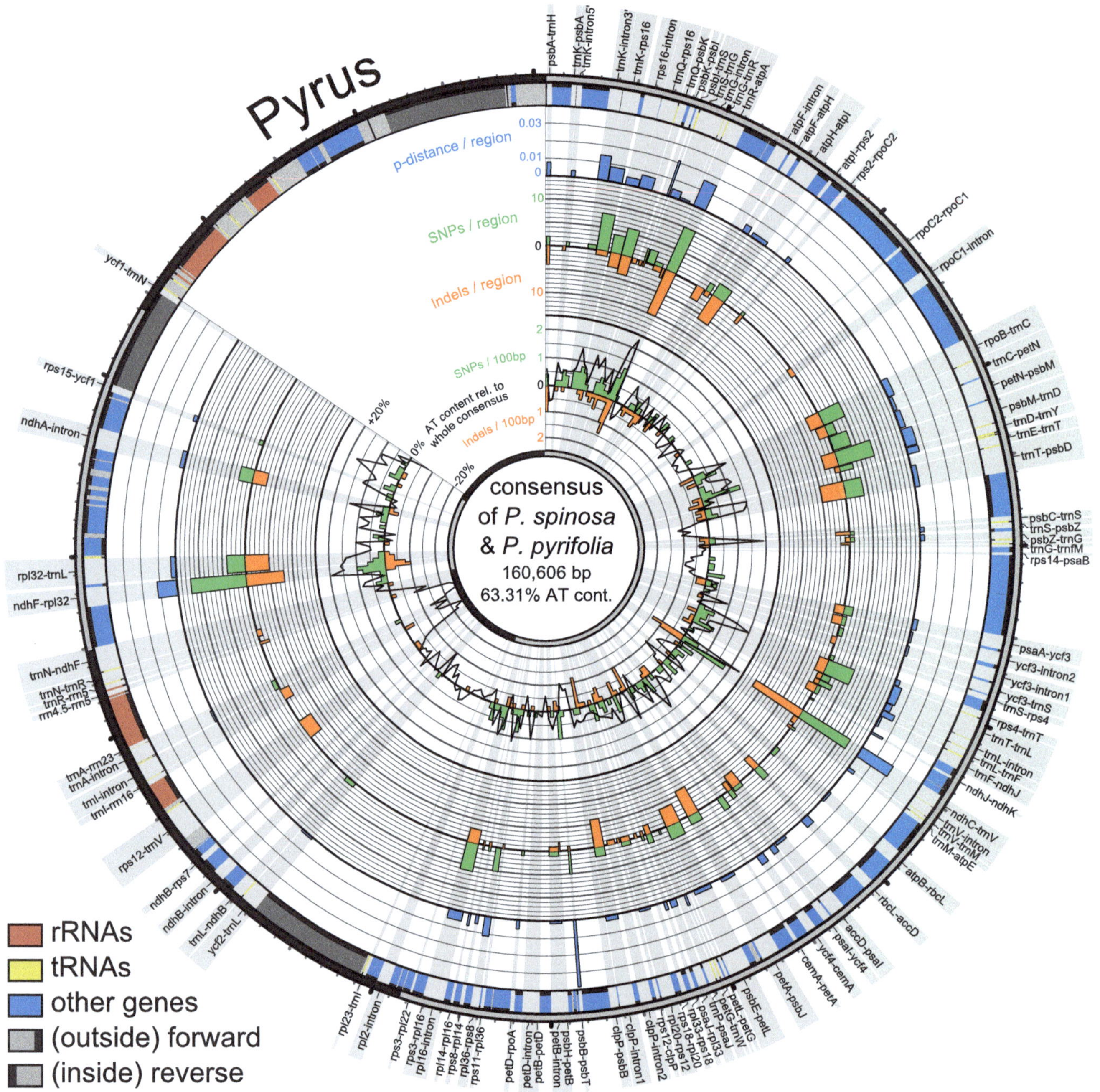

Figure 1. Circular representation of plastid genome pair in *Pyrus*. Shown are consensus sequences of compared species pairs of *Pyrus spinosa* and *P. pyrifolia* with their differing *p*-distances, numbers of SNPs and indels across the consensus. Radial grey highlights show the regions in focus of study with their names. Circular graphs from outside to inside: outermost circle with ticks for every 1,000 bp (small) and 10,000 bp (big) indicates part of genome, single copy regions in light grey and inverted repeats in dark grey; bands show locations of genes (blue), tRNAs (yellow) and rRNAs (red); the three outermost histograms display *p*-distances (blue), number of SNPs (green) and indels (orange) per spacer region; innermost graph shows number of SNPs (green histogram), indels (orange histogram), and AT content relative to the whole consensus (black line graph) of 500 bp long parts of the whole consensus.

amplification and sequencing strategies including primer binding sites, region size and the information content per primer read had to be considered in addition to a high rank in terms of variability. Furthermore, polyA/T stretches larger than seven nucleotides (microsatellites) had to be considered. Their presence usually require two primer reads for sequencing that start from both ends of the amplicon because slippage is likely to occur after the polyA/T stretch. Since a region >1000 bp usually requires two primers to sequence, one microsatellite was not considered a problem, while several microsatellites within the same region led to dismiss it. Considering that current technology generates reliable read lengths of 800–1000 bases, we selected fragments of 900–1300 bp in size _ a size range that can be easily amplified and then sequenced with a maximum of two primers.

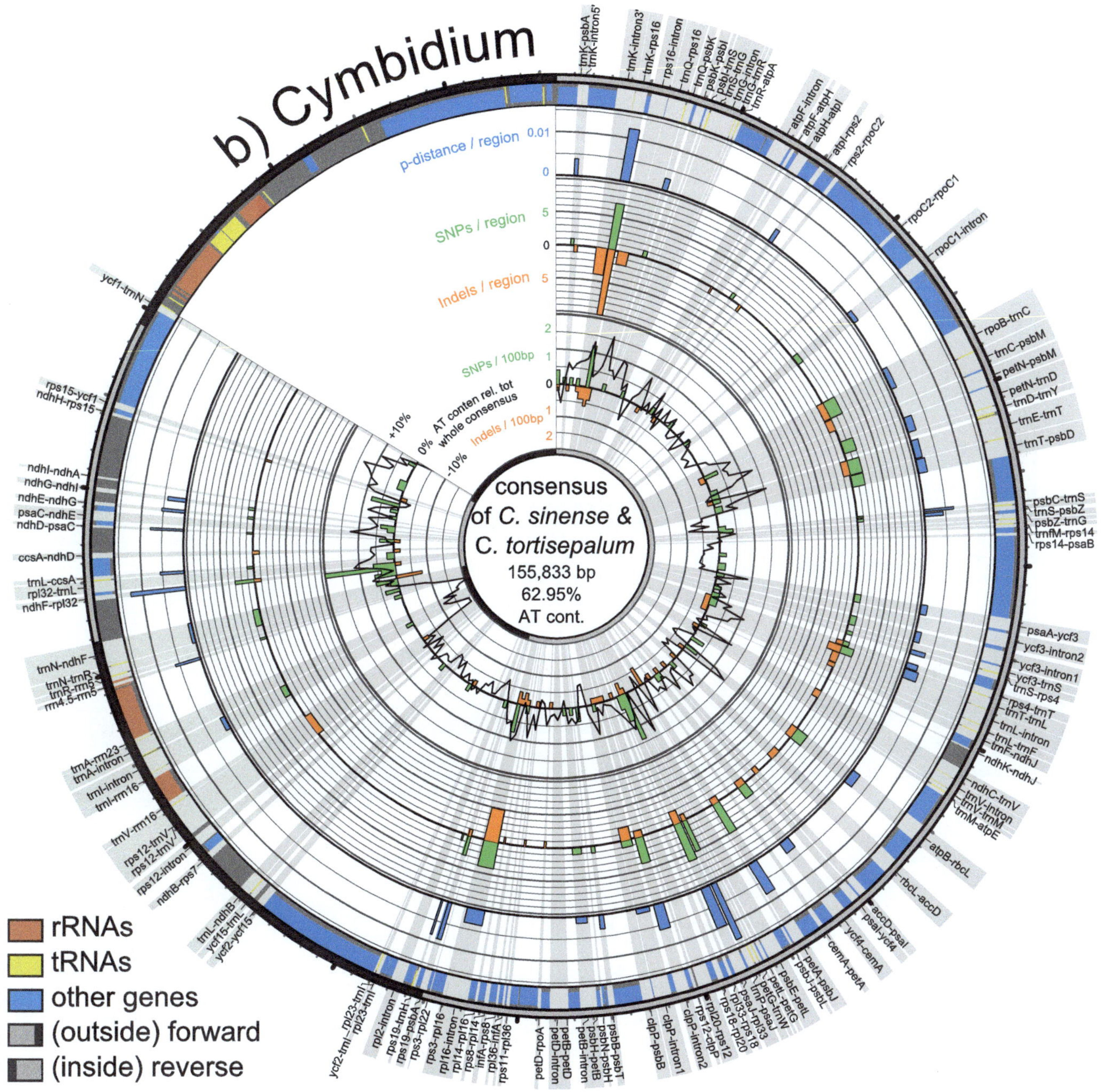

Figure 2. Circular representation of plastid genome pair in *Cymbidium*. Shown are consensus sequences of compared species pairs of *Cymbidium tortisepalum* and *C. sinense* with their differing *p*-distances, numbers of SNPs and indels across the consensus. Radial grey highlights show the regions in focus of study with their names. Circular graphs from outside to inside: outermost circle with ticks for every 1,000 bp (small) and 10,000 bp (big) indicates part of genome, single copy regions in light grey and inverted repeats in dark grey; bands show locations of genes (blue), tRNAs (yellow) and rRNAs (red); the three outermost histograms display *p*-distances (blue), number of SNPs (green) and indels (orange) per spacer region; innermost graph shows number of SNPs (green histogram), indels (orange histogram), and AT content relative to the whole consensus (black line graph) of 500 bp long parts of the whole consensus.

Results and Discussion

Size and structure of the *Pyrus* plastid genome

The plastid genome of *Pyrus spinosa* is 159,694 bp in length, and the inverted repeats (IRs) account for 26,396 bp. The large single-copy region (LSC) is 87,694 bp in length and the small single-copy region (SSC) 19,205 bp. The genome has a GC content of 36.6%. Gene content and order are identical to *Pyrus pyrifolia*, with 113 unique genes and 17 duplicates in the IR [30]. The extension of IRs is identical to *P. pyrifolia*, while a 137 bp gap in the LSC of *P. spinosa* directly adjacent to IRa leads to a different IR boundary. The *p*-distance between the two genomes is 0.00145 (Table 2). The consensus structure of the two *Pyrus* genomes and the variability between them is illustrated in Fig. 1. Most of the variation occurs in the noncoding parts, especially in intergenic spacers of the LSC region. The SSC is less variable and

Table 3. Ranking and comparison of *p*-distances and differences in the four plastid genome pairs.

	Pyrus				*Cymbidium*				*Oenothera*				*Olea*			
Rank	**Region**	**Aligned length [bp]**	**PICs (SNPs/ Indels)**	**p-distance [*10^{-3}]**	**Region**	**Aligned length [bp]**	**PICs (SNPs/ Indels)**	**p-distance [*10^{-3}]**	**Region**	**Aligned length [bp]**	**PICs (SNPs/ Indels)**	**p-distance [*10^{-3}]**	**Region**	**Aligned length [bp]**	**PICs (SNPs/ Indels)**	**p-distance [*10^{-3}]**
1	*psbB–psbT*	184	6 (5/1)	37.88	*trnP–psaJ*	366	6 (5/1)	14.04	*ycf1–ndhF*	381	15 (9/6)	36.73	*trnG–trnR*	170	5 (4/1)	23.81
2	*psbI–trnS*	149	4 (3/1)	22.06	*ndhF–rpl32*	259	4 (3/1)	13.04	*psbJ–psbL*	134	2 (2/0)	14.93	*psbC–trnS*	243	6 (5/1)	21.1
3	*ndhC–trnV*	760	24 (12/12)	20.34	*trnK–rps16*	613	17 (7/10)	12.15	*rps4–trnT*	332	5 (4/1)	12.16	*trnR–atpA*	112	3 (2/1)	18.02
4	*trnR–atpA*	909	20 (10/10)	13.61	*psaJ–rpl33*	629	8 (6/2)	9.93	*trnG–trnfM*	172	3 (2/1)	11.9	*trnS–trnG*	715	12 (10/2)	14.33
5	*ndhF–rpl32*	1078	20 (12/8)	11.41	*rps19–psbA*	345	4 (3/1)	8.7	*ndhG–ndhI*	408	5 (4/1)	9.9	*accD–psaI*	702	11 (10/1)	14.27
6	*rpl36–rps8*	459	5 (5/0)	10.89	*rps19–trnH*	122	1 (1/0)	8.2	*accD–psaI*	577	6 (5/1)	8.68	*ycf15–trnL*	354	5 (5/0)	14.12
7	*trnK–rps16*	974	9 (8/1)	8.38	*petA–psbJ*	635	6 (5/1)	7.9	*trnQ–psbK*	355	6 (3/3)	8.52	*psbA–trnH*	447	8 (6/2)	13.51
8	*trnQ–rps16*	905	10 (6/4)	8.3	*ndhD–psaC*	129	1 (1/0)	7.75	*ndhF–rpl32*	932	7 (7/0)	8.26	*rps4–trnT*	326	5 (4/1)	12.31
9	*psbA–trnH*	268	6 (2/4)	7.81	*psbC–trnS*	146	1 (1/0)	6.85	*trnQ–accD*	2615	23 (12/11)	5.59	*trnG–trnM*	174	2 (2/0)	11.49
10	*trnL–trnF*	403	4 (3/1)	7.59	*rrn4.5–rrn5*	168	1 (1/0)	5.95	*rps12–clpP*	397	4 (2/2)	5.13	*psbB–psbT*	186	2 (2/0)	10.75
11	*ndhJ–ndhK*	137	1 (1/0)	7.3	*clpP intron 2*	676	5 (4/1)	5.93	*rps16–rbcL*	976	8 (4/4)	5.06	*trnT–psbD*	1322	15 (14/1)	10.6
12	*rpl14–rpl16*	145	2 (1/1)	6.99	*trnL–ccsA*	180	1 (1/0)	5.56	*atpI–rps2*	216	1 (1/0)	4.63	*trnK–rps16*	899	12 (9/3)	10.06
13	*trnD–trnY*	448	4 (3/1)	6.79	*ndhE–ndhG*	185	1 (1/0)	5.41	*rps2–rpoC2*	219	2 (1/1)	4.59	*rps2–rpoC2*	209	4 (2/2)	9.66
14	*psbM–trnD*	1235	11 (8/3)	6.51	*trnS–psbZ*	230	1 (1/0)	4.35	*trnP–psaJ*	515	4 (2/2)	4.37	*trnS–rps4*	314	3 (3/0)	9.55
15	*trnW–trnP*	156	1 (1/0)	6.41	*ndhG–ndhI*	233	1 (1/0)	4.29	*trnI–ycf2*	462	2 (2/0)	4.33	*psbN–psbH*	105	1 (1/0)	9.52
16	*rpl16 intron*	1003	9 (6/3)	6.01	*trnK–psbA*	257	1 (1/0)	3.89	*atpH–atpI*	939	5 (4/1)	4.26	*trnD–trnY*	107	2 (1/1)	9.43
17	*ycf4–cemA*	526	3 (3/0)	5.7	*trnS–rps4*	287	1 (1/0)	3.48	*trnK intron 5'*	249	2 (1/1)	4.03	*psbM–trnD*	657	7 (6/1)	9.15
18	*rbcL–accD*	569	6 (3/3)	5.33	*rpl16 intron*	1191	9 (4/5)	3.46	*petN–psbM*	926	3 (3/0)	3.24	*trnQ–psbK*	335	5 (3/2)	9.01
19	*trnT–trnL*	1241	8 (6/2)	4.94	*trnT–trnL*	610	4 (2/2)	3.36	*psaA–ycf3*	669	3 (2/1)	3.01	*atpI–rps2*	222	2 (2/0)	9.01
20	*psaI–ycf4*	413	4 (2/2)	4.89	*atpI–rps2*	300	1 (1/0)	3.33	*trnS–psbZ*	348	1 (1/0)	2.87	*atpF intron*	697	7 (6/1)	8.62
21	*rps8–rpl14*	207	2 (1/1)	4.83	*psbB–psbT*	323	1 (1/0)	3.1	*trnK–rps16*	758	4 (2/2)	2.67	*petN–psbM*	1171	12 (10/2)	8.62
22	*rpl33–rps18*	218	2 (1/1)	4.67	*trnQ–psbK*	348	1 (1/0)	2.87	*petD intron*	761	3 (2/1)	2.65	*trnK–psbA*	236	2 (2/0)	8.47
23	*trnS–trnG*	651	3 (3/0)	4.62	*ycf4–cemA*	728	3 (2/1)	2.75	*trnS–trnG*	788	2 (2/0)	2.54	*ndhF–rpl32*	479	4 (4/0)	8.37
24	*rps16 intron*	909	7 (4/3)	4.48	*rps4–trnT*	367	2 (1/1)	2.74	*trnG intron*	804	4 (2/2)	2.5	*psbI–trnS*	120	1 (1/0)	8.33
25	*petD–rpoA*	225	1 (1/0)	4.48	*trnT–psbD*	947	2 (2/0)	2.11	*psaI–ycf4*	412	5 (1/4)	2.45	*petG–trnW*	121	1 (1/0)	8.26
26	*atpF–atpH*	451	3 (2/1)	4.44	*atpB–rbcL*	960	3 (2/1)	2.11	*psbE–petL*	984	4 (2/2)	2.05	*rps14–psaB*	122	1 (1/0)	8.2
27	*trnM–atpE*	242	2 (1/1)	4.29	*petN–trnD*	1020	2 (2/0)	1.96	*trnT–trnL*	1114	4 (2/2)	1.89	*psaA–ycf3*	741	6 (6/0)	8.1
28	*psaJ–rpl33*	472	4 (2/2)	4.29	*trnE–trnT*	1216	3 (2/1)	1.68	*trnT–psbD*	1441	6 (2/4)	1.43	*trnK intron 5'*	268	3 (2/1)	7.49
29	*rpoB–trnC*	1216	8 (5/3)	4.13	*psaA–ycf3*	638	1 (1/0)	1.57	*ycf3 intron 2*	720	1 (1/0)	1.39	*rpl32–trnL*	835	10 (6/4)	7.26
30	*trnL intron*	514	3 (2/1)	3.9	*rpoB–trnC*	1461	3 (2/1)	1.38	*petB intron*	775	4 (1/3)	1.34	*ndhC–trnV*	1119	12 (8/4)	7.24

The regions are sorted according to *p*-distances.

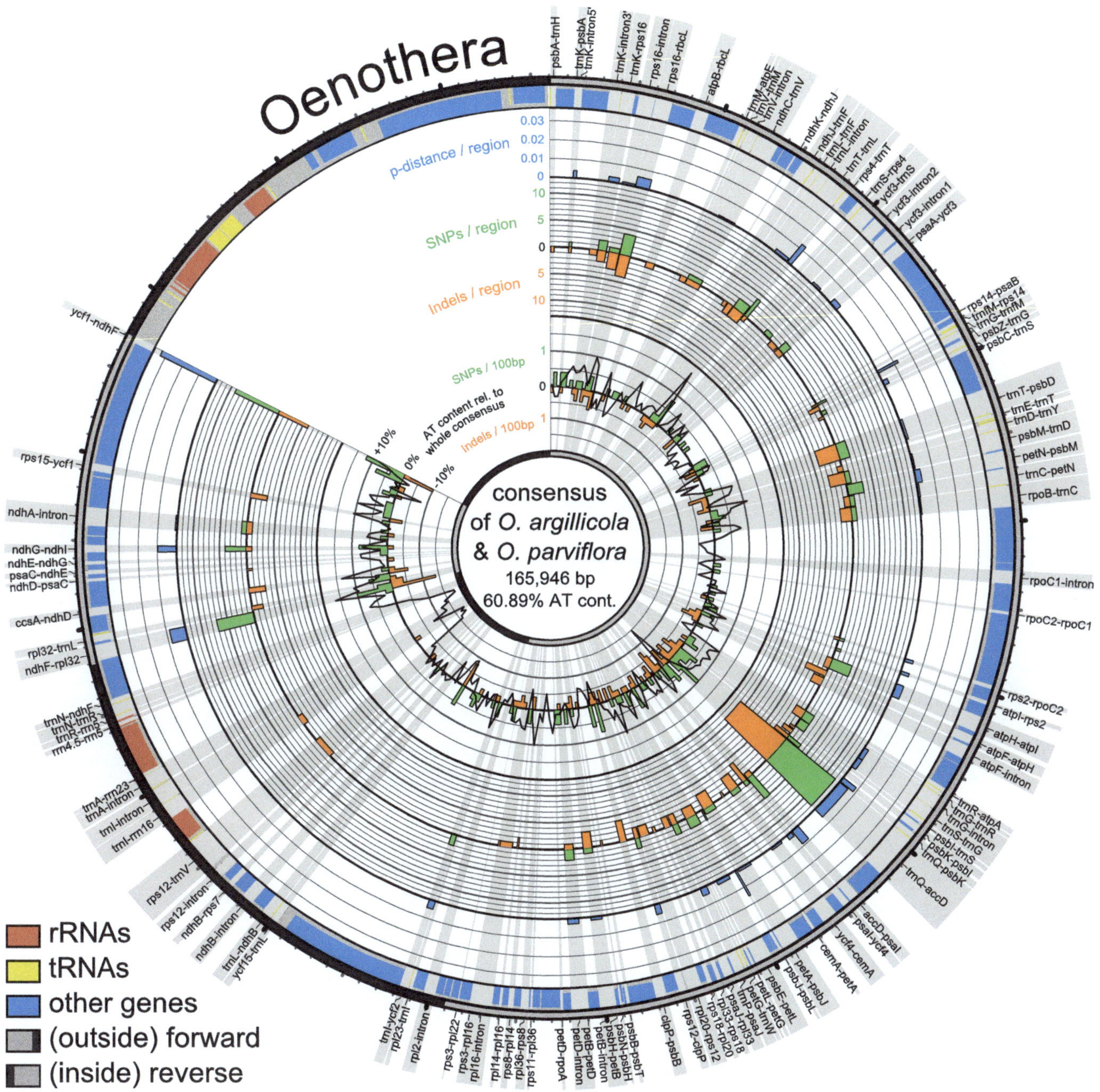

Figure 3. Circular representation of plastid genome pairs in *Oenothera*. Shown are consensus sequences of compared species pairs of *Oenonthera parviflora* and *O. argillicola* with their differing *p*-distances, numbers of SNPs and indels across the consensus. Radial grey highlights show the regions in focus of study with their names. Circular graphs from outside to inside: outermost circle with ticks for every 1,000 bp (small) and 10,000 bp (big) indicates part of genome, single copy regions in light grey and inverted repeats in dark grey; bands show locations of genes (blue), tRNAs (yellow) and rRNAs (red); the three outermost histograms display *p*-distances (blue), number of SNPs (green) and indels (orange) per spacer region; innermost graph shows number of SNPs (green histogram), indels (orange histogram), and AT content relative to the whole consensus (black line graph) of 500 bp long parts of the whole consensus.

almost no variation is found in the IRs. There are some genome parts with intergenic spacers alternating tRNA genes where variation appears to accumulate. This is especially the case in the region from *trnK* to *trnA* and from *rpoB* to *psbD* (Figs. 1, 2).

Finding the most variable regions of the *Pyrus* plastid genome. The five regions with the highest *p*-distances are the intergenic spacers *psbB–psbT*, *psbI–trnS*, *ndhC–trnV*, *trnR–atpA*, and *ndhF–rpl32*. Taking the PICs as a basis, the five top-ranked regions are *ndhC–trnV*, *trnR–atpA*, *ndhF–rpl32*, *psbM–trnD*, and *trnQ–rps16* (Table 3, Fig. 1–4).

Comparing our results with the ranking of Shaw et al. [9] it appears that 17 of our 30 top-ranked regions in *Pyrus* are also among the 32 top-ranked in their study. However, their ranks are different. For example, in Shaw et al. [8], the *rpl32–trnL* spacer

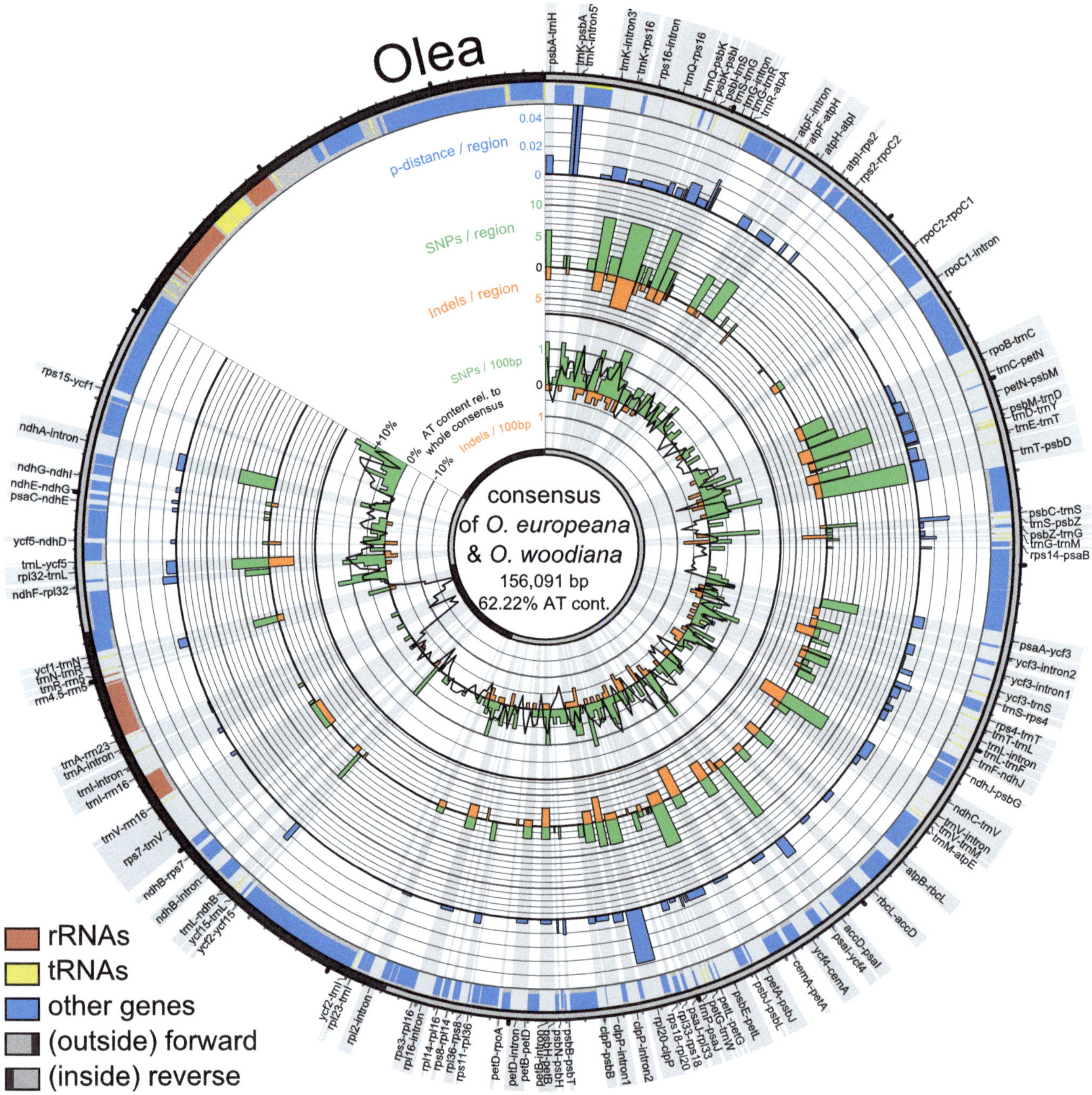

Figure 4. Circular representation of plastid genome pairs in *Olea*. Shown are consensus sequences of compared species pairs of *Olea europaea* and *O. woodiana* with their differing *p*-distances, numbers of SNPs and indels across the consensus. Radial grey highlights show the regions in focus of study with their names. Circular graphs from outside to inside: outermost circle with ticks for every 1,000 bp (small) and 10,000 bp (big) indicates part of genome, single copy regions in light grey and inverted repeats in dark grey; bands show locations of genes (blue), tRNAs (yellow) and rRNAs (red); the three outermost histograms display *p*-distances (blue), number of SNPs (green) and indels (orange) per spacer region; innermost graph shows number of SNPs (green histogram), indels (orange histogram), and AT content relative to the whole consensus (black line graph) of 500 bp long parts of the whole consensus.

has the highest number of PICs whereas it is only at rank 8 in *Pyrus*. The *trnR–atpA* spacer, which has the second-highest number of PICs in *Pyrus*, was not at all reported. However, the ranking of Shaw et al. may not be that comparable because the authors "normalized" their PICs with the aim to reduce the influence of different evolutionary rates or genetic distances. They divided the number of PICs within a region from a certain taxonomic lineage by the total sum of PICs within the same lineage. Therefore, their results do not directly show lineage-specific differences in marker variability, although the absolute variability of a given genomic region is the only relevant fact in any analysis.

Low genetic distances in *Pyrus* have been pointed out in two earlier studies of *Pyrus* plastid genomes [27,39]. These studies were motivated by the horticultural importance of *Pyrus*, and focused on Asian species and cultivars. Katayama and Uematsu [39] provided a physical map of the plastid genome of *Pyrus ussuriensis* var. *hondoensis* and ran an RFLP analysis on cpDNAs from 11 accessions of five *Pyrus* and two *Prunus* species. However, there were no sequence data to support their conclusions. Terakami et al. [27] aligned the three plastid genomes of *Pyrus pyrifolia*, *Malus* × *domestica*, and *Prunus persica*. The authors calculated the proportion of mutational events using the same formula as Shaw et al. [8] for 89 noncoding regions, and ranked the compared regions according to their variability comparing *Pyrus* with *Malus* and *Prunus* (ingroup and outgroup were not specifically defined). While the *ndhC–trnV* and *trnR–atpA* spacers depict the highest sequence divergence in both, Terakami et al. and our work presented here, the overall rankings are strongly different. Terakami et al. found the spacers *rpl33–rps18*, *psbI–trnS*, and *rpl14–rpl16* from the third to fifth rank. In our *Pyrus* ranking, these spacers are at positions 22, 2, and 12 (based on *p*-distances) and 43, 22, and 41 (based on PICs), respectively. These differences may be explained by the much greater distance between the *Pyrus* and *Malus* plastid genomes than our two *Pyrus* genomes. The crown group of *Pyrus* diversified 27–33 mya while the crown group of *Malus* was inferred to have diversified 34–46 mya [40].

Various plastid regions have also been sequenced for a large number of samples in *Pyrus*. Katayama et al. [41] sequenced the *rps16–trnQ* and *accD–psaI* spacers and reconstructed a network based on 25 different haplotypes including 21 species of *Pyrus* and multiple individuals of *P. pyrifolia* and *P. ussurienis*, respectively. The authors found both spacers to contain highly variable AT-rich mutational hotspots and concluded that these regions are "hypervariable", while their remaining *Pyrus* sequences showed hardly any variation. The authors argued that their results confirmed their earlier hypothesis of strong sequence conservation in the plastid genomes of *Pyrus* [39]. No explanation, however, was given why particularly the *rps16–trnQ* and *accD–psaI* spacers had been chosen and not one of the highest ranked ones in terms of variability. The authors noted that the frequency of microstructural mutations in both spacers studied was markedly higher than of substitutions and that haplotypes were mostly defined by indels. Such a dominance of microstructural mutations over substitutions is typical of AT-rich sequence elements that constitute terminal stem-loops of introns and transcribed spacers which are often unique to small lineages of plants [11]. At the same time such sequence elements often exhibit high levels of homoplasy. Thus, the exclusive application of these elements to calculate networks or trees may potentially lead to wrong conclusions. Wuyun et al. [42] sequenced the *rps16–trnQ* and *accD–psaI* spacers to reconstruct a phylogenetic network of *Pyrus ussuriensis* in China, which was largely based on the presence or absence of indels in the two spacers. Compared with our results, the two regions used by Katayama et al. [47] and Wuyun et al. [48] are also not the most variable plastid regions in *Pyrus*: the *trnQ–rps16* spacer ranks at place 24 for *p*-distances and at place 5 for PICs. The *accD–psaI* spacer ranks at place 18 for *p*-distances and at place 20 for PICs.

Plastid markers proposed for *Pyrus*

Four intergenic spacers of 900 to 1000 bp and the *rpl16* group II intron (ca. 1000 bp) are proposed here to be sequenced for evolutionary studies in *Pyrus* (Table 4). They were selected from the most variable genomic regions (Table 3) considering an efficient sequencing strategy (see methods section).

Among the regions with a minimum size of 500 bp, the *ndhC–trnV* and *trnR–atpA* spacers rank 3rd and 4th according to *p*-distances, and *ndhC–trnV* has the highest number of PICs. Both can be sequenced with just one primer (either forward or reverse). Thus, these spacers are especially useful if large sample numbers need to be analysed. The *ndhF–rpl32* spacer (ranked 3rd of the regions >500 bp in Table 4) was not considered further because there are two large microsatellites. This fragment can therefore not be sequenced with two primers. The same problem occurs in the *rps16–trnK* spacer (ranked 4th of the regions >500 bp in Table 4) where two poly G and one poly T are likely to cause sequencing problems with pherograms unreadable after the homonulceotide stretches. The *trnQ-rps16* and *psbM-trnD* spacers follow in the ranking. Both also have polyA/T microsatellites. While they can be covered with two primer reads that overlap at the microsatellite, they may not be as efficiently sequenced than the *ndhC–trnV* and *trnR–atpA* spacers for large sample numbers. The *rpl16* intron (ranked at 7th position of the regions >500 bp in Table 4), is particularly recommended because it was shown to also possess a high phylogenetic structure *R* in different angiosperm sequence data sets [43–45]. Multiple *rpl16* sequence alignments can therefore be expected to yield well-resolved and well-supported trees also in *Pyrus*. The intron can be co-amplified with the *rpl14–rpl16* spacer. The use of the reverse primer PYR-rpl16R (Table 4) will allow to sequence the whole intron with one read. The *rpl16* intron contains a polyA/T stretch of variable length in different species of *Pyrus* (see also Fig. 5c), what implies that an additional forward primer read may be necessary to cover the whole intron in some samples.

Primers were newly designed for *trnR-atpA* as this region to our knowledge has never been used in any evolutionary study so far. For *ndhC-trnV*, primers were available [46] but we designed a new *Pyrus*-specific reverse primer in order to completely cover the spacer-exon boundary. For *trnQ-rps16*, the universal primers designed by Shaw & al. [9] work for *Pyrus* as well. Available primers for *psbM-trnD* [47] were re-designed for *Pyrus* to avoid mismatches in the forward and then to obtain a similar melting temperature in the reverse primer. For the *rpl16* intron, primers were also adapted to *Pyrus* following the general amplification strategy of [43] and [44] with a forward primer that anneals to the *rps3* exon. This ensures that the *rpl16* intron can be amplified and sequenced completely. The universal reverse primer rpl16R [48] was replaced by a *Pyrus*-specific primer that anneals further downstream to cover the intron-exon boundary.

Comparison of plastid genomes with low *p*-distances in angiosperms. In addition to *Pyrus*, we explored variability patterns in plastid genome pairs of *Oenothera argillicola* and *O. parviflora* (Onagraceae), *Olea europaea* and *O. woodiana* (Oleaceae), and *Cymbidium sinense* and *C. tortisepalum* (Orchidaceae) which have comparable low *p*-distances (Table 2). The variability patterns of all four genome pairs are illustrated using a Circos-plot (Figs. 2–4). Each genome pair has different regions with highest *p*-distances and highest numbers of PICs, resulting in a genome pair-specific ranking (Table 3). The results of the pairwise comparisons of individual introns and spacers for each genome pair are provided in Table S2.

The SNPs and indels are almost evenly spread across the LSC and the SSCs in *Olea*. In *Cymbidium*, SNPs and indels are more clustered. The plastid genomes of *Pyrus* and *Oenothera* exhibit strong variation in certain areas, e.g. between *trnT* and *rpoB* (Figs. 1, 3) but alsoalso homogeneously distributed mutations across their genomes. The *Olea* genome stands out by many more SNPs than indels, while the other genomes have almost as many indels as SNPs.

Table 4. Genomic regions proposed for evolutionary analyses in *Pyrus* and primers for their amplification.

Region	Amplified fragment	Primer name	Primer sequence	Reference
ndhC–trnV	900 bp	ndhC-F	TGCCAAAATAGGAATAACAC	Goodson et al. [46]
		PYRtrnV-150R	CCACATAATGAATCAGAGCAC	this study
trnR–atpA	1000 bp	trnR-F	GTCTAATGGATAGGACAGAGG	this study
		atpA-180R	GGAACRAACGGYTATCTTGATTC	this study
psbM–trnD	1350 bp	PYRpsbM-F	CCTTGGCTGACTGTTTTTACG	this study
		PYRtrnD-R	GAGCACCGCCCTGTCAAGG	this study
trnQ–rps16	900 bp	trnQ (UUG)	GCGTGGCCAAGTGGTAAGGC	Shaw et al. [9]
		rps16x1	GTTGCTTTCTACCACATCGTTT	Shaw et al. [9]
rpl16 intron	1300 bp	PYR-rps3F	GATTATTGTTCCTATGCAG	this study
		PYR-rpl16R	GCTTGAAGAGCATATCTAC	this study

In our summary of the 30 most variable genomic regions including all four genome pairs,77 different regions appear in total (Table 3). It is noteworthy that only two spacers, *ndhF–rpl32* and *trnK–rps16*, are consistently placed among the 30 most variable regions. Eight spacers appear three times: *atpI–rps2*, *psaA–ycf3*, *psbB–psbT*, *rps4–trnT*, *trnQ–psbK*, *trnS–trnG*, *trnT–psbD*, and *trnT–trnL*.

Earlier comparisons of plastid genomes in angiosperms for marker selection. In an approach to explore hitherto unused plastid regions as phylogenetic markers, Shaw et al. [9] in 2007 compared whole plastid genomes in a comprehensive way. They analysed genome pairs from three different lineages of angiosperms [*Atropa* and *Nicotiana* (Solanaceae) for the asterids, *Lotus* and *Medicago* (Fabaceae) for the rosids, and *Oryza* and *Saccharum* (Poaceae) for the monocots]. They found nine previously unexplored plastid regions with high levels of variation based on the numbers of PICs: *rpl32–trnL*, *trnQ–rps16*, *ndhC–trnV*, *ndhF–rpl32*, *psbD–trnT*, *psbJ–petA*, *rps16–trnK*, *atpI–atpH*, and *petL–psbE*. As noted before, we were interested to compare the distance levels of these genomes to the genome pairs examined here, as we expected considerable differences. The *p*-distances were indeed much higher and are here calculated as follows: *Lotus japonicus*/*Medicago truncatula* $p = 0.17603$, *Nicotiana tabacum*/*Atropa belladonna* $p = 0.01363$, *Saccharum* hybrid/*Oryza sativa* $p = 0.04879$.

Another comparative study of plastid genomes was carried out by Dong et al. [13] five years later.They looked at 14 angiosperm genera for which more than one plastid genome was available, again with the goal of finding markers for phylogeny reconstruction and DNA barcoding. They concluded that *ycf1*, *psbA–trnH*, *rpl32–trnL*, *trnQ–rps16*, *ndhC–trnV*, *trnK/matK*, and *trnS–trnG* are best-suited.

Next generation sequencing has resulted in an increased availability of plastid genome data in recent years (Table 5) that were used to find markers for various phylogenetic analyses in certain angiosperm lineages, to recover promising regions for haplotype studies or to differentiate closely related species and cultivars [14,21,22,24–27,49–52]. None of the authors addressed more general patterns of plastid genome mutational dynamics and molecular evolution. As noted before, the studies span an enormous range of different genetic distances in the genomes compared. The compared economically important asterids (e.g., *Solanum*, *Nicotiana*, *Lactuca*) are well represented while studies on other taxa are still scarce. Moreover, the approaches and methods applied in these studies differ. Most of them calculated some kind of sequence variability, while others additionally or solely reconstructed phylogenetic trees based on small taxon sets to assess the phylogenetic utility of these regions. A spectrum of 37 plastid loci was reported as "highly variable" in the studies cited above. Most commonly mentioned were *rpl32–trnL* (7x), *trnQ–rps16* (5x) *trnK–rps16* (4x), and *ndhC–trnV* (4x). Nevertheless, the question remains how representative the earlier pairwise genome comparisons are, and to what extent their conclusions are also valid for other families and genera of flowering plants.

Shaw et al. [8] assumed a high universality of their results. But Daniell et al. [52], who compared plastid genomes of Solanaceae, found spacers with higher sequence divergence not mentioned in [8]. Timme et al. [49] analysed Asteraceae and indicated that their ranking of most variable regions barely overlapped with the ranking of Shaw et al., and suspected that "each family or major lineage will most likely have a unique set of variable regions" [43]. Shaw et al. [9] in 2007 found no less than 11 new highly variable markers not considered in their 2005 study therefore pointed to the need of a test-wise screening of the "universal" regions to find the most suitable one for a given lineage. Likewise, Dong et al. [13] stated that markers useful for one group may not be useful for another and recommended evaluating markers in detail before selecting them for further use. With the aim of resolving the species tree in the huge genus *Solanum*, Särkinen and George [14] found that the average amount of variable characters differs within subclades of the genus. In their view, the degree to which the utility of a marker can be extended to more inclusive clades would then also be clade-specific.

In summary, lineage specific differences in variability and phylogenetic utility of plastid genomic regions were reported in various cases in flowering plants although there was never any standardized comparative approach to better understand this issue. Moreover, none of the previous studies explicitly addressed phylogenetic signal as being different from similarity-based variability, or looked at any molecular evolutionary characteristics.

Molecular evolution and lineage specific variability of genomic regions. Lineage-specific differences in variability are often explained by patterns of molecular evolution. It has been exemplarily demonstrated for regions such as *psbA–trnH* [53] or *trnL–trnF* [54] that variability is strongly influenced by structural constraints. Empirical analysis of *petD* group II intron sequences has further shown that increased length correlates with increased AT strongly influenced byal constraints Empirical analysis of *petD* group II intron sequences has further shown that increased length correlates with increased AT content [12]. Figure 5 shows the AT

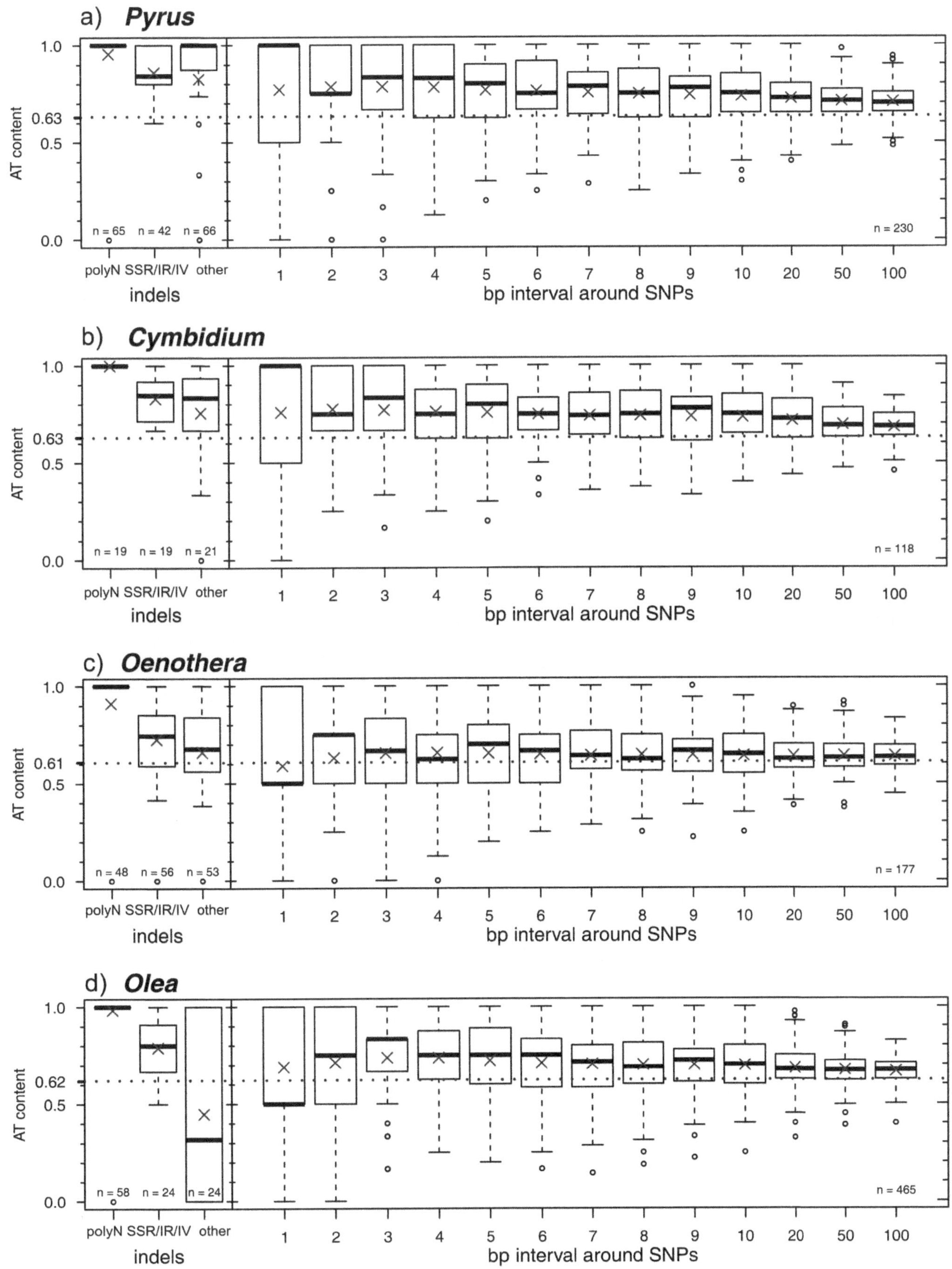
a) Pyrus
AT content
1.0
0.63
0.5
0.0
n = 65
n = 42
n = 66
n = 230
polyN SSR/IR/IV other
indels
1 2 3 4 5 6 7 8 9 10 20 50 100
bp interval around SNPs
b) Cymbidium
AT content
1.0
0.63
0.5
0.0
n = 19
n = 19
n = 21
n = 118
polyN SSR/IR/IV other
indels
1 2 3 4 5 6 7 8 9 10 20 50 100
bp interval around SNPs
c) Oenothera
AT content
1.0
0.61
0.5
0.0
n = 48
n = 56
n = 53
n = 177
polyN SSR/IR/IV other
indels
1 2 3 4 5 6 7 8 9 10 20 50 100
bp interval around SNPs
d) Olea
AT content
1.0
0.62
0.5
0.0
n = 58
n = 24
n = 24
n = 465
polyN SSR/IR/IV other
indels
1 2 3 4 5 6 7 8 9 10 20 50 100
bp interval around SNPs

Figure 5. Mutational dynamics in group II introns. a) Schematic consensus structure of plastid group II introns based on Michel et al. (1989). Roman numbers indicate the six domains. B) Alignment and predicted RNA secondary structure for domain IV of the *atpF* intron in *Cymbidium*, *Pyrus*, *Oenothera* and *Olea*. The apparently non-homologous sequence blocks are placed separately in the alignment. There are no substitutions or length mutations in *Pyrus* and *Cymbidium*, the structures shown are therefore identical in the two species compared. The shown secondary structures of *Oenothera* and *Olea* are consensus structures. Two conserved nucleotide blocks at the 3′ and 5′ ends, indicated by thick blue bars, are conserved across all taxa and homologous in primary sequence and secondary structure. These conserved sequence blocks form the stem of the domain while variation occurs in the terminal stem-loops part of the domain. c) Alignment and predicted secondary RNA structures of domain IV of the *rpl16* intron. For clarity, only the part of the domain with positions variable within genera are shown; "[-]" mark the omitted stem-loop elements. The apparently non-homologous sequence blocks are placed separately in the alignment. Those positions where variation occurs within a genus are marked with arrows. See text for more explanation.

contents of three types of indels (left side) and around SNPs (right side) in intervals of increasing size of each of our genome pairs. AT content distributions are displayed in boxplots with the cross showing the mean and the thick line referring to the median. Respective boxplots arranged along the *x*-axis then depict maximum distances of the intervals in each direction of the SNP. Apart from rare exceptions the surroundings of SNPs are distinctly more AT-rich than the whole genome (Fig. 6), indicating that substitutions occur predominantly in AT-rich stretches. The AT contents of the consensus sequences are displayed as dotted lines. Looking at indels, considerable differences are apparent in the frequency of different kinds among the four plant lineages. In *Olea*, length-variable polyA/T stretches are most common. In *Oenothera*, all three kinds of indels occur with almost equal frequency, while in *Cymbidium* and *Pyrus* indels without a clear motif predominate.

The AT content is significantly increased in sequence elements affected by microstructural changes (Fig. 6), both in SSRs and in the non-SSR indels. The SSRs are generally AT-rich, so the templates for these SSRs must be AT-rich as well. And therefore, their frequency is also significantly higher in AT-rich sequence elements. It can thus be suggested that mutational dynamics is increased in AT-rich sequence. A strong correlation between high AT content and high substitutional rates was also recently demonstrated in plastid genomes of Lentibulariaceae [55].

Comparative studies of the molecular evolution of group II introns showed substitutions, length-variable homonucleotide stretches and indels to predominantly occur in domains I, III and IV. These domains are also the most variable with respect to size and experience less strong functional constraints compared to the other domains [12,56,57]. Furthermore, considerable variation occurs in sequence elements that are unique to certain lineages, where they have evolved through stepwise insertion processes connected to the formation of stable helical elements [11]. In our data set, this is for example evident in the *petD* and *rpl16* introns. They appear at strikingly different positions in the rankings of the respective genome pairs (Table 3 and S2). In both introns the variation between the sequences of a genome pair is mostly caused by length variable polyA/T stretches or AT-rich indels.

Domain IV of the *atpF* intron belongs to a conserved group II intron (Fig. 5a) with no variation between the *Cymbidium* and *Pyrus* sequences, two substitutions in *Olea* and a length-variable polyA-stretch in *Oenothera* (Fig. 5b). The alignment (Fig. 5b) illustrates two conserved sequence blocks that are homologous and

Table 5. Identification of most variable plastid regions based on pairwise genome comparisons across angiosperms.

Reference	Taxa studied	Markers found as most variable
Daniell et al. [52]	Asterids: *Atropa belladonna, Nicotiana tabacum, Solanum bulbocastanum, S. lycopersicum* (Solanaceae)	*psbK–psbI, rps12–clpP, trnG–trnfM, trnK–rps16, trnQ–rps16*
Timme et al. [49]	Asterids: *Helianthus annuus, Lactuca sativa* (Asteraceae)	*ndhC–trnV, rpl32–trnL, rps12–clpP, trnE–rpoB, trnY–trnE*
Shaw et al. [9]	Angiosperms: Asterids: *Atropa belladonna, Nicotiana tabacum* (Solanaceae), Rosids: *Lotus, Medicago* (Fabaceae), Monocots: *Oryza, Saccharum* (Poaceae)	*rpl32–trnL trnQ–rps16 ndhC–trnV, ndhF–rpl32, psbD–trnT, psbJ–petA, rps16–trnK, atpI–atpH, petL–psbE*
Doorduin et al. [50]	Asterids: *Jacobaea vulgaris, Helianthus anuus, Lactuca sativa, Parthenium argentatum, Guizotia abyssinica* (Asteraceae)	*ndhC–trnV, ndhC–atpE, rps18–rpl20, clpP, psbM–trnD*
Gargano et al. [51]	Asterids: *Solanum tuberosum* subsp. *tuberosum, S. bulbocastanum* (Solanaceae)	*ndhA intron, petN–psbM, rpl32–trnL, rps2–rpoC2, trnQ–rps16*
Yang et al. [24]	Monocots: *Cymbidium* (Orchidaceae)	*cemA–petA, clpP–psbB, ndhF–rpl32, petA–psbJ, psbA–trnK, rpl32–trnL, trnE–trnT, trnK–rps16, trnL–ccsA, trnP–psaJ, trnT–trnL*
Dong et al. [13]	Angiosperms: *Acorus* (Acoraceae), *Aethionema* (Brassicaceae), *Calycanthus* (Calycanthaceae), *Chimonanthus* (Calycanthaceae), *Eucalyptus* (Myrtaceae), *Gossypium* (Malvaceae), *Nicotiana* (Solanaceae), *Oenothera* (Onagraceae), *Oryza* (Poaceae), *Paeonia* (Paeoniaceae), *Populus* (Salicaceae), *Solanum* (Solanaceae)	*ycf1, trnH–psbA, rpl32–trnL, trnQ–rps16, ndhC–trnV, trnK/matK, trnS–trnG*
Ku et al. [26]	Asterids: *Catharanthus roseus* (Apocynaceae), *Asclepias syriaca* (Apocynaceae), *Coffea arabica* (Rubiaceae), *Solanum lycopersicon* (Solanaceae)	*ndhF–rpl32, rpl32–trnL, rps16–trnQ, trnE–trnT, trnK–rps16*
Ku et al. [25]	Asterids: *Ardisia polysticta* (Primulaceae – Myrsinioideae) *Panax ginseng* (Araliaceae) *Sesamum indicum* (Pedaliaceae)	*ccsA–ndhD, ndhG–ndhI, rpl14–rpl16, rpl32–trnL, trnK–rps16*
Särkinen & George [14]	Asterids: *Solanum tuberosum, S. bulbocastanum, S. lycopersicum* (Solanaceae)	*atpB–rbcL, clpP–psbB, ndhF, ndhF–rpl32, petL–psaJ, petN–psbM, rpl32–trnL, rpoC1–rpoB, trnA–trnI, trnK–rps16, ycf1*

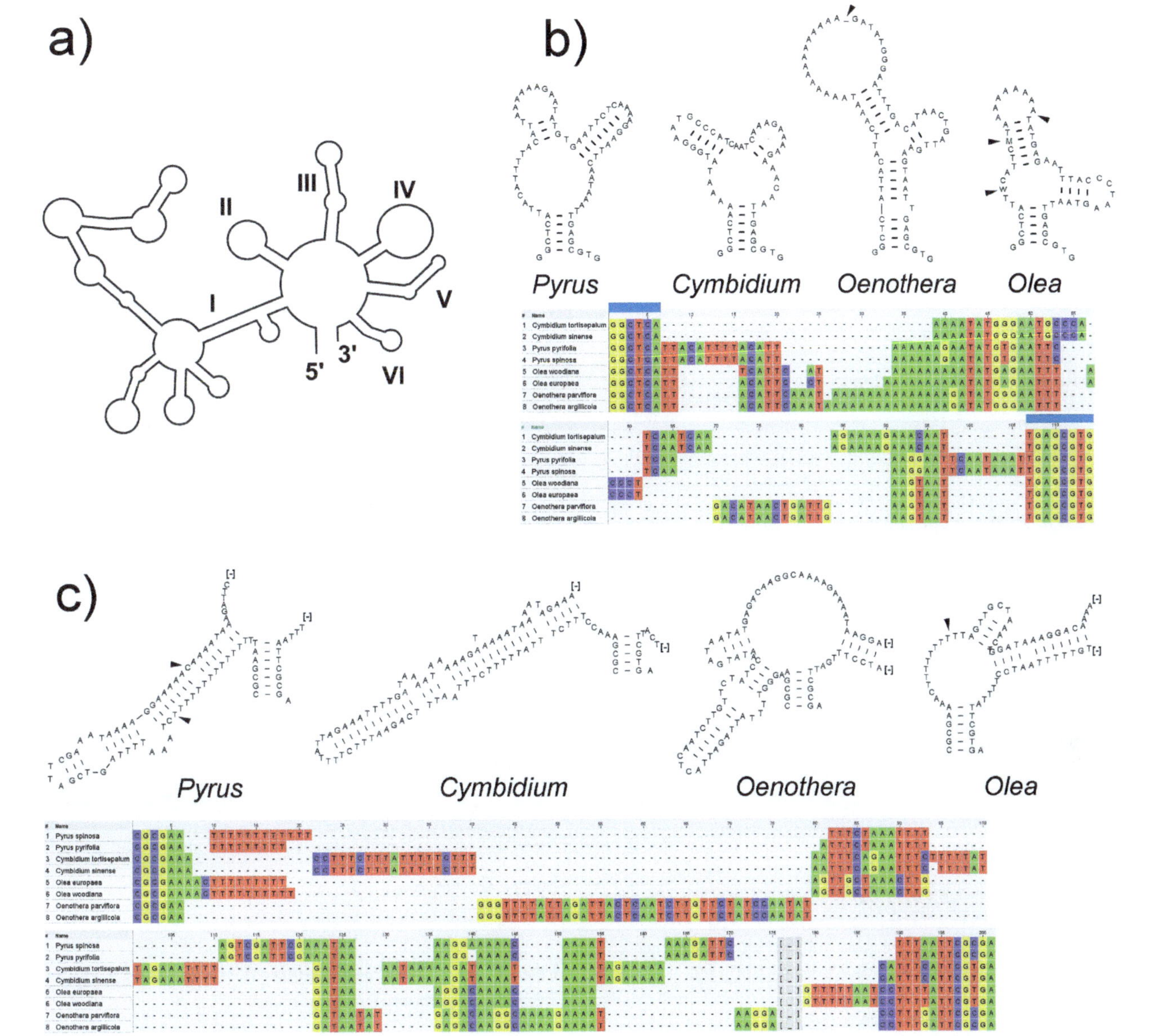

Figure 6. AT content of indels and areas around substitutions. Boxplot representation of the AT content in different types of indels (polyN, short sequence repeats (SSR) and other indels) on the left side and in areas with different sizes around all substitutions (SNPs) in the genome on the right side for a) *Pyrus spinosa* and *P. pyrifolia*), b) *Cymbidium tortisepalum* and *C. sinense*, c) *Oenonthera parviflora* and *O. argillicola* and d) *Olea europaea* and *O. woodiana*. The cross in each boxplot indicates the mean of the distribution, the thick line refers to the median. The dotted line shows the AT content of the whole consensus sequence.

conserved across all genera. They form the stem of the domain. Terminal parts of the domain such as the length-variable polyA-stretch in *Oenothera* have no structural constraints and therefore evolve rather freely. In *Olea*, there are two substitutions (indicated with ambiguity codes in the secondary structure) and one length variable polyA stretch. Again they occur in the terminal stem-loop and have no influence on the structure. The *rpl16* intron is more variable in *Pyrus* than in the other genome pairs. The polyT-stretch of *Olea* and *Pyrus* (beginning at position 10) is hypothesized as homologous in the alignment. But the predicted secondary structures (Fig. 5c) show that this polyT stretch forms different secondary structures caused by the different adjacent sequence elements. In *Olea*, it forms a bulge but in *Pyrus* it forms a stem-element together with a complementary 'AAAACACAAAAAA' motif [12,54].

Sequence variability versus phylogenetic signal. It is important to note that sequence variability as such does not necessarily correlate with the amount of hierarchical phylogenetic signal in a multiple sequence matrix. Thus, *p*-distances and PICs_which are both measures of sequence variability and describe the similarity of sequences_will not necessarily indicate the phylogenetically most informative regions. The phylogenetic utility of genomic regions depends on the distribution and kind of character state transformations throughout the evolutionary history of the sequences. Several statistics have been proposed to measure the hierarchical phylogenetic signal (referring to the phylogenetic structure in a data set) that take into account the number of resolved nodes and the statistical support for these nodes [58,59]. Specifically, the statistics *R*, *B*, and *C*, have been defined by Müller et al. [59]. The most important one, *R*, measures the proportion of resolved clades and their support in a tree inferred from a given data set relative to the maximum possible resolution and support. If all nodes have maximum support, *R* will get the value 1; if the phylogeny is completely unresolved (consists only of polytomies), *R* will have the value 0.

The empirical evaluation of phylogenetic structure in a genomic region generally requires a multiple sequence alignment of a representatively sampled clade. From the datasets that have been evaluated in detail using the *R* statistic [44,45,59], it is evident that at one hand higher variability often leads to more phylogenetic information (simply because there are more potentially informative characters). On the other hand, there are marked differences in the quality of hierarchical phylogenetic signal coming from the same number of variable positions in different kinds of genomic regions [45]. These can be explained by different molecular evolutionary patterns. The general trend across angiosperms is that high phylogenetic structure is found in intergenic spacers and group I and II introns, but not in protein-coding genes except *matK*. In our case of very closely related plastid genomes, the effects of multiple changes of the same site, eventually leading to saturation, or reversals, will probably not be very significant because these sequences are just starting to diverge. Nevertheless, it will be interesting to determine the phylogenetic structure in the top-ranked genomic regions in terms of variability once more extensive taxon sets will be available.

Moreover, highly variable regions will be needed to distinguish haplotypes (or species), even if they do not provide sufficient information about their phylogeny [44]. If haplotypes are used in the sense of individual alleles, the pure variability is most important. However, AT-rich sequence elements (often in stem-loops) can be highly homoplastic with respect to the evolution of microstructural mutations [60,61]. The most extreme causes of homoplasy are inversions [62,63]. Therefore, especially those markers that contain a single AT-rich mutational hotspot should be tested for congruence in signal with other plastid markers. Haplotype analyses often only use one or two markers, but experiences from other studies that have successfully reconstructed evolutionary relationships among closely related species indicate that the combination of four or five regions will be needed. An increased number of characters increases resolution and support also in network analyses [64,65].

Implications for plastid marker development in angiosperms. About 20–30 plastid spacers and introns are regularly sequenced for phylogenetic and haplotype analyses, for which universal amplification primers exist. Also, considerable progress has been made during recent years in predicting phylogenetic utility from molecular evolutionary patterns, revealing differences in phylogenetic structure of genes, group I and group II introns, and intergenic spacers [10–12,45,59]. In this way, markers with high versus low phylogenetic signal can be distinguished. For higher levels of genetic distance levels (e.g. distantly related species, genera, and families of flowering plants), a detailed evaluation of markers is therefore hardly necessary because sound predictions can be made. But is it worth to sequence whole plastid genomes when very closely related groups of species are to be studied?

Our comparison of genome pairs at comparable low distances shows that the mutational dynamics of plastid genomic regions may follow its own path in different lineages. While the variability in the respective unique sequence elements contributes the major proportion of the overall variability of a genomic region at that level, this contribution will be increasingly negligible at higher distance levels. The exploration of the plastid genome for the most variable and most suitable regions will therefore be a worthwhile investment when genetic distances are low.

It is of course possible to sequence all or at least most of the 30 promising plastid regions individually for a small taxon set in a given group. However, the effort needed is quite high. At least 60 individual fragments would need to be PCR-amplified and sequenced using many individual primers. Since only three to five loci are usually sequenced in evolutionary studies, a large part of these data would be wasted or deposited in GenBank as "unpublished". The sequencing and assembly of whole plastid genomes is still laborious, especially if critical areas of low coverage or homonucleotide stretches are verified by Sanger sequencing. Often overlooked costs have to be considered as well: this includes higher requirements for IT hardware and much increased time for sequence assembly and data management compared to traditional sequencing. Still, sequencing a complete plastid genome has many benefits over many single-marker PCRs. First, the complete genome sequence ensures that all genomic regions can be considered for marker development. And second, generating complete genomes allows for using the genome sequence for other studies, so that data are added in a complementary way to build proper information sources for the respective lineages (e.g., for comparative genomics, primer design, detection of plastid microsatellites, or extraction of regions for phylogenetic studies). We therefore conclude that whole plastid genome sequencing will remain a worthwhile approach for marker development in evolutionary studies of plants.

Supporting Information

Table S1 Primers for verification of sequence parts ambiguously read by the 454 sequencing.

Table S2 Ranking of all regions for the four genome pairs.

File S1 Pairwise alignment of the plastid genomes of *Pyrus spinosa* and *P. pyrifolia*.

File S2 Pairwise alignment of the plastid genomes of *Cymbidium tortisepalum* and *C. sinense*.

File S3 Pairwise alignment of the plastid genomes of *Oenothera parviflora* and *O. argillicola*.

File S4 Pairwise alignment of the plastid genomes of *Olea woodiana* and *O. europaea*.

Acknowledgments

The genomics work was done at the Berlin Center for Genomics in Biodiversity Research (BeGenDiv). We greatly acknowledge the lab assistance of Susan Mbedi and Virginia Duwe. We are also grateful to Felix Heeger for assistance with the genome assembly and Susan Wicke (Münster) for helpful advice on genome annotation.

This is publication number 11 of the Berlin Center for Genomics in Biodiversity Research (BeGenDiv). The study was carried out as part of the project "Developing tools for conserving the plant diversity of the Transcaucasus" funded by VolkswagenStiftung.

Author Contributions

Conceived and designed the experiments: NK TB. Performed the experiments: NK HTV LN. Analyzed the data: NK LN HTV MA TB. Contributed reagents/materials/analysis tools: LN MA. Wrote the paper: NK LN TB.

References

1. McCauley DE (1995) The use of chloroplast DNA polymorphism in studies of gene flow in plants. Trends Ecol Evol 10: 198–202.
2. Sang T, Crawford DJ, Stuessy TF (1997) Chloroplast DNA phylogeny, reticulate evolution, and biogeography of *Paeonia* (Paeoniaceae). Am J Bot 84: 1120–1136.
3. Sang T, Donoghue MJ, Zhang DM (1997) Evolution of alcohol dehydrogenase genes in peonies (*Paeonia*): Phylogenetic relationships of putative nonhybrid species. Mol Biol Evol 14: 994–1007.
4. Kurtto A (2009) Rosaceae (pro parte majore). Euro+Med Plantbase - the information resource for Euro-Mediterranean plant diversity. Available: http://www.emplantbase.org/home.html. Accessed February 2014 February 19.
5. Cuizhi G, Spongberg SA (2003) *Pyrus*. Flora of China: eFloras (2008). Published on the Internet http://www.efloras.org, Missouri Botanical Garden, St. Louis, MO & Harvard University Herbaria, Cambridge, MA. pp. 173–179.
6. Akopian JA (2007) O vidakh roda *Pyrus* L. (Rosaceae) v Armenii [On the *Pyrus* L. (Rosaceae) species in Armenia]. Fl Rast Rastitel Resurs Armenii 16: 15–26.
7. Fedorov AA (1954) Rod grusha *Pyrus* L. [The genus pear *Pyrus* L.]. In: Sokolov SJ, editor. Derev'a i kustarniki SSSR [The trees and shrubs of the USSR]. Moskva, Leningrad: Izdatel'stvo Akademii Nauk SSSR [National Science Academy of the U.S.S.R. publishing]. pp. 378–414.
8. Shaw J, Lickey EB, Beck JT, Farmer SB, Liu W, et al. (2005) The tortoise and the hare II: relative utility of 21 noncoding chloroplast DNA sequences for phylogenetic analysis. Am J Bot 92: 142–166.
9. Shaw J, Lickey EB, Schilling EE, Small RL (2007) Comparison of whole chloroplast genome sequences to choose noncoding regions for phylogenetic studies in angiosperms: the tortoise and the hare III. Am J Bot 94: 275–288.
10. Kelchner SA (2002) Group II introns as phylogenetic tools: Structure, function, and evolutionary constraints. Am J Bot 89: 1651–1669.
11. Borsch T, Quandt D (2009) Mutational dynamics and phylogenetic utility of noncoding chloroplast DNA. Plant Syst Evol 282: 169–199.
12. Korotkova N, Schneider J, Quandt D, Worberg A, Zizka G, et al. (2009) Phylogeny of the eudicot order Malpighiales: analysis of a recalcitrant clade with sequences of the *petD* group II intron. Plant Syst Evol 282: 201–228.
13. Dong W, Liu J, Yu J, Wang L, Zhou S (2012) Highly variable chloroplast markers for evaluating plant phylogeny at low taxonomic levels and for DNA barcoding. PLoS ONE 7: e35071.
14. Särkinen T, George M (2013) Predicting plastid marker variation: Can complete plastid genomes from closely related species help? PLoS ONE 8: e82266.
15. Jansen RK, Cai Z, Raubeson LA, Daniell H, dePamphilis CW, et al. (2007) Analysis of 81 genes from 64 plastid genomes resolves relationships in angiosperms and identifies genome-scale evolutionary patterns. Proc Natl Acad Sci USA 104: 19369–19374.
16. Moore MJ, Dhingra A, Soltis PS, Shaw R, Farmerie WG, et al. (2006) Rapid and accurate pyrosequencing of angiosperm plastid genomes. BMC Plant Biol 6.
17. Cronn R, Liston A, Parks M, Gernandt DS, Shen R, et al. (2008) Multiplex sequencing of plant chloroplast genomes using Solexa sequencing-by-synthesis technology. Nucleic Acids Res 36.
18. Parks M, Cronn R, Liston A (2009) Increasing phylogenetic resolution at low taxonomic levels using massively parallel sequencing of chloroplast genomes. BMC Biol 7.
19. Xu Q, Xiong G, Li P, He F, Huang Y, et al. (2012) Analysis of complete nucleotide sequences of 12 *Gossypium* chloroplast genomes: Origin and evolution of allotetraploids. PLoS ONE 7.
20. Njuguna W, Liston A, Cronn R, Ashman TL, Bassil N (2013) Insights into phylogeny, sex function and age of *Fragaria* based on whole chloroplast genome sequencing. Mol Phylogenet Evol 66: 17–29.
21. Mariotti R, Cultrera NG, Diez CM, Baldoni L, Rubini A (2010) Identification of new polymorphic regions and differentiation of cultivated olives (*Olea europaea* L.) through plastome sequence comparison. BMC Plant Biol 10: 211.
22. Ahmed I, Matthews PJ, Biggs PJ, Naeem M, McLenachan PA, et al. (2013) Identification of chloroplast genome loci suitable for high-resolution phylogeographic studies of *Colocasia esculenta* (L.) Schott (Araceae) and closely related taxa. Mol Ecol Resour: 929–937.
23. Jheng CF, Chen TC, Lin JY, Wu WL, Chang CC (2012) The comparative chloroplast genomic analysis of photosynthetic orchids and developing DNA markers to distinguish *Phalaenopsis* orchids. Plant Sci 190: 62–73.
24. Yang J-B, Tang M, Li H-T, Zhang Z-R, Li D-Z (2013) Complete chloroplast genome of the genus *Cymbidium*: lights into the species identification, phylogenetic implications and population genetic analyses. BMC Evol Biol 13: 84.
25. Ku C, Hu J-M, Kuo C-H (2013) Complete plastid genome sequence of the basal asterid *Ardisia polysticta* Miq. and comparative analyses of asterid plastid genomes. PLoS ONE 8: e62548.
26. Ku C, Chung W-C, Chen L-L, Kuo C-H (2013) The complete plastid genome sequence of Madagascar Periwinkle *Catharanthus roseus* (L.) G. Don: Plastid genome evolution, molecular marker identification, and phylogenetic implications in Asterids. PLoS ONE 8: e68518.
27. Terakami S, Matsumura Y, Kurita K, Kanamori H, Katayose Y, et al. (2012) Complete sequence of the chloroplast genome from pear (*Pyrus pyrifolia*): genome structure and comparative analysis. Tree Genet Genomes 8: 841–854.
28. Chevreux B, Wetter T, Suhai S (1999) Genome sequence assembly using trace signals and additional sequence information. Computer Science and Biology: Proceedings of the German Conference on Bioinformatics (GCB) 99: 45–56.
29. Wyman SK, Jansen RK, Boore JL (2004) Automatic annotation of organellar genomes with DOGMA. Bioinformatics 20: 3252–3255.
30. Müller J, Müller K, Neinhuis C, Quandt D (2005+) PhyDE: Phylogenetic Data Editor. – Available: www.phyde.de. Accessed 2014 March 3.
31. Huse SM, Huber JA, Morrison HG, Sogin ML, Welch DM (2007) Accuracy and quality of massively parallel DNA pyrosequencing. Genome Biol 8: R143.
32. Kelchner SA (2000) The evolution of non-coding chloroplast DNA and its application in plant systematics. Ann Missouri Bot Gard 87: 482–498.
33. Löhne C, Borsch T (2005) Molecular evolution and phylogenetic utility of the *petD* group II intron: A case study in basal angiosperms. Mol Biol Evol 22: 317–332.
34. Katoh K, Standley DM (2013) MAFFT multiple sequence alignment software version 7: improvements in performance and usability. Mol Biol Evol 30: 772–780.
35. Swofford DL (1998) PAUP*. Phylogenetic Analysis Using Parsimony (*and other Methods). Sunderland, Massachussets: Sinauer Associates.
36. Krzywinski MI, Schein JE, Birol I, Connors J, Gascoyne R, et al. (2009) Circos: An information aesthetic for comparative genomics. Genome Res.
37. Michel F, Umesono K, Ozeki H (1989) Comparative and functional anatomy of group II catalytic introns - a review. Gene 82: 5–30.
38. Mathews DH, Disney MD, Childs JL, Schroeder SJ, Zuker M, et al. (2004) Incorporating chemical modification constraints into a dynamic programming algorithm for prediction of RNA secondary structure. Proc Natl Acad Sci USA 101: 7287–7292.
39. Katayama H, Uematsu C (2003) Comparative analysis of chloroplast DNA in *Pyrus* species: physical map and gene localization. Theor Appl Genet 106: 303–310.
40. Lo EYY, Donoghue MJ (2012) Expanded phylogenetic and dating analyses of the apples and their relatives (Pyreae, Rosaceae). Mol Phylogenet Evol 63: 230–243.
41. Katayama H, Tachibana M, Iketani H, Zhang S-L, Uematsu C (2012) Phylogenetic utility of structural alterations found in the chloroplast genome of pear: hypervariable regions in a highly conserved genome. Tree Genet Genomes 8: 313–326.
42. Wuyun T, Ma T, Uematsu C, Katayama H (2013) A phylogenetic network of wild Ussurian pears (*Pyrus ussuriensis* Maxim.) in China revealed by hypervariable regions of chloroplast DNA. Tree Genet Genomes 9: 167–177.
43. Löhne C, Borsch T, Wiersema JH (2007) Phylogenetic analysis of Nymphaeales using fast-evolving and noncoding chloroplast markers. Bot J Linn Soc 154: 141–163.
44. Korotkova N, Borsch T, Quandt D, Taylor NP, Müller K, et al. (2011) What does it take to resolve relationships and to identify species with molecular markers? An example from the epiphytic Rhipsalideae (Cactaceae). Am J Bot 98: 1549–1572.
45. Barniske A-M, Borsch T, Müller K, Krug M, Worberg A, et al. (2012) Phylogenetics of early branching eudicots: Comparing phylogenetic signal across plastid introns, spacers, and genes. J Syst Evol 50: 85–108.
46. Goodson BE, Santos-Guerra A, Jansen RK (2006) Molecular systematics of *Descurainia* (Brassicaceae) in the Canary Islands: biogeographic and taxonomic implications. Taxon 55: 671–682.
47. Lee C, Wen J (2004) Phylogeny of *Panax* using chloroplast *trnC–trnD* intergenic region and the utility of *trnC–trnD* in interspecific studies of plants. Mol Phylogenet Evol 31: 894–903.
48. Campagna ML, Downie SR (1998) The intron in chloroplast gene *rpl16* is missing from the flowering plant families Geraniaceae, Goodeniaceae, and Plumbaginaceae. Trans Illinois State Acad Sci 91: 1–11.
49. Timme RE, Kuehl JV, Boore JL, Jansen RK (2007) A comparative analysis of the *Lactuca* and *Helianthus* (Asteraceae) plastid genomes: Identification of divergent regions and categorization of shared repeats. Am J Bot 94: 302–312.
50. Doorduin L, Gravendeel B, Lammers Y, Ariyurek Y, Chin-A-Woeng T, et al. (2011) The complete chloroplast genome of 17 individuals of pest species *Jacobaea vulgaris*: SNPs, microsatellites and barcoding markers for population and phylogenetic studies. DNA Res 18: 93–105.
51. Gargano D, Scotti N, Vezzi A, Bilardi A, Valle G, et al. (2012) Genome-wide analysis of plastome sequence variation and development of plastidial CAPS markers in common potato and related *Solanum* species. Genet Resour Crop Evol 59: 419–430.

52. Daniell H, Lee SB, Grevich J, Saski C, Quesada-Vargas T, et al. (2006) Complete chloroplast genome sequences of *Solanum bulbocastanum*, *Solanum lycopersicum* and comparative analyses with other Solanaceae genomes. Theor Appl Genet 112: 1503–1518.
53. Štorchová H, Olson MS (2007) The architecture of the chloroplast *psbA-trnH* non-coding region in angiosperms. Plant Syst Evol 268: 235–256.
54. Quandt D, Müller K, Stech M, Frahm J-P, Frey W, et al. (2004) Molecular evolution of the chloroplast *trnL-F* region in land plants. In: Goffinet B, Hollowell V, Magill R, editors. Molecular Systematics of Bryophytes. St Louis: Missouri Botanical Garden Press. pp. 13–37.
55. Wicke S, Schäferhoff B, dePamphilis CW, Müller KF (2014) Disproportional plastome-wide increase of substitution rates and relaxed purifying selection in genes of carnivorous Lentibulariaceae. Mol Biol Evol 31: 529–545.
56. Lehmann K, Schmidt U (2003) Group II introns: Structure and catalytic versatility of large natural ribozymes. Crit Rev Biochem Mol Biol 38: 249–303.
57. Pyle AM, Lambowitz AM (2006) Group II Introns: Ribozymes that splice RNA and invade DNA. In: Gesteland RF, Cech TR, Atkins JF, editors. The RNA World. 3rd ed. Cold Spring Harbor: Cold Spring Harbor Laboratory Press. pp. 449–505.
58. Källersjö M, Farris JS, Chase MW, Bremer B, Fay MF, et al. (1998) Simultaneous parsimony jackknife analysis of 2538 *rbcL* DNA sequences reveals support for major clades of green plants, land plants, seed plants and flowering plants. Plant Syst Evol 213: 259–287.
59. Müller K, Borsch T, Hilu KW (2006) Phylogenetic utility of rapidly evolving DNA at high taxonomical levels: Contrasting *matK*, *trnT-F*, and *rbcL* in basal angiosperms. Mol Phylogenet Evol 41: 99–117.
60. Tesfaye K, Borsch T, Govers K, Bekele E (2007) Characterization of *Coffea* chloroplast microsatellites and evidence for the recent divergence of *C. arabica* and *C. eugenioides* chloroplast genomes. Genome 50: 1112–1129.
61. Borsch T, Hilu KW, Wiersema JH, Lohne C, Barthlott W, et al. (2007) Phylogeny of *Nymphaea* (Nymphaeaceae): Evidence from substitutions and microstructural changes in the chloroplast *trnT-trnF* region. Int J Plant Sci 168: 639–671.
62. Quandt D, Müller K, Huttunen S (2003) Characterisation of the chloroplast DNA *psbT-H* region and the influence of dyad symmetrical elements on phylogenetic reconstructions. Plant Biol 5: 400–410.
63. Whitlock BA, Hale AM, Groff PA (2010) Intraspecific inversions pose a challenge for the *trnH-psbA* plant DNA barcode. PLoS ONE 5: e11533.
64. Fior S, Li M, Oxelman B, Viola R, Hodges SA, et al. (2013) Spatiotemporal reconstruction of the *Aquilegia* rapid radiation through next-generation sequencing of rapidly evolving cpDNA regions. New Phytol 198: 579–592.
65. Erixon P, Oxelman B (2008) Reticulate or tree-like chloroplast DNA evolution in Sileneae (Caryophyllaceae)? Mol Phylogenet Evol 48: 313–325.
66. Greiner S, Wang X, Herrmann RG, Rauwolf U, Mayer K, et al. (2008) The complete nucleotide sequences of the 5 genetically distinct plastid genomes of *Oenothera*, subsection *Oenothera*: II. A microevolutionary view using bioinformatics and formal genetic data. Mol Biol Evol 25: 2019–2030.
67. Greiner S, Wang X, Rauwolf U, Silber MV, Mayer K, et al. (2008) The complete nucleotide sequences of the five genetically distinct plastid genomes of *Oenothera*, subsection *Oenothera*: I. Sequence evaluation and plastome evolution. Nucleic Acids Res 36: 2366–2378.
68. Besnard G, Hernandez P, Khadari B, Dorado G, Savolainen V (2011) Genomic profiling of plastid DNA variation in the Mediterranean olive tree. BMC Plant Biol 11.

11

New Insights into Regulation of Proteome and Polysaccharide in Cell Wall of *Elsholtzia splendens* in Response to Copper Stress

Tingting Liu, Chaofeng Shen, Yi Wang, Canke Huang, Jiyan Shi*

Institute of Environmental Science and Technology, College of Environmental and Resource Sciences, Zhejiang University Hangzhou, P.R. China

Abstract

Background and Aims: Copper (Cu) is an essential micronutrient for plants. However, excess amounts of Cu are toxic and result in a wide range of harmful effects on the physiological and biochemical processes of plants. Cell wall has a crucial role in plant defense response to toxic metals. To date, the process of cell wall response to Cu and the detoxification mechanism have not been well documented at the proteomic level.

Methods: An recently developed 6-plex Tandem Mass Tag was used for relative and absolute quantitation methods to achieve a comprehensive understanding of Cu tolerance/detoxification molecular mechanisms in the cell wall. LC–MS/MS approach was performed to analyze the Cu-responsive cell wall proteins and polysaccharides.

Key Results: The majority of the 22 up-regulated proteins were involved in the antioxidant defense pathway, cell wall polysaccharide remodeling, and cell metabolism process. Changes in polysaccharide amount, composition, and distribution could offer more binding sites for Cu ions. The 33 down-regulated proteins were involved in the signal pathway, energy, and protein synthesis.

Conclusions: Based on the abundant changes in proteins and polysaccharides, and their putative functions, a possible protein interaction network can provide new insights into Cu stress response in root cell wall. Cu can facilitate further functional research on target proteins associated with metal response in the cell wall.

Editor: Raffaella Balestrini, Institute for Sustainable Plant Protection, C.N.R., Italy

Funding: The work was supported by the National Natural Science Foundation of China (11179025, 21177109) and Program for New Century Excellent Talents in University (NCET-11-0455). The funders had no role in study design, data collection and analysis, decision to publish, or preparation of the manuscript.

Competing Interests: The authors have declared that no competing interests exist.

* Email: shijiyan@zju.edu.cn

Introduction

Copper (Cu) is an essential micronutrient for plants because it is a structural and catalytic component of several proteins or enzymes involved in various physiological metabolisms [1]. However, excess Cu in plants is harmful; it induces phytotoxicity [2] and disturbs the physiological metabolisms, including carbohydrate metabolism, protein metabolism, mineral nutrition, cell respiration, and photosynthesis [3]. Given its redox properties, Cu^{2+} can catalyze the generation of reactive oxygen species (ROS) and Fenton reactions that damage cellular components or interfere with cellular transport processes [4]. Plants cannot avoid the stress conditions from Cu-contaminated areas. Thus, several resistance mechanisms for Cu tolerance have been developed [5], including exclusion, compartmentalization, chelation, and binding to organic ligands, which enable maintenance of Cu homeostasis for plant survival even in a highly contaminated environment [6].

Plant cell wall is the primary site for signal perception and defense response, which is significant in responding to environmental stresses [7]. The cell walls of plants that grow on heavy metal-contaminated areas provide physical barriers against toxic heavy metals and actively participate in plant defense response. The cell wall serves as a repository for high contents of metals; it senses stress signals and transmits them to the cell interior, thereby affecting the cell fate decision [8]. The main chemical components of a plant cell wall include polysaccharides and proteins. Polysaccharides represent up to 90% of the plant cell wall and constitute three different kinds of polymers, namely, cellulose, hemicelluloses, and pectins. Cell wall proteins (CWPs) of plants comprise less than 10% of the cell wall dry weight but have major roles in cell wall structure, metabolism, cell enlargement, signal transduction, defense responses, and many other physiological events [9].

Certain plants can accumulate high content of metal ions in their cell wall through various compounds. Cell wall polysaccharides possess crucial roles in heavy metal binding and accumulation. Binding of metal ions depends on the number of functional groups in the cell wall polysaccharides. Previous studies confirmed that the essential capacity of the cell wall for binding metal ions depends mainly on the abundance of polysaccharides in the carboxyl groups [10]. Current information about the regulation of cell wall polysaccharides under metal stress remains limited.

Abiotic stress induces expression of stress-responsive proteins, which lead to cellular adaptation in plant growth under stress conditions. Therefore, studies have been carried out to identify various CWPs involved in stress response, particularly those of regulatory or targeting function proteins [11–17]. Most reports have focused on leaf cell wall proteomes under abiotic stress, such as salt in tobacco [11]; wounding in *Medicago* [12]; dehydration in chickpea, rice, and maize [9,14,15]; zinc toxicity in *Hordeum* [18]; nickel toxicity in barley [19]; manganese toxicity in cowpea [20,21]; boron deficiency in *Lupinus albus* [22]; and flooding in soybean [23]. Few studies have focused on the root cell wall proteomes under abiotic stress [24–26]. All of these studies broadened our understanding of the complicated regulation of apoplast proteins. An in-depth proteome study of the cell wall can provide extensive information on the intrinsic mechanism of stress response attributed to a possible relationship between protein abundance and plant stress tolerance [27].

Elsholtzia splendens is a plant that can survive in a highly Cu-polluted area. This species can accumulate more than 1000 mg/kg of Cu in the body [28], and the major proportion (70%) of Cu ions is stored in the root cell wall [29]. Nevertheless, little is known about the defense and detoxification mechanisms in the root cell wall of *E. splendens*. A recently developed 6-plex Tandem Mass Tag (TMT) was used for relative and absolute quantitation methods to gain a comprehensive understanding of Cu tolerance/detoxification molecular mechanisms. LC–MS/MS approach was also conducted to analyze Cu-responsive CWPs and polysaccharides in this study. This proteomic study of the apoplast is the first in response to Cu. The results will significantly expand knowledge on the apoplast complexicity in metal-tolerant plants and develop a more comprehensive understanding of the involved cellular reactions in cell defense mechanisms.

Results

Isolation and purification of CWPs

Root CWPs from *E. splendens* were isolated with sucrose gradients and washed according to the method of Feiz et al. [30]. The activity of G'DH, a typical marker enzyme used to detect contamination of cytoplasmic proteins, was measured to validate the purification method [31]. The G6PDH activity in $CaCl_2$-extracted CWPs, NaCl-extracted CWPs, and cytosolic protein fractions were 8.36, 5.78, and 107.55 mU/mg protein, respectively (Figure 1). The G6PDH activity in $CaCl_2$- or NaCl-extracted CWPs was 10 times lower than that in cytosolic proteins ($p<0.01$). These results indicate that the contamination of intracellular proteins was minimal.

The identified proteins were classified into different gene ontology (GO) terms according to their cellular compartment(s), function(s), and biological processes (Table S1). Some of these identified proteins showed a compartment label (apoplast, cell wall, extracellular region, and plant-type cell wall) that is coherent with secreted proteins. The same proteins can also be attributed to various compartments in the GO classification. Three different computer programs (TargetP, Predotar, and WoLF PSORT) were used to predict the protein subcellular localization (Table S1). A protein was secreted when at least two out of the three algorithms predicted extracellular localization. Based on the results, 15 proteins were predicted to be secreted proteins in the fractions (Table S1).

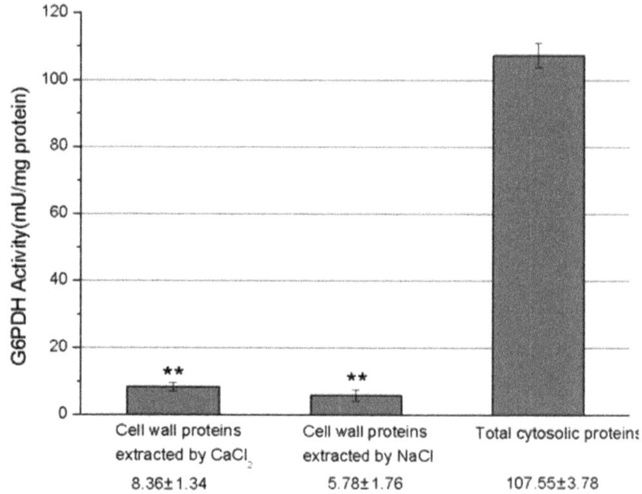

Figure 1. Evaluation of purity by the G6PDH activity in the cell wall proteins isolated from the roots of *Elsholtzia splendens*. The activity of G6PDH was assayed in $CaCl_2$-extracted cell wall proteins, NaCl-extracted cell wall proteins and total soluble proteins. One unit of G6PDH activity is defined as 1 μmol of NADPH turnover per min/mg protein. Results are presented as mean ±SE of G6PDH activity from three biological replicates. The asterisks indicate significant differences in the G6PDH activity of $CaCl_2$-extracted cell wall proteins, NaCl-extracted cell wall proteins compared with that of total soluble cytosolic proteins (**$p<0.01$).

Cu-induced changes in root CWP contents

Plants were grown on a medium supplemented with different $CuSO_4$ concentrations to determine changes in CWP abundance under Cu stress (Figure 2A). Cu toxicity led to a strong increase in CWP contents at low Cu concentrations. The CWP contents increased 1.39-fold at 25 μM Cu and 1.96-fold at 50 μM Cu. The highest content of the CWPs was observed at 50 μM Cu concentrations. The plant root began to decompose at 100 μM Cu (Figure 2C). The increase of Cu contents in the cell wall could proportionally increase with the Cu supply in the medium (slope = 0.3, $R^2 = 0.98$) at Cu concentrations below 75 μM. The Cu content in the root cell wall continued to increase significantly (slope = 0.45, $R^2 = 0.99$) above 75 μM Cu. The highest value of Cu content in the root cell wall reached 1.41 mg/g with the root cell wall dry weight at 100 μM Cu. These findings indicate an increase in Cu binding to cell walls with Cu supply in the medium (Figure 2A). Polypeptide separation by SDS-PAGE (Figure 2B) revealed about 30 polypeptide bands. Band intensities increased with the increase in Cu concentration below 50 μM (Figure S1). The staining intensity was the lowest among all the cell wall polypeptides at 100 μM Cu concentration.

Cu-induced remodeling in root cell wall polysaccharide

Evidence generally suggests that cell wall polysaccharides are crucial sites for Cu retention in plants. Therefore, changes in the contents of root cell wall polysaccharide of *E. splendens* were measured. Treatments with 25, 50, 75, and 100 μM Cu significantly increased the expression of CWPs and polysaccharides compared with controls (Figure 3). Cu supply enhanced the total sugar contents in various cell wall fractions. Sugar contents were higher in hemicellulose than in pectin and cellulose. Sugar contents in pectin increased 1.22-fold with 25 μM Cu. The major components of cell wall fractions are galacturonic acids. The content of galacturonic acids was considerably higher than that in

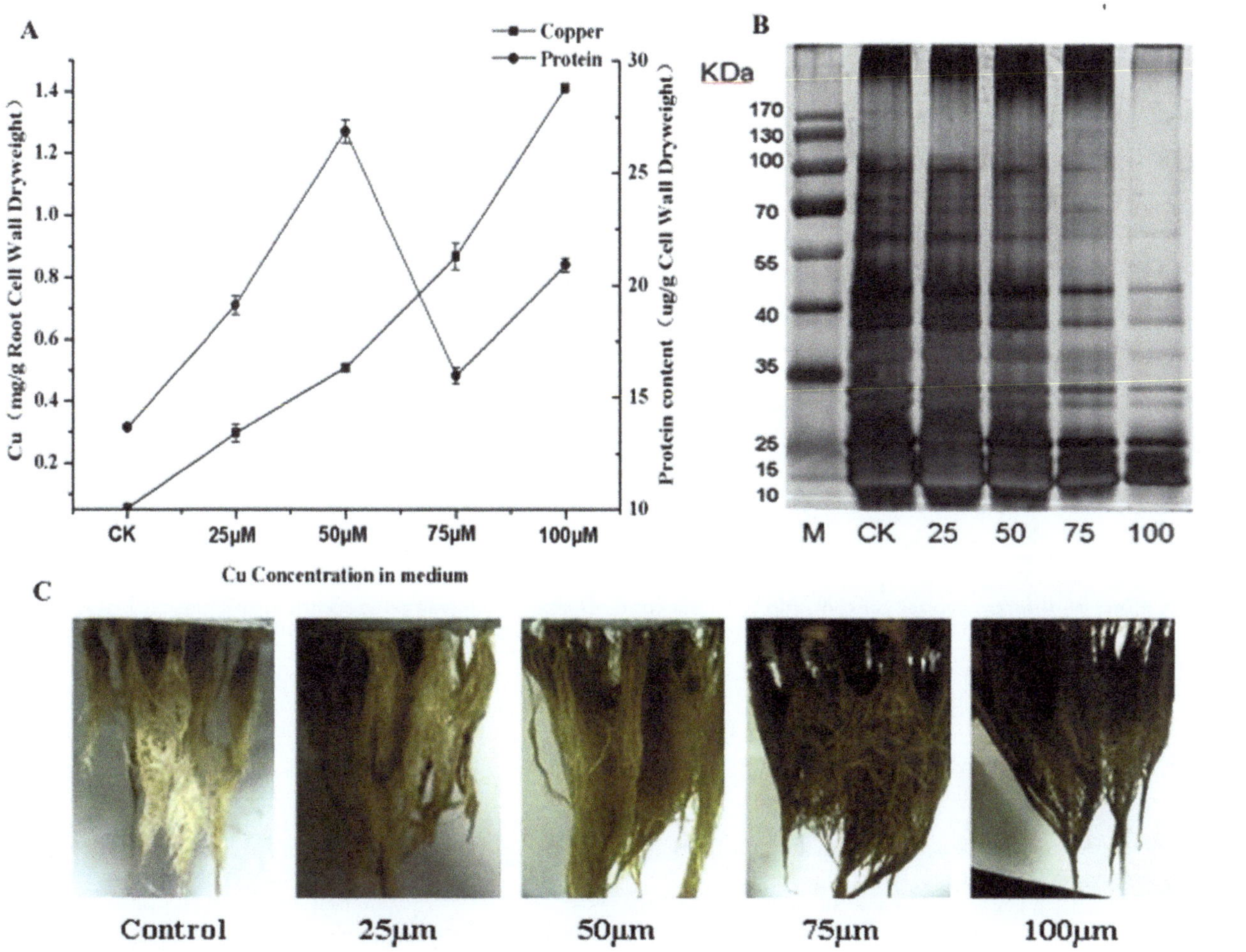

Figure 2. Effect of different copper concentration on the cell wall protein and copper ions in *Elsholtzia splendens*'s root cell. A: The content of copper and protein every root cell wall dry weight. B: SDS-PAGE of root cell wall protein under different copper stress. C: Root growth patterns of control and copper-stressed plants during different copper concentration. Data presented are mean ±SE (n = 20),*Significant mean differences from control at $p = 0.05$ in multiple comparison by LSD test.

the control during Cu treatment, particularly at higher Cu supply. The galacturonic acid values rapidly increased in pectin (1.28-fold), hemicellulose (1.22-fold), and cellulose (1.79-fold) at 100 μM Cu. A conserved disaccharide residue known as 2-keto-3-deoxyoctonic acid (KDO) is attached to C-3 of the backbone of rhamnogalacturonan II in pectin. The KDO concentration in pectin increased during Cu exposure. Thus, the highest concentration of KDO (1.49-fold) was found in the 100 μM Cu-treated group.

The results indicated that excess Cu affected the intracellular distribution of polysaccharides in the root cell wall. Given this reason, we focused on the relationship between wall polysaccharide composition and CWPs with increasing Cu intensity. The dendrogram proved the connection between the contents of polysaccharides and proteins as matched with different Cu concentrations (Figure 4). This finding suggests that Cu induced changes in the contents of root cell wall polysaccharides and proteins, which altered the intracellular distribution patterns in assigning effects on cellular activities.

Cu-responsive CWP identification using TMT LC-MS/MS

This study aimed to gain a better understanding of the metabolic processes and molecular mechanisms involved in Cu-responsive CWPs. A TMT-based shotgun quantitation approach was used to obtain an overall view of the proteome changes associated with Cu stress in the root cell wall of *E. splendens*. Unique proteins were successfully identified based on homology searching with a common protein confidence cutoff of 95% and relative quantitative information in control and Cu-treated samples, as indicated by the peak areas of the different TMT tags (Table 1, Table S2). A threshold of 1.5-fold change and a $p < 0.05$ were set as stringent criteria for significant differences between control and treatment. Fifty five proteins were classified as differentially expressed upon Cu stress using the criteria; 22 and 33 of these proteins showed increasing trends and decreasing abundance, respectively (Table 1, Table 2).

Functional classification of identified CWPs

Database search using Uni-Prot/NCBI accession numbers was conducted to obtain detailed information of differentially expressed proteins. These proteins were categorized into biological process and molecular function using an in-house Perl script according to the extracted GO terms from InterPro (www.ebi.ac.uk/interpro/) or Pfam (pfam.sanger.ac.uk/) (Figure 5). Functional classifications of the 55 differentially expressed proteins (Table 1, Table 2) were classified based on the GO terms that correspond to

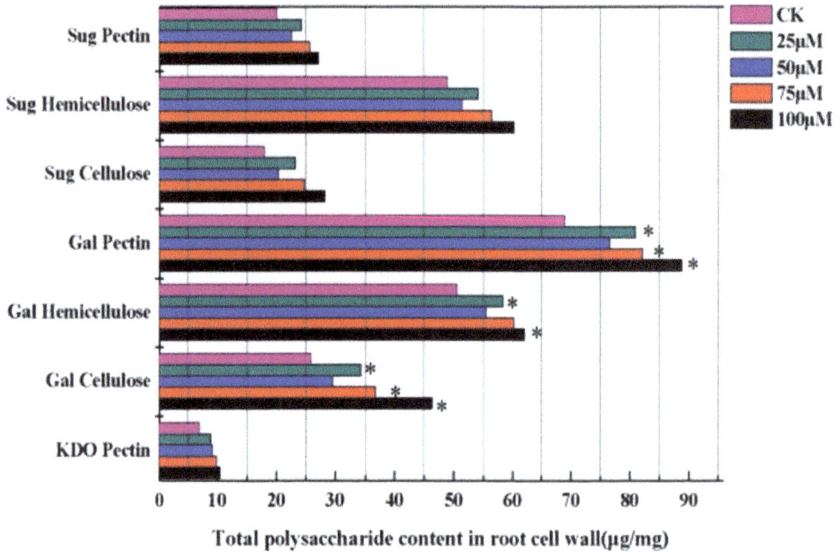

Figure 3. Effect of different copper concentrations on the contents of root cell wall polysacchride. Sug: sugar, Gal: galacturonic acids, KDO: 2-Keto-3-deoxyoctonic acid. Data presented are mean ±SE (n = 3). *Significant mean difference from control at $p = 0.05$ in multiple comparison by LSD test.

the biological process. The top three categories were those involved in metabolic processes (23.24%), cellular processes (16.20%), and response to stimuli (14.79%) (Figure 5A). These three categories were classified according to the GO molecular annotation into binding (41.38%), catalytic activity (31.03%), and transport activity (11.49%) (Figure 5B). The observed diversity in biological function and processes of the differentially expressed CWPs demonstrated that the response of the root cell wall of *E. splendens* to Cu stress was a complex process. Many physiological and biochemical changes were altered to counteract the adverse conditions.

Specific enzymes constituted by some small differentially expressed CWPs participated in various metabolic processes to counteract Cu stress. Classification pathways according to the KEGG database searches categorized the cell wall enzymes into 12 classes (Table 3, Table S2). The more important category pathways were involved in oxidative phosphorylation, phenylalanine biosynthesis and metabolism, ascorbate and aldarate metabolism, and citrate cycle. All these pathways suggested the relatively altered abundance of the enzymes in various metabolic pathways that probably re-optimized the metabolic processes to combat against Cu stress.

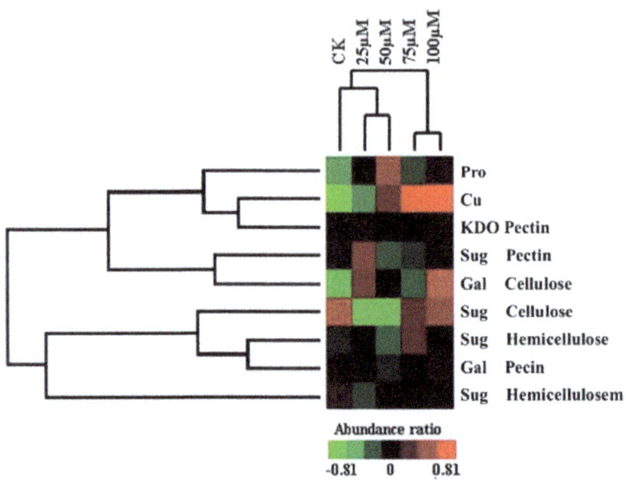

Figure 4. Hierarchical cluster result of cell wall polysaccharide abundance ratio using the average linkage distance between clusters is shown. The color weighting represents normalized levels of each variable from the high (red) to the low (green).

Discussion

Cell wall polysaccharide remodeling in root under Cu stress

Cell walls provide a physical barrier to plants grown in Cu-contaminated soil, and cell wall polysaccharides have a crucial role in metal binding and accumulation [7]. The binding ability depends on the number of functional groups in polysaccharides, particularly for abundant carboxyl groups [7]. In the present study, the contents of the various cell wall polysaccharides significantly increased with the increase in Cu concentration. The expression of cell wall polysaccharides, which contain more carboxyl groups, provided more binding sites for Cu ions. Excess Cu also affected the intracellular distribution of polysaccharide in the root cell walls (Figure 3). Similar situations were also described in rice roots under Cd stress [45] and *Arabidopsis* roots under Al stress [46]. The starch and sucrose metabolism-related proteins (Table 3) – Chloroplast NAD-MDH (No. 1, Table 1) – were also markedly up-regulated in response to Cu stress based on our observations. These findings suggest that Cu or Cu-induced oxidative stress may activate some crucial enzymes in the glyconeogenesis pathway, which can alter the contents of cell wall polysaccharides.

Increasing physiological, biochemical, and molecular evidence also showed that the remodeling of the binding properties of root apoplasts was attributed to Cu resistance. Enhancement at the level of low-methylesterified pectins, where the polysaccharides bound more trivalent metal ions, is one of the most remarkable alterations. Pectin methylesterases (PMEs), known as pectinesterases, catalyzed

Table 1. Root cell wall proteins with significant increased in expression level under 50 μM Cu treatment identified by LC-ESI-MS/MS-based proteomics using SIEVE ($p<0.05$ and fold chang >1.5).

Serial No.	Protein name	gi number[a]	Percent coverage [b] (%)	Fold Change[c]	*p* value[d]	Species identified[e]	MW(kDa)	p*I*
1	Chloroplast NAD-MDH	3256066	8.68	1.50	0.00000	Arabidopsis thaliana	49.9	8.2
2	D-glyceraldehyde-3-phosphate Dehydrogenase(GAPDH)	51849658	17.26	1.52	0.00000	Periploca sepium	25.3	6.2
3	40S ribosomal protein S19	74231038	9.32	1.52	0.00271	Solanum peruvianum	15.7	5.5
4	Phosphoglycerate kinase	21272	5.08	1.53	0.00963	Spinacia oleracea	55.1	6.1
5	Putative malate dehydrogenase	50080249	15.29	1.55	0.00346	Oryza sativa Japonica Group	41.6	5.0
6	Cyc07	1856971	6.12	1.56	0.02389	Catharanthus roseus	35.6	6.0
7	2,3-bisphosphoglycerate-independent phosphoglycerate mutase	168588	1.61	1.57	0.00000	Zea mays	68.8	5.2
8	Non-symbiotic hemoglobin	11095158	13.75	1.58	0.01599	Medicago sativa	23.0	9.0
9	P0 ribosomal protein	1143507	6.83	1.59	0.00000	Lupinus luteus	41.0	9.8
10	Putative 40S ribosomal protein s12	643074	7.75	1.60	0.03853	Fragaria x ananassa	21.2	4.7
11	Unknown	255648032	11.18	1.62	0.02313	Glycine max	21.3	6.4
12	Mitochondrial chaperonin	415733	7.16	1.68	0.01830	Brassica napus	74.5	5.4
13	Maturase K	33332509	4.13	1.69	0.01270	Pitcairnia xanthocalyx	48.8	6.9
14	Pectin methylesterase	1222552	2.23	1.70	0.00633	Solanum lycopersicum	75.5	9.5
15	Porin-like protein	47847590	3.61	1.71	0.01348	Oryza sativa Japonica Group	36.1	6.4
16	Pollen allergen Que a 1 isoform	167472851	7.50	1.72	0.00062	Quercus alba	21.3	8.8
17	Putative 1,4-benzoquinone reductase	53749369	5.37	1. 73	0.00000	Oryza sativa Japonica Group	49.9	8.2
18	Thioredoxin h	13624884	9.32	1.89	0.02335	Pisum sativum	25.3	6.2
19	Predicted protein	162680354	5.16	1.91	0.00523	Physcomitrella patens subsp	24.8	8.9
20	Hypothetical protein OsI_31140	125563499	2.51	1.99	0.00035	Oryza sativa Indica Group	16.1	4.7
21	Hypothetical protein OsI_07053	218190658	1.47	2.70	0.00000	Oryza sativ a Indica Group	37.3	5.8
22	60s acidic ribosomal protein P2	551267	7.89	3.31	0.00000	Parthenium argentatum	56.7	5.4

[a] gi no., gene identification number as in GenBankTM.
[b] Coverage., sequence coverage.
[c] Fold change.
[d] p value, indicates the significance of up- or down- regulation of spots according to the F-test through analysis of variance ($p<0.05$).
[e] Species identified by Mascot search using NCBI.

Table 2. Root cell wall proteins with significant decreased in expression level under 50 μM Cu treatment identified by LC-ESI-MS/MS-based proteomics using SIEVE ($p<0.05$ and fold chang >1.5).

Serial No.	Protein name	gi number[a]	Percent coverage [b] (%)	Fold Change[c]	*p* value[d]	Species identified[e]	MW(kDa)	p*I*
23	Predicted protein	162671613	1.87	−2.68	0.00000	Physcomitrella patens subsp	71.6	4.9
24	Peroxidase	17066703	2.71	−2.59	0.00000	Capsicum annuum	39.5	6.9
25	Os07g0223100	113610742	18.75	−2.28	0.03704	Oryza sativa Japonica Group	6.1	6.0
26	Mitochondrial phosphate translocator	1842188	2.47	−2.25	0.00466	Betula pendula	45.3	9.4
27	Predicted protein	549049082	3.53	−2.23	0.01662	Populus trichocarpa	30.5	8.9
28	Actin	288300146	16.80	−2.17	0.00166	Cleistogenes songorica	46.2	5.5
29	Ubiquitin fusion protein	149391425	36.04	−2.08	0.02095	Oryza sativa Indica Group	16.6	10.0
30	Heat shock protein Hsp70	124360342	2.59	−1.98	0.02106	Medicago truncatula	80.9	6.4
31	Putative nascent polypeptide associated complex alpha chain	20160782	6.93	−1.97	0.00728	Oryza sativa Japonica Group	26.9	4.4
32	Hypothetical protein SORBIDRAFT_09g020360	241945088	8.50	−1.88	0.00003	Sorghum bicolor	49.1	5.2
33	Plasma membrane-type calcium ATPase	3335060	1.08	−1.86	0.03138	Arabidopsis thaliana	125.3	8.9
34	Cytochome b5	2695711	11.94	−1.83	0.01678	Olea europaea	18.3	5.5
35	Unknown	118482646	3.48	−1.82	0.00000	Populus trichocarpa	41.8	4.9
36	ATP synthase beta subunit	8452631	4.35	−1.80	0.04389	Cichorium intybus	18.1	6.2
37	Peptidyl-prolyl cis-trans isomerase	118104	7.56	−1.76	0.00000	Zea mays L	21.8	9.0
38	Ferredoxin–NADP+ reductase-like protein	7267299	8.61	−1.73	0.00081	Arabidopsis thaliana	48.3	8.9
39	Putative protein	5262759	0.82	−1.71	0.04231	Arabidopsis thaliana	124.3	8.5
40	Conserved hypothetical protein	223517971	3.65	−1.71	0.00000	Ricinus communis	24.2	4.2
41	Unnamed protein product	219742878	7.50	−1.70	0.01691	Glycine max	38.2	5.6
42	Unknown	255634488	7.50	−1.69	0.01879	Glycine max	38.0	5.4
43	Ferredoxin-NADP Reductase	6634773	7.59	−1.69	0.00157	Oryza sativa	50.9	9.2
44	Predicted protein	222844395	1.20	−1.68	0.00000	Populus trichocarpa	72.6	9.2
45	60S ribosomal protein L25	310935	14.94	−1.66	0.00002	Nicotiana tabacum	24.8	8.8
46	Vacuolar H(+)-ATPase subunit-like protein	7801655	2.27	−1.64	0.00000	Arabidopsis thaliana	58.0	5.9
47	Small G protein	974780	4.06	−1.61	0.01530	Beta vulgaris subsp. vulgaris	26.3	10.1
48	Elongation factor 1 subunit alpha	162036292	8.22	−1.58	0.02150	Cenchrus americanus	30.2	6.6
49	2-dehydro-3-deoxy phosphor heptonate aldolase	2546988	5.61	−1.52	0.01476	Morinda citrifolia	65.5	9.3
50	RAS-related GTP-binding protein	20756	5.34	−1.52	0.00000	Pisum sativum	26.7	8.3
51	Aspartic proteinase nepenthesin-1 precursor, putative	223535787	3.05	−1.52	0.00000	Ricinus communis	46.6	6.5
52	Hypothetical protein	147774267	0.73	−1.52	0.00000	Vitis vinifera	24.1	4.9
53	Hypothetical protein VITISV_025412	147822728	8.42	−1.52	0.00000	Vitis vinifera	21.9	7.3
54	Predicted protein	222855760	6.33	−1.51	0.00185	Populus trichocarpa	41.8	5.0
55	Hypothetical protein VITISV_023718	147801802	1.87	−1.51	0.00000	Vitis vinifera	71.1	7.8

[a] gi no., gene identification number as in GenBankTM.
[b] Coverage., sequence coverage.
[c] Fold change.
[d] p value, indicates the significance of up- or down- regulation of spots according to the F-test through analysis of variance ($p<0.05$).
[e] Species identified by Mascot search using NCBI.

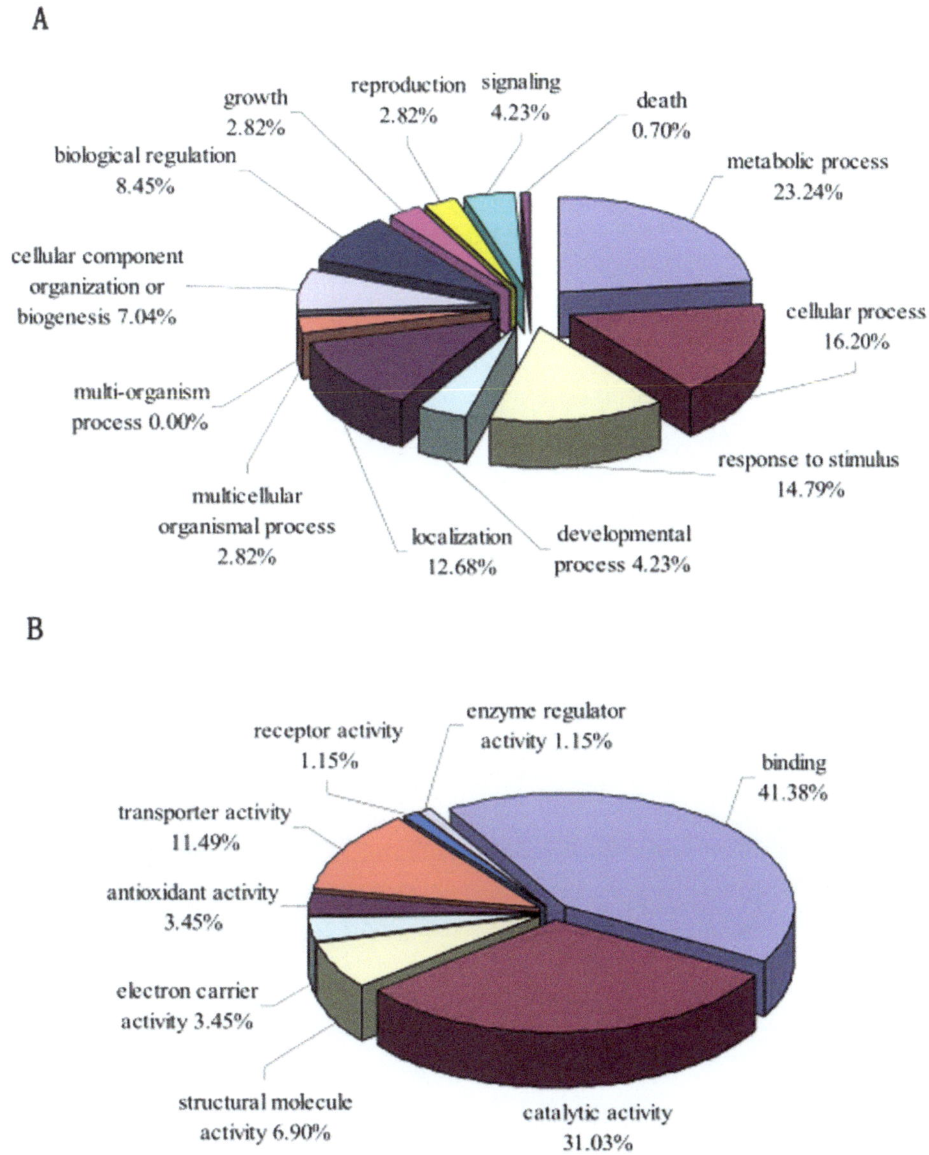

Figure 5. Functional cataloguing of 55 differentially expressed cell wall proteins in *Elsholtzia splendens*'s root based on GO annotation. The pie charts show the distribution of 55 differentially expressed cell wall proteins on of the Cu -responsive proteins into their functional classes in percentage. A: Biological Process Ontology, B: Molecular Function Ontology.

the demethylesterification of the homogalacturonan pectin domain in the cell wall. Demethylesterification of the pectin increases the abundance of free carboxylic acid groups on the galacturonic acid residue [47]. A flax PME has been recently implicated in wall remodeling following Cd treatment [48]. Interestingly, this study identified PMEs (No. 14, Table 1) as Cu-induced differentially expressed proteins; the protein had changes in abundance of 1.7-fold ($p \leq 0.5$). Catalyzed ROS by excess Cu^{2+} can also damage or cause degradation of essential complex molecules in cell wall polysaccharide. Generating hydroxyl radicals from H_2O_2 possesses a direct role in cell wall loosening through polysaccharide cleavage [49,50]. Hydroxyl radicals (OH) may cause non-enzymic scission of polysaccharides in vivo (e.g., in plant cell walls) [50]. Basing on these findings, we concluded that the chemical composition and distribution of plant cell wall polysaccharides are a factor in the outcome of plant–metal interaction, which elucidated a possible role in a novel Cu-resistance mechanism.

Cu-induced oxidative stress and antioxidant defense in cell wall

Cu is a redox active metal that catalyzes ROS production, such as superoxide ($\dot{O}_2$), hydrogen peroxide (H_2O_2), and hydroxyl radicals (OH·) through Habere Weiss and Fenton reactions [51]. ROS can act as signaling molecules for stress response. However, ROS can cause damage to many cellular components above a certain threshold. Most Cu-tolerance mechanisms are primarily involved in protecting the cellular structure. An important method is the control of the level of ROS or the limitation of damage caused by ROS. In the current study, significant changes in the abundance of some antioxidant and defense-related proteins suggested that ROS can be involved in a Cu-induced oxidative

Table 3. Category of differentially expressed cell wall proteins refers to the entry on the classification of pathways from KEGG database.

No.	Pathway	Enzyme	Enzyme ID	Change Folds	Protein Name
1	Phenylpropanoid biosynthesis??? phenylalanine metabolism	Lactoperoxidase	ec:1.11.1.7	−2.59	Peroxidase
				−1.69	Ferredoxin-NADP Reductase
				−1.66	Predicted protein
2	Phenylalanine, tyrosine and tryptophan biosynthesis	Synthase	ec:2.5.1.54	−1.52	Hypothetical protein VITISV_025412
				2.70	Hypothetical protein OsI_07053
3	Nitrogen metabolism	(6S)-tetrahydrofolate aminomethyltransferase dehydratas	ec:2.1.2.10	1.72	Pollen allergen Que a 1 isoform
			ec:4.2.1.1	−1.83	Cytochome b5
4	One carbon pool by folate	(6S)-tetrahydrofolate aminomethyltransferase	ec:2.1.2.10	1.72	Pollen allergen Que a 1 isoform
5	Starch and sucrose metabolism	Synthase	ec:2.4.1.34	1.50	Chloroplast NAD-MDH
6	Oxidative phosphorylation	ATPase	ec:3.6.3.6	−1.76	Peptidyl-prolyl cis-trans isomerase
		Dehydrogenase	ec:1.3.5.1	1.53	Phosphoglycerate kinase
7	Ascorbate and aldarate metabolism	Oxidase	ec:1.1.3.8	−1.64	Vacuolar H(+)-ATPase subunit-like protein
8	Glyoxylate and dicarboxylate metabolism	(Si)-synthase	ec:2.3.3.1	−1.86	Plasma membrane-type calcium ATPase
9	Sphingolipid metabolism	Phosphodiesterase	ec:3.1.4.12	−1.52	2-dehydro-3-deoxyphosphoheptonate aldolase
10	Citrate cycle (TCA cycle)	(Si)-synthase	ec:2.3.3.1	−1.86	Plasma membrane-type calcium ATPase
		dehydratas	ec:1.3.5.1	1.53	Phosphoglycerate kinase
11	Glycine, serine and threonine metabolism	(6S)-tetrahydrofolate aminomethyltransferase	ec:2.1.2.10	1.72	Pollen allergen Que a 1 isoform
12	Methane metbolisma	Lactoperoxidase	ec:1.11.1.7	−2.59	Peroxidase
				−1.69	Ferredoxin-NADP Reductase
				−1.66	Predicted protein

stress response. Analysis of *E. splendens* root CWPs confirmed the associated gene expression for enzymes involved in ROS scavenging and oxidative phosphorylation (Table 3), such as peroxidase (No. 24, Table 2), peptidyl-prolyl cis-trans isomerase (No. 37, Table 2), and phosphoglycerate kinase (No. 4, Table 1). Among these proteins, peptidyl-prolyl cis-trans isomerase, as a protein chaperone, possess complementary and sometimes overlapping roles in protecting the proteins [15].

Excess Cu generates oxidative stress, thereby hindering some important metabolic process, such as up-regulation of antioxidant and stress-related regulatory proteins, that help maintain cellular homeostasis [52]. Maturase K (No. 13, Table 1) changed in abundance of 1.69-fold. Maturase K has an important role in plant growth, cell division, and expansion as well as in protecting metabolic processes against H_2O_2 and other toxic derivatives of oxygen. Hydroquinone formation was accompanied by the oxidation of two moles of NADPH and the presence of an inducible 1,4-benzoquinone reductase. Thus, putative 1,4-benzoquinone reductase (No. 17, Table 1) plays an essential role in antioxidant defense pathway [53]. Studies also reported about wall-bound malate dehydrogenase (No. 5, Table 1) [54,55], which can regenerate NAD(P)H that is needed by cell wall peroxidases for free radical generation associated with lignin polymerization. Thioredoxin (No. 18, Table 1) is also involved in redox regulation by reducing disulfides on the target protein for detoxifying lipid hydroperoxides or repairing oxidized proteins and relaying the signal to mitogen-activated protein kinase pathway of stress signaling [56]. Phosphoglycerate kinase (No. 4, Table 1) can also interact with cytosolic catalase and has a role in relieving oxidative stress. Interestingly, down-regulation of key metabolic enzymes revealed that oxidant protection conferred by these proteins was also regulated during Cu treatment. Ferredoxin-NADP reductase (No. 43, Table 2) and ferredoxin-NADP reductase-like protein (No. 38, Table 2), known to sequester highly reactive Fe^{3+} and prevent formation of toxic˙ OH species, were also identified [57]. Vacuolar H(+)-ATPase subunit-like protein (No. 46, Table 2) was involved in ascorbate and aldarate metabolism. L-ascorbic acid is characterized by plant tissues, and ascorbate is one of the most important antioxidant molecules [58]. Furthermore, 2-dehydro-3-deoxy-phosphorheptonate aldolase synthase (No. 49, Table 2) may be involved in aromatic pathways in the secondary metabolites, which are known to act as defense responses to abiotic stress [59].

Cu stress-activated signaling pathways

The communication between the cytoskeleton and the CWPs is one of the most characterized features of cellular mechanisms that enable cells to respond effectively to various extracellular signals. Several candidate components involved in signal transduction were identified in this study, including Hsp70, small G protein, and cytochrome. The biosynthesis and accumulation of HSP proteins (No. 30, Table 2) can generally contribute to the protection and repair of cells under stress [60]. These proteins may be involved in cell wall biogenesis [61]. Small G proteins (No. 47, Table 2) transduced signals from receptors to control a wide range of cellular functions, particularly for regulating Ca^{2+} channel expression at the cell surface [62]. These proteins are clustered into distinct families but all act as molecular switches, which are active in their GTP-bound form but inactive when GDP-bound [63]. RAS-related GTP-binding protein (No. 50, Table 2) also has a primarily role in modulating cellular functions that involve actin cytoskeleton (No. 28, Table 2), such as establishing cell polarity and morphology [64]. Cytochrome has been identified in the stem cell wall of *Medicago sativa* [65] and germinating embryos of *Oryza sativa* [66]. This first proteomic study showed that cytochromes (No. 6, Table 1; No. 34, Table 2) were Cu-responsive proteins in the root cell wall of *E. splendens*. More preliminary evidence suggested that porins can form aqueous transmembrane channels for transporting solutes and macromolecules across the extracellular surface [67] [68]. These pores are regulated by ATP and GTP with a gating mechanism that modulates the pore size and ion selectivity [54]. In this case, Cu resistance mechanism may be induced by active apoplastic permeability because of up-regulated porin-like proteins (No. 15, Table 1).

Cu stress-activated energy pathways and protein synthesis

ATP is an essential metabolite in cell walls associated with energy conversion [69]. The plasma membrane calcium ATPase (PMCA) (No. 33, Table 2) or Ca^{2+} pump transports Ca^{2+} ions out of the cells using the stored energy in ATP. Control of Ca^{2+} concentration is significant in the cytosol [70]. PMCA down-regulation at the protein levels was observed during the experiment, which may represent an adaptive mechanism to facilitate removal of Ca^{2+} in the maintenance of calcium homeostasis in abiotic stress. The vacuolar H(+)-ATPase (No. 46, Table 2) functions as a primary proton pump that generates electrochemical gradients of protons across the transmembrane region [71], which provides the primary driving force for transporting numerous ions and metabolites against their electrochemical gradients [72]. Glyceraldehyde-3-phosphate dehydrogenase (GAPDH) (No. 2, Table 1), as a classical cellular enzyme involved in glycolysis, is also differentially expressed in response to Cu exposure, where GAPDH is specifically targeted at the cell wall [73]. Mitochondrial phosphate translocator (No. 26, Table 2), involved in ATP/ADP transportation, is also identified in the cell wall. Amino acids and amino acid-derived molecules can be chelated with metal ions by high affinity ligands [74]. Pollen allergen Que a 1 isoform (No. 16, Table 1), which is involved in glycine, serine, and threonine metabolism (Table 3), was up-regulated in the root cell wall of *E. splendens*, whereas aspartic proteinase nepenthesin-1 precursor (No. 51, Table 2) was down-regulated by Cu stress. Different expression levels of amino acid-related proteins in the cell wall elucidated their possible role in Cu tolerance.

Other Cu stress-responsive proteins and their potential functions

The 2,3-bisphosphoglycerate independent phosphoglycerate mutase (No. 7, Table 1) was up-regulated in the study, which was also differentially expressed in rice anthers under cold stress [75]. Previous evidence indicated that the expression of non-symbiotic hemoglobins (nsHbs) from cotton is induced with Verticillium wilt fungus [76]. The proteins also had a role in the defense responses against pathogen invasions in *Arabidopsis* [76]. Our studies showed that the nsHbs (No. 8, Table 1) were up-regulated in the cell wall of *E. splendens* root under Cu stress, although their physiological function is yet to be determined. A significant amount of the unidentified proteins were classified as hypothetical or predictable. These proteins were annotated in databases as unknown, hypothetical, or putative proteins because of the theoretical translation of open reading frame sequences [77]. The roles of these proteins remain to be investigated.

Overview of regulated CWP functions

This study provides insights into the functional role of the cell wall of *E. splendens* under Cu stress, where several CWPs were identified with possible roles in Cu tolerance/detoxification. Cu stress conditions can alter the composition (polysaccharide and protein) of the cell wall both qualitatively and quantitatively. Approximately 40% of the differentially expressed CWPs showed higher abundance in response to Cu stress involved in antioxidant defense, cell wall polysaccharide remodeling, and metabolism process. Up to 60% of the CWPs were in low abundance in response to Cu stress that is involved in signal, energy, and protein synthesis. Polysaccharide analysis confirmed the cell wall remodeling under Cu stress. The amount, composition, and distribution of the cell wall polysaccharides are consequential for plant adaptation to enhance Cu ion levels. Proteome analysis of cell wall confirmed that most proteins were associated with an antioxidant defense response. Hsp 70, small G protein, and RAS-related GTP-binding protein also have essential roles in signal transduction across the cell wall and through the cytoskeleton. Literature provides fundamental information about the role of polysaccharide composition of plant cell wall in metal tolerance and complementary evidence on continuous crosstalk between CWPs and the cytoskeleton. Therefore, knowledge has been expanded on plant stress-related signaling pathways in the cell wall. Cu regulation of these proteins may also not solely respond on abundance changes. Post-translational modification and dynamics are interesting subjects for future investigation. The present study is the first cell wall proteome and polysaccharide investigation of plants in response to Cu and important for the understanding of the plant cell wall response to environmental heavy metal stresses. We propose a possible protein interaction network (Figure 6) that provides new insights into Cu stress response in the root cell wall and facilitates further functional research of target proteins that are associated with Cu response based on the abundant changes in these proteins, as well as polysaccharides and their putative functions.

Materials and Methods

Ethics statement

No specific permissions were required for collecting *E. splendens* seeds from deposited Cu-mining soil in Zhuji County, Zhejiang Province, China. *E. splendens* is neither endangered nor protected. Authors maintained the population at sustainable levels. All plant work was conducted according to relevant national and international guidelines.

Plant material and Cu treatment

E. splendens seeds were collected from plants that grew on deposited Cu-mining soil in Zhuji County, Zhejiang Province, China. The seeds were washed with deionized water and soaked in distilled water to germinate in a controlled dark condition (25°C)

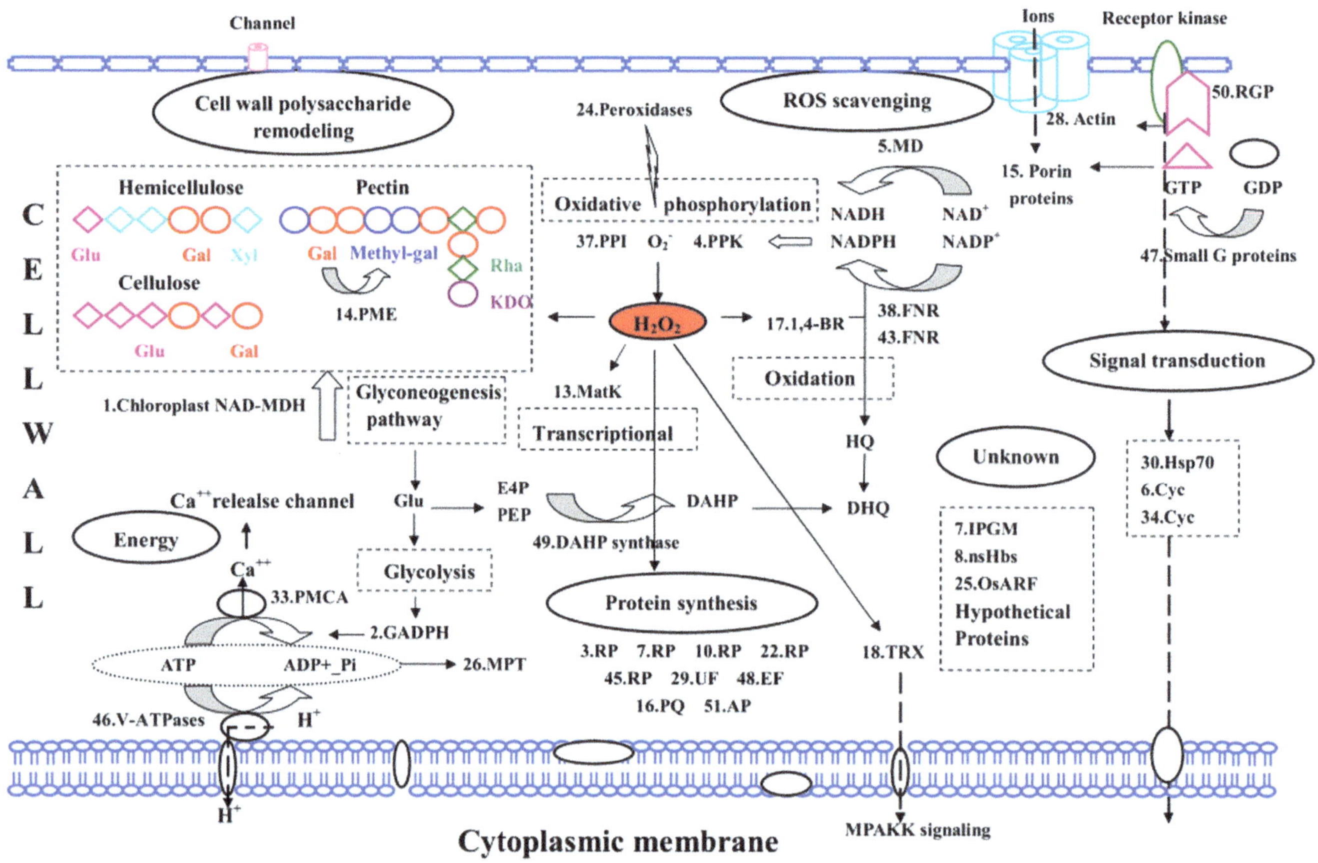

Figure 6. Pathways involved in cell defense, signaling, and cell wall remodeling under copper stress in the root cell wall of *Elsholtzia splendens*. Proteins identified in this study are displayed on the corresponding metabolic pathways and the number indicates the protein identification number. Gal, galacturonic acids; Methyl-gal, methylated galacturonic acids; KDO, 2-keto-3-deoxyoctonic acid; Rha, rhamnose; PME, Pectin methylesterase; Xyl, xylanase; Glu, glucose; MD, malate dehydrogenase; Trx, thioredoxin; MAPKK, mitogen-activated protein kinase kinase; PPI, peptidyl-prolyl cis-trans isomerase; PPK, phosphoglycerate kinase; MatK, maturase K; 1,4-BR, putative 1,4-benzoquinone reductase; oxidation; FNR, ferredoxin-NADP reductase; HQ, hydroquinone; DHQ, Durohydroquinone; DAHP, 3-deoxy-D-arabino-heptulosonic acid-7-phosphate; E4P, perythrose-4-phosphate; PEP, phosphoenolpyruvate; RGP, RAS-related GTP-binding protein; PLP, porin-like proteins; Cyc, cytochrome; PMCA, plasma membrane calcium ATPase; V-ATPases, vacuolar H(+)-ATPases; MPT, mitochondrial phosphate translocator; GADPH glyceraldehyde-3-phosphate dehydrogenase; RP, ribosomal protein; UF, ubiquitin fusion protein; EF, elongation factor 1 subunit alpha; PQ, Pollen allergen Que a 1 isoform; AP, aspartic proteinase nepenthesin-1 precursor; nsHbs. non-symbiotic hemoglobin; OsARF, Os070223100; iPGM, 2,3-bisphosphoglycerate-independent phosphoglycerate mutase.

for 48 h. The solution was changed to one-quarter-strength complete nutrient solution after germination. At the fourth leaf stage, uniform seedlings were transferred into vessels filled with full-strength nutrient solution that contained macronutrients (in mM): 1.0 $Ca(NO_3)_2$, 0.5 $MgSO_4$, and 0.5 K_2HPO_4; and micronutrients (in μmol/L): 27.0 Fe(III)–EDTA, 23.0 H_3BO_3, 0.8 $CuSO_4$, 0.5 Na_2MoO_4, 0.5 $ZnSO_4$, and 4.5 $MnSO_4$. The pH was adjusted to 5.8 with 0.1 M HCl and NaOH. Nutrient solutions were renewed every 3 d and aerated continuously through 0.2 μm filters. Plants were grown in a hydroponic nutrient solution chamber with a 16 h, 25°C day and an 8 h, 20°C night regimen at 60% to 70% relative humidity. Light conditions in the growth chamber were fixed at 5 μmol to 10 μmol photons m^{-2} s^{-1}. Plants were exposed to various concentrations of Cu (25, 50, 75, and 100 μM) after 28 d of growth, which was added as sulfate for 48 h. Each treatment (15 plants) was conducted in triplicate, and the control plants (CK) were grown in pure nutrient solution for comparison. Plant roots were rinsed with distilled water at harvest and then immersed in 5 mmol/L of $Ca(NO_3)_2$ for 20 min to remove the putative adsorbed Cu^{2+} [32]. Roots were separated, pooled, and rinsed with deionized water; plant parts were washed, immediately frozen in liquid nitrogen, and stored at −80°C for analysis [33].

Cell wall preparation

Cell wall purification and CWP extraction were prepared as described by Feiz et al. [34] with slight modifications. All procedures were conducted at 4°C unless mentioned otherwise. Fifteen independent root preparations were pooled to yield one biological replicate. Root tissues (4 g) were homogenized in precold extraction buffer [5 mM acetate, 0.4 M sucrose, 1 mM phenylmethylsulfonyl fluoride (PMSF), pH 4.6, 125 mL] with a chilled mortar and pestle. After adding 0.4 g of polyvinylpolypyrrolidone (PVPP), the mixture was incubated at 4°C for 30 min with stirring. Cell walls were separated from soluble cytoplasmic fluid by centrifuging the homogenate at 1,000×*g* for 15 min. The pellet was washed by suspension in 125 mL of 5 mM acetate buffer at pH 4.6 with 0.6 M sucrose. The mixture was centrifuged at 1,000×*g* for 15 min. The pellet was further purified by resuspension in 125 mL of 5 mM acetate buffer at pH 4.6 with

1 M sucrose and centrifuged at 1,000×*g* for 15 min. The residue was washed thoroughly with 750 mL of 5 mM acetate buffer at pH 4.6 while filtered on a layer of Miracloth (Merck, Darmstadt, Germany). The supernatant was discarded, and the final pellet was freeze-dried overnight. The freeze-dried cell wall materials were then stored at −80°C for further use. Most intracellular proteins were removed from the cell walls given the advantage of sucrose gradients and extensive washing with low ionic strength acidic buffer. PVPP was treated with acid to increase polymerization and to remove metal ions and contaminants.

Cell wall composition extraction and analysis

Isolation of cell wall material. Cell wall materials were extracted according to Zhong and Lauchli (1993) [35]. The pectin fraction was extracted twice with 0.5% ammonium oxalate buffer that contained 0.1% $NaBH_4$ (pH 4) in boiling water bath for 1 h each and pooled the supernatants. Pellets were subsequently subjected to triple extractions with 4% KOH that contained 0.1% $NaBH_4$ at room temperature for 24 h, followed by similar extraction with 24% KOH that contained 0.1% $NaBH_4$. The supernatants from the 4% and 24% KOH extractions were collected and thus yielded the hemicellulose fractions. The remaining pellet from the 24% KOH extraction was then lyophilized, weighed, and considered to be the cellulose fraction [36].

Protein extraction and digestion. A portion (2 g) of the roots was homogenized with 1 mL of PBS (pH 7.6) that contained 65 mM K_2HPO_4, 2.6 mM KH_2PO_4, 400 mM NaCl, and 3 mM NaN_3 in a mortar and pestle. The homogenate was centrifuged twice at 15,000×*g* for 10 min, and the supernatant was collected as the total soluble protein. CWPs were extracted from the root cell wall fraction in two successive steps. The first step used $CaCl_2$ solution (5 mM sodium acetate buffer, pH 4.6, 0.2 M $CaCl_2$, and 1 mM PMSF; $CaCl_2$ can efficiently extract CWPs that exert weak electrostatic interactions with other cell wall components [37]). The second step involved two extractions with NaCl solution (5 mM sodium acetate buffer, pH 4.6, 1 M NaCl, and 1 mM PMSF; NaCl was also used to extract the strong ionically bound proteins) [34]. For each extraction, the sample was incubated with vortexing at 4°C, and the supernatant was collected after centrifugation at 4,000×*g* for 15 min. The supernatants from all the extraction steps were pooled and concentrated to 4 mL by centrifugation at 1,500×*g* at 10°C using a 3 kDa molecular weight cut-off Amicon Spin Tube (Millipore, MA). The protein sample was buffer-exchanged with ultrapure water.

CWP extracts were dissolved in 100 μL of lysis buffer (7 M urea, 2 M thiourea, 5 mM EDTA, 10 mM DTT, and 1 mM PMSF). The supernatant was transferred to a new tube, reduced with 10 mM DTT for 1 h at 56°C, and alkylated with 55 mM iodoacetamide for 45 min at room temperature in darkness. The protein was precipitated with four volumes of pre-chilled acetone for 30 min at −20°C. The pellet was dissolved in 0.5 M TEAB after centrifugation and sonicated for 5 min. The centrifugation step was repeated, and the supernatant was collected. Approximately 100 μg of proteins from each sample was digested with trypsin (Promega) overnight at 37°C in a 1:20 trypsin-to-protein mass ratio.

Analysis of Cu, protein, and cell wall polysaccharide content. Cell wall fraction was dried at 70°C to a constant weight and then digested at 145°C for 24 h in an acid mixture of HNO_3:$HClO_4$ (3:1, v:v). Cu concentrations were measured by Elmer flame atomic absorption spectrometry (AAS-3600).

Protein contents were determined by the Bradford method [38] with bovine serum albumin as the standard. CWPs were dissolved with 400 μL of SDS sample lysis buffer by boiling for 5 min and then loaded onto a 12% acrylamide mini-gel (5 cm to 8 cm) for 1D SDS-PAGE.

The contents of total sugars, galacturonic acids, and KDO were determined colorimetrically by sulfate-phenol [39], hydroxydiphenyl [40], and thiobarbituric acid [41], respectively.

Labeling TMT reagents

Peptide was desalted by Strata XC18 SPE column (Phenomenex) and vacuum-dried after trypsin digestion. Peptide was reconstituted in 0.5 M TEAB and processed according to the protocol for 6-plex TMT reagent kits. In brief, two units of TMT reagent (defined as the amount of reagent required to label 100 μg of protein) was thawed and reconstituted in 41 μL of acetonitrile. Peptide samples from 50 μM Cu were treated for 48 h, and CK samples were labeled with TMT tags 126, 128, and 130 as well as TMT tags 127, 129, and 131. Both samples were incubated at room temperature for 2 h. The peptide mixtures were then pooled, desalted, and dried by vacuum centrifugation.

LC–ESI–MS/MS analysis by Q Exactive

The labeled peptide was resuspended in buffer A (2% ACN, 0.1% FA) and centrifuged at 20,000×*g* for 2 min. The supernatant was transferred into a sample tube and loaded onto an Acclaim PepMap 100 C18 trap column (75 μm×2 cm; Dionex) by EASY nLC1000 nano UPLC (Thermo). The peptide was then eluted onto an Acclaim PepMap RSLC C18 analytical column (50 μm×15 cm; Dionex). An 85 min gradient program was run at 300 nL/min, which started from 3% to 35% B (80% ACN, 0.1% FA), followed by 5 min linear gradient to 90% B, and maintained at 90% B for 5 min.

The peptides were subjected to NSI source, followed by tandem mass spectrometry (MS/MS) in Q Exactive (Thermo) coupled online to the UPLC. Intact peptides were detected in the Orbitrap at a resolution of 70,000. Peptides were selected for MS/MS using 27% NCE with 12% stepped NCE; ion fragments were detected in the Orbitrap at a resolution of 17,500. A data-dependent procedure that alternated between one MS scan followed by 20 MS/MS scans was applied to the top 20 precursor ions above a threshold ion count of 3E4 in the MS survey scan with 5.0 s dynamic exclusion. The applied electrospray voltage was 1.8 kV. Automatic gain control was used to prevent overfilling of the ion trap; 1E5 ions were accumulated for generation of MS/MS spectra. The m/z scan range was 350 Da to 2000 Da for MS scans.

Proteomic database search

The instrument data file (.raw) was merged and transformed to an.mgf file by Proteome Discoverer (ver. 1.3.0.339; Thermo). Peptide and protein identifications were performed using the Mascot search engine (ver. 2.3.02; Matrix Science). Derived protein sequences from plants in the NCBI were collected, and a database containing 1,596,443 sequences was created. In the Mascot search engine version 2.3.02 software, all parameters were set as follows. Database searching was restricted to tryptic peptides. Carbamidomethyl (C), TMT 6-plex (N-term), and TMT 6-plex (K) were selected as fixed, and deamidated (NQ), Gln→pyro-Glu (N-term Q), and oxidation (M) were selected as variable modifications, where two missed cleavages were allowed with precursor error tolerance at 10 ppm and fragment deviation at 0.02 Da. The Mascot search results were quantified using Mascot 2.3.02 with the following criteria: protein ratio type = median, minimum unique peptides = 1, peptide threshold type = at least homolog. Peptides were not quantified if peptide score was too low or the deviation was too large. The final ratios of protein were then normalized by taking the median of all the proteins quantified. The complete list of identified peptides was encoded in an Excel (Microsoft) database for grouping of results into proteins and calculation of ratios and coefficient of variation.

To obtain the differentially expressed proteins, we selected the proteins according to the following principles. (a) The expression change of protein folds is more than 1.5. (b) The protein quantitative T-test p value is less than 0.05. A total of 55 differentially expressed proteins were obtained in our bioinformatics analysis data.

Bioinformatics studies and statistical analysis

Data were the average of at least three independent experimental replicates. One-way ANOVA (Duncan's test) and LSD test with $p<0.05$ as the significance level were performed to analyze the differences between the control and treatments. All data were presented as the mean value ± standard error and were analyzed using SPSS statistical software package (version 16.0).

The clustering of content abundance profiles was performed using Cluster 3.0. Euclidean metric was used for computing the distance between points and cluster centroids, which is a typical choice for K-means clustering analysis. The output was visualized using TreeView after hierarchical clustering.

GO information was used to categorize the biological processes of identified proteins from TMT data sets. GO annotations were extracted from the Uni-Prot database and were matched with corresponding gene locus identifiers embedded in the NCBI RefSeq database. Proteins were then classified based on their biological process using the Web Gene Ontology Annotation Plot [42] (http://wego.genomics.org.cn/cgi-bin/wego/index.pl). Functionally analyzed Cu stress-induced CWPs in biosynthesis pathways were identified through the Kyoto Encyclopedia of Genes and Genomes (KEGG) [43]. These pathways were classified into hierarchical categories according to the KEGG website (http://www.genome.jp/kegg/). Subcellular locations were predicted using TargetP (http://www.cbs.dtu.dk/services/TargetP/), Predotar (http://urgi.versailles.inra.fr/predotar/predotar.html), and WoLF PSORT (http://wolfpsort.org/) (http://www.genscript.com/psort/wolf_psort.html)software applications as previously described [44].

Supporting Information

Figure S1 SDS–PAGE of root cell wall proteins under different copper stress.

Table S1 Protein subcellular localization and go annotation of 55 significantly differentially expressed proteins.

Table S2 Details of 55 significantly differentially expressed proteins identified by TMT using ANOVA including their *p* –value.

Author Contributions

Conceived and designed the experiments: TTL JYS. Performed the experiments: TTL CFS. Analyzed the data: TTL CKH JYS. Contributed reagents/materials/analysis tools: YW. Wrote the paper: TTL.

References

1. Maksymiec W (1998) Effect of copper on cellular processes in higher plants. Photosynthetica 34: 321-342.
2. Teisseire H, Guy V (2000) Copper-induced changes in antioxidant enzymes activities in fronds of duckweed (*Lemna minor*)Plant science 153: 65-72.
3. Vinit-Dunand F, Epron D, Alaoui-Sossé B, Badot PM (2002) Effects of copper on growth and on photosynthesis of mature and expanding leaves in cucumber plants. Plant science 163: 53-58.
4. Bona E, Marsano F, Cavaletto M, Berta G (2007) Proteomic characterization of copper stress response in *Cannabis sativa* roots. Proteomics 7: 1121-1130.
5. Hall J (2002) Cellular mechanisms for heavy metal detoxification and tolerance. Journal of Experimental Botany 53: 1-11.
6. Cobbett C, Goldsbrough P (2002) Phytochelatins and metallothioneins: roles in heavy metal detoxification and homeostasis. Annual review of plant biology 53: 159-182.
7. Krzesłowska M (2011) The cell wall in plant cell response to trace metals: polysaccharide remodeling and its role in defense strategy. Acta Physiologiae Plantarum 33: 35-51.
8. Jarvis MC (2009) Plant cell walls: supramolecular assembly, signalling and stress. Structural Chemistry 20: 245-253.
9. Zhu J, Chen S, Alvarez S, Asirvatham VS, Schachtman DP, et al. (2006) Cell wall proteome in the maize primary root elongation zone. I. Extraction and identification of water-soluble and lightly ionically bound proteins. Plant physiology 140: 311-325.
10. Vorwerk S, Somerville S, Somerville C (2004) The role of plant cell wall polysaccharide composition in disease resistance. Trends in plant science 9: 203-209.
11. Dani V, Simon WJ, Duranti M, Croy RR (2005) Changes in the tobacco leaf apoplast proteome in response to salt stress. Proteomics 5: 737-745.
12. Soares NC, Francisco R, Vielba JM, Ricardo CP, Jackson PA (2009) Associating wound-related changes in the apoplast proteome of Medicago with early steps in the ROS signal-transduction pathway. Journal of Proteome Research 8: 2298-2309.
13. Jamet E, Canut H, Boudart G, Pont-Lezica RF (2006) Cell wall proteins: a new insight through proteomics. Trends in plant science 11: 33-39.
14. Bhushan D, Pandey A, Choudhary MK, Datta A, Chakraborty S, et al. (2007) Comparative proteomics analysis of differentially expressed proteins in chickpea extracellular matrix during dehydration stress. Molecular & Cellular Proteomics 6: 1868-1884.
15. Pandey A, Rajamani U, Verma J, Subba P, Chakraborty N, et al. (2010) Identification of extracellular matrix proteins of rice (*Oryza sativa* L.) involved in dehydration-responsive network: a proteomic approach. Journal of Proteome Research 9: 3443-3464.
16. Floerl S, Druebert C, Majcherczyk A, Karlovsky P, Kües U, et al. (2008) Defence reactions in the apoplastic proteome of oilseed rape (*Brassica napus var. napus*) attenuate Verticillium longisporum growth but not disease symptoms. Bmc Plant Biology 8: 129.
17. Goulet C, Goulet C, Goulet MC, Michaud D (2010) 2-DE proteome maps for the leaf apoplast of Nicotiana benthamiana. Proteomics 10: 2536-2544.
18. Brune A, Urbach W, Dietz K-J (1994) Zinc stress induces changes in apoplasmic protein content and polypeptide composition of barley primary leaves. Journal of Experimental Botany 45: 1189-1196.
19. Blinda A, Koch B, Ramanjulu S, Dietz K (1997) De novo synthesis and accumulation of apoplastic proteins in leaves of heavy metal-exposed barley seedlings. Plant, Cell & Environment 20: 969-981.
20. Fecht-Christoffers MM, Braun H-P, Lemaitre-Guillier C, VanDorsselaer A, Horst WJ (2003) Effect of manganese toxicity on the proteome of the leaf apoplast in cowpea. Plant physiology 133: 1935-1946.
21. Fecht-Christoffers MM, Führs H, Braun H-P, Horst WJ (2006) The role of hydrogen peroxide-producing and hydrogen peroxide-consuming peroxidases in the leaf apoplast of cowpea in manganese tolerance. Plant physiology 140: 1451-1463.
22. Alves M, Francisco R, Martins I, Ricardo C (2006) Analysis of *Lupinus albus* leaf apoplastic proteins in response to boron deficiency. Plant and Soil 279: 1-11.
23. Komatsu S, Kobayashi Y, Nishizawa K, Nanjo Y, Furukawa K (2010) Comparative proteomics analysis of differentially expressed proteins in soybean cell wall during flooding stress. Amino Acids 39: 1435-1449.
24. Kataoka T, Furukawa J, Nakanishi T (2003) The decrease of extracted apoplast protein in soybean root tip by aluminium treatment. Biologia plantarum 46: 445-449.
25. Kong F-J, Oyanagi A, Komatsu S (2010) Cell wall proteome of wheat roots under flooding stress using gel-based and LC MS/MS-based proteomics approaches. Biochimica et Biophysica Acta (BBA)-Proteins and Proteomics 1804: 124-136.
26. Zhou L, Bokhari SA, Dong C-J, Liu J-Y (2011) Comparative proteomics analysis of the root apoplasts of rice seedlings in response to hydrogen peroxide. Plos One 6: e16723.
27. Hossain Z, Nouri M-Z, Komatsu S (2011) Plant cell organelle proteomics in response to abiotic stress. Journal of Proteome Research 11: 37-48.

28. Yang MJ, Yang XE, Römheld V (2002) Growth and nutrient composition of *Elsholtzia splendens Nakai* under copper toxicity. Journal of plant nutrition 25: 1359-1375.
29. Lou L, Shen Z, Li X (2004) The copper tolerance mechanisms of *Elsholtzia haichowensis*, a plant from copper-enriched soils. Environmental and Experimental Botany 51: 111-120.
30. Feiz L, Irshad M, Pont-Lezica RF, Canut H, Jamet E (2006) Evaluation of cell wall preparations for proteomics: a new procedure for purifying cell walls from Arabidopsis hypocotyls. Plant Methods 2: 10.
31. Kwon H-K, Yokoyama R, Nishitani K (2005) A proteomic approach to apoplastic proteins involved in cell wall regeneration in protoplasts of Arabidopsis suspension-cultured cells. Plant and Cell Physiology 46: 843-857.
32. Harrison S, Lepp N, Phipps D (1979) Uptake of Copper by excised roots II. Copper desorption from the free space. Zeitschrift für Pflanzenphysiologie 94: 27-34.
33. Li F, Shi J, Shen C, Chen G, Hu S, et al. (2009) Proteomic characterization of copper stress response in *Elsholtzia splendens* roots and leaves. Plant Molecular Biology 71: 251-263.
34. Feiz L, Irshad M, Pont-Lezica RF, Canut H, Jamet E (2006) Evaluation of cell wall preparations for proteomics: a new procedure for purifying cell walls from Arabidopsis hypocotyls. Plant Methods 2.
35. Zhong H, Lauchli A (1993) Changes of cell wall composition and polymer size in primary roots of cotton seedlings under high salinity. Journal of Experimental Botany 44: 773-778.
36. Yu M, Shen R, Liu J, Chen R, Xu M, et al. (2009) The role of root border cells in aluminum resistance of pea (*Pisum sativum*) grown in mist culture. Journal of Plant Nutrition and Soil Science 172: 528-534.
37. Jamet E, Boudart G, Borderies G, Charmont S, Lafitte C, et al. (2008) Isolation of plant cell wall proteins. 2D PAGE: Sample Preparation and Fractionation: Springer. pp. 187-201.
38. Bradford MM (1976) A rapid and sensitive method for the quantitation of microgram quantities of protein utilizing the principle of protein-dye binding. Analytical biochemistry 72: 248-254.
39. Dubois M, Gilles KA, Hamilton JK, Rebers P, Smith F (1956) Colorimetric method for determination of sugars and related substances. Analytical chemistry 28: 350-356.
40. Dronnet V, Renard C, Axelos M, Thibault J-F (1996) Heavy metals binding by pectins: selectivity, quantification and characterisation. Progress in Biotechnology 14: 535-540.
41. Mohnen D (2008) Pectin structure and biosynthesis. Current opinion in plant biology 11: 266-277.
42. Ye J, Fang L, Zheng H, Zhang Y, Chen J, et al. (2006) WEGO: a web tool for plotting GO annotations. Nucleic acids research 34: W293-W297.
43. Kanehisa M, Goto S (2000) KEGG: kyoto encyclopedia of genes and genomes. Nucleic acids research 28: 27-30.
44. Day A, Fénart S, Neutelings G, Hawkins S, Rolando C, et al. (2013) Identification of cell wall proteins in the flax (*Linum usitatissimum*) stem. Proteomics.
45. Xiong J, An L, Lu H, Zhu C (2009) Exogenous nitric oxide enhances cadmium tolerance of rice by increasing pectin and hemicellulose contents in root cell wall. Planta 230: 755-765.
46. Yang JL, Zhu XF, Peng YX, Zheng C, Li GX, et al. (2011) Cell wall hemicellulose contributes significantly to aluminum adsorption and root growth in Arabidopsis. Plant physiology 155: 1885-1892.
47. Pelloux J, Rusterucci C, Mellerowicz EJ (2007) New insights into pectin methylesterase structure and function. Trends in plant science 12: 267-277.
48. Paynel F, Schaumann A, Arkoun M, Douchiche O, Morvan C (2009) Temporal regulation of cell-wall pectin methylesterase and peroxidase isoforms in cadmium-treated flax hypocotyl. Annals of Botany 104: 1363-1372.
49. Liszkay A, Kenk B, Schopfer P (2003) Evidence for the involvement of cell wall peroxidase in the generation of hydroxyl radicals mediating extension growth. Planta 217: 658-667.
50. Fry S, Dumville J, Miller J (2001) Fingerprinting of polysaccharides attacked by hydroxyl radicals in vitro and in the cell walls of ripening pear fruit. Biochem J 357: 729-737.
51. Thounaojam TC, Panda P, Mazumdar P, Kumar D, Sharma G, et al. (2012) Excess copper induced oxidative stress and response of antioxidants in rice. Plant Physiology and Biochemistry 53: 33-39.
52. Ahsan N, Lee D-G, Lee S-H, Kang KY, Lee JJ, et al. (2007) Excess copper induced physiological and proteomic changes in germinating rice seeds. Chemosphere 67: 1182-1193.
53. Spain JC, Wyss O, Gibson DT (1979) Enzymatic oxidation of p-nitrophenol. Biochemical and Biophysical Research Communications 88: 634-641.
54. Gross GG (1977) Cell wall-bound malate dehydrogenase from horseradish. Phytochemistry 16: 319-321.
55. Li Z-C, McClure JW, Hagerman AE (1989) Soluble and bound apoplastic activity for peroxidase, β-D-glucosidase, malate dehydrogenase, and nonspecific arylesterase, in barley (*Hordeum vulgare* L.) and oat (*Avena sativa* L.) primary leaves. Plant physiology 90: 185-190.
56. Apel K, Hirt H (2004) Reactive oxygen species: metabolism, oxidative stress, and signal transduction. Annu Rev Plant Biol 55: 373-399.
57. Jensen KA, Houtman CJ, Ryan ZC, Hammel KE (2001) Pathways for extracellular Fenton chemistry in the brown rot basidiomycete Gloeophyllum trabeum. Applied and Environmental Microbiology 67: 2705-2711.
58. Howard L, Talcott S, Brenes C, Villalon B (2000) Changes in phytochemical and antioxidant activity of selected pepper cultivars (Capsicum species) as influenced by maturity. Journal of Agricultural and Food Chemistry 48: 1713-1720.
59. Keith B, Dong X, Ausubel FM, Fink GR (1991) Differential induction of 3-deoxy-D-arabino-heptulosonate 7-phosphate synthase genes in Arabidopsis thaliana by wounding and pathogenic attack. Proceedings of the National Academy of Sciences 88: 8821-8825.
60. Timperio AM, Egidi MG, Zolla L (2008) Proteomics applied on plant abiotic stresses: role of heat shock proteins (HSP). Journal of proteomics 71: 391-411.
61. Wang SB, Hu Q, Sommerfeld M, Chen F (2004) Cell wall proteomics of the green alga *Haematococcus pluvialis* (Chlorophyceae). Proteomics 4: 692-708.
62. Béguin P, Nagashima K, Gonoi T, Shibasaki T, Takahashi K, et al. (2001) Regulation of Ca2+ channel expression at the cell surface by the small G-protein kir/Gem. Nature 411: 701-706.
63. Rittinger K, Walker PA, Eccleston JF, Nurmahomed K, Owen D, et al. (1997) Crystal structure of a small G protein in complex with the GTPase-activating protein rhoGAP. Nature 388: 693-697.
64. Hall A (1994) Small GTP-binding proteins and the regulation of the actin cytoskeleton. Annual review of cell biology 10: 31-54.
65. Watson BS, Lei Z, Dixon RA, Sumner LW (2004) Proteomics of *Medicago sativa* cell walls. Phytochemistry 65: 1709-1720.
66. Zhang H, Lian C, Shen Z (2009) Proteomic identification of small, copper-responsive proteins in germinating embryos of *Oryza sativa*. Annals of Botany 103: 923-930.
67. Odou M, Singer E, Romond M, Dubreuil L (1998) Isolation and characterization of a porin-like protein of 45 kilodaltons from Bacteroides fragilis. Fems Microbiology Letters 166: 347-354.
68. Massari P, Ho Y, Wetzler LM (2000) Neisseria meningitidis porin PorB interacts with mitochondria and protects cells from apoptosis. Proceedings of the National Academy of Sciences 97: 9070-9075.
69. Chivasa S, Ndimba BK, Simon WJ, Lindsey K, Slabas AR (2005) Extracellular ATP functions as an endogenous external metabolite regulating plant cell viability. The Plant Cell Online 17: 3019-3034.
70. Carafoli E, Lim D (2009) Plasma Membrane Calcium ATPase. Handbook of Neurochemistry and Molecular Neurobiology: Springer. pp. 581-596.
71. Nelson H, Nelson N (1990) Disruption of genes encoding subunits of yeast vacuolar H (+)-ATPase causes conditional lethality. Proceedings of the National Academy of Sciences 87: 3503-3507.
72. Sze H, Ward JM, Lai S (1992) Vacuolar H+-translocating ATPases from plants: structure, function, and isoforms. Journal of bioenergetics and biomembranes 24: 371-381.
73. Petersen J, Brinkmann H, Cerff R (2003) Origin, evolution, and metabolic role of a novel glycolytic GAPDH enzyme recruited by land plant plastids. Journal of molecular evolution 57: 16-26.
74. Sharma SS, Dietz K-J (2006) The significance of amino acids and amino acid-derived molecules in plant responses and adaptation to heavy metal stress. Journal of Experimental Botany 57: 711-726.
75. Imin N, Kerim T, Rolfe BG, Weinman JJ (2004) Effect of early cold stress on the maturation of rice anthers. Proteomics 4: 1873-1882.
76. Qu Z-L, Zhong N-Q, Wang H-Y, Chen A-P, Jian G-L, et al. (2006) Ectopic expression of the cotton non-symbiotic hemoglobin gene GhHbd1 triggers defense responses and increases disease tolerance in Arabidopsis. Plant and Cell Physiology 47: 1058-1068.
77. Zhou X, Li J, Cheng W, Liu H, Li M, et al. (2010) Gene structure analysis of rice ADP-ribosylation factors (OsARFs) and their mRNA expression in developing rice plants. Plant Molecular Biology Reporter 28: 692-703.

12

Morphological and Physiological Responses of Cotton (*Gossypium hirsutum* L.) Plants to Salinity

Lei Zhang, Huijuan Ma, Tingting Chen, Jun Pen, Shuxun Yu*, Xinhua Zhao*

State Key Laboratory of Cotton Biology, Institute of Cotton Research of CAAS, Anyang, P. R. China

Abstract

Salinization usually plays a primary role in soil degradation, which consequently reduces agricultural productivity. In this study, the effects of salinity on growth parameters, ion, chlorophyll, and proline content, photosynthesis, antioxidant enzyme activities, and lipid peroxidation of two cotton cultivars, [CCRI-79 (salt tolerant) and Simian 3 (salt sensitive)], were evaluated. Salinity was investigated at 0 mM, 80 mM, 160 mM, and 240 mM NaCl for 7 days. Salinity induced morphological and physiological changes, including a reduction in the dry weight of leaves and roots, root length, root volume, average root diameter, chlorophyll and proline contents, net photosynthesis and stomatal conductance. In addition, salinity caused ion imbalance in plants as shown by higher Na^+ and Cl^- contents and lower K^+, Ca^{2+}, and Mg^{2+} concentrations. Ion imbalance was more pronounced in CCRI-79 than in Simian3. In the leaves and roots of the salt-tolerant cultivar CCRI-79, increasing levels of salinity increased the activities of superoxide dismutase (SOD), ascorbate peroxidase (APX), and glutathione reductase (GR), but reduced catalase (CAT) activity. The activities of SOD, CAT, APX, and GR in the leaves and roots of CCRI-79 were higher than those in Simian 3. CAT and APX showed the greatest H_2O_2 scavenging activity in both leaves and roots. Moreover, CAT and APX activities in conjunction with SOD seem to play an essential protective role in the scavenging process. These results indicate that CCRI-79 has a more effective protection mechanism and mitigated oxidative stress and lipid peroxidation by maintaining higher antioxidant activities than those in Simian 3. Overall, the chlorophyll a, chlorophyll b, and Chl (a+b) contents, net photosynthetic rate and stomatal conductance, SOD, CAT, APX, and GR activities showed the most significant variation between the two cotton cultivars.

Editor: Wagner L. Araujo, Universidade Federal de Vicosa, Brazil

Funding: This work was supported by the National Natural Science Foundation of China (31301262) and China Postdoctoral Science Foundation (Grant no. 2013M540169). The funders had no role in study design, data collection and analysis, decision to publish, or preparation of the manuscript.

Competing Interests: The authors have declared that no competing interests exist.

* Email: yu@cricaas.com.cn (SY); zhaoxinhua1968@126.com (XZ)

Introduction

The proportion of agricultural land that is negatively affected by high salinity is increasing worldwide, owing to natural causes and agricultural practices [1]. This problem has been aggravated by the development of more recent agricultural practices such as irrigation. Approximately 20% of the world's cultivated lands and more than half of all irrigated lands are affected by salinity [2]. High salt concentrations in the soil cause various events that negatively impact agricultural production, such as delays in plant growth and development, inhibition of enzymatic activities and reductions in photosynthetic rates [3]. Therefore, before attempts can be made to introduce genetic and environmental factors to alleviated salt stress, it is critical to elucidate the morphological and physiological responses of particular crops and cultivars to salinity.

In general, salt stress causes an imbalance of the cellular ions resulting in ion toxicity, osmotic stress and production of reactive oxygen species (ROS) [4], thus affecting plant growth, morphology, and survival [5]. High concentrations of NaCl outside the roots reduce the water potential and make it more difficult for the root to extract water. On the other hand, high concentrations of Na^+ and Cl^- ions inside plant cells are inhibitory to many enzyme processes. Salt-tolerant plants can not only regulate ion and water movements more efficiently but should also have a better antioxidant system for effective removal of ROS [6]. Salt stress causes excessive generation of ROS such as superoxide anions (O_2^-), hydrogen peroxide (H_2O_2), and hydroxyl radicals (OH•) [7]. To mitigate the oxidative damage initiated by ROS formed under salt stress, plants possess a complex antioxidant system, including non-enzymatic antioxidants such as ascorbic acid, glutathione (GSH), tocopherols, and carotenoids; antioxidant enzymes such as superoxide dismutase (SOD, EC 1.15.1.1), catalase (CAT, EC 1.11.1.6), glutathione peroxidase (EC 1.11.1.9), and peroxidases (POD, EC 1.17.1.7); and enzymes of the so-called ascorbate-glutathione cycle, including ascorbate peroxidase (APX, EC 1.11.1.11) and glutathione reductase (GR, EC 1.6.4.2). These components of the antioxidant system act in concert to alleviate the cellular damage accumulated under conditions of oxidative stress [8], [9].

SOD is generally considered as the first line of the antioxidant defense system, as it catalyzes the dismutation of O_2^- into H_2O_2 and O_2 in the cytosol, chloroplasts, and mitochondria [10]. POD is mainly located in the apoplastic space and vacuoles, where it plays an important role in catalyzing the conversion of H_2O_2 to H_2O and O_2 [11]. H_2O_2 is scavenged by CAT and APX. CAT dismutates H_2O_2 to H_2O and O_2, whereas APX, together with monodehydroascorbate reductase, dehydroascorbate reductase, and GR, converts H_2O_2 to H_2O via the ascorbate-glutathione pathway. APX is the first enzyme in this pathway, which eliminates H_2O_2 by using ascorbate as an electron donor in an oxidation-reduction reaction [12]. GSH is the final enzyme in this pathway and functions to protect plants from oxidative stress by maintaining GSH in the reduced state [13]. Under salt stress, malondialdehyde (MDA), the decomposition product of the polyunsaturated fatty acids of biomembranes, tends to accumulate [14]. Accordingly, cell membrane stability has been widely used to differentiate between stress-tolerant and -susceptible cultivars of some crops [15], in some cases, higher membrane stability could be correlated with abiotic stress tolerance.

In most plants, higher levels of the activity of the above-mentioned antioxidant enzyme are considered as a salt tolerance mechanisms [9]. Indeed, previous studies have shown that within the same species, salt-tolerant cultivars generally have enhanced or higher constitutive antioxidant enzyme activity under salt stress when compared with sensitive-cultivars. This has been demonstrated in numerous plant species such as cotton [16], rice [17], and pea [18]. Moreover, the response of plant antioxidant enzymes to salinity has been shown to vary among plant species, tissues, and subcellular localizations [19]. Several studies have demonstrated that salt-tolerant species show increased antioxidant enzyme activities and antioxidant contents in response to salt stress, whereas salt-sensitive species fail to do so [20]. Thus, the evidence accumulated to data indicates that intrinsic antioxidant resistance mechanisms of plants may provide a strategy to enhance salt tolerance. However, to achieve efficient selection of genetically-transformed salt-tolerant plants, the mechanisms underlying the effects of salt on the morphology, physiology, growth, and antioxidative responses of plants must first be identified [21].

Salt may affect plant growth indirectly by decreasing the rate of photosynthesis. Indeed, under saline conditions, substantial reduction in photosynthesis has been associated with a decrease in total chlorophyll content and distortion in chlorophyll ultrastructures [22]. Although the factors that limit photosynthesis in salt-stressed plants have been investigated for a number of species, the mechanistic pattern of inhibition remains unclear [23].

In addition, the relationships between ion accumulation, morphological and physiological changes, salt stress, and resulting plant injury are poorly understood. A number of conflicting views have been proposed in the literature over the toxic ions and enzymatic protection and activity involved in the response to salinity [24], [25]. Oxygen radicals are generated during plant metabolism that need to be scavenged by antioxidant systems to maintain normal growth; therefore, determining any adverse effects on these antioxidant systems due to salinity is an important consideration for appropriate cultivar selection.

Cotton is one of the most economically important crops in China. Although it is classified as a salt-tolerant crop, this tolerance is not only limited but also varies according to the growth and developmental stages of the plant [26]. Several studies have been conducted to assess the effect of salinity on the germination, vegetative growth, or yield of cotton [27–29]. However, the interactions between growths rates, ionic content, enzymatic activity, and oxidation reactions are likely to be complex and perhaps vary significantly between cultivars; therefore, such interactions deserve more detailed investigation.

In this study, the effect of NaCl on the growth behavior of two cotton cultivars that differ in their tolerance to salt was investigated. Changes in growth, ion concentration, pigment contents, and photosynthesis were assessed and linked to differences in the antioxidant system observed during salt stress. Since no detailed investigations have been conducted on this responses to data, the information presented here will not only provide criteria for improving salt-tolerant cotton specifically but also for selecting other salt-tolerant species and cultivars.

Materials and Methods

Experimental design

Seeds of two cotton cultivars, CCRI-79 (salt tolerant) and Simian 3 (salt sensitive), were obtained from the National Medium-Term Gene Bank of Cotton in China and soaked in sterile deionized water at 28°C for 6 h. They were then transferred to two sheets of sterile filter paper moistened with deionized water and placed in plastic trays for germination at 28°C for 72 h in the dark. The seeds were then sown into pots filled with perlite and grown under controlled conditions (light/dark regime of 16/8 h at 23°C, relative humidity of 60%–70%, photosynthetic photon flux density of 350 $\mu mol \cdot m^{-2} \cdot s^{-1}$). Germinated seeds were sown into holes in Styrofoam boards that were placed in deionized water and grown hydroponically in a growth room for 3 weeks under fluorescent and incandescent lights.

After 3 weeks, healthy and uniform seedlings were transplanted to 4-L plastic pots (10 plants per pot) filled with an aerated Hoagland nutrition solution (pH 5.2). The nutrition solution was aerated constantly and replaced twice a week throughout the experiment. Plants were cultured under non-saline conditions for 10 d to ensure full establishment before starting the salinity treatments. Salt stress treatment was initiated by providing the plants with full-strength Hoagland's solutions containing 0, 80, 160, or 240 mM NaCl. To avoid osmotic shock, salt concentrations were increased daily by 40 mM NaCl, until reaching the required concentration. After 1 week, the plants were harvested, cleaned and their fresh weights were measured.

Leaf photosynthesis measurements

Net photosynthetic rate (Pn) and stomatal conductance (gs) of leaves were measured in three plants per cultivar per treatment with a photosynthesis system (Li-6400, Li-COR Inc., NE, USA) under 1500 $\mu mol \cdot m^2 \cdot s^2$ light intensity, 65%±5% relative humidity, 32°C±2°C leaf temperature, and 380 $\mu mol \cdot mol^{-1}$ CO_2 concentrations at 9:30–11:00 AM.

Growth parameter measurements

From each treatment group, 10 plants were randomly selected and separated into leaf and root fractions. The leaf area of the youngest fully developed leaf of each plant was measured. Root samples were placed in a rectangular glass dish with a thin layer of water (4–5 mm depth) to allow all roots to spread appropriately. Entire roots were scanned with an EPSON Transparency unit (Epson Perfection V700 Photo; Indonesia), and then analyzed with WinRHIZO software version 5.0 (Regent Instruments, Inc.; Canada) to calculate the total root length, total root surface area, total root volume, and average root diameter. Leaves and roots were washed with deionized water and dried at 70°C for 48 h to determine dry weight before being ground to determine the ion contents.

In addition, another six plants per replicate of each of the four treatments were harvested. Fresh roots of seedlings were separated and frozen immediately in liquid nitrogen before being stored at −80°C pending further analysis.

Chlorophyll and carotenoid content measurements

To determine chlorophyll a (Chl a), chlorophyll b (Chl b), and carotenoids levles, 3–5 discs (0.8-cm diameter) were cut from the upper-most fully expanded leaves randomly selected from five plants per replicate. Discs were homogenized with 2 mL of acetone (80%) and washed twice with an additional 2 mL of acetone. The absorbance of the pooled extracts was measured using a spectrophotometer at 480 nm, 645 nm, and 663 nm. Contents of Chl a, Chl b, and carotenoids in the extracts was determined using MacKinney equations [30]:

$$Chla(mg\ g^{-1}DW) = (12.72 \times A663) - (2.58 \times A645)$$

$$Chlb(mg\ g^{-1}DW) = (22.87 \times A645) - (4.67 \times A663)$$

$$Carotenoids(mg\ g^{-1}DW) = (0.114 \times A663) + A480 - (0.638 \times A645)$$

Ion analyses

The concentration of calcium (Ca), magnesium (Mg), potassium (K), and sodium (Na) were analyzed in subsamples of dried plant materials, which were finely ground in a mill grinder. Approximately 0.5 g of finely ground plant samples were placed into digesting tubes, to which 10 mL of concentrated nitric acid and 3 mL perchlorate acid were added. All the samples were soaked for 12 h and then burned at 300°C for 3 h. The residue was transferred to a 50-mL volumetric flask, which was topped up to 50 mL with distilled water. The cation content was then measured using an atomic absorption spectrophotometer (TAS-986; Persee; China) [31]. For the determination of Cl^- content, leaf samples (0.1 g) were extracted in 10 mL of distilled water by heating at 80°C for 3 h [32]. The Cl^- content in the extracts was analyzed by ion chromatography (DX-300; Sunnyvale, CA, USA) [33].

Determination of lipid peroxidation

Frozen leaf and root segments (0.5 g) were homogenized in a 0.1% (w/v) trichloroacetic acid (TCA) solution. The homogenate was centrifuged at 15,000× *g* for 10 min and 1 mL of the supernatant was added to 4 mL 0.5% (w/v) 2-Thiobarbituric acid (TBA) in 20% (w/v) TCA. The mixture was incubated at 90°C for 30 min, and the reaction was stopped by placing the reaction tubes in an ice water bath. Samples were centrifuged at 10,000× *g* for 5 min and the absorbance of the supernatant was read at 532 nm. The value for non-specific absorption at 600 nm was subtracted from the measured values. The concentration of MDA was calculated using an extinction coefficient of 155 $mM^{-1}\cdot cm^{-1}$.

H_2O_2 determination

H_2O_2 content was estimated according to the methods of Bernt and Bergmeyer [34]. Approximately 0.5 g of root and leaf samples from control and treatment groups were homogenized with liquid nitrogen, and the powders were suspended in 1.5 mL of 100 mM potassium phosphate buffer (pH 6.8). The suspensions were then centrifuged at 18,000× *g* for 20 min at 4°C. The enzymatic reaction was initiated with 0.25 mL supernatant and 1.25 mL peroxidase reagent, consisting of 83 mM potassium phosphate buffer (pH 7.0), 0.005% (w/v) *O*-dianizidine, and 40 μg peroxidase/mL at 30°C. The reaction was stopped after 10 min by adding 0.25 mL of 1 N perchloric acid and the reaction mixture was centrifuged at 5000× *g* for 5 min. The absorbance of the supernatant was read at 436 nm, and the amount of H_2O_2 was determined using an extinction coefficient of 39.4 $mM^{-1}\cdot cm^{-1}$.

Extraction and assay of antioxidant enzyme assay

For enzyme extractions, frozen root and leaf samples (0.3 g) were ground into a fine powder by using a mortar that was placed in an ice bath and a pestle that was pre-cooled with liquid nitrogen, and homogenized in 50 mM potassium phosphate buffer (pH 7.8) containing 1 mM ascorbate and 2% (w/v) polyvinylpolypyrrolidone. Homogenates were then centrifuged at 20,000× *g* for 30 min at 4°C.

SOD activity was determined according to the methods of Foster and Hess [35]. The reaction was performed in a total volume of 1 mL containing 50 mM potassium phosphate buffer (pH 7.8, containing 0.1 mM ethylenediaminetetraacetic acid [EDTA]), 0.1 mM cytochrome c, 0.1 mM xanthine, enzyme extract, and 0.3 U/mL of xanthine oxidase. The reaction was initiated by the addition of xanthine oxidase and absorbance was measured at 560 nm. One unit of SOD activity was defined as the amount of enzyme that inhibits the rate of cytochrome c reduction by 50%.

Total CAT activity was measured according to the method reported by Beers and Sizer [36], with minor modifications. The reaction mixture (1.5 mL) consisted of 100 mM phosphate buffer (pH 7.0), 0.1 mM EDTA, 20 mM H_2O_2, and 50 μL enzyme extract. The reaction was initiated by the addition of the enzyme extract. The decrease in H_2O_2 was monitored at 240 nm and was quantified by its molar extinction coefficient (36 $M^{-1}\cdot cm^{-1}$).

Peroxidase activity was analyzed in 2.9 mL of 0.05 M phosphate buffer, containing 1.0 mL of 0.05 M guaiacol and 1.0 mL of 2% H_2O_2 [37]. Increases in absorbance at 470 nm were recorded after adding 2.0 mL of 20% chloroacetic acid.

APX activity was determined according to Nakano and Asada [38] by following the decline in absorbance at 290 nm as ascorbate was oxidized. The oxidation rate of ascorbate was estimated between 1 and 60 s after starting the reaction with the addition of H_2O_2. The 1-mL reaction mixture contained 50 mM HEPES-NaOH (pH 7.6), 0.22 mM ascorbate, 1.0 mM H_2O_2, and an enzyme extract. Corrections were made for the low, non-enzymatic oxidation of ascorbate in the absence of H_2O_2.

GR activity was measured as described by Foyer et al. [39]. The assay medium contained 1 mM EDTA in 50 mM potassium phosphate buffer (pH 7.8), 0.1 mM NADPH, enzyme extract, and 0.1 mM glutathiol (GSSG) in a total volume of 1 mL. The reaction was initiated by adding GSSG and the NADPH oxidation rate was monitored at 340 nm. GR activity was calculated using an extinction coefficient of 6.2 $mM^{-1}\cdot cm^{-1}$ for NADPH, and one unit of enzyme was defined as the amount of enzyme required to oxidize 1 μmol NADPH per minute. The specific enzyme activity for all enzymes was expressed as units/mg protein.

Statistical analysis

The experiments were set up as a completely randomized design, including two cotton cultivars and four salinity levels. All data obtained were subjected to ANOVA, and the mean difference was compared by the LSD test at 95% or 99% levels of probability. In all figures, the spread of values is shown as error bars representing standard errors of the means.

Results and Discussion

Growth parameters

Salinity exposure can lead to various physiological and biochemical changes within plant cells causing numerous changes in their structure and function [40]. In the present study, salt-induced changes in the growth and antioxidant profile of cotton plants were evaluated. Increasing NaCl concentration, up to 240 mM, gradually reduced leaf and root growth (Fig. 1). In general, both cultivars showed decreased growth rates of the roots and leaves with increasing salt concentration, but there was no significant variation in the leaf dry weight of Simian 3 when

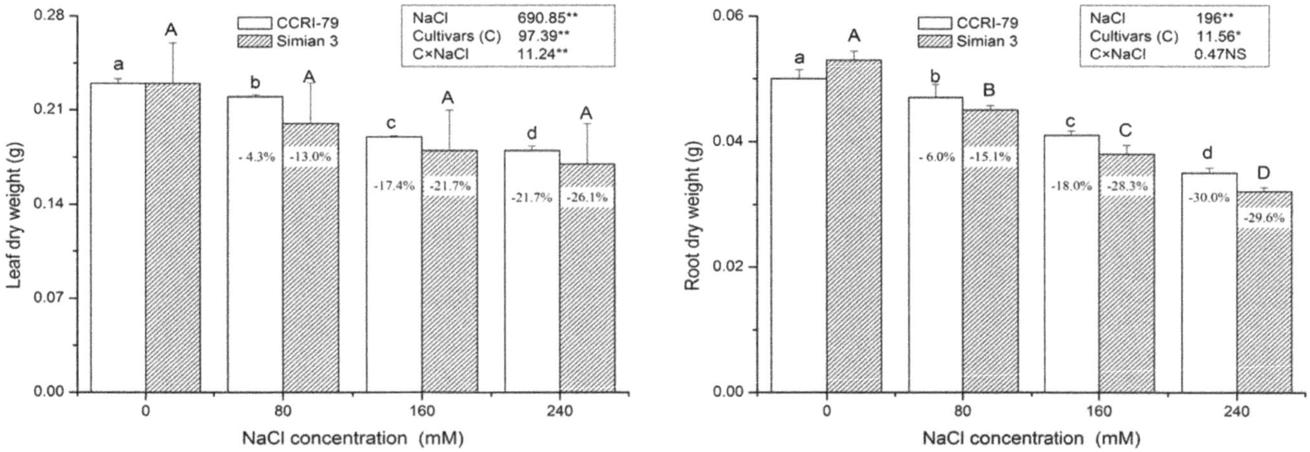

Figure 1. Effect of NaCl salinity on the leaf and root dry weight of two cotton cultivars. Vertical bars represent ± standard error (n = 3). Bars labeled with the different lowercase letters on open square bars or uppercase letters on closed square bars are significant difference ($P<0.05$). *, ** Significant at $P<0.05$ and $P<0.01$, respectively. NS, not significant. Figures in column indicate± increase/decrease under salinity stress over control.

subjected to salt stress. However, the percentage reduction in leaf and root dry weights due to salinity over control was lower in CCRI-79 as compared with Simian 3, indicating that CCRI-79 is a more salt-tolerant cultivar. Inhibition of growth due to NaCl stress in CCRI-79 is comparable to the observations of Takemura et al. [41]. This reduction in growth may be due to osmotic injury or specific ion toxicity caused by the uptake of salt [42]. However, the differential response of growth to salinity observed between CCRI-79 and Simian 3 could be due to genotypic differences, which have also been reported by Qidar and Shams [26]. An increase in the tissue maintenance process (through respiration) is believed to be the primary cause of growth decline during salinity stress, and could represent a mechanism of adaptation to salinity [25]. The sacrifice of leaf photosynthetic tissue during salt adaptation may serve to conserve energy that can then be redirected to maintaining leaf multiplication and growth, indicating successful use of the tissue maintenance process. Thus, the fact that the leaf dry weight of the salt-sensitive cultivar Simian 3 was maintained indicates that it is not reallocating its energy reserves when faced with high-salinity conditions as compared to the more salt-tolerant cultivar CCRI-79.

When plants are grown under conditions of salt stress, the immediate response is a cessation in the expansion of the leaf surface [43]. Several authors have reported the phenomenon of leaf expansion in response to salinity in halophytes as well as in glycophytes [44]. Similarly, in our study, leaf area was highest in the control group (0 mM NaCl), whereas it decreased continuously as salinity increased (Table 1). One possible reason for this decline might be related to salt osmotic effects, which affect cell turgor and expansion [45].

Compared with the no-salt control treatment, salt stress significantly ($P<0.01$) reduced the length, surface area, volume,

Table 1. Morphological parameters of cotton at different NaCl concentrations and results of ANOVA (F-ratios).

Cultivar	NaCl (mM)	Leaf area (cm^2)	Root length (cm)	Surface area (cm^2)	Volume (cm^3)	Average diameter (mm)
CCRI-79	0	36.19±0.91a	370.86±3.45a	52.07±0.94a	0.57±0.014a	0.54±0.016a
	80	35.18±1.06a	320.33±2.53b	44.04±1.29b	0.48±0.016b	0.45±0.014b
	160	34.43±0.46a	192.72±4.38c	32.45±1.67c	0.42±0.009c	0.43±0.009b
	240	34.07±0.12a	161.28±4.88d	28.03±0.71d	0.34±0.014d	0.35±0.007c
Simian 3	0	35.13±0.85a	302.58±3.02a	49.07±0.78a	0.62±0.041a	0.52±0.005a
	80	33.99±1.07a	260.86±8.67b	35.71±1.28b	0.39±0.043b	0.49±0.018a
	160	33.46±1.36a	188.32±14.26c	31.12±0.98c	0.38±0.023b	0.40±0.014b
	240	32.98±0.63a	138.58±1.05d	24.19±0.60d	0.30±0.008c	0.31±0.004c
NaCl		4.31NS	673.99**	338.36**	75.49**	182.17**
Cultivars(C)		5.73*	134.92**	50.99**	4.60 NS	3.12 NS
C×NaCl		0.01 NS	20.54**	6.70*	5.15*	9.41**

Values are the mean of three replicates ± S.E. Means followed by a different letter within a column for each cotton cultivar are significantly different at $P<0.05$ according to the Student's LSD test.
*, ** Significant at $P<0.05$ and $P<0.01$, respectively.
NS, not significant.

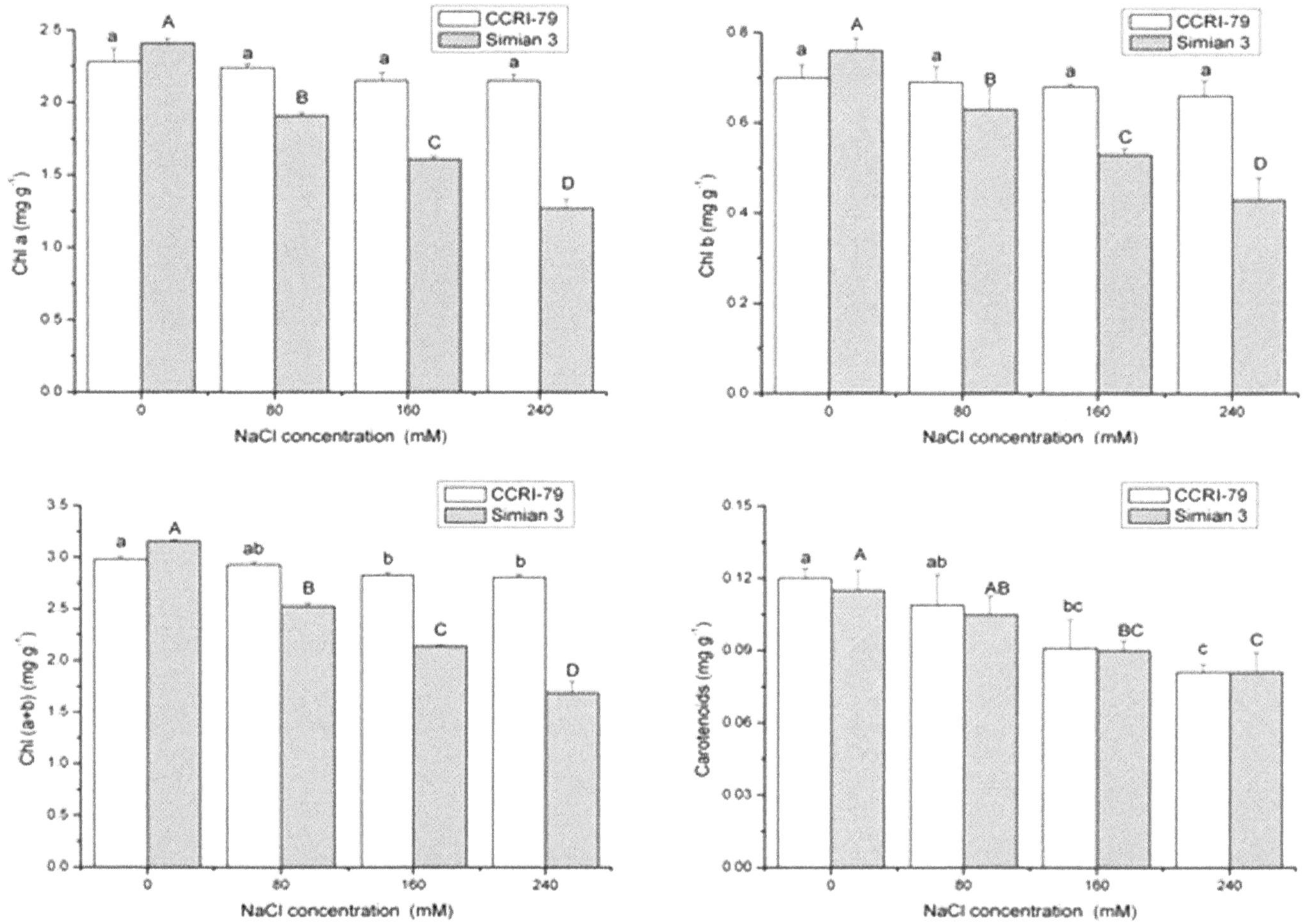

Figure 2. Concentrations of chlorophylls and carotenoids in cotton grown at different NaCl concentrations. Vertical bars represent ± standard error (n = 3). Bars labeled with the different lowercase letters on open square bars or uppercase letters on closed square bars are significant difference (P<0.05).

and average diameter of the roots in both cotton cultivars (Table 1). Significant differences were observed when NaCl concentration was increased from 80 to 240 mM. There are several reasons for the reduced root length, including cell growth restriction due to low water potential of the external medium, interference caused by ions, or the toxicity of accumulated ions [46]. Our findings are consistent with the results of Siroka et al. [47], who reported that salinity suppressed the development of maize roots cell. The inhibition of root growth in terms of root length, surface area, volume, and average diameter can be attributed to the inhibition of mitosis, reduced synthesis of cell wall components, damage to the Golgi apparatus, and changes in polysaccharide metabolism [48]. However, this decrease was more

Table 2. *F* values of ANOVA of the effects of NaCl, cultivars, and their interaction for chlorophylls, carotenoids content, net photosynthetic rate (Pn), and stomatal conductance (gs).

Item	NaCl	Cultivars(C)	C×NaCl
Chl a	108.52**	245.65**	66.29**
Chl b	45.49**	61.62**	26.86**
Chl (a+b)	186.62**	380.12**	113.08**
Carotenoids	13.62**	0.28 NS	0.10 NS
Pn	98.15**	200.15**	51.5**
gs	56.31**	77.17**	23.44**

*, ** Significant at P<0.05 and P<0.01, respectively.
NS, not significant.

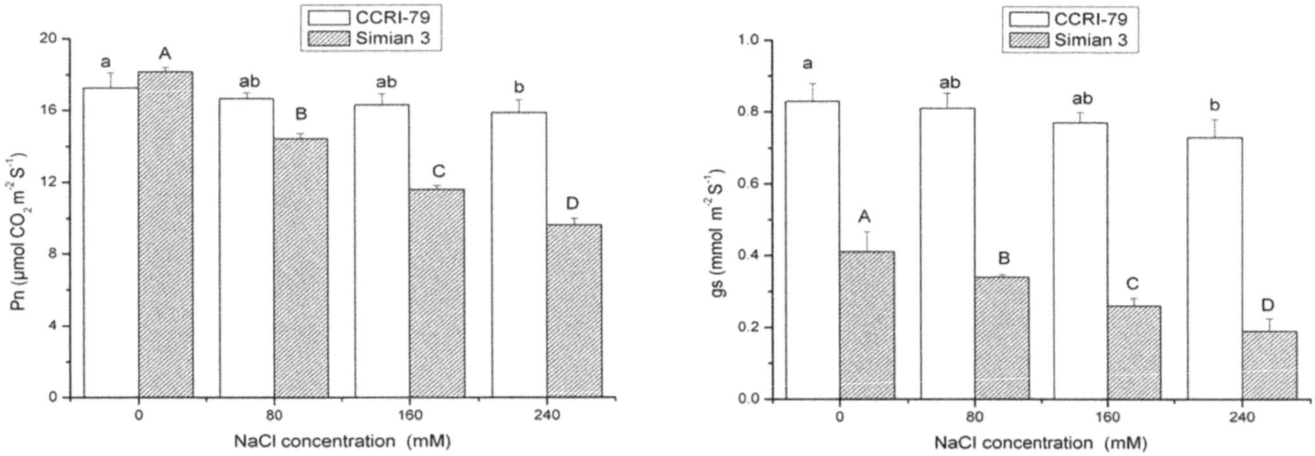

Figure 3. Net photosynthetic rate (Pn) and stomatal conductance (gs) of two cotton cultivars as affected by different NaCl concentrations. Vertical bars represent ± standard error (n = 3). Bars labeled with the different lowercase letters on open square bars or uppercase letters on closed square bars are significant difference ($P<0.05$).

predominant in Simian 3 than in CCRI-79, indicating that CCRI-79 was more tolerant to salinity than Simian 3.

Chlorophyll and carotenoid contents and photosynthesis

In general, the reduction in growth and productivity when plants are grown under conditions of salt stress is accompanied by as strong reduction in the rate of photosynthesis owing to severe impairments in photosynthetic activities and the photosynthetic apparatus, the degree of which depends on the varieties of species considered [49]. As shown in Fig. 2, the contents of Chl a, Chl b, and Chl (a+b) in the plants significantly ($P<0.01$) decreased as salinity increased in Simian 3, but only marginal changes were observed in CCRI-79 (Table 2). Maintenance of chlorophyll content has been reported in other salt-tolerant crops such as durum wheat and legume species [50]. One possible mechanisms of salt tolerance in these species may be the possession of a salt exclusion and/or sequestration trait that prevents leaf injury, thus maintaining chlorophyll content. Since chlorophyll content is directly correlated with growth and development of the plant [51], the decrease in chlorophyll content in Simian 3 suggested substantial damage to the photosynthetic mechanism, as reported previously in salt-treated rice and sorghum plants [52], [53]. The inhibitory effects of salt on chlorophylls could be due to the suppression of specific enzymes responsible for the synthesis of chlorophyll. Our findings regarding total chlorophyll content are comparable with the observations of Meloni et al. [54], who demonstrated that Guazuncho, another salt-sensitive cotton cultivar, showed a 35% reduction in total chlorophyll content after 21 days of salinity treatment.

Carotenoids are reported to play an important role in ROS scavenging, thereby protecting membranes from salt stress [55]. However, we did not observe changes in carotenoid contents in response to salinity treatments in either cultivar, suggesting that these pigments are not involved in the response to salt stress in cotton. Although the rate of change was slower in carotenoids than in chlorophylls, carotenoid content also showed a decreasing trend with increasing salt stress, indicating that this trait could also serve as a useful indicator of NaCl stress in cotton.

Since plant growth is dependent on photosynthesis, environmental stresses affecting growth will also affect photosynthesis [56]. In the present study, the net photosynthetic rate (Pn) and stomatal conductance (gs) of both cotton cultivars were inhibited by salinity due to NaCl (Fig. 3, Table 2). However, the net photosynthetic rate and stomatal conductance were significantly lower for Simian 3 than for CCRI-79 under conditions of salt stress. Compared with the control treatment, the net photosynthetic rate and stomatal conductance of Simian 3 significantly decreased with increasing salinity, whereas there was no significant difference in either trait in CCRI-79.

Photosynthesis was reduced in both cotton cultivars in response to salt stress, which was likely caused by the reduction in stomatal conductance. Indeed, parallel decreases in stomatal conductance and net photosynthesis due to NaCl salinity have previously been reported for cotton [57]. Our results suggest that stomatal closure limited the leaves' photosynthetic capacity in the NaCl-treated plants of both cultivars, Although only Simian 3 showed a significant decline in the leaf chlorophyll contents due to NaCl stress for 7 days. Similarly, Delfine et al. reported no changes in the chlorophyll content in spinach plants (*Spinacia oleracea* L.) exposed to salt stress for 20 days [58]. Our results suggest that the greater reduction in stomatal conductance accompanied by decreased leaf chlorophyll content could have contributed to the higher reduction in the leaf photosynthetic rate of Simian 3 when compared with that of CCRI-79.

Further examination showed that the decrease in the Chl a, Chl b, Chl (a+b), net photosynthetic rate, and stomatal conductance levels with increasing salt concentrations occurred more rapidly compared with the rate of decrease in carotenoid content; this trend was more conspicuous in Simian 3 than in CCRI-79.

Ion concentrations

High external salt concentration causes an ion imbalance or disturbance in ion homeostasis [43]. In our experiments, the leaves and roots of both cultivars had higher levels of Na^+ and Cl^- ions under salt stress due to nonspecific ion uptake and/or membrane leakage. However, as the NaCl concentration increased, the levels of Na^+ and Cl^- also increased further, suggesting that these cultivars may not differ in terms of Na^+ uptake and its transportation to leaves, and thus the increase observed can not be explained by an ionic exclusion mechanism (Fig. 4). In addition, Na^+ ion concentrations in leaves were higher than those in the roots of both cultivars at all salinity levels, which indicated

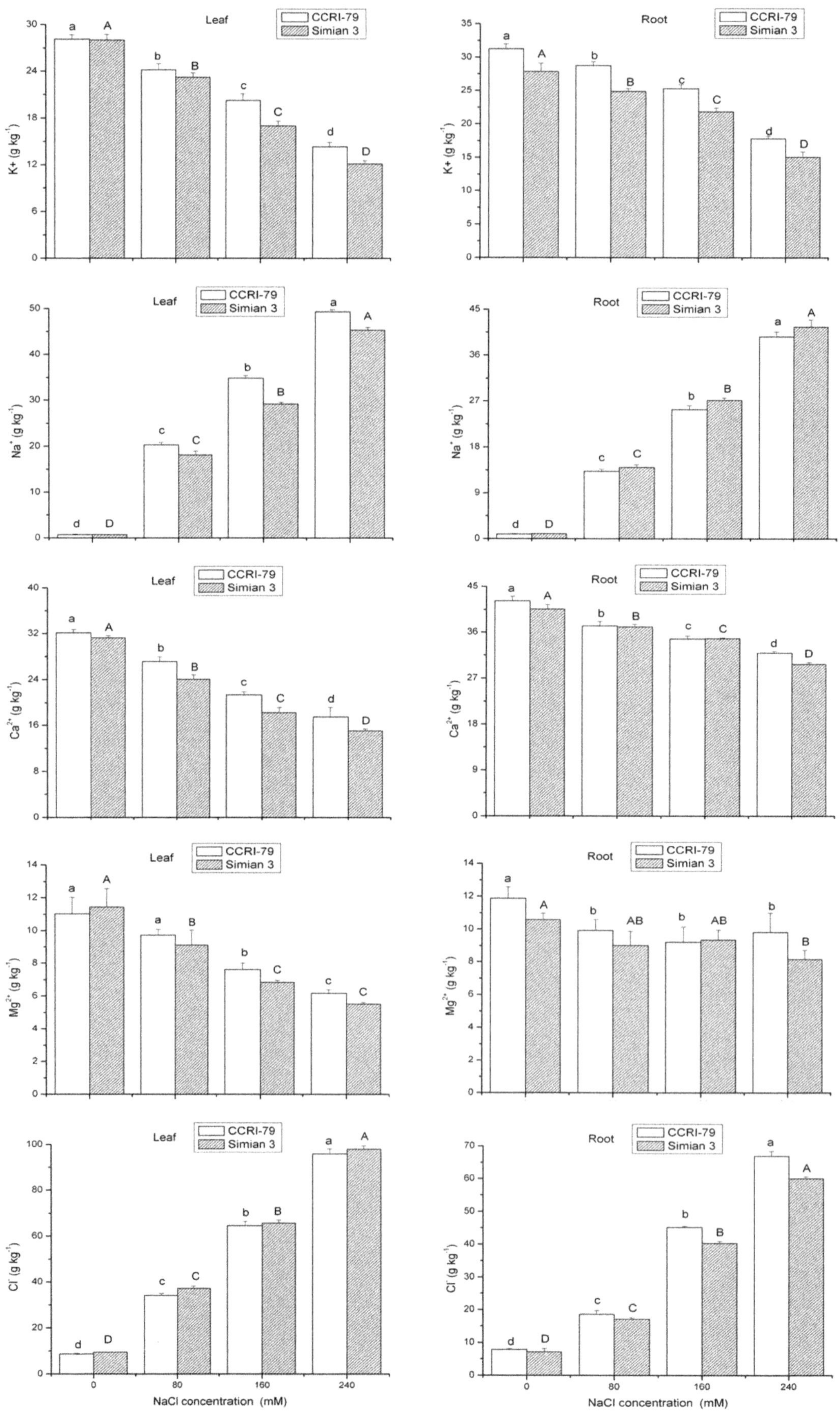
Leaf
Root
CCRI-79
Simian 3
K+ (g kg^{-1})
Na^+ (g kg^{-1})
Ca^{2+} (g kg^{-1})
Mg^{2+} (g kg^{-1})
Cl^- (g kg^{-1})
0
80
160
240
NaCl concentration (mM)

Figure 4. Effect of NaCl salinity on the concentrations of K^+, Na^+, Ca^{2+}, Mg^{2+} and Cl^- in the roots and leaves of cotton. Vertical bars represent ± standard error (n = 3). Bars labeled with the different lowercase letters on open square bars or uppercase letters on closed square bars are significant difference ($P<0.05$).

the inability of these cultivars to prevent Na^+ ion transportation from the roots to leaves. Chlorine ions showed a similar distribution pattern to Na^+ ions, despite being at higher salinity levels. Na^+ and Cl^- are highly water-soluble and are readily taken up by plants and transported into leaves; these ions most likely act as osmotica, but only moderate concentrations can be tolerated before growth and photosynthesis are reduced.

High Na^+ and Cl^- absorption competes with the uptake of other nutrient ions such as K^+, Mg^{2+}, and Ca^{2+}, leading to a deficiency of these ions and an imbalance among cations [59]. During the same period, K^+ and Ca^{2+} content in the leaves and roots of both cultivars were significantly reduced, and a further decrease in these ions was observed with increasing NaCl concentration (Fig. 4, Table 3). Salt tolerance involves not only adaptation to Na^+ influx but also acquisition of K^+, the uptake of which is adversely affected by high external Na^+ concentration due to the chemical similarity of these two ions [60], [61]. Indeed, selective uptake of K^+ as opposed to Na^+ is considered to be one of the key physiological mechanisms contributing to salt tolerance in many plant species [62]. K^+ efflux has already been used as an indicator of cellular toxicity for a range of toxic compounds, and losses of K^+ and Ca^{2+} have already been documented during salinity stress [63], [64]. A large and permanent efflux of K^+ and Ca^{2+} usually indicates damage to the limiting membranes.

On the other hand, salt treatments induced a significant decrease in Mg^{2+} concentrations in cotton leaves in the present study. The significantly lower levels of Mg^{2+} in the leaves under salinity conditions are probably related to the lower levels of chlorophylls present in the NaCl-treated leaves. However, in the roots of NaCl-treated plants, Mg^{2+} concentrations were close to or lower than those observed in control plants, even at the highest salt dose (240 mM NaCl). This was in contrast to the Ca^{2+} content, which was significantly lower in salt-treated plants. Our findings are similar to those of Khan [65], who reported that NaCl treatment induced a decline in Ca^{2+} and Mg^{2+} levels in *Ceriops tagal* plants.

In Addition, there were no significant differences between the two cotton cultivars in the variation of ion content (including K^+, Na^+, Mg^{2+}, Ca^{2+} and Cl^-) under NaCl stress. This similarity in ionic levels may be a consequence of the shoot culture method used, which would not involve a selective mechanisms of ion transport. For example, there was likely to be no regulation between the xylem parenchyma and xylem interface, in which ion selection and reabsorption from the medium may be regulated. Previous results reported for other species are consistent with these observations [66]. However, the restriction of entry of ions into metabolically active areas of cells in the more-tolerant CCRI-79 cultivar can not be ruled out as a mechanism to maintain ionic equilibrium when ions are highly concentrated in the external environment [67].

Lipid peroxidation and proline content

Salt stress is known to result in extensive lipid peroxidation, which has often been used as an indicator of salt-induced oxidative damage in membranes [68]. The MDA content increased with increasing salinity in the leaves and roots of both cotton cultivars (Fig. 5, Table 4), indicating cell membrane damage in both cotton cultivars. However, as the salinity increased, the accumulation of MDA was higher in Simian 3 as compared to CCRI-79, indicating a higher degree of lipid peroxidation in Simian 3 due to salt stress. Lipid peroxidation could be a result of light-dependent formation of singlet oxygen during stress conditions [69]. The low values of MDA content obtained with CCRI-79 might account for the lower lipid peroxidation levels observed and the reduced effect on membrane permeability. Low levels of lipid peroxidation may have contribute to the observed tolerance of CCRI-79 plants exposed to high levels of salinity. Similar results for lipid peroxidation have been reported by other researchers in barley [70].

Many plants accumulate proline as a nontoxic and protective osmoylte under saline conditions [55]. In this study, the levels of proline continued to increase in both of the cultivars as the NaCl concentration increased (Fig. 5). The proline concentration of

Table 3. *F* values of ANOVA of the effect of NaCl, cultivars, and their interaction for K^+, Na^+, Ca^{2+}, Mg^{2+}, and Cl^- contents.

Tissue	Item	Cultivars (C)	NaCl	C×NaCl
Leaf	K^+	22.47**	349.67**	4.05 NS
	Na^+	131.7**	5755.89**	22.16**
	Ca^{2+}	27.27**	220.03**	1.28 NS
	Mg^{2+}	2.09 NS	69.77**	0.94 NS
	Cl^-	7.03*	3175.89**	0.58 NS
Root	K^+	72.66**	204.40**	0.34 NS
	Na^+	21.19**	4546.89**	3.12 NS
	Ca^{2+}	15.30**	311.67**	4.96*
	Mg^{2+}	5.19NS	6.21*	0.91 NS
	Cl^-	65.95**	3497.79**	11.22**

*, ** Significant at $P<0.05$ and $P<0.01$, respectively.
NS, not significant.

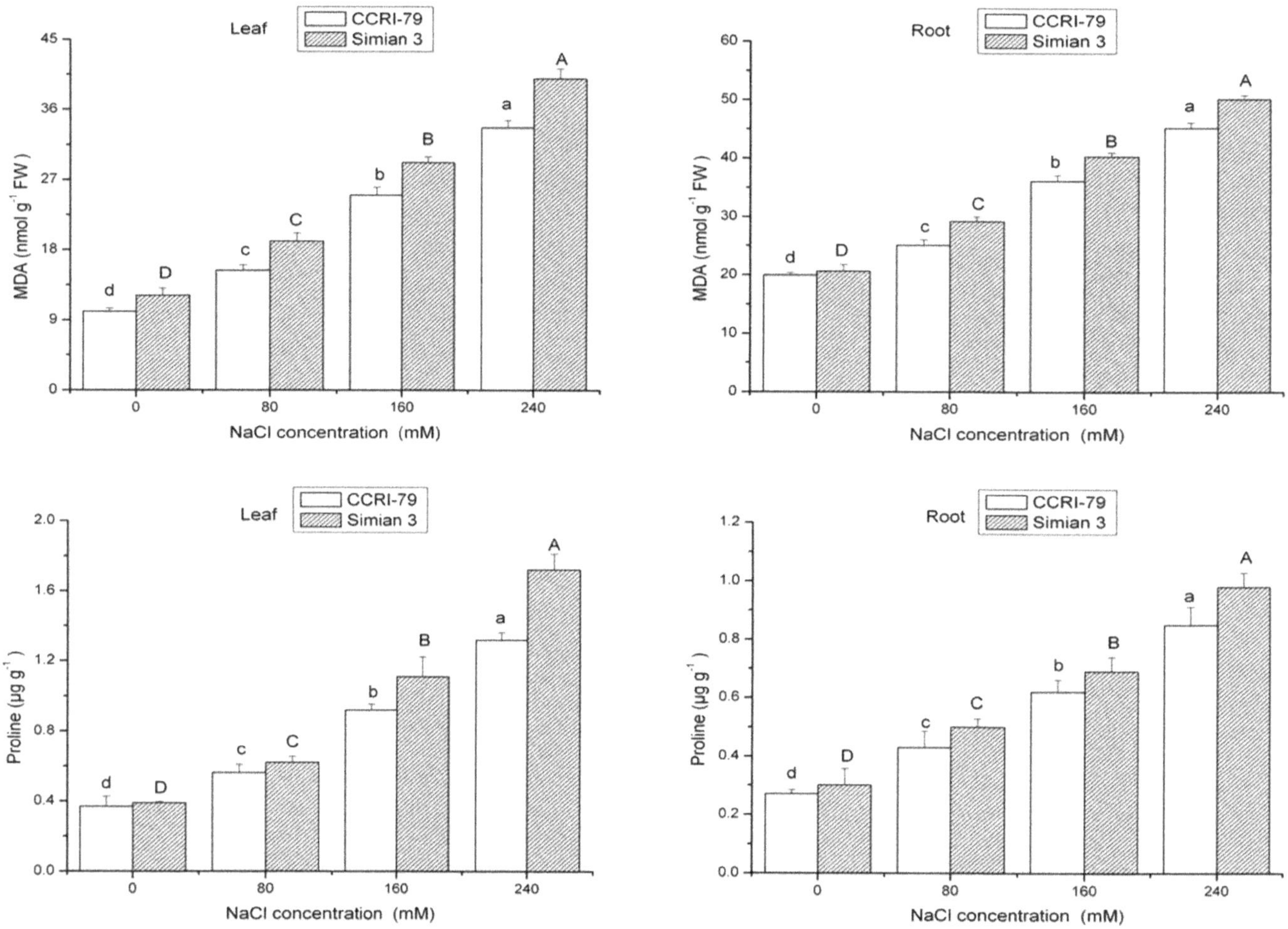

Figure 5. Effect of NaCl salinity on the concentrations of malondialdehyde (MDA) and proline in the roots and leaves of cotton. Vertical bars represent ± standard error (n = 3). Bars labeled with the different lowercase letters on open square bars or uppercase letters on closed square bars are significant difference ($P<0.05$).

CCRI-79 plants was lower than that of Simian 3 plants, especially at the highest salinity level, which could be attributable to the greater salt resistance of the CCRI-79 cultivar, i.e., less injury was induced by the salt [71]. Similar trends were observed by Rabie and Almadini in broad bean plants [72]. Moreover, in our study, proline accumulation in the leaves was higher than that in the roots, which is similar to the findings of Sharma and Dietz [73], indicating that proline plays a more important protective role in the leaves of cotton seedlings than in the roots under salinity stress.

H_2O_2 content

Stress conditions enhance H_2O_2 production in different compartments of plants cells through both enzymatic and non-enzymatic processes [74]. In our study, both genotypes had similar levels of H_2O_2 levels at 0 mM NaCl treatment. However, salinity treatments caused a marked increase in H_2O_2 content, and the Simian 3 cultivar had a higher H_2O_2 content than did CCRI-79 after NaCl treatment (Fig. 6, Table 4), which could be mainly due to the decreased H_2O_2-scavenging activity in the salt-sensitive cultivar. These results are comparable to those reported in a previous study [75], where accumulation of H_2O_2 in the roots of the salt-tolerant rice cultivar FL478 was significantly higher than that of the salt-sensitive rice cultivar IR-29 in response to moderate salt stress applied for 12 days. Simultaneously, the H_2O_2 content were markedly lower in the leaves than in the roots, regardless of NaCl dose, which contrasts with the findings of Lee et al. [76] in rice. This discrepancy in the effects of salt on H_2O_2 content between studies may be related to technical difficulties, and therefore these results should be evaluated with caution [77]. Furthermore, H_2O_2 has been shown to induce cytosolic APX [78]; therefore, accumulation of H_2O_2 under high salinity conditions may be a signal to initiate an adaptive response to the stress [79]. Although differences in salt tolerance among different cultivars are not necessarily related to differences in the ability to detoxify ROS, many comparative studies using salt-tolerant and salt-sensitive genotypes have shown a correlation between salt tolerance and increased activity of antioxidant enzymes [80].

Activities of antioxidant enzymes

Environmental stresses that limit photosynthesis can increase oxygen-induced cellular damage due to increased ROS generation [81]. Therefore, salt stress resistance may depend, at least in part, on the enhancement of the antioxidative defense system, which involves antioxidant compounds and several antioxidant enzymes. In the present study, the responses of SOD, POD, CAT, APX, and GR enzyme activities suggested that oxidative stress is an important component of salt stress in cotton plants.

Table 4. *F* values of ANOVA of the effect of NaCl, cultivars, and their interaction for malondialdehyde (MDA), proline, H_2O_2 content, and antioxidant enzyme activities.

Tissue	Item	Cultivars (C)	NaCl	C×NaCl
Leaf	MDA	78.15**	608.62**	3.65 NS
	Proline	25.01**	230.23**	6.58*
	H_2O_2	74.86**	1357.74**	11.54**
	SOD	315.65**	213.58**	63.01**
	CAT	874.75**	445.12**	43.46**
	POD	0.47 NS	5.52*	0.49 NS
	APX	1.36 NS	5.93*	0.08 NS
	GR	28.88**	194.18**	15.77**
Root	MDA	60.16**	708.42**	4.47*
	Proline	9.10*	119.51**	0.72 NS
	H_2O_2	174.18**	4622.30**	40.92**
	SOD	8559.93**	1510.48**	1357.86**
	CAT	111.07**	1.92 NS	28.60**
	POD	50.42**	322.16**	2.08 NS
	APX	37.10**	208.76**	5.83*
	GR	17.47**	82.69**	1.45 NS

*, ** significant at P = 0.05 and P = 0.01 levels, respectively.
NS, not significant.
H_2O_2, hydrogen peroxide; SOD, superoxide dismutase; CAT, catalase; POD, peroxidase; APX, ascorbate peroxidase; GR, glutathione reductase.

Because SOD can catalyze the dismutation of superoxide to molecular oxygen and H_2O_2, this enzyme is considered the most effective intracellular enzymatic antioxidant. Indeed, it has been suggested that SOD plays an important role in plant stress tolerance and provides the first line of defense against the toxic effects of elevated levels of ROS [82]. In this study, salt stress increased SOD activity in the leaves of both cultivars and in the roots of CCRI-79 only (Fig. 7, Table 4). However, increased SOD activity in both the leaves and roots was more conspicuous in the salt tolerant cultivar CCRI-79 than in the salt-sensitive cultivar Simian 3, suggesting that the salt-tolerant genotype has a more efficient $O_2^{\bullet-}$ radical-scavenging ability. Similar results have also been shown in both the leaves and roots of cotton and pea plants [14], [83]. In plants, high induction of SOD activity can lead to H_2O_2 accumulation as well as lipid peroxidation [84], which could contribute to the increased H_2O_2 content observed in the roots than in the leaves of cotton seedlings exposed to salinity.

POD is the primary enzyme that detoxifies H_2O_2 in the chloroplasts and cytosol of plant cells [85]. CAT plays an important role in the antioxidant system because it converts H_2O_2 into oxygen and water [86]. These two enzymes constitute the main H_2O_2-scavenging systems in cells. The present data showed that the roots had higher POD activity compared to the leaves in both cultivars; however, the enzyme activity in the roots and leaves responded differently to incremental levels of salinity. In the roots of both cultivars, there was a significant decline in POD activity with an increase in salinity levels, whereas there was no significant difference in the leaf of Simian 3 across salt treatments. However, POD activity in the leaf of CCRI-79 showed no significant difference when subjected to 0 to 160 mM NaCl concentrations. Conversely, the 240 mM NaCl concentration induced a significant decrease in the leaf POD activity of CCRI-79 when compared to the control (0 mM NaCl concentration). Amor et al. reported that H_2O_2 accumulation under salinity stress was related to a decrease in CAT activity in *C. maritima* [87]. In our study, CAT activity in the leaves and roots of Simian 3 declined with an increase in NaCl concentration (Fig. 7, Table 4). However, although the CAT activity in CCRI-79 increased with salinity in the roots, it decreased with increasing salinity in the leaves. CAT activity in the leaves and roots of CCRI-79 was significantly higher than that in Simian 3 leaves and roots for all salinity treatments. This indicates that the scavenging of H_2O_2 is more effective in CCRI-79 than in Simian 3. The roots of both the CCRI-79 and Simian 3 cultivars showed higher CAT activities than did the leaves. The stimulation of POD and CAT suggests that these enzymes are important in the detoxification of H_2O_2 in plant seedlings under salinity stress [88]. In contrast, POD and CAT activities were found to be reduce in response to excess salinity in the leaves and roots of various other plants species [52], [89]. Thus, the effect of salinity on antioxidant enzyme activities varies among plant species, organs, and even treatment concentrations.

H_2O_2 scavenging is also accomplished by the glutathione-ascorbate cycle, a series of coupled redox reactions involving three enzymes, APX, GR and monodehydroascorbate reductase [90]. APX plays a key role in protecting the plant against oxidative stress by scavenging H_2O_2 in different cell compartments. It also has a higher affinity for H_2O_2 than POD and CAT and, as such, may play a more crucial role in the management of ROS during stress [82]. As shown in Fig. 7, there was a significant difference in the effect of salt on APX activity between the leaves and roots of the two cultivars. As salinity increased, APX activity in the root increased in CCRI-79 but decreased in Simian 3. However, there was no significant difference in the leaf APX activity of plants subjected to 0 to 160 mM NaCl concentrations in both cultivars. Conversely, the 240 mM NaCl concentration induced a significant

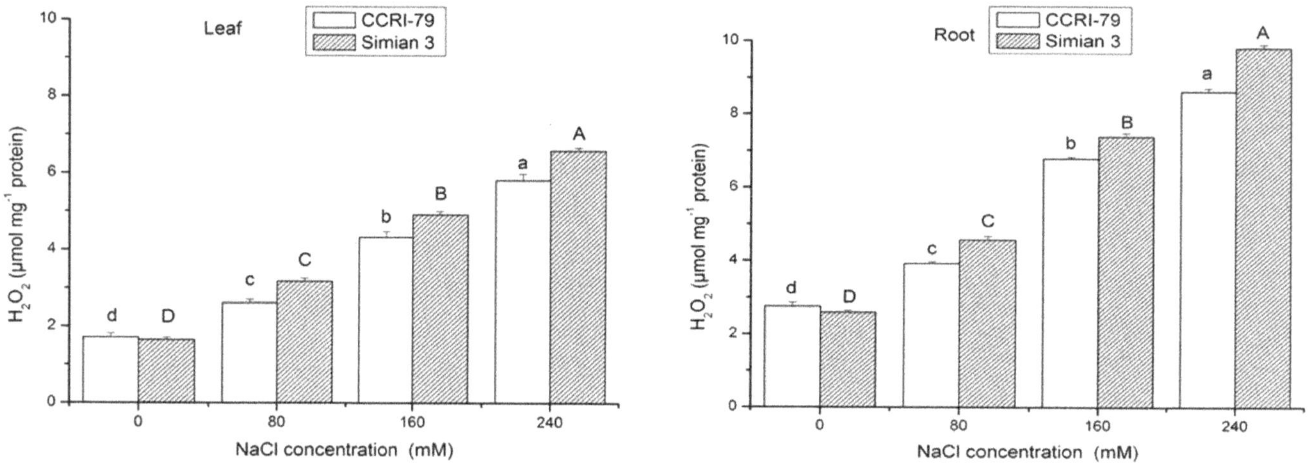

Figure 6. Effect of NaCl salinity on the concentrations of H_2O_2 in the roots and leaves of cotton. Vertical bars represent ± standard error (n = 3). Bars labeled with the different lowercase letters on open square bars or uppercase letters on closed square bars are significant difference ($P< 0.05$).

decrease in the leaf APX activity of both cultivars when compared to the control (0 mM NaCl concentration). We also found that higher APX activities were accompanied by higher CAT activities in CCRI-79, implying that CCRI-79 has a more effective H_2O_2-scavenging mechanism than Simian 3.

The role of GSH and GR in H_2O_2 scavenging in plant cells has been well established in the Halliwell-Asada enzyme pathway. GR catalyzes the rate-limiting and last step of the ascorbate-glutathione pathway [91]. In our study, GR activities increased in the leaves of both cotton cultivars and in the roots of CCRI-79 only in response to salt stress. On the other hand, salinity significantly reduced GR activity in the roots of Simian 3. Several studies investigating salt-tolerant and salt-sensitive cultivars have suggested that the salt tolerance trait is related to increased GR activity in salt-tolerant cultivars [83], [92], which is similar to our results. The elevated levels of GR activity may increase the rate of NADPH oxidation to $NADP^+$, thereby ensuring the availability of $NADP^+$ to accept electrons from the photosynthetic electron transport chain. Under such conditions, the flow of electrons to O_2^-, and therefore the formation of O_2^- can be minimized. However, in the roots of Simian 3, the reduction in GR activity suggests a decrease in the GSH turnover rate. Considering that salinity also reduced APX activity in the roots of Simian 3, these results suggests that salt-sensitive cultivars exhibit a less-active ascorbate-glutathione cycle in the roots, which may be a key enzyme for the development of salt-tolerance in cotton plants.

Furthermore, the variation in SOD, CAT, APX, and GR activities differed significantly between CCRI-79 and Simian 3 under the NaCl concentrations tested. The increased salinity resistance of CCRI-79 was associated with its ability to maintain higher activity of these antioxidant enzymes, which resulted in lower H_2O_2 production, lipid peroxidation, and higher membrane stability. This provides further evidence that the H_2O_2-scavenging mechanisms were more effective in CCRI-79. By contrast, the relatively lower CAT, APX, and GR activities in salt-stressed Simian 3 compared to control plants indicated that H_2O_2 scavenging was less effective in this cultivar. This excess of H_2O_2 may be the main contributor to the extensive lipid peroxidation and growth inhibition observed in Simian 3.

When compared with the other scavenging enzymes tested, CAT and APX had much higher H_2O_2-scavenging activities in the leaves and roots of both cotton cultivars in our study. The importance of these two enzymes in H_2O_2 scavenging has been demonstrated in several previous studies, in which increased CAT and APX activities were correlated with tolerance to salt and other environmental stresses [93]. These enzymes were also shown to be important in salt tolerance of barley and mulberry [94], [95]. Therefore, the results suggest that the coordination of CAT and APX with SOD activity could comprise an additional constituent in the enzymatic antioxidant mechanism of cotton plants against oxidative stress.

Conclusions

In this study, we compared the response of two cotton cultivars that differ with respect to salt tolerance to increasing NaCl concentrations. Overall, salinity significantly reduced the leaf and root dry weights, root volume, root length, root surface area, root average diameter, chlorophyll content and photosynthesis in the cotton plants of both cultivars. In contrast, antioxidant enzyme activity and proline and MDA contents increased in response to salinity. The salt-tolerant cultivar CCRI-79 showed evidence of possessing a more efficient antioxidant defense system against oxidative stress and lipid peroxidation by maintaining higher SOD, CAT, APX, and GR activities than those in Simian 3 during salt stress. The differences in the antioxidant enzyme activity of the leaves and roots may, at least in part, explain the greater tolerance to salt stress exhibited by CCRI-79 compared to that exhibited by Simian 3. Besides differences in antioxidant enzyme activities, the two cotton cultivars also showed marked variation in Chl a, Chl b, and Chl (a+b) contents, net photosynthetic rate, and stomatal conductance in response to NaCl stress. Therefore, acquisition of tolerance to salt may not only involve improved resistance to oxidative stress owing to enzymes that primarily function to protect membranes and tissues from such damage, but might also involve improvement in the biosynthetic pathway of photosynthetic pigments to maintain higher rates of photosynthesis in the face of stress. However, it should be noted that salt stress was only assessed at concentrations of 0, 80, 160, and 240 mM NaCl; therefore, further studies should be conducted to verify and screen the selection criteria for salt-tolerant species and cultivars. Nonetheless, the data presented

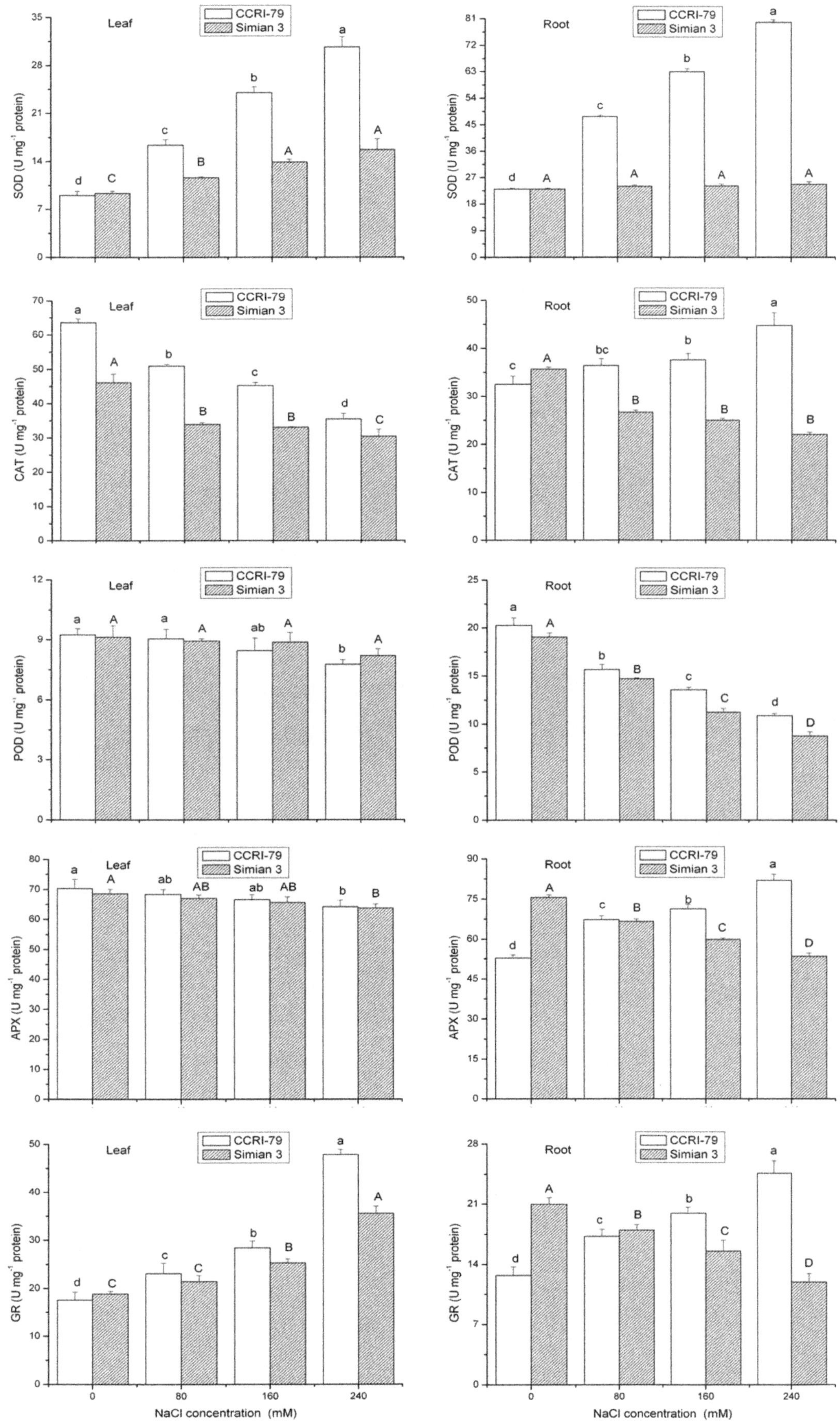
Leaf
Root
CCRI-79
Simian 3
SOD (U mg-1 protein)
CAT (U mg-1 protein)
POD (U mg-1 protein)
APX (U mg-1 protein)
GR (U mg-1 protein)
0
80
160
240
NaCl concentration (mM)

Figure 7. Effect of NaCl salinity on the antioxidant enzyme activities in the roots and leaves of cotton. Vertical bars represent ± standard error (n = 3). Bars labeled with the different lowercase letters on open square bars or uppercase letters on closed square bars are significant difference (P<0.05). SOD, superoxide dismutase; CAT, catalase; POD, peroxidase; APX, ascorbate peroxidase; GR, glutathione reductase.

herein provide novel information on the mechanisms and traits involved in salt tolerance, which could be exploited for cultivar selection and breeding to increase crop production in the face of increased salinity stress.

Acknowledgments

The authors thank Dr. Zhiguo Zhou for critically reading this manuscript and technical assistance.

Author Contributions

Conceived and designed the experiments: SXY XHZ. Performed the experiments: LZ JP HJM TTC. Analyzed the data: LZ. Contributed reagents/materials/analysis tools: LZ JP TTC. Wrote the paper: LZ. Final approval of the version to be published: XHZ.

References

1. Munns R, Tester M (2008) Mechanisms of salinity tolerance. Annu Rev Plant Biol 59: 651–681.
2. Arzani A (2008) Improving salinity tolerance in crop plants: a biotechnological view. In Vitro Cell Dev Biol Anim 44: 373–383.
3. Gaber MA (2010) Antioxidative defense under salt stress. Plant Signaling &Behaviour 5: 369–374.
4. Khan MA, Ungar IA, Showalter AM (2000) Effects of salinity on growth, water relations and ion accumulation in the subtropical perennial halophyte, *Atriplex griffithii var. stocksil*. Ann Bot 85: 225–232.
5. Locy RD, Chang CC, Neilson BL, Singh NK (1996) Photosynthesis in salt-adapted heterotrophic tobacco cells and regenerated plants. Plant Physiol 110: 321–328.
6. Rout NP, Shaw BP (2001) Salt tolerance in aquatic macrophytes: possible involvement of the antioxidative enzymes. Plant Sci 160: 415–423.
7. Zheng C, Jiang D, Liu F, Dai T, Jing Q, et al. (2009) Effects of salt and water logging stresses and their combination on leaf photosynthesis, chloroplast ATP synthesis, and antioxidant capacity in wheat. Plant Sci 176: 575–582.
8. Foyer CH, Halliwell B (2000) Oxygen processing in photosynthesis: regulation and signaling. New Phytol 146: 359–388.
9. Ashraf M (2009) Biotechnological approach of improving plant salt tolerance using antioxidants as markers. Biotechnol Adv 27: 84–93.
10. Sigaud-Kutner TSC, Pinto E, Okamoto OK, Latorre LR, Colepicolo P (2002) Changes in superoxide dismutases activity and photosynthetic pigment content during growth of marine phytoplankters in batch-cultures. Plant Physiol 114: 566–571.
11. Gratao PL, Polle A, Lea PJ, Azevedo RA (2005) Making the life of heavy metal-stressed plants a little easier. Funct Plant Biol 32: 481–494.
12. Noctor G, Foyer CH (1998) Ascorbate and glutathione: keeping active oxygen under control. Annu Rev Plant Phys 49: 249–279.
13. Blokhina O, Virolainen E, Fagestedt KV (2002) Antioxidants, oxidative damage, and oxygen deprivation stress: A review. Ann Bot 91:179–194.
14. Gosset DR, Millhollon EP, Lucas MC (1994) Antioxidant response to NaCl stress in salt-tolerant and salt-sensitive cultivar of cotton. Crop Sci. 34: 706–714.
15. Blum A, Ebercon A (1981) Cell membrane stability as a measure of drought and heat tolerance in wheat. Crop Sci 21: 43–47.
16. Gosset DR, Millhollon EP, Lucas MC (1994) Antioxidant response to NaCl stress in salt-tolerant and salt-sensitive cultivars of cotton. Crop Sci 34: 706–714.
17. Dionisio-Sese ML, Tobita S (1998) Antioxidant response of rice seedlings to salinity stress. Plant Sci 135: 1–9.
18. Hernandez JA, Jimenez A, Mullineaux P, Sevilla F (2000) Tolerance of pea (*Pisum sativum* L.) to long-term salt stress is associated with induction of antioxidant defenses. Plant Cell Environ 23: 853–862.
19. Mittova V, Tal M, Volokita M, Guy M (2003) Up-regulation of the leaf mitochondrail and peroxisomal antioxidative systems in response to salt-induced oxidative stress in the wild salt-tolerant tomato species *Lycopersicon pennellii*. Plant Cell Environ 26: 845–856.
20. Meneguzzo S, Navario-Izzo F, Izzo R (1999) Antioxidative responses of leaves and roots of wheat to increasing NaCl concentrations. J Plant Physiol 155: 274–280.
21. Xiong L, Zhu JK (2002) Molecular and genetic aspects of plant responses to osmotic stress. Plant Cell Environ 25: 131–139.
22. Meng HB, Jiang SS, Hua SJ, Lin XY, Li YL, et al. (2011) Comparison between a tetraploid turnip and its diploid progenitor (*Brassica rapa* L.): the adaptation to salinity stress. ASC 10: 363–375.
23. Steduto P, Albrizio R, Giorio P (2000) Gas exchange response and stomatal and non-stomatal limitations to carbon assimilation of sunflower under salinity. Environ Exp Bot 44: 243–255.
24. Fedina IS, Tsoner T, Guleva EI (1993) The effect of pretreatment with proline on the response of *Pisum sativum* to salt stress. Photosynthetica 29: 521–527.
25. Flowers TJ, Yeo AR (1995) Breeding for salinity resistance in crop plants: Where next? Aust J Plant Physiol 22: 875–884.
26. Qidar M, Shams M (1997) Some agronomic and physiological aspects of salt tolerance in cotton (*Gossypium hirsutum* L.). Journal of Agronomy and Crop Science 179: 101–106.
27. Guo WX, Mass SJ, Bronson KF (2012) Relationship between cotton yield and soil electrical conductivity, topography, and Landsat imagery. Precision Agronomy 13(2): 678–692.
28. Zhang L, Zhang GW, Wang YH, Zhou ZG, Meng YL, et al. (2013) Effect of soil salinity on physiological characteristics of functional leaves of cotton plants. J Plant Res 126(2): 293–304.
29. Ahmads S, Khan N, Iqbal MZ (2002) Salt tolerance of cotton (*Gossypium hirsutum* L.). Asian Journal of Plant Science 1(6): 715–719.
30. Sestak K, Catsky J, Jarvis PG (1971) Plant photosynthetic production. Dr. W Junk N.V publishers The Hague.
31. Zheng Y, Wang Z, Sun X, Jia A, Jiang G, et al. (2008) Higher salinity tolerance cultivars of winter wheat relieved senescence at reproductive stage. Environ Exp Bot 62(2): 129–138.
32. Ashraf M, Orooj A (2006) Salt stress effects on growth, ion accumulation and seed oil concentration in an arid zone traditional medicinal plant ajwain (*Trachyspermum ammi* [L.] Sprague). J Arid Environ 64(2): 209–220.
33. Liu J, Guo WQ, Shi DC (2010) Seed germination, seedling survival, and physiological response of sunflowers under saline and alkaline conditions. Photosynthetica 48(2): 278–286.
34. Bernt E, Bergmeyer HU (1974) Inorganic peroxidases. In: Bergmeyer HU (ed) methods of enzymatic analysis, vol 4. Academic press, New York: 2246–2248.
35. Foster JG, Hess JL (1980) Responses of superoxide dismutase and glutathione reductase activities in cotton leaf tissue exposed to an atmosphere enriched in oxygen. Plant Physiol 66: 482–487.
36. Beers RF, Sizer IW (1952) A spectrophotometric method of measuring the breakdown of hydrogen peroxide by catalase. J Biol Chem 195: 133–140.
37. Tan W, Liu J, Dai T, Jing Q, Cao W, et al. (2008) Alterations in photosynthesis and antioxidant enzyme activity in winter wheat subjected to post-anthesis waterlogging. Photosynthetica 46: 21–27.
38. Nakano Y, Asada K (1981) Hydrogen peroxide is scavenged by ascorbate-specific peroxidase in spinach chloroplasts. PCPhy 22(5): 867–880.
39. Foyer CH, Halliwell B (1976) The presence of glutathione and glutathione reductase in chloroplasts: a proposed role in ascorbic acid metabolism. Planta 133: 21–25.
40. Takemura T, Hanagata N, Sugihara K, Baba S, Karube I, et al. (2000) Physiological and biochemical responses to salt stress in the mangrove, *Bruguiera gymnorrhiza*. Aquat Bot 68: 15–28.
41. Garratt LC, Janagoudar BS, Lowe KC, Anthony P, Power JB, et al. (2002) Salinity tolerance and antioxidant status in cotton cultures. Free Radical Biol Med 33(4): 502–511.
42. Meloni DA, Oliva MA, Ruiz HA, Martinez CA (2001) Contribution of proline and inorganic solutes to osmotic adjustment in cotton under salt stress. J Plant Nutr 24: 599–612.
43. Parida AK, Das AB (2005) Salt tolerance and salinity effects on plants: a review. Ecotoxicol Environ Saf 60: 324–349.
44. Curtis PS, Läuchli A (1986) The role of the leaf area development and photosynthetic capacity in determining growth of Kenaf under moderate salt stress. Aust J Plant Physiol 18: 553–565.
45. Thiel G, Lynch J, Läuchli A (1988) Short-term effects of salinity stress on the turgor and elongation of growing barley leaves. J Plant Physiol 132: 38–44.
46. Cuartero J, Fernandez-Munoz R (1999) Tomato and salinity. HortScience 78: 83–125.
47. Siroka B, Huttova J, Tamas L, Simonoviva M, Mistrik I (2004) Effect of cadmium on hydrolytic enzymes in maize root and celeopptile. Biologia 59: 513–517.
48. Berkelaar E, Beverley H (2000) The relationship between morphology and cadmium accumulation in seedlings of two durum wheat cultivars. Canadian Journal of Botany 78: 381–387.

49. Singh AK, Dubey RS (1995) Changes in chlorophyll a and b contents and activities of photosystems I and II in rice seedlings induced by NaCl. Photosynthetica 31: 489–499.
50. Munns R, James RA, Lauchli A (2006) Approaches to increasing the salt tolerance of wheat and other cereals. J Exp Bot 57: 1025–1043.
51. Zhang Q, Alfarra MR, Worsnop D, Allan JD, Coe H, et al. (2005) Deconvolution and quantification of hydrocarbon-like and oxygenated organic aerosols based on aerosol mass spectrometry. Environment Science and Technology 39: 4938–4952.
52. Lee MH, Cho EJ, Wi SG, Bae H, Kim JE, et al. (2013) Divergences in morphological changes and antioxidant responses in salt-tolerant and salt-sensitive rice seedlings after salt stress. Plant Physiol Biochem 70: 325–335.
53. de Lacerda CF, Cambraia J, Oliva MA, Ruiz HA, Prisco JT (2003) Solute accumulation and distribution during shoot and leaf development in two sorghum genotypes under salt stress. Environ Exp Bot 49(2): 107–120.
54. Meloni DA, Oliva MA, Martinez CA, Cambraia J (2003) Photosynthesis and activity of superoxide dismutase, peroxidase and glutathione reductase in cotton under salt stress. Environ Exp Bot 49(1): 69–76.
55. Asharf M, Foolad MR (2005) Pre-sowing seed treatment- a shotgun approach to improve generation, plant growth, and crop yield under saline and non-saline conditions. Advances in Agronomy 88: 223–271.
56. Dubey RS (1997) Photosynthesis in plants under stressful conditions. In: Pessarakli, M. (Ed.), Handbook of Photosynthesis. Marcel Dekker, New York,: 859–975.
57. Brugnoli E, Lauteri M (1991) Effects of salinity on stomatal conductance, photosynthetic capacity and capacity and carbon isotope discrimination of salt-tolerant (*Gossypium hirsutum* L.) and salt-sensitive bean (*Phaseilus vulgaris* L.) C3 non-halophytes. Plant Physiol 95: 658–635.
58. Delfine S, Alvino A, Villani MC, Loreto F (1999) Restrictions to carbon dioxide conductance and photosynthesis in spainach leaves recovering from salt stress. Plant Physiol 119: 1101–1106.
59. Parida A, Das AB, Mittra B (2004) Effects of salt on growth, ion accumulation, photosynthesis and leaf anatomy of the mangrove, *Bruguiera parviflora*. Trees: Structure and Function 18: 167–174.
60. Kaya C, Tuna AL, Asharf M, Altunlu H (2007) Improved salt tolerance of molon (*Cucumis melo* L.) by the addition of proline and potassium nitrate. Environ Exp Bot 60: 397–403.
61. Borsani O, Valpuesta V, Botella MA (2003) Developing salt tolerant plants in a new century: a molecular biology approach. Plant Cell Tissue and Organ Culture 73: 101–115.
62. Asgari HR, Cornelis W, Damme PV (2012) Salt stress on wheat (*Triticum aestivum* L.) growth and leaf ion concentrations. International Journal of Plant Production 6: 195–208.
63. Taleisnik E, Grunberg K (1994) Ion balance in tomato cultivars differing in salt tolerance. I. Soidium and potassium accumulation and fluxes under moderate salinity. Plant Physiol 92: 528–534.
64. Graifenberg A, Giustiniani L, Temperini O, Paola ML (1995) Allocation of Na, Cl, K and Ca within plant tissues in globe artichoke (*Cynara scolimus* L.) under saline-sodic conditions. Scientia Horticulturae 63: 1–10.
65. Khan MA (2001) Experimental assessment of salinity tolerance of *Ceriops tagal* seedlings and saplings from the *Indus delta*. Pakistan Journal of Acquatic Botany 70: 259–268.
66. Munns R, Schachtman DP, Condon AG (1995) The significance of a two-phase growth response to salinity in wheat and barely. Aust J Plant Physiol 22: 561–569.
67. Niu X, Bressan RA, Hasegawa PM, Pardo JM (1995) Ion homeostasis in NaCl stress environments. Plant Physiol 109: 735–742.
68. Hernandez JA, Jimenez A, Mullineaux P, Sevilla F (2002) Tolerance of pea (*Pisum sativum* L.) to long-term salt stress is associated with induction of antioxidant defenses. Plant Cell Environ 23: 853–862.
69. Boo YC, Jung J (1999) Water deficit-induced oxidative stress and antioxidative defense in rice plants. Plant Physiol 155: 255–261.
70. Seckin B, Turkan I, Sekmen AH, Ozfidan C (2010) The role of antioxidant defense systems at differential salt tolerance of *Hordeum marinum* Hds. (sea barlygrass) and *Hordeum vulgare* L. (cultivated barley). Environ Exp Bot 69: 76–85.
71. Ruiz-Lozano JM, Azcon R (1997) Effect of calcium application on the tolerance of mycorrhizal lettuce plants to polyethylene glycol-induced water stress. Symbiosis 23: 9–21.
72. Rabie GH, Almadini AM (2005) Role of bioinoculants in development of salt-tolerance of *Vicia faba* plants under salinity stress. Africa Journal of Biotechnology 4: 210–220.
73. Sharma SS, Dietz KJ (2006) The significance of amino acids and amino acid derived molecules in plant responses and adaptation to heavy metal stress. J Exp Bot 57: 711–726.
74. Foyer CH, Lopez-Delgado H, Dat JF, Scott IM (1997) Hydrogen peroxide and glutathione-associated mechanism of acclimatory stress tolerance and signaling. Plant Physiol 100: 241–254.
75. Senadheera P, Tirimanne TLS, Maathuis FJM (2012) Long term salinity stress reveals variety specific differences in root oxidative stress response. Rice Sci 19: 36–43.
76. Lee DH, Kim YS, Lee CB (2001) The inductive responses of the antioxidant enzymes by salt stress in the rice (*Oryza sativa* L.). J Plant Physiol 158(6): 737–745.
77. Queval G, Hager J, Gakiere B, Noctor G (2008) Why are literature data for H_2O_2 contents so variable? A discussion of potential difficulties in the quantitative assay of leaf extracts. J Exp Bot 59: 135–146.
78. Morita S, Kaminaka H, Masumura K, Tanaka K (1999) Induction of rice cytosolic ascorbate peroxidase mRNA by oxidative stress: the involvement of hydrogen peroxide in oxidative stress signaling. Plant Cell Physiol 40: 417–422.
79. Breusegem FV, Vranova E, Dat JF, iNZE D (2001) The role of active oxygen species in plant signal transduction. Plant Sci 161: 405–414.
80. Sekmen AH, Turkan I, Takio S (2007) Differential responses of antioxidative enzymes and lipid peroxidation to salt stress in salt-tolerant Plantago maritima and salt-sensitive Plantago media. Plant Physiol 131: 399–411.
81. Mittler R (2002) Oxidative stress, antioxidants and stress tolerance. Trends in Plant Science 7(9): 405–410.
82. Gill SS, Tuteja N (2010) Reactive oxygen species and antioxidant machinery in abiotic stress tolerance in crop plants. Plant Physiol Biochem 48: 909–930.
83. Henandez JA, Jimenez A, Mullineaux P, Sevilla F (2000) Tolerance of pea to long-term salt stress in associated with induction of antioxidant defenses. Plant Cell Environ 23: 853–862.
84. Laspina NV, Groppa MD, Tomaro ML, Benavides MP (2005) Nitric oxide protects sunflower leaves against Cd-induced oxidative stress. Plant Sci 169: 323–330.
85. Zhang Q, Zhang JZ, Chow WS, Sun LL, Chen JW, et al. (2011) The influence of low temperature on phtosynthesis and antioxidant enzymes in sensitive banana and tolerant plantain (*Musa sp.*) cultivars. Photosynthetica 49: 201–208.
86. Asada K (2006) Production and scavenging of reactive oxygen species in chloroplasts and their functions. Plant Physiol 141: 391–396.
87. Amor NB, Jimenez A, Megdiche W, Lundqvist M, Sevilla M, et al. (2007) Kinetics of the antioxidant response to salinity to the halophyte *Cakile maritime*. J Integrative Plant Biol 126: 446–457.
88. Ben Amor N, Ben Hamed K, Debez A, Grignon C, Abdelly C (2005) Physiological and antioxidant responses of the perennial halophyte *Crithmum maritimum* to salinity. Plant Sci 168(4): 889–899.
89. Xu R, Yamada M, Fujiyama H (2013) Lipid peroxidation and antioxidative enzymes of two turgrass species under salinity stress. Pedosphere 23(2): 213–222.
90. Nakano Y, Asada K (1981) Hydrogen peroxide is scavenged by ascorbate-specific peroxidase in spinsth chloroplasts. PCPhy 22: 867–880.
91. Bower C, Montagu MV, Inze D (1992) Superoxide dismutase and stress tolerance. Annu Rev Plant Physiol Plant Mol Biol 43: 81–116.
92. Demiral T, Turkan I (2005) Comparative lipid peroxidation, antioxidant defense systems and proline content in roots of two rice cultivars differing in salt tolerance. Environ Exp Bot 53: 247–257.
93. Locato V, Gadaleta C, Gara I, Pinto MC (2008) Production of reactive species and modulation of antoxidant network in response to heat shock: a critical balance for cell fate. PCPhy 31: 1606–1619.
94. Liang Y, Chen Q, Liu Q, Zhang W, Ding R (2003) Exogenous silicon (Si) increases antioxidant enzyme activity and reduces lipid peroxidation in roots of salt-stressed barley (*Hordeum vulgare* L.). J Plant Physiol 160: 1157–1164.
95. Sudhakar C, Lakshmi A, Giridarakumar S (2001) Changes in the antioxidant enzyme efficiency in two high yielding genotypes of mulberry (*Morus alba* L.) under NaCl salinity. Plant Sci 161: 613–619.

13

Reduction of the Cytosolic Phosphoglucomutase in Arabidopsis Reveals Impact on Plant Growth, Seed and Root Development, and Carbohydrate Partitioning

Irina Malinova[1,2], Hans-Henning Kunz[3,5], Saleh Alseekh[4], Karoline Herbst[1], Alisdair R. Fernie[4], Markus Gierth[5], Joerg Fettke[1,2]*

1 Plant Physiology, University of Potsdam, Potsdam-Golm, Germany, 2 Biopolymers analytics, University of Potsdam, Potsdam-Golm, Germany, 3 Plant Physiology, Washington State University, Pullman, Washington, United States of America, 4 Max-Planck-Institute of Molecular Plant Physiology, Potsdam-Golm, Germany, 5 Department of Botany II, University of Cologne, Cologne, Germany

Abstract

Phosphoglucomutase (PGM) catalyses the interconversion of glucose 1-phosphate (G1P) and glucose 6-phosphate (G6P) and exists as plastidial (pPGM) and cytosolic (cPGM) isoforms. The plastidial isoform is essential for transitory starch synthesis in chloroplasts of leaves, whereas the cytosolic counterpart is essential for glucose phosphate partitioning and, therefore, for syntheses of sucrose and cell wall components. In Arabidopsis two cytosolic isoforms (PGM2 and PGM3) exist. Both PGM2 and PGM3 are redundant in function as single mutants reveal only small or no alterations compared to wild type with respect to plant primary metabolism. So far, there are no reports of Arabidopsis plants lacking the entire cPGM or total PGM activity, respectively. Therefore, *amiRNA* transgenic plants were generated and used for analyses of various parameters such as growth, development, and starch metabolism. The lack of the entire cPGM activity resulted in a strongly reduced growth revealed by decreased rosette fresh weight, shorter roots, and reduced seed production compared to wild type. By contrast content of starch, sucrose, maltose and cell wall components were significantly increased. The lack of both cPGM and pPGM activities in Arabidopsis resulted in dwarf growth, prematurely die off, and inability to develop a functional inflorescence. The combined results are discussed in comparison to potato, the only described mutant with lack of total PGM activity.

Editor: Miyako Kusano, RIKEN Center for Sustainable Resource Science, Japan

Funding: The authors acknowledge financial support by the Deutsche Forschungsgemeinschaft to J.F. (DFG-Az. FE 1030/1-1 and FE 1030/2-1). H.-H.K. was supported by the Human Frontier Science Program Long-Term fellowship and an Alexander von Humboldt Feodor Lynen fellowship. The funders had no role in study design, data collection and analysis, decision to publish, or preparation of the manuscript.

Competing Interests: Joerg Fettke is still a PLOS ONE Editorial Board member. This does not alter the authors' adherence to PLOS ONE Editorial policies and criteria.

* Email: fettke@uni-potsdam.de

Introduction

Phosphoglucomutase (PGM) catalyzes the reversible interconversion of glucose 6-phosphate (G6P) and glucose 1-phosphate (G1P). In higher plants PGM activity is verifiable in two compartments, the plastidial stroma and the cytosol. The plastidial isoform is essential for the formation of glucose 1-phosphate a substrate of ADPglucose pyrophosphorylase and, therefore, for starch synthesis. Lack of this isoform results in dramatically diminished starch levels [1,2]. Furthermore, mutants lacking the ability to form starch displayed a higher amount of soluble sugars, like glucose and sucrose [3,4]. The latter carbohydrate is the main transport form in higher plants and supplies non-photosynthetic tissues and organs of the plant with energy and carbon. Sucrose is formed in the light from triose-phosphates exported from the chloroplasts. During the formation of sucrose the cytosolic PGM (cPGM) is essential as it converts G6P into G1P, which is the substrate for the UDPglucose pyrophosphorylase.

Also in the dark, when the photosynthetic driven export of carbon from the chloroplast is absent, the formation of sucrose is dependent on cPGM activity [5,6]. Furthermore, this pathway is linked to starch breakdown products. By the action of various enzymes, in most cases hydrolyzing enzymes, the transitory starch is degraded and the major carbohydrates released from the chloroplasts are glucose and maltose [5,7,8]. Starch derived maltose enters the cytosol via maltose exporter 1 (MEX1; [9]) and is further metabolized by disproportionating enzyme 2 (DPE2; [10,11,12]). DPE2 transfers one of the glucosyl residues (the non-reducing) of maltose on cytosolic heteroglycans and releases the second as free glucose. The glucosyl residues of the cytosolic heteroglycans can be released as G1P by the action of the cytosolic phosphorylase (AtPHS2; [13,14]). However, the starch derived glucose is exported from the chloroplast via pGlcT [15,16]. Both the exported glucose and the glucose released by the action of DPE2 are thought to be immediately converted into G6P by the action of hexokinase [5]. The cPGM controls partitioning of both sugar phosphates in the cytosol. G6P is used primarily in respiratory pathways, whereas G1P is linked to sucrose metabolism and in addition to cell wall synthesis. *Arabidopsis thaliana*, tobacco and maize contain one plastidial and two cytosolic isoforms; for potato and spinach only one plastidial and one cytosolic isoform were reported [17,18,19,20,21]. Recently, potato plants with antisense repression of cytosolic phosphoglucomutase were analyzed. These plants displayed a stunted phenotype, diminished root growth and reduced tuber yield [20]. Antisense plants were also characterized by reduced rates of photosynthesis and dramatic reduction in nucleotide level compared to the wild type [22]. Moreover, transgenic lines with altered cPGM activity revealed alterations in starch-related cytosolic heteroglycans. From these results it was concluded that elevated levels of cPGM activity favor the cytosolic phosphorylase-mediated conversion of glucosyl residues from the cytosolic heteroglycans into the cytosolic hexose-phosphate pools during starch degradation [23].

The two genes encoding cytosolic phosphoglucomutase activities in *Arabidopsis thaliana* At1g23190 (PGM 3) and At1g70730 (PGM2) [24,17] reveal high sequence homology as well as possess similar exon/intron structures. Indeed, they encode two isoforms with 91% sequence identity at the amino acid level. Egli *et al.* [24] reported that *pgm2* and *pgm3* mutants deficient in one of the cytosolic isoforms grown under standard 12 h light/12 h dark regime displayed phenotypes similar to that of wild type. The authors suggested that under these conditions the functions of the isoforms were redundant to one another and the loss of one isoform did not affect plant metabolism. Unfortunately, the generation of double mutants was unsuccessful, as formation of homozygous seeds was prevented. Therefore, it was concluded that an absolute lack of cPGM activity compromises gametophyte development [24].

Not so long ago, transgenic potato lines with strongly decreased total PGM activities were identified. Transgenic plants were reduced in growth, tuber yield, and revealed lower levels of starch and sucrose in leaves compared to wild type [25]. Interestingly, rate of starch synthesis was similar to the wild type [26]. A possible explanation for this phenotype is a direct G1P transport over the plastidial membranes, which has been verified for both potato and Arabidopsis [27,1].

However, until now no *A. thaliana* transgenic plants with a strong reduction of both cPGM isoforms or the simultaneous reduction of plastidial and cytosolic phosphoglucomutases have been reported. For this reason, we generated and analyzed Arabidopsis lines with *amiRNA* (artificial micro RNA) repression of both cPGMs. Furthermore, the cPGM *amiRNA* construct was introduced into *pgm1* mutants by Agrobacterium mediated transformation to explore whether a similar bypass to that observed in potato also occurred in Arabidopsis. In order to test this, the generated plants were assessed at the level of isoform specific activity as well as carbohydrate and metabolite content and phenotypic characterization of vegetative growth and propagative development. Results are discussed in the context of current understanding of the importance of the reactions catalyzed by phosphoglucomutase.

Materials and Methods

Plant material and growth conditions

The *pgm1* mutants were as described in [17]. The *pgm2* [SALK_068481 (AR)] and *pgm3* [SALK_023069 (AZ)] mutants were ordered from NASC. Mutants were identified by PCR amplification using the primers presented in Table S1 in File S1.

pgm2 pgm1 and *pgm3 pgm1* were generated by crosses between individual homozygous mutants and the resulting F1 generations were allowed to self-pollinate. Double mutants were identified in the F2 generation by native PAGE and PGM activity staining (see below). For further analyses plants of F3 or F4 generation were used. Plants were grown either in 14, 10, 8 or 7 h light (110 μmol $m^{-2}s^{-1}$, 22°C; dark, 18°C, humidity 60%) or in a 12 h diurnal cycle (12 h light [110 μmol $m^{-2}s^{-1}$], 20°C; 12 h dark, 16°C, humidity 60%). For all Arabidopsis lines used, the genetic background was Col-0.

Generation of *amiRNA* c-pgm plants

The PGM2/3-specific *amiRNA* (tctgttaagataaatgcgcct) was designed and amplified by three consecutive PCR reactions according to guidelines found at http://wmd3.weigelworld.org using the vector pRS300 as template [28] (Table S1 in File S1). The final PCR product including the cPGM-specific *amiRNA* was subcloned into the pENTR/D-TOPO vector and sequence identity was verified. Subsequently, the *amiRNA* was recombined by L/R reaction into pGWB2 [29] to obtain the binary expression plasmid *p35S:amiRNA cPGM*. The binary vector was transformed into *Argobacterium tumefaciens* strain GV3101 and used for plant transformation.

Plant transformation was performed using the floral dip method [30]. Agrobacterium strains were grown in 1 L of LB medium containing antibiotics rifampicin (100 mg/L), kanamycin (50 mg/L), gentamycin (25 mg/L), hygromycin (50 mg/L) at 28°C for 24 h. Cells were collected by centrifugation at 4,000 ***g*** for 15 min at room temperature (RT) and gently resuspended in 1 L of freshly made 5% [w/v] sucrose solution containing 0.02% [v/v] Silwet L-

Table 1. Carbohydrate content.

carbohydrate	growth photoperiod [h light]	Col-0	*pgm3*	*pgm2*
starch [mg/g FW]	7 h	6.1±0.3	5.4±0.6	6.1±0.3
	8 h	9.2±0.4	7.7±0.3	9.4±0.2
	10 h	6.9±0.4	5.0±0.1	6.1±0.2
	14 h	7.0±0.7	6.7±0.6	6.4±0.6
sucrose content [μmol/g FW]	7 h	1.55±0.11	1.45±0.03	1.81±0.03*
	8 h	1.17±0.04	1.04±0.05	1.87±0.33*
	10 h	1.86±0.08	1.96±0.09	2.71±0.05*
	14 h	2.58±0.19	2.46±0.21	2.90±0.03**

Leaves were harvested one hour before beginning of the dark phase. Values are means of four replicates representing a mix of 7–10 plants ± SD. Asterisks denote the significance levels as comparing mutants to Co1-0 : * $p \leq 0.01$;** $p \leq 0.05$.

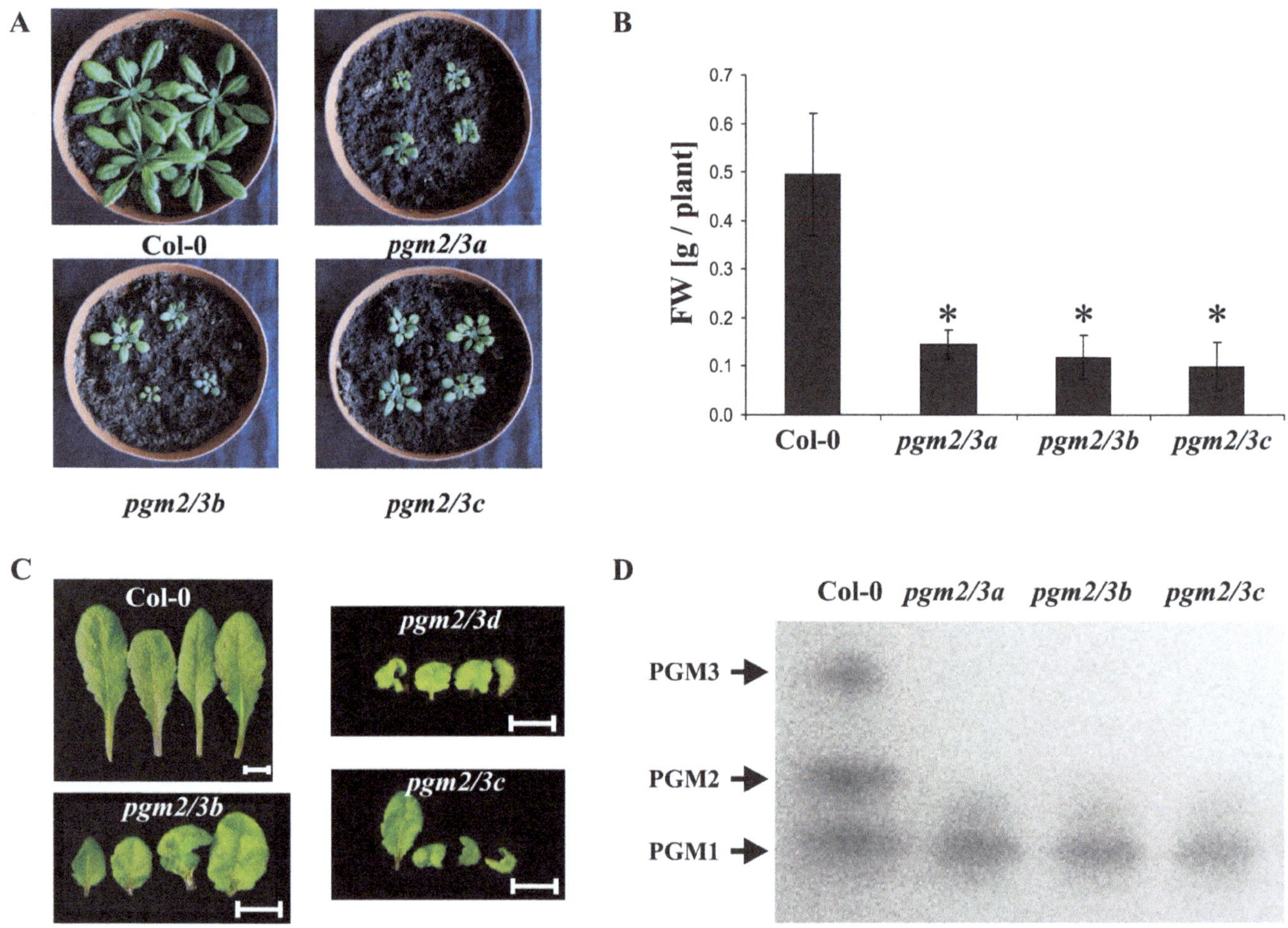

Figure 1. Phenotype of Col-0 and *pgm2/3* plants in 12 h light/12 h dark regime. A, Growth phenotypes. Photographs were taken of six-week-old plants. Bar = 1 cm. B, Fresh weight of plant rosettes. Values are means ± SD (n = 29–30). Plants were harvested after five weeks. Asterisks indicate significant difference from Col-0 (Student Test, P≤0.01). C, Leaf form from Col-0 and transgenic plants. Leaves were harvested from the middle of rosettes from six-week-old plants. Bar = 1 cm. D, Phosphoglucomutase activity in Col-0 and *pgm2/3* plants. Crude extracts were subjected to native PAGE and subsequent PGM activity staining. Separation gel was 7.5% [T] and 25 µg protein was loaded per lane.

77 (Lehle Seeds, USA). Col-0 and *pgm1* plants (approximately four to five weeks after germination) were used for transformation. On reaching the mature stage plants were transferred to a 14 h light/10 h dark regime until mature silique stage.

Screening of *amiRNA* plants

Dry seeds from transformed plants were collected and sterilized. Seeds were immersed in 70% [v/v] ethanol for 5 min, followed by a 20 min soaking in 2.4% [w/v] sodium hypochlorite, 0.02% [v/v] Triton X-100. Seeds were rinsed six times with sterile water and dried under sterile conditions. Seeds were screened on MS-plates with sucrose (4.3 g/L MS salt (Duchefa, Haarlem, Netherlands), 2.5 mM MES, pH 5.7 (NaOH), 1% [w/v] sucrose, 0.8% [w/v] Agar-agar) except where indicated. Selective antibiotics were added: hygromycin (50 mg/L), kanamycin (50 mg/L). Plates were placed in growth chambers and plants were germinated under 12 h light/12 h dark, except otherwise stated. Transformants with well developed leaves (four leaves stage) and roots were planted in soil and grown under standard conditions (12 h light/12 h dark). Seeds of at least four plants were harvested separately and used for generation of four plant lines (*pgm2/3 a* to *d*). Analyses were performed with the F3 to F5 generation of the respective lines.

Phosphoglucomutase assay and PGM activity staining

Buffer-soluble proteins were extracted as described elsewhere [12]. Phosphoglucomutase activity measurement was performed as described [23]. However, in the reaction mixture soluble starch and rabbit muscle phosphorylase were omitted. Measurement was started by addition of 17.5 mM G1P to the reaction mixture. Native PAGE and PGM activity staining were performed according to Fettke *et al.* [23].

Carbohydrate quantification

Starch was extracted and measured as described [1]. Monosaccharides, disaccharides and sugar phosphates were determined according to Stitt *et al.* [31].

Isolation and analysis of cell wall matrix polysaccharides

Leaf material, frozen in liquid nitrogen, was homogenized and resuspended in ice-cold 20% [v/v] ethanol, mixed thoroughly, and centrifuged for 10 min at 20,000 ***g*** (4°C). Pellets were washed with 20% [v/v] ethanol two times, finally resuspended in 70% [v/v] ethanol and centrifuged (as above). Subsequently, pellets were resuspended in chloroform/methanol (1:1 [v/v]) and incubated for 20 min under continuous stirring followed by centrifugation (as

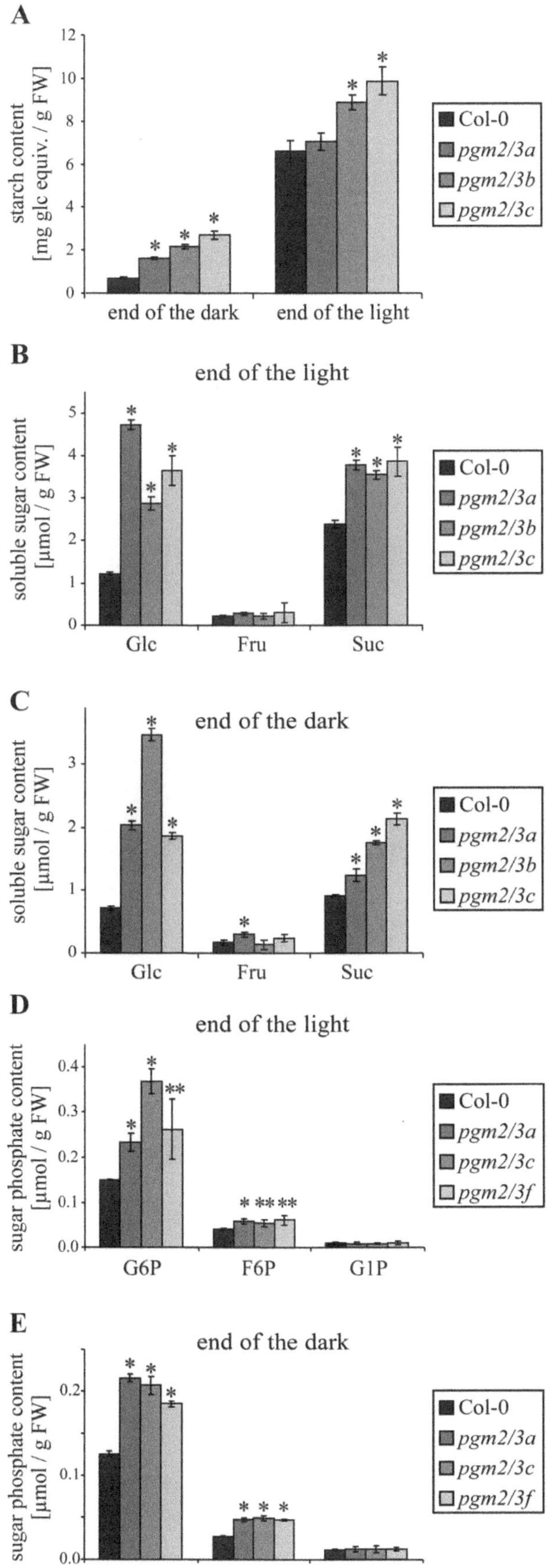

Figure 2. Carbohydrate analysis of Col-0 and *pgm2/3* plants. A–E, Plants were grown under 12 h light/12 h dark conditions and after five weeks 7–8 plants were collected and homogenized per line. Values are means of four technical replicates (A–C), and three technical parallels (D–E) ± SD, respectively. A, Starch content. B–C, Soluble sugar content. D–E, Sugar phosphate content. Asterisks denote the significance levels comparing *pgm2/3* mutants to Co1-0: * $p \leq 0.01$;** $p \leq 0.05$.

above). The resulting pellets were completely destained by washing with acetone followed by water. Then pellets were resolved in 0.1 M sodium acetate buffer (pH 5.0) and incubated for 20 min at 80°C. The suspension was cooled to RT and residual starch was removed by treatment with 25 U of α-amylase (from *Basillus sp*. Typ II-A, Sigma-Aldrich, Germany) and 7 U pullulanase (from *Klebsiella planticola*, Macerozyme, Ireland) as described elsewhere [32]. The residual pellet was washed at least five times with water and subjected to TFA hydrolysis (2 M final concentration) for 3 h at 100°C. After that samples were centrifuged and the supernatants were collected. Pellets were washed two times with water and supernatants pooled together. Collected supernatant represents matrix polysaccharides of the cell wall. Following lyophilization, samples were dissolved in water and monomer content was estimated [33] (glucose was used as a standard). Aliquots were subjected to HPAEC-PAD for monosaccharide separation (as described elsewhere [12]).

Isolation and quantification of crystalline cellulose

Residual pellets from cell wall matrix isolation were subjected to hydrolysis in Updegraff reagent (8:1:2 of concentrated acetic acid:concentrated nitric acid:water) [34] for 30 min at 100°C. Crystalline cellulose was separated, completely hydrolyzed into glucose, and determined as described elsewhere [35].

Metabolic Profiling

For GC-MS analyses, Col-0 and transgenic lines were grown in 12 h light/12 h dark regime and harvested at the end of the light and at the end of the dark. Plants were five-week-old. Leaves from several plants per line were pooled together and processed as previously described [36].

Trypan blue staining

Trypan blue (Sigma-Aldrich, Germany) staining was performed as described [37]. Leaves were boiled 1 min at 100°C with lactophenol-trypan blue solution (10 mL lactic acid, 10 mL glycerol, 10 g phenol, 10 mL 0.1% [w/v] trypan blue solution) and decolorized with chloral hydrate (2.5 g mL^{-1} distilled water) overnight.

Statistical analysis

Statistical analysis (Student's t-test [two-sided]) was performed using MS Excel 2010 (Microsoft Corporation, Washington, USA).

Results

Elimination of one cPGM isoform in Arabidopsis has no significant effect on starch metabolism

In native PAGE the total PGM activity was resolved in three distinct bands of activity, the fastest moving band represented the plastidial PGM (PGM1), whereas the slowest moving band represented PGM3 (At1g23190) and the intermediate band PGM2 (At1g70730). Both PGM2 and PGM3 are cytosolic isoforms [23,24]. The localization of the three isoforms was further confirmed by non-aqueous fractionation [38]. All three

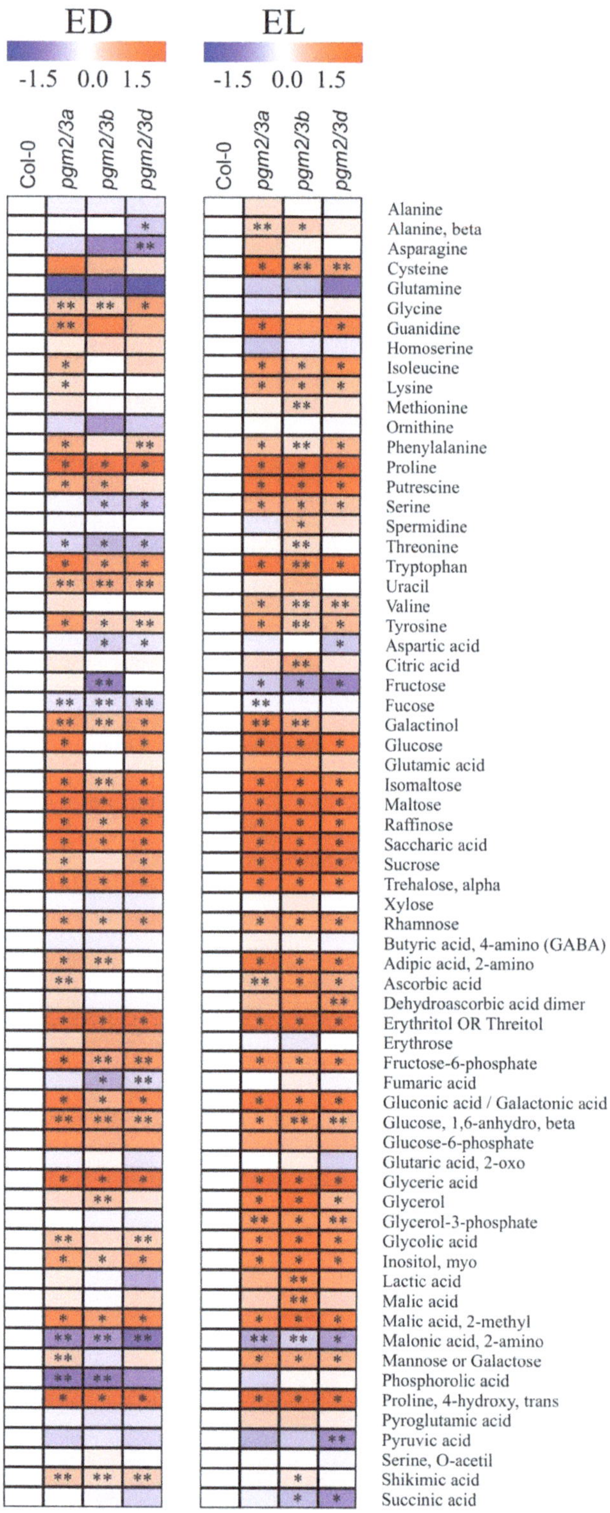

Figure 3. Overlay heat map of the metabolite changes in *pgm2/3* mutants in comparison with control (Co1-0) using false-color scale. Red or blue indicate that the metabolite content is increased or decreased, respectively. Five-week-old plants were grown under 12 h light/12 h dark conditions and harvested at the end of light phase (EL) and dark phase (ED), and three replicates represented 3–4 plants were analyzed (two technical replicates each); asterisks denote the significance levels as comparing *pgm2/3* mutants to Co1-0 : * $p \leq 0.01$;** $p \leq 0.05$.

isoforms were detected in various organs (Fig. S1A in File S1). PGM activity was analyzed in leaves of different Arabidopsis accessions (Fig. S1B in File S1). Results indicate a wide diversity of cytosolic PGM isoforms. Consistent with previously published data [24], Cvi-0 was the single accession which displayed only one cytosolic isoform.

Two mutants lacking an isoform of cytosolic PGM (*pgm2, pgm3*) were previously analyzed [24]. No substantial differences compared to the wild type were observed even when various parameters like starch and soluble sugar content as well as root and shoot growth were examined. However, we here generated independent homozygous T-DNA mutant lines. The total reduction in PGM activity was determined to be 23% in *pgm3* plants and 35% in *pgm2* plants compared to control Col-0. These results were consistent with the PGM activity staining analysis (Fig. S1B in File S1), since the PGM2 band had a higher intensity than PGM3.

Additionally, PGM2 and PGM3 proteins from *A. thaliana* have previously been cloned and expressed in *Escherichia coli* and the recombinant proteins were analyzed for substrate specificity and affinity. However, no differences between PGM2 and PGM3 were observed [39].

In order to analyze the influence of different growth conditions on *pgm2* and *pgm3* mutants, plants were cultivated under various light/dark conditions (light phase: 7 h, 8 h, 10 h or 14 h). Still both mutants revealed a similar growth phenotype (data not shown) and starch content compared to the Col-0. The *pgm2* plants displayed an increased level of sucrose under different growth conditions but this was not observed for *pgm3* (Table 1). Most likely, PGM2 has a higher impact on glucose-phosphate turnover. However, no significant differences in steady-state levels of sugar phosphate contents (F6P, G1P, G6P) were observed (data not shown).

As the cytosolic pools of sugar phosphates are linked to starch metabolism via the action of two transglucosidases (DPE2 and AtPHS2), the activity of both enzymes and the composition of soluble heteroglycans (SHG_L) were analyzed. However, neither differences in enzyme activities nor composition of SHG_L were observed (Fig. S2 in File S1).

Thus, it seems likely that PGM2 and PGM3 could substitute for one another since the residual PGM activity in either mutant is relatively abundant.

Simultaneous reduction of PGM2 and PGM3 activities affect plants growth and carbohydrate partitioning

Given that single *pgm2* and *pgm3* mutants do not reveal significant changes in e.g. starch metabolism, generation of double mutants is essential to clarify the role of cPGM for plant metabolism. An *amiRNA* cPGM construct was therefore transformed into Col-0 plants and four independent lines were generated. Transgenic *pgm2/3* lines were strongly retarded in growth and revealed diminished fresh weight compared to Col-0 (Fig. 1A–B). Additionally, *pgm2/3* leaves revealed small and abnormally curled leaves (Fig. 1C) and slightly elevated chlorophyll levels (Table S2 in File S1). Protein crude extracts of Col-0 and *pgm2/3* leaves were subjected to native PAGE and PGM activity staining (Fig. 1D). In all *pgm2/3* lines the two bands of cPGM activity were below the limit of detection (cPGM activity was not observed, even if 75 μg of protein crude extracts were loaded on the gel; data not shown). In addition, PGM activities in protein crude extracts were measured (Fig. S3A in File S1). In all three transgenic lines a strong reduction in total PGM activity was observed (residual activity 30–34%, [wt = 100%]). Furthermore, analyses of gene expression revealed that PGM2 and PGM3 were

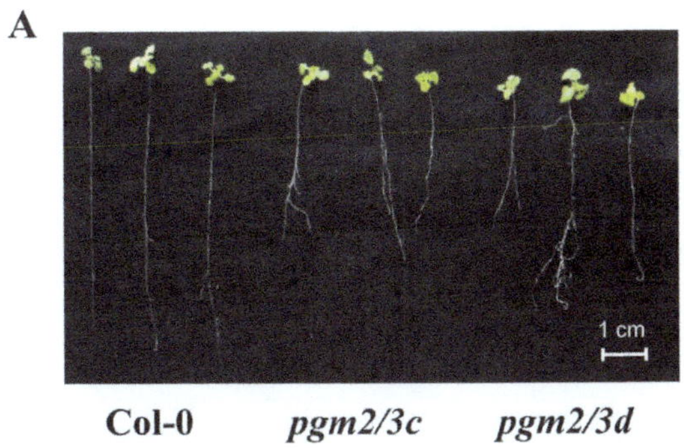

genotype	length of roots [mm]
Col-0	70 ± 11
pgm2/3c	37 ± 9*
pgm2/3d	51 ± 10*

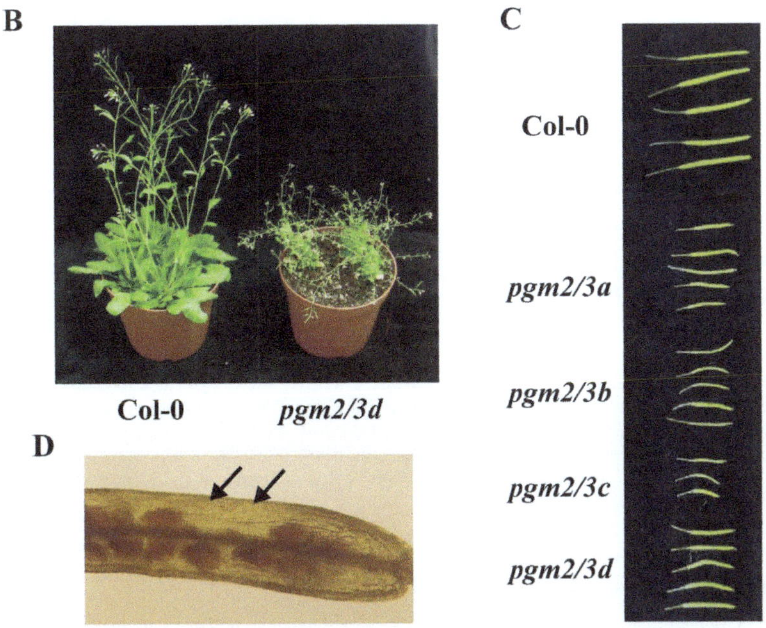

Figure 4. Roots and siliques of Col-0 and *pgm2/3* plants. A, Root length and morphology of Col-0 and *pgm2/3* lines. Plants were grown on vertical MS plates without any external sugar and antibiotics under long day conditions (16 h light/8 h dark). Plants were two-week-old. Length of central roots was measured. Values are means ± SD (n = 26−35). Asterisks indicate significant difference from Col-0 (Student Test, P≤0.01) B, Mature Col-0 and *pgm2/3d* plants. Col-0 and *pgm2/3* plants were six and 11- week-old, respectively. C, Morphology of siliques of Col-0 and *pgm2/3* lines. D, *pgm2/3d* silique. Siliques were destained in chloral hydrate solution (2.5 g in 1 mL distilled water). Black arrows indicate absence of seeds. C–D, Plants were grown under 14 h light/10 h dark regime.

strongly down-regulated in *pgm2/3* lines. In contrast PGM1 was somewhat up-regulated (Fig. S3B in File S1).

However, transgenic *pgm2/3* plants grown under prolonged day conditions (14 h light/10 h dark) revealed similar results with transgenic plants being significantly smaller than Col-0, but larger as compared to the 12 h light/12 h dark grown plants (Fig. S3C in File S1).

With respect to metabolites all *pgm2/3* lines showed increased starch content at the end of the dark phase compared to Col-0 (Fig. 2A). The increased starch content was also detected at the end of the light phase except for *pgm2/3a*. Similarly, starch content was significantly increased in *pgm2/3* lines compared to Col-0 when grown in 14 h light/10 h dark regime (data not shown). Transgenic *pgm2/3* lines displayed increased levels of glucose and sucrose on a fresh weight basis. In contrast the amount of fructose was comparable in the transgenic lines and Col-0 (Fig. 2B–C). Similar results were also obtained, if metabolite content was evaluated on a dry weight basis (data not shown).

Given that PGMs catalyze the interconversion of G1P and G6P, levels of sugar phosphates were determined. The *pgm2/3* plants displayed increased levels of G6P and fructose 6-phosphate (F6P) but G1P levels were similar to those in Col-0 (Fig. 2D–E). Nevertheless, further enzymes involved in the metabolism (DPE2 and phosphorylases) were not affected (Fig. S3D in File S1). In addition metabolic profiling was performed, revealing that numerous metabolites were increased both at the end of light and dark phase. At the end of the light period clear increases were seen in a range of sugars including maltose, glucose, trehalose, isomaltose and raffinose as well as the sugar alcohols galactinol, inositol and erythritol or threitol but fructose was unchanged or even decreased. Similarly, a large number of amino and organic acids were increased in the transgenic lines including tryptophan, proline, galacturonic acid, malate and shikimate (Fig. 3, Table S3 in File S1). By contrast, relatively few metabolites were consistently decreased in the transgenic lines at this time point those that were included were ornithine, phosphoric acid, asparagine, glutamine, and malonate. Consistent with these global effects on the primary

Table 2. Amount of crystalline cellulose and of cell wall matrix in Col-0 and *pgm2/3*.

genotype	crystalline cellulose [mg/g FW]	cell wall matrix [mg/g FW]
Col-0	5.17±0.42	4.73±0.01
pgm2/3a	6.24±0.11*	7.42±0.85*
pgm2/3b	5.80±0.06**	6.28±0.33*
pgm2/3c	5.43±0.24	6.63±0.58*

Plants were grown under 12 h light/12 h dark regime and harvested at the end of the light phase (six-week-old). Values are means of four replicates representing a mix of 7–10 plants ± SD. Asterisks denote the significance levels as comparing *pgm2/3* mutants to Co1-0 : * $p \le 0.01$;** $p \le 0.05$.

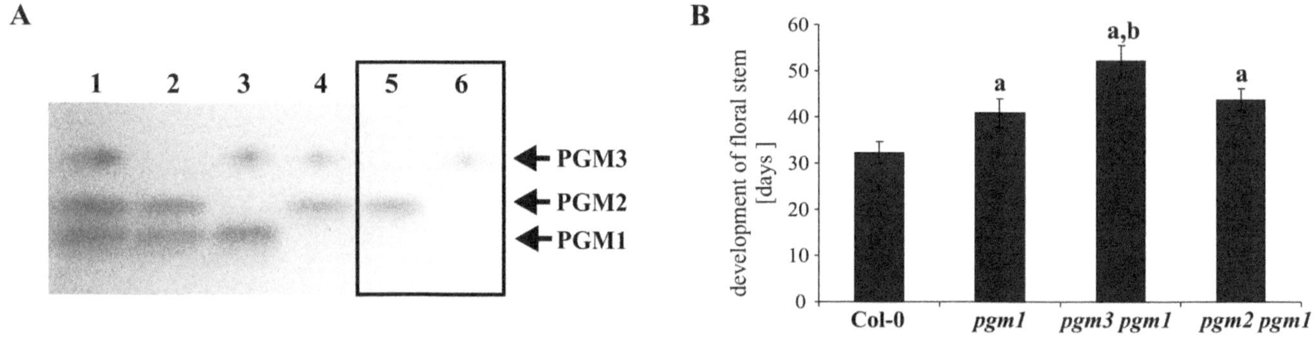

Figure 5. Characterization of knock-out mutants lacking one cytosolic and the plastidial PGM. A, Analysis of PGM activity in the Col-0 and *pgm3 pgm1* and *pgm2 pgm1* mutants using native PAGE and PGM activity staining. Separation gel 7.5% [T]. 35 µg proteins were loaded per lane. 1- Col-0, 2- *pgm3*, 3- *pgm2*, 4- *pgm1*, 5- *pgm3 pgm1*, 6- *pgm2 pgm1*. B, Analysis of floral stems development in Col-0 and different PGM knock-out plants. Plants were grown under long day conditions (14 h light/10 h dark). Days after germination were registered, when plants developed floral stems 1 cm long. Values are means ± SD (n = 24). a - significant difference from Col-0 (Student Test, $p \leq 0.01$), b - significant difference from *pgm1* (Student Test, $p \leq 0.01$).

metabolome being strongly influenced by the sugar status and more specifically by a likely inhibition of sucrose export, they became considerably stronger and more consistent by the end of the night. At this time point all three transgenic lines display alterations including maltose, glucose, trehalose, isomaltose, raffinose, galactinol, inositol, and erythritol or threitol, fructose 6-phosphates, tryptophan, proline, galacturonic acid, malate, and shikimate, which were also elevated in the day. Additionally, the levels of amino adipic acid, guanadine, glutamate, glycolate, lactate, and the branched chain amino acid increased in the dark. As for the situation observed in the light this is most likely the result of inhibition of sucrose export from the leaves. By contrast, at the end of the night the levels of malonate, pyruvate, glutamine and to a lesser extent succinate were significantly decreased in the transgenic lines. The exact reasons underlying these decreases are, however, unclear from the current study.

As G1P is strictly connected with formation of UDP-glucose in the cytosol, which acts as a major substrate for synthesis of cell wall constituents [40], crystalline cellulose and matrix component were analyzed. The *pgm2/3* lines displayed increased amounts of cell wall matrix components and in two of the lines the crystalline cellulose amount was altered (Table 2). Additionally, samples of cell wall matrix were hydrolyzed and the monomer composition was analyzed using HPAEC-PAD. The transgenic lines were characterized by an increased amount of all analyzed monosaccharides and changes in the arabinose/galactose ratio in comparison to Col-0 (Fig. S3E in File S1). For analyses of the impact of cPGM on roots Col-0 and two *pgm2/3* lines were grown on vertical MS plates. *amiRNA pgm2/3* plants carry antibiotic resistance markers, kanamycin and hygromicin. However, it was reported that hygromycin is toxic even to resistant plants during long exposure, which may cause their abnormal development [41]. Indeed, when *pgm2/3* plants were grown in the presence of antibiotics, roots of *pgm2/3* transgenic lines were much shorter and more branched as compared to Col-0 cultivated without antibiotics (data not shown). To avoid such effects, Col-0 and *pgm2/3* seeds were sown on vertical MS plates without antibiotics. After two weeks plants were gently removed from plates and the length of main root was measured (Fig. 4A). Additionally, the lack of cytosolic PGM activity was confirmed in these plants using native PAGE. The root length of transgenic plants was increased on plates without antibiotics (compared to MS plates containing antibiotics), which confirmed that the antibiotics might affect the root growth of the transgenic plants. However, even without antibiotics the root length of transgenic plants was significantly decreased in comparison to Col-0 (Fig. 4A).

Furthermore, it was observed that *pgm2/3* lines were delayed in silique development, as compared to Col-0, independent of growth conditions (short day, long day) (Fig. 4B). The *pgm2/3* transgenic lines develop mature siliques approximately after 10–11 weeks under long day conditions (14 h light/10 h dark regime), whereas Col-0 achieves this after five to six weeks. Siliques from *pgm2/3* lines are much smaller (Fig. 4C) and possess a lower number of seeds compared to Col-0 (data not shown). In addition missing seeds were observed in the siliques of the transgenics (Fig. 4D).

Impact of simultaneous reduction of cytosolic and plastidial phosphoglucomutase activities on Arabidopsis plants

Action of the plastidial phosphoglucomutase (PGM1) is an essential step in starch synthesis. Arabidopsis mutants lacking PGM1 are strongly reduced in starch content [1,2]. In order to analyze the influence of single PGM2 or PGM3 mutation in the *pgm1* background, *pgm2* and *pgm3* mutants were crossed with *pgm1*. Both *pgm2 pgm1* and *pgm3 pgm1* are similar in growth compared to *pgm1*, under long day conditions (Fig. S4 in File S1). Crude extracts from double mutants were subjected to native PAGE and PGM activity staining (Fig. 5A). Both double mutants possess one band of cPGM activity each. Total PGM activity was reduced to 38±2% for *pgm3 pgm1* mutants and 36±2% for *pgm2 pgm1* plants (wt = 100%; n = 3).

Both double mutants possess very low yet still detectable amounts of starch (Table 3). *pgm3 pgm1* mutants revealed an elevated starch amount both in the light and in the dark compared to *pgm1*. However, when plants were grown under 12 h light/12 h dark or 16 h light/8 h dark, these results were not reproduced, as starch content was similar in *pgm1* and both double mutants under these photoperiod regimes (data not shown).

Furthermore, *pgm1* and both double mutants displayed elevated levels of soluble sugar compared to Col-0 (Table 3). Additionally, it was consistently observed that the double knock-out mutants flowered significantly later compared to Col-0 (data not shown). Therefore, floral stem development was investigated. *pgm1* mutants were delayed in floral stem development compared to Col-0, which is consistent with a previous report [42]. The *pgm2 pgm1* mutant displayed a floral stem development time similar to

Figure 6. Growth phenotype of *cp-pgm* plants. A, Seeds were sowed on MS medium containing sucrose and antibiotics (kanamycin [50 μg/mL], hygromycin [50 μg/mL]). Plants were grown under long day conditions (16 h light/8 h dark) and were two-week-old. Bar = 1 cm. B, *cp-pgm* plant before trypan blue staining. C, Col-0 and *cp-pgm* plants after trypan blue staining. The *cp-pgm* plant was five- week-old, germinated on MS plate (as above) and the two last weeks grown under continuous illumination. Leave of Col-0 from three-week-old plant grown under 12 h light/12 h dark conditions. Bars = 1 cm. D–F, Phenotype of cp-*pgm* plants under continuous illumination. Seeds were germinated on MS medium containing sucrose with antibiotics (kanamycin [50 μg/mL], hygromycin [50 μg/mL]). After four weeks plants were transferred to soil and grown further under continuous illumination. D, Plant was six-week-old. Bar = 1 cm. E–F, Flower buds of *cp-pgm* transgenic plants. Plant was six-week-old (E) and seven-week-old (F). Bars = 1 mm.

that of *pgm1*, by contrast *pgm3 pgm1* plants were significantly delayed (Fig. 5B). Although, *pgm1*, *pgm2 pgm1*, and *pgm3 pgm1* plants contained very low amounts of starch, they were not strongly compromised in growth under long day conditions and were able to develop normal flowers and seeds. By contrast, plants with reduced cPGM activity are strongly diminished in growth and seed development (Fig. 4). Therefore, transgenic Arabidopsis lines with a substantial reduction of total PGM were generated by introducing the cPGM *amiRNA* construct into *pgm1* mutants by Agrobacterium mediated transformation (*cp-pgm* plants). Seeds were germinated on MS medium supplemented with sucrose and antibiotics and transformants with well developed leaves and roots were identified (Fig. 6A). It was noted that sucrose is essential for *cp-pgm* seed germination, as seeds sown on sucrose-free MS medium with appropriate antibiotics were not able to germinate.

In order to prove that the transgenic lines are strongly reduced in total PGM activity, protein crude extracts were subjected to native PAGE and PGM activity staining. The *cp-pgm* plants did not display any residual PGM activity (Fig. S5 in File S1). As a control the same crude extracts were used for phosphorylase activity staining, revealing activities comparable to Col-0 for both the cytosolic and plastidial phosphorylase isoforms (data not shown). After approximately three weeks *cp-pgm* plants were transferred to soil at different light/dark conditions: 12 h light/12 h dark, 14 h light/10 h dark and continuous illumination. Independent of growth conditions, plants were very tiny and

Table 3. Starch and soluble sugar content in Col-0 and PGM knock-out mutants.

genotype	starch content [mg glc equiv./g FW]		soluble sugars content (7 h in the light) [μmol/g FW]		
	7 h in the light	3.5 h in the dark	glucose	fructose	sucrose
Col-0	2.930±0.303	3.738±0.196	1.03±0.20	0.28±0.03	1.88±0.28
pgm1	0.012±0.003	0.010±0.001	4.23±0.65	1.04±0.21	2.69±0.11
pgm3 pgm1	0.025±0.005*	0.023±0.004*	4.91±0.59	0.94±0.04	2.70±0.17
pgm2 pgm1	0.015±0.003	0.016±0.003	4.67±0.51	0.87±0.11	2.74±0.31

Plants were grown under long day conditions (14 h light/10 h dark). Plants were five-week-old. Values are means of three biological replicates (two technical replicates each) ± SD. Asterisks indicate values significantly different from *pgm1* and *pgm2 pgm1* (Student Test, $p \leq 0.05$).

rapidly became chlorotic and dry (Fig. 6B). However, under prolonged light conditions and continuous illuminations plants stayed green longer. Nevertheless, trypan blue which selectively stains dead tissue revealed that the plants are not longer vital (Fig. 6C; [37]). That said, some transgenic *cp-pgm* plants were even able to develop normal looking flowering buds under continuous illumination (Fig. 6D–E), but further development of flowers failed as buds shriveled within one week (Fig. 6F). Even if plants were supplied for the entire growth period with exogenous sugars (MS medium+sucrose) they failed to grow to maturity (data not shown). Thus, significant reduction of total PGM activity leads to a dramatic dwarf phenotype and inability to develop functional flowers and seeds. Therefore, *cp-pgm* plants showed a more severe phenotype compared with transgenic potato plants reduced in total PGM activity [24]. Additionally, the phenotype exhibited by the lack of total PGM activity was corroborated by crossing *pgm2/3d* with *pgm1* (named *pgm2/3d pgm1* plants) which displayed the same phenotype as *cp-pgm* plants (data not shown). Despite of the tiny amount of available leaf material, initial analysis of the starch content in *pgm2/3d pgm1* was performed revealing that *pgm2/3d pgm1* plants possess very low amounts of starch (0.21±0.02 μmol glc. equiv./g FW), similar to *pgm1* (0.25±0.06 μmol glc. equiv./g FW) at the middle of the day.

Discussion

Analyses of single knock-out mutants of both cytosolic phosphoglucomutase isoforms (*pgm2* and *pgm3*) confirmed that the isoforms are redundant and expressed at a sufficient abundance to substitute for one another. Even the additional lack of PGM1 reveals only small alterations in metabolism and development in comparison to *pgm1* (Table 3, Fig. S4 in File S1). Furthermore, investigations with purified recombinant Arabidopsis enzymes, reveal that the kinetic properties of both cytosolic isoforms are very similar (for example the K_M's, using G1P as substrate, are PGM3 3.7±0.5 mM; PGM2 4.2±0.4 mM; [39]). The possible difference in substrate specificity observed for phosphoglucomutases of *Pseudomonas aeruginosa* [43] or *Giardia lamblia* [44], which show the additional interconversion of mannose 1-phosphate to mannose 6-phosphate, was not observed here. In competition experiments, where recombinant PGM2 or PGM3 were used with different amounts of mannose 1-phosphate in the presence of glucose 1-phosphate, no alteration in glucose 6-phosphate formation were observed. Furthermore, no formation of mannose 6-phosphate was detectable using HPAEC-PAD [39].

In contrast the *pgm2/3* lines reveal a very considerable phenotype. Even when the cytosolic phosphoglucomutase activity was below the detection limit, there was still a slight residual expression of both cytosolic isoforms (Fig. S3B in File S1). This is likely the reason for the severe yet not lethal phenotype. Thus, also the formation of seeds, albeit reduced or in some cases completely inhibited, could be explained and is in agreement with previous reports [24]. Furthermore, *pgm2/3* reveals alterations in cell wall composition, which were not previously detected in transgenic potato plants with strong reduction of cPGM [22].

Surprisingly in the *pgm2/3* lines a strong increase in sucrose, as well as the starch breakdown derived maltose, was observed. A significant increase in sucrose was additionally detected in the single knock-out line *pgm2* (Table 1). The formation of sucrose in the light is dependent on cPGM activity, as G1P is essential for the formation of UDPglucose via both routes of sucrose synthesis. However, several pathways for formation of G1P and thereby sucrose remain in the cytosol of *pgm2/3* plants: (i), the conversion via the mentioned residual cPGM activity in the plants, (ii), the formation of G1P in the night by the pathway of starch derived maltose, disproportionating enzyme 2, cytosolic heteroglycans, and the cytosolic phosphorylase [12,13], (iii), the direct transport of G1P from the chloroplasts into the cytosol as demonstrated from isolated chloroplasts [1]. That said on the basis of our results flux through all of these routes can be anticipated to be relatively minor since formation via the starch degradation pathway is restricted to the night period, and it was shown that the G1P transport rate across the chloroplast membrane is minor in Arabidopsis in comparison to situation observed in potato [27]. Furthermore, it has been demonstrated that G1P that is taken up by the Arabidopsis chloroplast is directly converted into starch via ADPglucose pyrophosphorylase pathway, indicating that free G1P is immediately metabolized thus reducing the possibility of the G1P export [1]. It is possible that the observed elevation of the expression of PGM1 (Fig. S3B in File S1) in the transgenic lines is an effort to overcome this limitation. Additionally, preliminary experiments point to an increased G1P transport rate in *pgm2/3* plants compared to Col-0 (more than 20%) when measuring G1P uptake with isolated chloroplasts (data not shown).

However, it is not possible to explain the increase of sucrose in *pgm2/3* compared to Col-0 merely in terms of its rate of synthesis. It would seem more likely to be the consequence of the reduced sink capacity in the heterotrophic tissues and, therefore, a reduced export from the leaves of these lines. When sink capacity is reduced, feedback to the autotrophic tissues occurs culminating in the high starch and maltose levels observed in these lines. Moreover, metabolic profiling reveals a massive effect on the entire plant metabolism. Furthermore, taking into account the carbohydrate partitioning between sucrose and starch, the increase in both is not unexpected. Sucrose is catabolized either by sucrose synthase or invertase. It is proposed that invertase rather than sucrose synthase might be the dominant route for sucrose catabolism in *A. thaliana* [45]. Consequently, products of sucrose catabolism would enter the hexose phosphate pool as G6P or F6P but not as G1P. Thus, it would appear that cPGM is essential for G1P production.

A strong reduction of G1P is also anticipated to affect the entire nucleotide sugar metabolism [40], resulting in reduced growth and altered cell wall formation. As shown for *pgm2/3* the composition of the cell wall is altered and the root length is reduced. This phenotype was also observed for plants deficient in cytosolic invertase (*cinv1*) revealing reduced cell wall flexibility, inhibited root cell elongation and shorter roots [46]. Furthermore, mutants lacking two isoforms of cytosolic invertase (*cinv/cinv2*) are drastically reduced in root growth [45].

Additionally, a development of curly leaves was described in plants exhibiting reduced expression of SUT1 [47,48] or plants expressed yeast derived invertase [49,50,51]. This leaf phenotype was postulated to be due to osmotic problems associated with carbohydrate accumulation, which is similar to the situation observed for *pgm2/3*. However, it is important to note that in some cases plants with alteration in cell wall synthesis, downstream of G1P, also display such curled leaves [52].

The tiny *cp-pgm* plants reveal an even more severe phenotype. Indeed under normal growth conditions these perturbations are lethal. Germination was only observed, when sucrose was supplemented, but also under these conditions complete formation of inflorescence and seeds were inhibited. As the expected residual cPGM activity is similar to the parental *pgm2/3* lines (not detectable), this is a strong indication that the glucose-phosphate interconversion via PGM1 and formation of G1P via the starch degradation pathway are essential in *pgm2/3* plants for the creation of the residual levels of G1P. The observed phenotype is much more severe than that observed for transgenic potato lines lacking both cPGM and pPGM activities [25]. The strongest reduced line was reported to have decreased leaf fresh weight of up to 33 percent. One explanation for the less distinct phenotype for potato is that in these plants a residual activity of both the pPGM and cPGM was still detectable (both 4%, [26]). However, also a second point is to mention, that the transport rate for G1P over the plastidial membranes seems to be much higher in potato compared to Arabidopsis [1,27]. Thus, the possible bypass of the PGM lack via G1P transport is minor in Arabidopsis and therefore results in the observed more pronounced phenotype. Nevertheless, the higher transport rate of G1P observed for potato tuber is insufficient to completely overcome the limitations by lacking PGMs, especially in heterotrophic tissues, as the reduction in tuber fresh weight is far more pronounced with up to 75% reduction [25]. Overall, this points to a more flexible metabolism related to alternative carbon fluxes in potato then in Arabidopsis in respect to starch/sucrose turn-over.

Supporting Information

File S1 Supporting Information containing Tables S1–S3 and Figures S1–S5. Table S1. Primers used for PCR and qPCR analysis. **Table S2**. Chlorophyll content of Col-0 and *pgm2/3* plants. **Table S3.** Values of the metabolic profiling used for the generation of the heat map. **Figure S1**. Phosphoglucomutase activity in Arabidopsis leaves. **Figure S2.** Analysis of single knock-out lines *pgm2* and *pgm3* and Col-0 under long day conditions (14 h light/10 h dark). **Figure S3.** Characterization of Col-0 and *pgm2/3* plants. **Figure S4.** Growth phenotypes of Col-0 and PGM knock-out mutants. **Figure S5.** Phosphoglucomutase activity in Col-0 and PGM transgenic plants.

Acknowledgments

The authors gratefully thank Ulrike Matthes and Jessica Alpers for excellent technical assistants and Tom Orawetz for help screening the various transgenic lines and Sebastian Mahlow for help during preparation of the figures (all University of Potsdam). The authors also thank Julia Vogt and Anke Koch (both University of Potsdam) for help performing the qPCR experiments.

Author Contributions

Conceived and designed the experiments: IM HHK MG JF. Performed the experiments: IM HHK SA KH JF. Analyzed the data: IM HHK SA KH MG ARF JF. Contributed reagents/materials/analysis tools: IM HHK SA KH MG ARF JF. Contributed to the writing of the manuscript: IM HHK MG ARF JF.

References

1. Fettke J, Malinova I, Albrecht T, Hejazi M, Steup M (2011) Glucose-1-phosphate transport into protoplasts and chloroplasts from leaves of Arabidopsis. Plant Physiol 155: 1723–1734.
2. Streb S, Egli B, Eicke S, Zeeman SC (2009) The debate on the pathway of starch synthesis: a closer look at low-starch mutants lacking plastidial phosphoglucomutase supports the chloroplast-localized pathway. Plant Physiol 151: 1769–1772.
3. Kofler H, Häusler RE, Schulz B, Gröner F, Flügge U-I, et al. (2000) Molecular characterization of a new mutant allele plastid phosphoglucomutase in Arabidopsis, and complementation of the mutant with the wild-type cDNA. Mol Gen Genet 263: 978–986.
4. Gibon Y, Bläsing OE, Palacios-Rojas N, Pankovic D, Hendriks JHM, et al. (2004) Adjustment of diurnal starch turnover to short days: depletion of sugar during the night leads to a temporary inhibition of carbohydrate utilization, accumulation of sugars and post-translational activation of ADP-glucose pyrophosphorylase in the following light period. Plant J 39: 847–862.
5. Fettke J, Hejazi M, Smirnova J, Höchel E, Stage M, et al. (2009) Eukaryotic starch degradation: integration of plastidial and cytosolic pathways. J Exp Bot 60: 2907–2922.
6. Fettke J, Fernie AR, Steup M (2012) Transitory starch and its degradation in higher plant cells. In: Tetlow IJ, editor. Essential reviews in experimental biology: starch: origins, structure and metabolism, Vol. 5. London, UK: SEB. 309–372.
7. Weise SE, Weber APM, Sharkey TD (2004) Maltose is the major form of carbon exported from the chloroplast at night. Planta 218: 474–482.
8. Stitt M, Zeeman SC (2012) Starch turnover: pathways, regulation and role in growth. Curr Opin Plant Biol 15: 282–292.
9. Niittylä T, Comparot-Moss S, Lue WL, Messerli G, Trevisan M, et al.(2006) Similar protein phosphatases control starch metabolism in plants and glycogen metabolism in mammals. J Biol Chem 281: 11815–11818.
10. Lu Y, Sharkey TD (2004) The role of amylomaltase in maltose metabolism in the cytosol of photosynthetic cells. Planta 218: 466–473.
11. Chia T, Thorneycroft D, Chapple A, Messerli G, Chen J, et al. (2004) A cytosolic glucosyltransferase is required for conversion of starch in Arabidopsis leaves at night. Plant J 37: 853–863.
12. Fettke J, Chia T, Eckermann N, Smith A, Steup M (2006) A transglucosidase necessary for starch degradation and maltose metabolism in leaves at night acts on cytosolic heteroglycans (SHG). Plant J 46: 668–684.
13. Fettke J, Eckermann N, Poeste S, Pauly M, Steup M (2004) The glycan substrate of the cytosolic (Pho 2) phosphorylase isoform from *Pisum sativum* L.: identification, linkage analysis, and subcellular localization. Plant J 39: 933–946.
14. Fettke J, Poeste S, Eckermann N, Tiessen A, Pauly M, et al. (2005) Analysis of heteroglycans from leaves of transgenic potato (*Solanum tuberosum* L.) plants that under- or overexpress the Pho2 phosphorylase isozyme. Plant Cell Physiol 46: 1987–2004.
15. Cho MH, Lim H, Shin DH, Jeon JS, Bhoo SH, et al. (2011) Role of the plastidic glucose transporter in the export of starch degradation products from the chloroplasts in *Arabidopsis thaliana*. New Phytol 190: 101–112.
16. Weber A, Servaites JC, Geiger DR, Kofler H, Hille D, et al. (2000) Identification, purification, and molecular cloning of a putative plastidic glucose translocator. Plant Cell 12: 787–801.
17. Caspar T, Huber SC, Somerville C (1985) Alterations in growth, photosynthesis, and respiration in a starchless mutant of *Arabidopsis thaliana* (L.) deficient in chloroplast phosphoglucomutase activity. Plant Physiol 79: 11–17.
18. Hanson KR, McHale NA (1988) A starchless mutant of *Nicotiana sylvestris* containing a modified plastid phosphoglucomutase. Plant Physiol 88: 838–844.

19. Manjunath S, Kenneth LCH, Winkle PV, Bailey-Serres J (1998) Molecular and biochemical characterisation of cytosolic phosphoglucomutase in maize. Expression during development and in response to oxygen deprivation. Plant Physiol 117: 997–1006.
20. Fernie AR, Tauberger E, Lytovchenko A, Roessner U, Willmitzer L, et al. (2002) Antisense repression of cytosolic phosphoglucomutase in potato (*Solanum tuberosum*) results in severe growth retardation, reduction in tuber number and altered carbon metabolism. Planta 214: 510–520.
21. Mühlbach H, Schnarrenberger C (1978) Properties and intracellular distribution of two phosphoglucomutases from spinach leaves. Planta 141: 65–70.
22. Lytovchenko A, Sweetlove L, Pauly M, Fernie AR (2002) The influence of cytosolic phosphoglucomutase on photosynthetic carbohydrate metabolism. Planta 215: 1013–1021.
23. Fettke J, Nunes-Nesi A, Alpers J, Szkop M, Fernie AR, et al. (2008) Alterations in cytosolic glucose-phosphate metabolism affect structural features and biochemical properties of starch-related heteroglycans. Plant Physiol 148: 1614–1629.
24. Egli B, Kölling K, Köhler C, Zeeman SC, Streb S (2010) Loss of cytosolic phosphoglucomutase compromises gametophyte development in Arabidopsis. Plant Physiol 154: 1659–1671.
25. Lytovchenko A, Fernie AR (2003) Photosynthetic metabolism is severely impaired on the parallel reduction of plastidial and cytosolic isoforms of phosphoglucomutase. Plant Phys and Biochemistry 41: 193–200.
26. Fernie AR, Swiedrych A, Tauberger E, Lytovchenko A, Trethewey RN, et al. (2002) Potato plants exhibiting combined antisense repression of cytosolic and plastidial isoforms of phosphoglucomutase surprisingly approximate wild type with respect to the rate of starch synthesis. Plant Physiol Biochem 40: 921–927.
27. Fettke J, Albrecht T, Hejazi M, Mahlow S, Nakamura Y, et al. (2010) Glucose 1-phosphate is efficiently taken up by potato (*Solanum tuberosum*) tuber parenchyma cells and converted to reserve starch granules. New Phytol 185: 663–675.
28. Schwab R, Ossowski S, Riester M, Warthmann N, Weigel D (2006) Highly specific gene silencing by artificial microRNAs in Arabidopsis. Plant Cell 18: 1121–1133.
29. Nakagawa T, Kurose T, Hino T, Tanaka K, Kawamukai M, et al. (2007) Development of series of gateway binary vectors, pGWBs, for realizing efficient construction of fusion genes for plant transformation. J Biosci Bioeng 104: 34–41.
30. Zhang X, Henriques R, Lin SS, Niu QW, Chua NH (2006) *Agrobacterium*-mediated transformation of *Arabidopsis thaliana* using the floral dip method. Nature Protocols. 1: 1–6.
31. Stitt M, McC Lilley R, Gerhardt R, Heldt HW (1989) Metabolite levels in specific cells and subcellular compartments of plant leaves. Methods in Enzymology 174: 518–552.
32. Foster CE, Martin TM, Paul M (2010) Comprehensive compositional analysis of plant cell walls (Lignocellulosic biomass) Part II: Carbohydrates. Journal of Visualized Experiments. 37. Available : http://www.jove.com/index/Details.stp?ID=1837, doi: 10.3791/1837.
33. Waffenschmidt S, Jaenicke L (1987) Assay of reducing sugars in the nanomole range with 2,2'-bicinchoninate. Anal Biochem 165: 337–340.
34. Updegraf DM (1969) Semimicro determination of cellulose in biological materials. Anal Biochem 32: 420–424.
35. Fettke J, Eckermann N, Tiessen A, Geigenberger P, Steup M (2005) Identification, subcellular localization and biochemical characterization of water-soluble heteroglycans (SHG) in the leaves of *Arabidopsis thaliana* L.: distinct heteroglycans reside in the cytosol and in the apoplast. Plant J 43: 568–585.
36. Fettke J, Nunes-Nesi A, Fernie AR, Steup M (2011) Identification of a novel heteroglycan-interacting protein, HIP 1.3, from *Arabidopsis thaliana*. J Plant Physiol 168: 1415–1425.
37. Koch E, Slusarenko A (1990) Arabidopsis is susceptible to infection by a downy mildew fungus. Plant Cell 2: 437–445.
38. Fettke J (2006) Stärkerelevante cytosolische Heteroglycane: Identifizierung und funktionelle Analyse. PhD thesis, University of Potsdam, Potsdam, Germany.
39. Herbst K (2011) Cytosolische Phosphoglucomutase in *Arabidopsis thaliana*: Heterologe Expression and Analyse. Diploma thesis. University of Potsdam, Potsdam, Germany.
40. Seifert GJ (2004) Nucleotide sugar interconversions and cell wall biosynthesis: how to bring the inside to outside. Curr Opin Plant Biol 7: 277–284.
41. Matsui M, Nakazawa M (2003) Selection of hygromycin-resistant Arabidopsis seedling. Biotechniques 34: 28–30.
42. Yu TS, Lue WL, Wang SM, Chen J (2000) Mutation of Arabidopsis plastid phosphoglucose isomerase affect leaf starch synthesis and floral initiation. Plant Physiol 123: 319–325.
43. Ye RW, Zielinski NA, Chakrabarty AM (1994) Purification and characterization of phosphomannomutase/phosphoglucomutase from *Pseudomonas aeruginosa* involved in biosynthesis of both alginate and lipopolysaccharide. Journal Bacteriol. 176: 4851–4857.
44. Mitra S, Cui J, Robbins PW, Samuelson J (2010) A deeply divergent phosphoglucomutase (PGM) of *Giardia lamblia* has both PGM and phosphomannomutase activities. Glycobiology 20: 1233–1240.
45. Barratt DH, Derbyshire P, Findlay K, Pike M, Wellner N, et al. (2009) Normal growth of Arabidopsis requires cytosolic invertase but not sucrose synthase. Proc Natl Acad Sci USA 106: 13124–13129.
46. Lou Y, Gou JY, Xue HW (2007) PIP5K9, an Arabidopsis phosphatidylinositol monophosphate kinase, interacts with a cytosolic invertase to negatively regulate sugar-mediated root growth. Plant Cell 19: 163–181.
47. Riesmeier JW, Wilmitzer L, Frommer W (1994) Evidence for a role of the sucrose transporter in phloem loading and assimilate partitioning. EMBO J 13: 1–7.
48. Bürkle L, Hibberd JM, Quick WP, Kuehn C, Hirner B, et al. (1998) The H+-sucrose co-transporter NtSUT1 is essential for sugar export from tobacco leaves. Plant Physiol 118: 59–68.
49. Heineke D, Sonnewald U, Büssis D, Günter G, Leidreiter K, et al. (1992) Apoplastic expression of yeast-derived invertase in potato: effects on photosynthesis, leaf solute composition, water relations, and tuber composition. Plant Physiol 100: 301–308.
50. Dickinson C, Altabella T, Chrispeels M (1991) Slow-growth phenotype of transgenic tomato expressing apoplastic invertase. Plant Physiol 95: 420–425.
51. von Schaewen A, Stitt M, Schmidt R, Sonnewald U, Willmitzer L (1990) Expression of a yeast-derived invertase in the cell wall of tobacco and Arabidopsis plants leads to accumulation of carbohydrate and inhibition of photosynthesis and strongly influences growth and phenotype of transgenic tobacco plants. EMBO J. 9: 3033–3044.
52. Reiter WD, Chapple C, Somerville CR (1997) Mutants of *Arabidopsis thaliana* with altered cell wall polysaccharide composition. Plant J 12: 335–345.

14

Comparative Analysis of Predicted Plastid-Targeted Proteomes of Sequenced Higher Plant Genomes

Scott Schaeffer[1,2], Artemus Harper[1], Rajani Raja[3], Pankaj Jaiswal[3], Amit Dhingra[1,2]*

1 Department of Horticulture, Washington State University, Pullman, WA, United States of America, **2** Molecular Plant Science Graduate Program, Washington State University, Pullman, WA, United States of America, **3** 2082 Cordley Hall, Department of Botany and Plant Pathology, Oregon State University, Corvallis, OR, United States of America

Abstract

Plastids are actively involved in numerous plant processes critical to growth, development and adaptation. They play a primary role in photosynthesis, pigment and monoterpene synthesis, gravity sensing, starch and fatty acid synthesis, as well as oil, and protein storage. We applied two complementary methods to analyze the recently published apple genome (Malus × domestica) to identify putative plastid-targeted proteins, the first using TargetP and the second using a custom workflow utilizing a set of predictive programs. Apple shares roughly 40% of its 10,492 putative plastid-targeted proteins with that of the Arabidopsis (Arabidopsis thaliana) plastid-targeted proteome as identified by the Chloroplast 2010 project and ~57% of its entire proteome with Arabidopsis. This suggests that the plastid-targeted proteomes between apple and Arabidopsis are different, and interestingly alludes to the presence of differential targeting of homologs between the two species. Co-expression analysis of 2,224 genes encoding putative plastid-targeted apple proteins suggests that they play a role in plant developmental and intermediary metabolism. Further, an inter-specific comparison of Arabidopsis, Prunus persica (Peach), Malus × domestica (Apple), Populus trichocarpa (Black cottonwood), Fragaria vesca (Woodland Strawberry), Solanum lycopersicum (Tomato) and Vitis vinifera (Grapevine) also identified a large number of novel species-specific plastid-targeted proteins. This analysis also revealed the presence of alternatively targeted homologs across species. Two separate analyses revealed that a small subset of proteins, one representing 289 protein clusters and the other 737 unique protein sequences, are conserved between seven plastid-targeted angiosperm proteomes. Majority of the novel proteins were annotated to play roles in stress response, transport, catabolic processes, and cellular component organization. Our results suggest that the current state of knowledge regarding plastid biology, preferentially based on model systems is deficient. New plant genomes are expected to enable the identification of potentially new plastid-targeted proteins that will aid in studying novel roles of plastids.

Editor: Steven M. Theg, University of California – Davis, United States of America

Funding: This research and support for AH was funded in part by United States Department of Agriculture National Institute of Food and Agriculture Grant # 2009-05031 to AD. Support for SS was provided by National Institutes of Health Training Grant number T32 GM008336 to AD as co-applicant. The funders had no role in study design, data collection and analysis, decision to publish, or preparation of the manuscript.

Competing Interests: Co-authors Pankaj Jaiswal and Amit Dhingra are members of the PLOS ONE Editorial Board. The authors would like to state that this does not alter their adherence to PLOS ONE Editorial policies and criteria.

* Email: adhingra@wsu.edu

Introduction

The plastid is an intracellular organelle derived from an endosymbiotic event wherein a free-living autotrophic photosynthetic bacterium was phagocytized by a separate heterotrophic organism [1]. These organelles have since become essential to plant survival and have been documented to participate in numerous biological processes including photosynthesis, storage of oils, and proteins, pigment synthesis and storage, monoterpene synthesis [2], gravity sensing [3], and starch and fatty acid synthesis [4]. Over an extensive period of evolution, large parts of the plastid genome are hypothesized to have integrated into the nuclear genome [5]. In higher plants, the vast majority of proteins constituting the plastid proteome are encoded by genes physically resident in the nuclear genome, with about 120 genes retained in the plastid genome, a number which varies between species [6]. Comparative genomic analysis between Arabidopsis (*Arabidopsis thaliana*) and cyanobacteria indicates that 18% of the Arabidopsis protein–coding genes were derived from events involving transfer of genetic material from the plastid to the nucleus [7]. In part, exchange of genetic material and related biological functionality has necessitated an orchestration of processes between the plastid and nucleus where the nucleus actively exerts control on all aspects of plastid function.

Plant cells have developed intricate mechanisms to import nuclear-encoded proteins to or across the three plastid membranes (outer, inner plastid envelope, thylakoid). The presence of multiple protein transport pathways have been shown to play a role in aiding protein transport across the inner and outer plastid envelopes; however, the vast majority of the plastid proteome is transported via the tic/toc pathway [8]. In order to utilize this pathway, most proteins possess a signal peptide which interacts with chaperones and is later cleaved. Stromal-targeting peptide sequences, while not conserved, possess some similarities in amino acid composition. These targeting peptides are typically comprised of a relatively high abundance of serine and threonine residues [9,10] and are positively charged [11]. There are also some proteins that do not have any canonical signaling peptides and yet localize to plastids [12,13,14]. Therefore, the signaling prediction programs provide a good reference point to initiate an understanding of the plastid-targeted proteome for any new species, but as predictions, they do require experimental validation.

Due largely to the technical complexity with whole plastid proteome characterization, transcriptome or genome sequences have become a widely used dataset to predict plastid-targeted motifs. Such an approach also enables the identification of plastid-targeting proteins in a spatial and temporal context. Prediction of subcellular localization has been reported to be performed with software such as PCLR [15], iPSORT [16], TargetP [17,18], and PREDOTAR [19] amongst many other programs. Most prediction methods exploit the presence of an N-terminal signal sequence to predict cellular localization. Of these, TargetP was recommended to be most successful in prediction and was comparable to PCLR, with each having sensitivity values of 0.72 [20].

The Rosaceae family represents a unique diversity in fruit development, which is unrealized in the many model plants whose genomes have been sequenced [21,22,23,24,25,26]. Pomes (apples and pears), stone fruits (cherry and peach), and aggregate fruits (strawberry and raspberry) display a diversity that suggests the presence of novel metabolic processes, and is supported by a large number of genes which far exceeds the number of genes in Arabidopsis. While fruit development in these Rosaceae species differs vastly, the ubiquitous process of the plastidial transition from a chloroplast to chromoplast is often assumed to be conserved. Within Rosaceae fruit, plastids play extremely important roles in determining fruit quality and organoleptic appeal as they are the site for synthesis of carotenoids [27,28], monoterpenes [2], fatty acids [4] and aromatic amino acids. Many of these compounds have been linked to human health and nutrition [29,30]. Plastids are also important in converting starch into various types of carbohydrate and sugars in developing fruits [31].

The plastid structure has also been reported to differ between different tissues of the fruit. Phan [32] reported the presence of a large single granum comprised only of stacked thylakoid membranes in the plastids of the endocarp tissue of apple. In addition, chloroplasts with leaf-like thylakoid and grana organization in the outer six cell layers of mature apple fruit and presence of globular chromoplasts in epidermal cells were described. It is expected that differences in structure, physiology and biochemistry in the Rosaceae fruit plastids as well as other non-model systems will assist in identifying novel processes associated with plastids in plants.

In this study we tested the three primary hypotheses using a bioinformatics approach, (1) The total number and composition of the plastid-targeted protein coding genes in apple, a model representative of Rosaceae, that is taxonomically different from Arabidopsis, (2) The plastid-targeted protein coding genes are under transcriptional control during apple fruit development and (3) There is a subset of unique plastid-targeted protein coding genes that are unique and novel to each plant species.

In order to test the first hypothesis, we performed an in-depth computational analysis predicting the plastid-targeted proteome of apple and compared it with Arabidopsis resulting in the identification of a much larger number of plastid-targeted genes with nearly 4000 plastid-targeted protein coding genes being unique to apple. The second hypothesis was tested by reanalyzing publically available apple transcriptome data which revealed the presence of co-expression profiles of plastid-targeted genes and their association to development and metabolism. Finally, the third hypothesis was tested by extending the custom analysis workflow to an inter-genera comparison between six published genomes: *Arabidopsis thaliana*, *Vitis vinifera*, *Prunus persica*, *Populus trichocarpa*, *Fragaria vesca*, and *Solanum lycopersicum* resulting in the identification of plastid-targeted proteins unique to each species. A core set of 737 *Arabidopsis thaliana* proteins, highly enriched in photosynthesis and primary metabolism gene ontology (GO) terms, were identified to have homologous plastid-targeted proteins in all investigated species.

Materials and Methods

TargetP-based prediction of *Malus* × *domestica* plastid proteome

The *Malus* × *domestica* predicted protein set was obtained from the apple genome sequencing project [26]. Protein sequences were analyzed using TargetP using plant networks with default parameters [17,18]. All sequences with predicted chloroplast transit peptides were compiled into a new dataset and were sorted based on length using USEARCH [33].

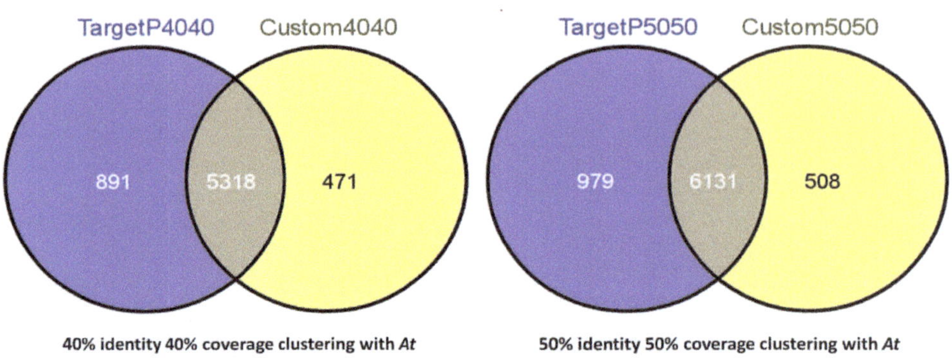

Figure 1. Venn diagrams displaying the predicted plastid-targeting proteins unique to apple compared to *Arabidopsis*. Two plastid-targeting methods, TargetP and a custom analysis method, were used to predict genes encoding plastid localized proteins. Sequences in these data sets were compared to *Arabidopsis* plastid-targeted proteins from the Chloroplast2010 project using USEARCH. Genes not clustered to *Arabidopsis* were compared between prediction methods displaying a high agreement between the methods. Venn diagrams were constructed using Venny (Oliveros, 2007).

Table 1. Enriched GO terms associated with co-expressed genes encoding plastid-targeted proteins in developing apple.

	GO term	Ontology	Description	p-value	FDR
Cluster 1	GO:0015979	P	photosynthesis	7.80E-08	2.60E-05
	GO:0009579	C	thylakoid	5.50E-16	9.10E-14
	GO:0009536	C	plastid	4.70E-06	3.90E-04
	GO:0016020	C	membrane	1.10E-05	6.20E-04
	GO:0043231	C	intracellular membrane-bounded organelle	3.20E-05	1.10E-03
	GO:0043227	C	membrane-bounded organelle	3.20E-05	1.10E-03
	GO:0043229	C	intracellular organelle	6.90E-05	1.60E-03
	GO:0043226	C	organelle	6.90E-05	1.60E-03
	GO:0005623	C	cell	8.30E-05	1.70E-03
	GO:0044464	C	cell part	1.30E-04	2.30E-03
	GO:0044424	C	intracellular part	6.50E-04	1.10E-02
	GO:0044444	C	cytoplasmic part	7.00E-04	1.10E-02
	GO:0005622	C	intracellular	1.60E-03	2.20E-02
Cluster 2	GO:0006629	P	lipid metabolic process	8.40E-10	2.60E-07
	GO:0019748	P	secondary metabolic process	8.70E-05	1.30E-02
	GO:0009058	P	biosynthetic process	1.60E-04	1.60E-02
	GO:0009056	P	catabolic process	4.40E-04	3.40E-02
	GO:0006810	P	transport	9.20E-04	4.00E-02
	GO:0051234	P	establishment of localization	9.20E-04	4.00E-02
	GO:0051179	P	localization	9.20E-04	4.00E-02
	GO:0003824	F	catalytic activity	5.60E-04	3.80E-02
	GO:0005737	C	cytoplasm	7.00E-12	1.40E-09
	GO:0044464	C	cell part	1.80E-11	1.80E-09
	GO:0005623	C	cell	4.90E-11	3.20E-09
	GO:0044444	C	cytoplasmic part	7.30E-11	3.50E-09
	GO:0044424	C	intracellular part	1.40E-10	4.50E-09
	GO:0005622	C	intracellular	1.30E-10	4.50E-09
	GO:0043229	C	intracellular organelle	1.40E-09	3.10E-08
	GO:0009536	C	plastid	1.30E-09	3.10E-08
	GO:0043226	C	organelle	1.40E-09	3.10E-08
	GO:0043231	C	intracellular membrane-bounded organelle	1.80E-09	3.20E-08
	GO:0043227	C	membrane-bounded organelle	1.80E-09	3.20E-08
Cluster 3	No significant term				
Cluster 4	No significant term				
Cluster 5	No significant term				
Cluster 6	No significant term				
Cluster 8	No significant term				
Cluster 9	No significant term				
Cluster 10	No significant term				
Cluster 11	No significant term				

Expression data from Janssen et al. [49] was mined to identify the expression profiles for any genes encoding plastid-targeted protein. Clustering of genes was performed based upon expression profile revealing a set of 11 significant clusters. GO term enrichment was performed to identify significantly enriched terms associated with each cluster. Gene Ontology terms are provided for biological process (P), molecular function (F), and cellular component (C).

Custom protein targeting analysis

A part of the functional annotation pipeline was applied to identify organelle 'plastid' targeted gene products encoded by the apple genome [26]. The peptide sequences were analyzed first through InterProScan [34] results provided by the genome consortium [26], followed by in-house analysis using the SignalP [17], Predotar [19] and TMHMM [35]. InterPro provided the domain annotations, and any genes/peptides with transposable element/domain annotations were filtered out for further analysis. The next steps of the pipeline employed: (1) SignalP to predict localization to the mitochondrial or plastid or secretion pathway, plus providing signal peptide cleavage sites, (2) Predotar to predict localization to either or both the mitochondrion or plastid, and (3)

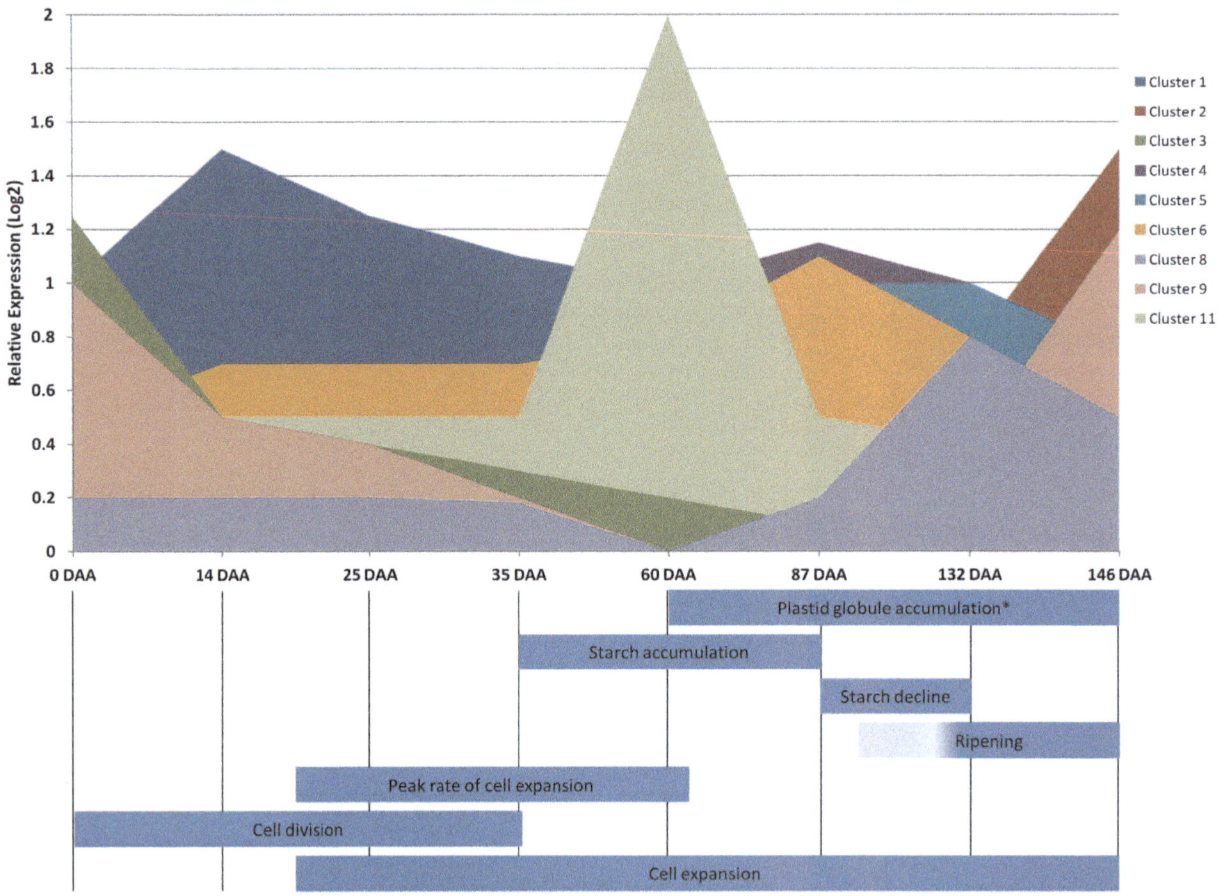

Figure 2. Overlay of apple gene expression clusters with distinct fruit development events. Relative expression of the co-expressed genes encoding plastid-targeted proteins were merged and displayed along significant events occurring within the developmental continuum of apple fruits, adapted from Janssen et al. 2008. An additional event unique to apple fruit plastids, globule accumulation was added as described by Clijsters in 1969.

TMHMM to identify predicted transmembrane domains in the protein sequences. After collecting these annotations, standardized protocols for assigning the annotations were adopted [24]. The higher quality scores with reviewed after computational analyses (RCA) were selected if the scores of 0.75 and greater were predicted for TargetP and Predotar and two or more transmembrane annotations were predicted by the TMHMM. The parameters selected for inferred by electronic annotation (IEA)

Table 2. Results of putative plastid-targeted protein prediction.

Species	Reference	Unique predicted protein-coding genes	Predicted to be plastid-targeted
Arabidopsis thaliana	[40]	27,416	5,382* (19.6%)
Malus × domestica (Apple)	[26]	57,386	10,492 (18.3%)
Vitis vinifera (Grapevine)	[22]	26,346	3,015 (11.4%)
Prunus persica (Peach)	[21]	28,689	3,860 (13.5%)
Populus trichocarpa (Black cottonwood)	[25]	45,555	4,515 (9.9%)
Fragaria vesca (Woodland strawberry)	[24]	34,809	4,922 (14.1%)
Solanum lycopersicum (Tomato)	[23]	33,926	4,009 (11.8%)

Predicted transcripts were selected from 7 sequenced genomes representing both model organisms as well as agriculturally important fruit-producing crops. Translated sequences were analyzed using TargetP to predict cellular localization. Sequences were clustered with 40% coverage and 40% identity to predict proteins unique to the plastid proteome of each species. *Represents sequences from Chloroplast 2010 project and sequences associated with predicted plastid-targeted embryo lethal mutant.

include the scores of 0.5–0.749 for TargetP and Predotar and one/single transmembrane domain suggested by the TMHMM. The majority of these annotations were IEA evidence codes. If the annotations overlapped for gene products that had plastid-targeting predicted from TargetP and Predotar and membrane spanning domains identified by the TMHMM, then the suggested location of the targeted protein was 'plastid membrane'.

The Inparanoid algorithm [36] was used to find orthologous genes and paralogous genes that arise by duplication events. The pipeline was discussed in the *Fragaria vesca* genome paper [24]. For this study, the analysis included the peptide sequences from 22 species, including *Arabidopsis thaliana, Brachypodium distachyon, Caenorhabitis elegans, Chlamydomonas reinhardtii, Danio rerio, Eschericia coli, Fragaria vesca, Glycine max, Homo sapiens sapiens, Zea mays, Malus × domestica, Mus musculus, Neurospora crassa, Oryza sativa, Physcomitrella patens, Populus trichocarpa, Saccharomyces cerevisiae and pombe, Selaginella moellendorffii, Sorghum bicolor, Synechosystis, and Vitis vinifera* to cover the tree of life with emphasis on fully/nearly complete and published genomes. The peptide sequences were downloaded from Phytozome.net for grapevine, *Selaginella, Physcomitrella, Chlamydomonas, Glycine, Populus, and Malus* from the genome portal [26], and *Gramene* [37] for rice, sorghum, maize and Arabidopsis. The remaining sequences were downloaded from Ensembl [38,39].

Identification of sequences unique to apple datasets compared to Arabidopsis

The *Arabidopsis thaliana* plastid-targeted gene set was obtained from the Chloroplast 2010 project website (www.plastid.msu.edu) [40]. Arabidopsis embryo lethal mutants were analyzed using TargetP [17,18] and any chloroplast targeted proteins were added to the aforementioned dataset, as these were omitted from the Chloroplast 2010 database. Proteins predicted to target the apple plastid were then compared to plastid-targeted proteins from *Arabidopsis thaliana* using USEARCH [33]. Predicted plastid-targeted proteins were compared using two conditions: first, a global USEARCH was performed using 40% amino acid identity and 40% coverage (40/40), and a second global comparison was performed using 50% amino acid identity with 50% coverage (50/50). Header files of proteins unique to the *M.* × *domestica* dataset were compared between the TargetP-based method as well as the custom analysis to investigate any bias associated with either respective prediction technique.

USEARCH-based multispecies comparative analysis of predicted plastid-targeted proteomes

Predicted coding sequences were collected from the genomes of *Fragaria vesca* (Woodland Strawberry) [24], *Vitis vinifera* (Grapevine) [22], *Solanum lycopersicum* ITAG1 release (Tomato) [23], *Prunus persica* (Peach) [21] and *Populus trichocarpa* (Black Cottonwood) [25]. Protein sequences were analyzed with TargetP [17,18] using default parameters to predict localization. Sequences predicted to be plastid-targeted were organized into new files for each species. Comparisons were performed for each plastid-targeted dataset using USEARCH [33] 40/40 and 50/50 global parameters against the *Arabidopsis thaliana* plastid-targeted dataset, the entire Arabidopsis TAIR V10 protein set (Arabidopsis.org) [41], the predicted proteins from *Solanum lycopersicum*, as well as a file comprised of the sequences of the predicted plastid-targeted protein sequences of the other six species. All datasets were first sorted by length using USEARCH.

Further analysis was performed with USEARCH to identify those proteins present in the predicted plastid proteomes of all investigated species. To perform this analysis, the *Arabidopsis thaliana* putative plastid-targeted protein set was compared using USEARCH 40/40 global parameters separately against the plastid-targeted proteins from woodland strawberry, grapevine, tomato, peach, black cottonwood, and apple. Output files were then analyzed to identify those Arabidopsis sequences which had a match in the plastid-targeted proteomes of all species.

UCLUST-based multispecies comparative analysis of predicted plastid-targeted proteomes

A second comparative analysis was performed using the clustering feature of the USEARCH package, UCLUST. In this analysis, the plastid-targeted protein sequences from the seven examined species were compiled into a single file and sorted by length. UCLUST was performed at 50% identity. The output was parsed to identify protein clusters with members from all seven species, as well as those clusters containing sequences from only one species.

Determination of Jaccard's similarity coefficients

Two separate techniques were used to create the similarity matrices based upon Jaccard's coefficients. To calculate the value of an individual cell (the distance between species A and species B) we first determined if two genes were considered homologous. If a

Table 3. Clustering of predicted plastid proteomes with *Arabidopsis* datasets.

Species	Predicted plastid-targeted proteins	Percent clustered with *At* chloroplast (40/40)	Percent clustered with all *At* proteins (40/40)	Percent clustered with *At* chloroplast (50/50)	Percent clustered with all *At* proteins (50/50)
Arabidopsis thaliana	5,382*	100%	100%	100%	100%
Malus × domestica	10,492	40.7%	56.7%	32.2%	44.5%
Vitis vinifera	3,015	57.3%	77.0%	49.5%	65.8%
Prunus persica	3,860	59.9%	76.9%	52.0%	67.3%
Populus tichocarpa	4,515	51.8%	68.3%	44.5%	58.4%
Fragaria vesca	4,922	40.9%	56.1%	33.3%	45.0%
Solanum lycopersicum	4,009	55.2%	69.7%	45.7%	57.0%

Predicted plastid-targeted proteins unique to each species were clustered against the peptide sequences derived from the Chloroplast 2010 dataset with the addition of embryo lethal sequences (*At* chloroplast) as well as the entire *Arabidopsis* protein set from TAIR V10 (all *At* proteins) using USEARCH. Clustering was performed using the parameters of 40% coverage with 40% identity (40/40) as well as 50% coverage with 50% identity (50/50).

Table 4. USEARCH-based matching of unique plastid-targeted proteins.

Species	Predicted plastid-targeted Proteins	Number Clustered with *At* Chloroplast (40/40)	Number Clustered with *Sl* Chloroplast (40/40)	Proteins clustered with Cp Proteins from 6 Other Species	Unique Plastid-targeted Proteins
Arabidopsis thaliana	5,382*	5,382 (100%)	2,503 (46.5%)	4,099 (76.2%)	1,446 (26.9%)
Malus × domestica	10,492	4,265 (40.7%)	3,442 (32.8%)	6,235 (59.4%)	4,257 (40.6%)
Vitis vinifera	3,015	1,729 (57.3%)	1,396 (46.3%)	2,265 (75.1%)	750 (24.9%)
Prunus persica	3,860	2,312 (59.9%)	1,857 (48.1%)	3,232 (83.7%)	628 (16.3%)
Populus tichocarpa	4,515	2,340 (51.8%)	1,949 (43.2%)	3,207 (71.0%)	1,308 (29.0%)
Fragaria vesca	4,922	2,014 (40.9%)	1,584 (32.2%)	2,880 (58.5%)	2,042 (41.5%)
Solanum lycopersicum	4,009	2,213 (55.2%)	4,009 (100%)	2,799 (69.8%)	1,210 (30.2%)

Datasets comprised of putative plastid-targeted proteins unique to each species were clustered against the predicted plastid-targeted protein sequences of *Arabidopsis thaliana (At)* from Chloroplast2010 and TAIR V10, *Solanum lycopersicum (Sl)*, and a database consisting of the plastid-targeted proteins from all 6 species using USEARCH. Sequences were clustered globally with 40% coverage with 40% identity (40/40). Results suggest that a large portion of the plastid-targeted proteins may be unique to each respective species.

gene from the species A matches with another gene in species B, then both A and B are included in the intersection set. The result in the cell is the count of the intersection set divided by the sum of all the genes in both species. For the USEARCH-based approach, two genes match if they align to each other using 40/40 parameters. Alternatively, in the UCLUST-based approach two genes match if they belong to the same cluster.

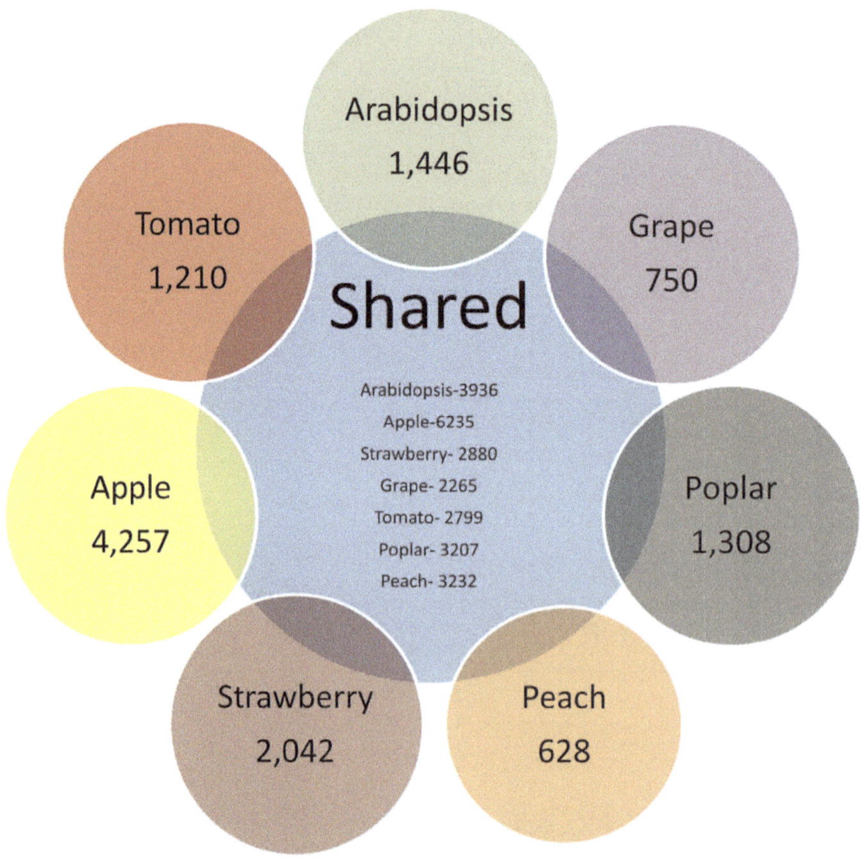

Figure 3. Comparison of predicted plastid-targeted proteomes across seven sequenced genomes. Protein sequences generated from the genomes of *Arabidopsis thaliana, Vitis vinifera, Populus trichocarpa, Prunus persica, Fragaria vesca, Malus × domestica*, and *Solanum lycopersicum* were clustered at 40 percent identity and 40 percent coverage using USEARCH. Comparison reveals large subsets for each species appearing to be unique to each species' respective plastids.

Table 5. Uniquely targeted plastid-targeted protein sequences.

Species	Predicted plastid-targeted Proteins	Unique to Species (USEARCH 40/40)	Unique to Species (UCLUST50)	Number Clustered with All Proteins from 6 Other Species (USEARCH 40/40)
Arabidopsis thaliana	5,382*	1,446 (26.9%)	2,154 (40.0%)	4,928 (91.6%)
Malus × domestica	10,492	4,257 (40.6%)	4,787 (45.6%)	7,911 (76.2%)
Vitis vinifera	3,015	750 (24.8%)	976 (32.3%)	2,561 (84.9%)
Prunus persica	3,860	628 (16.3%)	795 (20.6%)	3,506 (90.8%)
Populus tichocarpa	4,515	1,308 (29.0%)	1,732 (38.4%)	3,630 (80.4%)
Fragaria vesca	4,922	2,042 (41.5%)	2,305 (46.8%)	3,338 (67.8%)
Solanum lycopersicum	4,009	1,210 (30.2%)	1,654 (41.3%)	3,122 (77.9%)

A USEARCH comparison of plastid-targeted protein datasets was performed at 40% identity and 40% coverage against a database containing the chloroplast protein sequences from all six other species investigated in this study. A second comparison was performed against a database containing the entire protein set from the other species. An increase in matching suggests the presence of differentially localized homologues in other systems. Additionally, these results suggest a sizeable number of plastid-targeted proteins may be unique to each species.

Blast2GO Gene Ontology analysis and GO term enrichment analysis

Sequences for all genes encoding unique or shared plastid-targeted proteins in the investigated apple, Arabidopsis, grapevine, peach, strawberry, black cottonwood, and tomato datasets were analyzed via Blast2GO [42,43]. BLASTP was performed using the NCBI nr database with Blast2GO default parameters. Gene ontology mapping and annotation were also performed using default parameters with the August 2012 database. Following GO annotation, an Interpro scan [34,44]was performed and results were merged with the GO annotations. Annotation augmentation was performed using ANNEX [45], followed by GO-slim with the goslim_plant.obo database. Kyoto Encyclopedia of Genes and Genomes (KEGG) information was downloaded from the KEGG Pathway Database [46,47]. Datasets comprising those unique to each plastid-targeted proteome, as well as those shared between all seven species were investigated using Single Enrichment Analysis with agriGO [48] to identify enriched GO terms. Analysis was performed using the Fisher test for significance and adjusted using the Yekutieli multi test adjustment with the minimum mapping entries set to three. A significance level was set at 0.01 and all terms GO-terms with p-values lower than this cutoff were reported as enriched.

Analysis of apple fruit gene expression

In order to ascertain if genes encoding plastid-targeted proteins in apple are expressed in fruits, as well as to identify co-expressed gene sets, microarray data from a previously published experiment were used [49]. The Janssen study measured the relative expression of about 13,000 features designed from apple fruit expressed sequence tags (ESTs) at 8 time points ranging from 0 days after anthesis (DAA) to 146

Table 6. GO terms enriched in uniquely targeted proteins as determined by USEARCH 40/40 method.

	GO term	Ontology	Description	p-value	FDR
Apple	GO:0006259	P	DNA metabolic process	8.30E-60	3.30E-57
	GO:0044260	P	cellular macromolecule metabolic process	4.20E-14	8.40E-12
	GO:0006807	P	nitrogen compound metabolic process	1.60E-12	1.60E-10
	GO:0006139	P	nucleobase, nucleoside, nucleotide and nucleic acid metabolic process	1.60E-12	1.60E-10
	GO:0043170	P	macromolecule metabolic process	2.90E-11	2.30E-09
	GO:0003676	F	nucleic acid binding	1.30E-53	1.60E-51
	GO:0005488	F	Binding	2.80E-24	1.70E-22
	GO:0003677	F	DNA binding	8.30E-05	3.30E-03
Arabidopsis	GO:0005634	C	Nucleus	8.30E-16	1.70E-13
Grapevine	No significant term				
Peach	GO:0016787	F	hydrolase activity	1.60E-06	1.10E-04
Black cottonwood	GO:0003676	F	nucleic acid binding	2.90E-06	3.40E-04
Strawberry	No significant term				
Tomato	No significant term				

GO terms were determined for all predicted plastid-targeted proteomes. Enrichment of GO terms was determined with agriGO for the subset of plastid-targeted proteins unique to each species compared to the entire plastid proteome. Gene Ontology terms are provided for biological process (P), molecular function (F), and cellular component (C).

DAA. All EST sequences utilized in the microarray experiment were retrieved from NCBI and a BLASTX was performed against the predicted apple protein set generated from the apple genome [26]. The EST expression data were then assigned to the top protein hit. Sequences which were previously found to be plastid-targeted were extracted and their respective expression data were analyzed by determining relative expression to the lowest measured mean expression value. The Log2 of relative expression data were imported into and analyzed with Multi-Experiment Viewer [50,51]. Sequences were clustered using Cluster Affinity Search Technique [52] using Pearson Correlation and a threshold of 0.8. Blast2GO [42] was used to assign annotation to those proteins with associated gene expression data. Single Enrichment Analysis was performed with AgriGO [48] as previously described, however a chi-square test was used instead to determine statistical significance.

Results

Predicted Plastid-targeted Proteomes of *Malus* × *domestica*

The apple genome has a total of 57,386 predicted genes [26] nearly 30,000 more genes than Arabidopsis [41,53]. We analyzed the complete apple gene set for cellular localization using two approaches, namely TargetP [17,18] and a custom prediction method (see materials and methods section for details). TargetP predicted the presence of 10,492 plastid-targeted proteins in the apple genome, while the custom gene ontology-based analysis predicted 9,882 genes, with an overlap of 9,256. Each data set was then clustered with the Arabidopsis plastid-targeted protein set using USEARCH [33] with 40% identity and 40% coverage (40/40 parameters) to identify homologous protein sequences. The TargetP method and custom analysis predicted 6,209 and 5,789 plastid-targeted proteins respectively to be unique to the apple dataset. The two methods agreed upon 5,318 proteins (86% and 92% respectively) uniquely targeted to apple chloroplasts and absent from those of *Arabidopsis thaliana* (Figure 1). Alternative clustering using 50% identity and 50% coverage (50/50 parameters) resulted in less clustering with *Arabidopsis* sequences and, consequently, increased the number of proteins predicted to be unique to apple. Using these parameters 7,110 sequences were predicted to be unique to the apple plastid proteome by TargetP, 6,639 with the custom analysis, and a set of 6,131 agreed upon by the two methods.

In order to identify prediction biases between the custom analysis and TargetP, an agriGO [48] GO term enrichment was performed on the proteins predicted to be differentially targeted. No significant GO terms were found to be enriched in the 1,236 proteins predicted to target the plastid with TargetP. However, agriGO identified the GO terms oxygen binding (GO:0019825, p-value 3.2e-05), hydrolase activity (GO:0016787 p-value 6e-05) and catalytic activity (GO:0003824 p-value 3.3e-04) are enriched in the plastid-targeted-proteins unique to the custom analysis.

Expression analysis of genes encoding plastid-targeted proteins in *Malus* × *domestica*

In order to test the hypothesis that plastid-targeted protein coding genes are under transcriptional control as the apple fruit develops, we reanalyzed data from a previously published microarray-based analysis of developing apple fruit [49]. Of the 13,000 unigene microarray probe sets studied, 2,698 were determined to map back to putative plastid-targeted proteins identified in this study, and represent a total of 2,224 unique sequences. Clustering of expression data using MultiExperiment Viewer [50,51] identified 92 different expression clusters, however, only 64 of these had 5 or more members. Over 50% of the genes fit into 9 co-expression clusters. These co-expressed genes were annotated using Blast2GO to infer their functions. Expression data for each cluster are provided in File S1 as well as their associated GO term information (File S2).

Of the 11 main clusters investigated, only the two most populous clusters of co-expressed genes contained GO terms which were determined to be significantly enriched by agriGO analysis (Table 1). Cluster 1 had a single enriched biological process GO term of photosynthesis (GO:0015979) along with 12 enriched cellular component GO terms with thylakoid (GO:0009579) having the lowest p-value. Cluster 2 was enriched in the biological process GO terms lipid metabolic process (GO:0006629), secondary metabolic process (GO:0019748), biosynthetic process (GO:0009058), catabolic process (GO:0009056), transport (GO:006810), establishment of localization (GO:0051234), and localization (GO:0051179). Additionally, one molecular function GO term was enriched in cluster 2, catalytic activity (GO:0003824), along with 11 cellular component GO terms.

In order to determine if genes encoding plastid-targeted proteins were indeed expressed within the fruit of apple, data from a previous study were analyzed [49]. The initial microarray experiment was a large scale analysis representing 13,000 of the ~57,000 apple genes, and was designed around many significant physiological events occurring during apple fruit development. These 13,000 genes were compared to the genes encoding predicted plastid targeted proteins described earlier in this study. About 20% of the genes (2,224 genes) encoding predicted plastid-targeted proteins mapped back to genes represented in the Janssen study. Analysis with MultiExperiment Viewer revealed that the majority of these genes were co-expressed in 9 clusters. To show how these expression profiles may relate to important fruit developmental events, expression profiles for the co-expressed genes were overlaid with those events described in Janssen et al. (Figure 2). An additional event, plastid globule accumulation, was also added, as it was noted in developing apple fruits alongside the unstacking of photosynthetic membranes [54]. As seen in Figure 2, the gene expression of these clusters and their GO terms coincide to some extent with the processes occurring within the apple fruits. Many of the biological process GO terms and KEGG pathways associated with each gene expression cluster suggest that expression of genes encoding plastid-targeted proteins may coincide with these important events. The expression of Cluster 1 greatly mirrors the photosynthetic activity of apple fruit tissue, with highest expression occurring in young, photosynthetically-capable fruit, and expression lowering as the fruit matures and has a reduction in photosynthetic capabilities. Additionally, the expression of those genes in Cluster 2 appear to mirror the development of carotenoids, volatile compounds, and maturation of fruit, with expression lowest in young fruit and increasing as the fruit reaches maturity. In particular the expression of genes whose products are involved in lipid metabolic processes, secondary metabolic processes, biosynthetic processes, and catabolic processes, as determined via GO term enrichment would be great candidates for further study in their participation in apple fruit volatile production. Cluster 11 is particularly interesting as it is comprised of genes whose expression peaks at a single time point (60 DAA), however, the associated KEGG pathways and GO terms do not suggest a connection to the significant fruit processes of cell expansion and starch accumulation occurring at that time point. Blast2GO analysis revealed that 15.2% of the entire plastid-targeted proteome of apple lacked GO term information.

Table 7. GO terms enriched in uniquely plastid-targeted proteins identified with UCLUST 50% method.

	GO term	Ontology	Description	p-value	FDR
Apple	GO:0006259	P	DNA metabolic process	4.60E-08	1.80E-05
	GO:0016265	P	death	2.90E-04	3.80E-02
	GO:0008219	P	cell death	2.90E-04	3.80E-02
	GO:0003676	F	nucleic acid binding	3.90E-15	4.90E-13
	GO:0003677	F	DNA binding	2.30E-12	1.40E-10
	GO:0005488	F	binding	2.20E-10	9.30E-09
Arabidopsis	GO:0030528	F	transcription regulator activity	7.20E-06	3.60E-04
	GO:0003700	F	transcription factor activity	6.90E-06	3.60E-04
	GO:0003677	F	DNA binding	9.10E-05	3.00E-03
	GO:0005634	C	nucleus	1.30E-25	2.70E-23
	GO:0005654	C	nucleoplasm	6.60E-06	6.70E-04
	GO:0031981	C	nuclear lumen	3.90E-05	1.30E-03
	GO:0031974	C	membrane-enclosed lumen	3.90E-05	1.30E-03
	GO:0043233	C	organelle lumen	3.90E-05	1.30E-03
	GO:0070013	C	intracellular organelle lumen	3.90E-05	1.30E-03
	GO:0044428	C	nuclear part	1.10E-04	2.60E-03
	GO:0044446	C	intracellular organelle part	1.10E-04	2.60E-03
	GO:0044422	C	organelle part	1.10E-04	2.60E-03
Grapevine	No significant term				
Peach	GO:0016787	F	hydrolase activity	1.20E-04	8.90E-03
Black cottonwood	GO:0003676	F	nucleic acid binding	3.60E-04	4.40E-02
Strawberry	No significant term				
Tomato	GO:0003677	F	DNA binding	8.60E-05	6.60E-03
	GO:0030528	F	transcription regulator activity	2.60E-04	1.00E-02
	GO:0003700	F	transcription factor activity	4.70E-04	1.20E-02

Blast2GO was used to determine GO terms associated with all predicted plastid-targeted proteins. Enrichment analysis was performed with agriGO to identify significant enriched GO terms. Gene Ontology terms are provided for biological process (P), molecular function (F), and cellular component (C).

However, the set of 2,224 genes represented in this study reveals that this subset is better characterized as it contains only 78 (3.5%) sequences with no associated GO terms. Of course the mere expression of a gene does not indicate that a functional protein is present within the fruit plastids as this process could be affected or controlled at a number of levels including translation, interaction with chaperone proteins, redox state of the plastid, presence of appropriate translocation proteins, protein and mRNA stability and turnover, and likely many other factors. Regardless, the data presented in this study indicate that the expression of genes encoding plastid-targeted proteins is dynamic in the fruit of *Malus*

Table 8. Percentage of unique plastid proteome containing GO information.

Species	Plastid-targeted Proteins	Number w/o GO information	Unique to Species (USEARCH)	Number w/o GO information	Unique to Species (UCLUST)	Number w/o GO information
Arabidopsis thaliana	5,382	10 (0.2%)	1,446	6 (0.4%)	2,154	8 (0.4%)
Malus × domestica	10,492	1,592 (15.2%)	4,257	1,348 (31.7%)	4,817	1,386 (28.8%)
Vitis vinifera	3,015	542 (18.0%)	750	415 (55.3%)	976	455 (46.6%)
Prunus persica	3,860	719 (18.6%)	628	379 (60.4%)	795	450 (56.6%)
Populus tichocarpa	4,515	527 (11.7%)	1,308	425 (32.5%)	1,732	462 (26.6%)
Fragaria vesca	4,922	1,539 (31.3%)	2,042	1,350 (66.1%)	2,305	1,385 (60.1%)
Solanum lycopersicum	4,009	873 (21.8%)	1,210	693 (57.3%)	1,654	750 (45.3%)

Plastid-targeted proteins were analyzed using Blast2GO to identify GO terms associated with each protein sequence. With the exception of *Arabidopsis*, significant proportions of chloroplast-targeted proteins datasets lack GO term information. This further increases in the datasets comprised of chloroplast-targeted proteins unique to each investigated species.

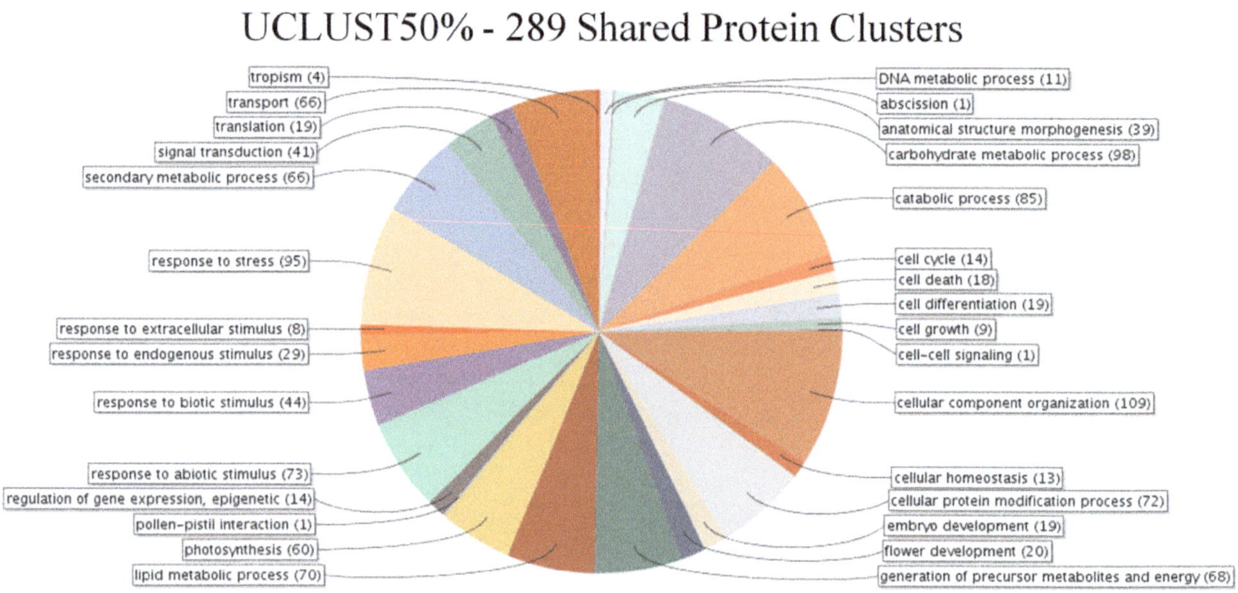

Figure 4. Biological process GO term composition of proteins shared between the 7 investigated plastid-targeted proteomes using two separate techniques. Two separate analyses were performed to identify proteins within the predicted plastid-targeted proteomes of seven species. The first, UCLUST with 50% identity, generated 15,750 clusters, 289 of which contained a member from all 7 species. USEARCH comparison performed at 40% identity and 40% coverage identified 737 sequences in the Arabidopsis thaliana putative plastid-targeted dataset which had a match in a protein sequence from the plastid-targeted sequences from all of the other 6 species. Sequences were analyzed via Blast2GO to determine the biological processes in which they partake.

× *domestica* and may play key roles in the development and quality of apple fruit.

Prediction and comparative analysis of plastid-targeted proteomes

In order to identify plastid targeted-proteins in seven species of interest (*Arabidopsis thaliana*, *Prunus persica*, *Malus* × *domestica*, *Populus trichocarpa*, *Fragaria vesca*, *Solanum lycopersicum*, and *Vitis vinifera*), plastid-targeting predictions were primarily performed using Target P. TargetP was selected in order to be consistent with previously published work in this area and because previous studies have found TargetP to be the most reliable single prediction program [17,20]. TargetP analysis revealed a large variance in the percentage of total transcripts encoding putative plastid-targeted proteins between the investigated species. The largest of these datasets belonged to *Malus* × *domestica*with 18.3% of its nuclear-encoded proteins predicted to be plastid-targeted, while the lowest was that of *Populus trichocarpa* with only 9.9% (Table 2). Header information for the predicted plastid-targeted datasets is provided in File S3.

Comparison of plastid-targeted proteomes with model systems

The predicted plastid proteomes for *Malus* × *domestica*, *Fragaria vesca*, *Populus trichocarpa*, *Prunus persica*, *Vitis vinifera*, and *Solanum lycopersicum* were independently compared with the *Arabidopsis* plastid proteome dataset as well as the entire *Arabidopsis* protein set using USEARCH. In *Arabidopsis*, the Chloroplast2010 project (www.plastid.msu.edu) [40] identified 5,181 unique genes encoding plastid-targeted proteins using software predictions and direct experimental evidence. Since this dataset did not represent embryo lethal mutants, an additional 201 genes predicted to encode embryo lethal plastid-targeted proteins from the SeedGenes database [55] were added to the dataset used in this study for comparative analysis to bring it to a total of 5,382 sequences. Comparison at 40% identity and 40% coverage (40/40) reveals that about 50% of each predicted plastid proteome has a likely homolog in the *Arabidopsis* plastid-targeted proteome subset, while about 60–70% of the proteins have likely homologs in the entire *Arabidopsis* protein subset (Table 3). Further comparison with 50/50 clustering parameter lowers these estimates significantly.

Additional comparison of putative plastid-targeted protein sequences for all six species was performed against the plastid-targeted proteome of *Solanum lycopersicum*, another model system for plastid biology research. This analysis showed that smaller proportions of the plastid proteomes had homologs in the tomato plastid proteome (ranging from 32–48%) than they had in the Arabidopsis proteome (40–60%) (Table 4). Strawberry and apple had the lowest similarity with the predicted plastid-proteomes of both *Arabidopsis* and tomato, while peach had the highest.

Identification of unique plastid-targeted proteins

Two separate analyses were performed to identify the plastid-targeted proteins in each of the seven species examined. The first consisted of a USEARCH-based comparison of predicted plastid-targeted proteins against the plastid-targeted protein sequences from the other six species. A second comparison utilized a clustering technique with UCLUST [33]. In this analysis the plastid-targeted proteins from all species were clustered together and clusters containing singletons or sequences from a single species were identified and further analyzed. In both the USEARCH and UCLUST-based analyses, a significant proportion of each predicted plastid proteome was found to be unique to that species (Table 5). The proportion of uniquely targeted proteins ranges from 16.3% in *Prunus persica* to 41.5% in *Fragaria vesca* in the USEARCH method and 20.6% in *Prunus persica* to 46.8% in *Fragaria vesca* in the UCLUST method.

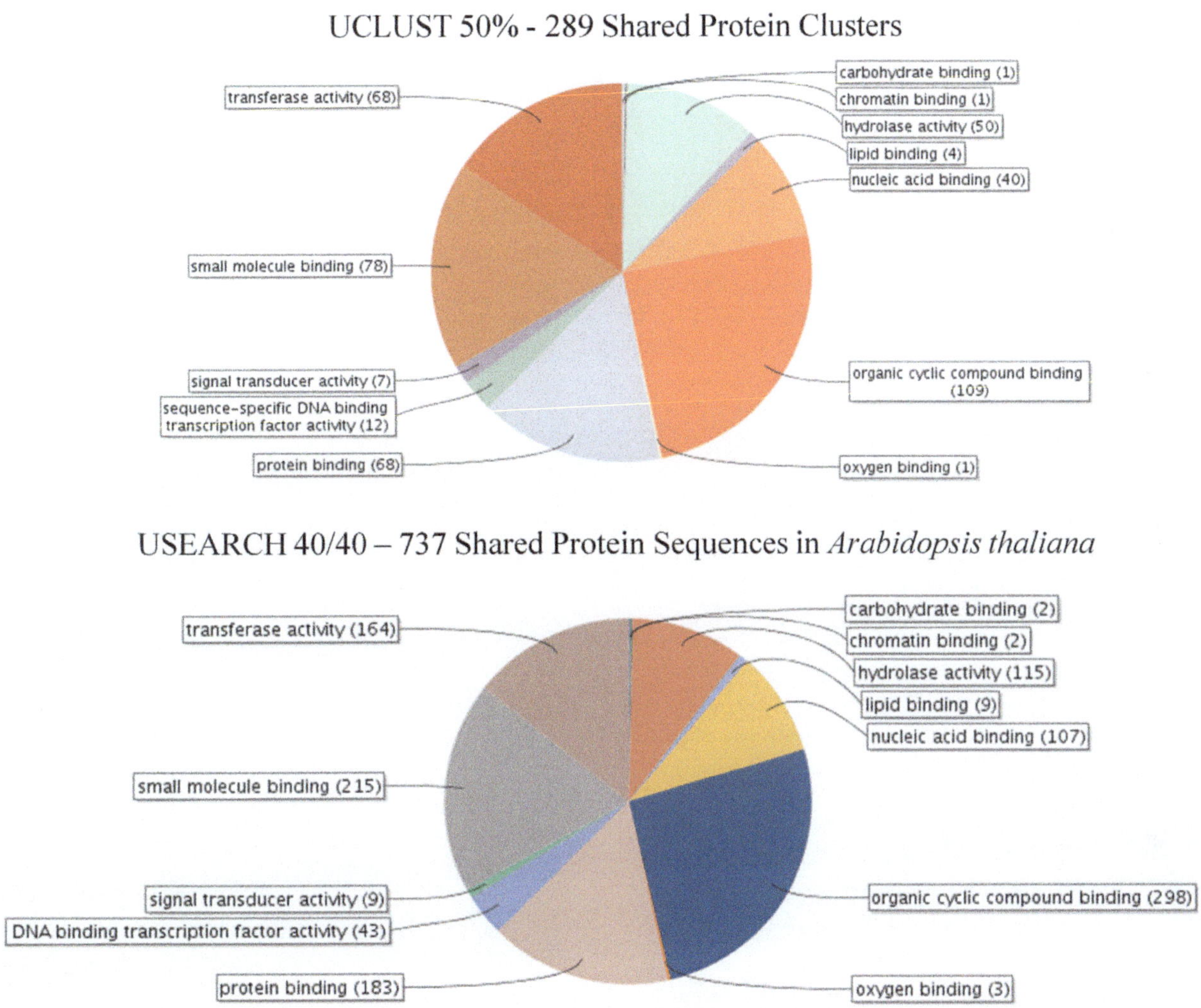

Figure 5. Molecular Function GO term composition of proteins shared between the 7 investigated plastid-targeted proteomes using two separate techniques. Two separate analyses were performed to identify proteins within the predicted plastid-targeted proteomes of seven species. The first, UCLUST with 50% identity, generated 15,750 clusters, 289 of which contained a member from all 7 species. USEARCH comparison performed at 40% identity and 40% coverage identified 737 sequences in the Arabidopsis thaliana putative plastid-targeted dataset which had a match in a protein sequence from the plastid-targeted sequences from all of the other 6 species. Sequences were analyzed via Blast2GO to determine the molecular functions of proteins at level 3.

However, the majority of the protein sequences have a homolog in at least one of the investigated species (Figure 3). In order to determine if uniquely-targeted proteins were in fact completely unique to each species, as opposed to alternatively targeted, a comparison with USEARCH 40/40 parameters was performed against datasets comprising the entire predicted proteomes of the other 6 species (Table 5). This investigation revealed thata significant number of proteins that lacked homology with interspecies plastid-targeted proteins, in fact have alternatively targeted homologs. This difference was most significant in *Arabidopsis* and *Malus* × *domestica* increasing the percentage of homology by 15.3% and 16.8%, respectively. Roughly 7–10% more proteins had homologs when using this matching scheme in the other investigated species. Information on these uniquely plastid-targeted proteins is provided in File S4.

GO term enrichment analysis was performed on the species-specific plastid-targeted protein sequences for both UCLUST and USEARCH-based analyses. Comparisons were performed using a Fisher's test with the entire predicted plastid proteome as a reference. This analysis revealed significantly enriched GO terms for both comparative techniques. In the USEARCH based technique, the majority of enriched GO terms were present in the unique apple plastid-targeted sequences with the most significant GO terms were DNA metabolic process (GO:006259, p-value 8.30E-60) and cellular macromolecule metabolic process (GO:0044260) (Table 6). Additionally, the GO term nucleic acid binding (GO:0003676) was enriched in poplar as well as apple. No significant GO terms were found for grape, strawberry or tomato. More GO terms were found to be enriched in the UCLUST-based comparison (Table 7). In this analysis transcription regulator activity (GO:0030528) and transcription factor activity (GO:0003700) were enriched in the *Arabidopsis* and tomato datasets. DNA binding (GO:0003677) was enriched in apple, *Arabidopsis* and tomato. Cell death (GO:0008219) and death (GO:0016265) were also enriched in the apple UCLUST50 dataset.

While a significant amount of functional annotation was performed with Blast2GO, a large proportion of the proteins

Table 9. GO terms enriched in *Arabidopsis thaliana* members of the 289 plastid-targeted protein clusters shared between all species investigated.

GO term	Ontology	Description	Number in UCLUST50	Number in At Cp-targeted Proteome	p-value	FDR
GO:0015979	P	photosynthesis	96	362	1.30E-22	5.10E-20
GO:0005975	P	carbohydrate metabolic process	167	911	3.70E-19	7.40E-17
GO:0043412	P	macromolecule modification	138	718	2.10E-17	2.10E-15
GO:0006464	P	protein modification process	138	718	2.10E-17	2.10E-15
GO:0008152	P	metabolic process	419	3515	1.90E-16	1.50E-14
GO:0006091	P	generation of precursor metabolites and energy	108	515	2.40E-16	1.60E-14
GO:0044238	P	primary metabolic process	388	3152	4.60E-16	2.60E-14
GO:0009987	P	cellular process	436	3796	9.90E-15	4.90E-13
GO:0044267	P	cellular protein metabolic process	157	925	1.40E-14	6.70E-13
GO:0016265	P	death	54	203	7.10E-13	2.60E-11
GO:0008219	P	cell death	54	203	7.10E-13	2.60E-11
GO:0019748	P	secondary metabolic process	112	624	8.60E-12	2.80E-10
GO:0009628	P	response to abiotic stimulus	148	940	6.80E-11	2.10E-09
GO:0019538	P	protein metabolic process	178	1238	8.30E-10	2.40E-08
GO:0006629	P	lipid metabolic process	116	703	1.00E-09	2.70E-08
GO:0016043	P	cellular component organization	181	1274	1.50E-09	3.30E-08
GO:0006519	P	cellular amino acid and derivative metabolic process	130	824	1.50E-09	3.30E-08
GO:0044281	P	small molecule metabolic process	130	824	1.50E-09	3.30E-08
GO:0050896	P	response to stimulus	220	1657	3.90E-09	8.10E-08
GO:0009058	P	biosynthetic process	293	2394	4.40E-09	3.80E-08
GO:0044237	P	cellular metabolic process	316	2637	4.20E-09	3.80E-08
GO:0006950	P	response to stress	177	1274	1.60E-08	2.80E-07
GO:0023052	P	signaling	99	636	5.40E-07	9.40E-06
GO:0007165	P	signal transduction	86	534	8.80E-07	1.30E-05
GO:0050794	P	regulation of cellular process	86	534	8.80E-07	1.30E-05
GO:0009056	P	catabolic process	146	1056	8.80E-07	1.30E-05
GO:0065007	P	biological regulation	135	959	1.00E-06	1.50E-05
GO:0023046	P	signaling process	87	545	1.10E-06	1.50E-05
GO:0023060	P	signal transmission	87	545	1.10E-06	1.50E-05
GO:0050789	P	regulation of celbiological process	102	674	1.40E-06	1.80E-05
GO:0044260	P	cellular macromolecule metabolic process	194	1530	2.60E-06	3.30E-05
GO:0006810	P	transport	136	1010	1.10E-05	1.30E-04
GO:0051234	P	establishment of localization	136	1010	1.10E-05	1.30E-04
GO:0051179	P	localization	136	1010	1.10E-05	1.30E-04
GO:0009605	P	reponse to external stimulus	57	342	3.50E-05	4.00E-04
GO:0006807	P	nitrogen compound metabolic process	203	1686	4.70E-05	5.00E-04
GO:0006139	P	nucleobase, nucleoside, nucleotide and nucleic acid metabolic process	203	1686	4.70E-05	5.00E-04
GO:0009607	P	response to biotic stimulus	88	618	1.10E-04	1.10E-03
GO:0043170	P	macromolecule metabolic process	216	1842	1.20E-04	1.20E-03
GO:0009719	P	reponse to endogenous stimulus	69	484	7.40E-04	7.40E-03
GO:0003824	F	catalytic activity	291	2207	4.00E-13	4.40E-11
GO:0000166	F	nucleotide binding	139	934	1.80E-08	1.00E-06
GO:0016740	F	transferase activity	115	760	2.20E-07	4.80E-06
GO:0016301	F	kinase activity	56	286	1.80E-07	4.80E-06

Table 9. Cont.

GO term	Ontology	Description	Number in UCLUST50	Number in At Cp-targeted Proteome	p-value	FDR
GO:0016772	F	transferase activity, transferring phosphorus-containing groups	56	286	1.80E-07	4.80E-06
GO:0005488	F	binding	327	2878	9.30E-07	1.70E-05
GO:0016817	F	hydrolase activity, acting on acid anhydrides	11	31	1.30E-04	1.30E-03
GO:0016818	F	hydrolase activity, acting on acid anhydrides, in phosphorus-containing anhydrides	11	31	1.30E-04	1.30E-03
GO:0003774	F	motor activity	11	31	1.30E-04	1.30E-03
GO:0016462	F	pyrophosphatase activity	11	31	1.30E-04	1.30E-03
GO:0017111	F	nucleoside-triphosphatase activity	11	31	1.30E-04	1.30E-03
GO:0009579	C	thylakoid	111	524	3.50E-17	6.90E-15
GO:0009536	C	plastid	337	2749	1.10E-11	1.10E-09
GO:0016020	C	membrane	228	1712	1.10E-09	7.10E-08
GO:0044444	C	cytoplasmic part	394	3569	4.70E-08	2.30E-06
GO:0005737	C	cytoplasm	409	3823	1.10E-06	4.40E-05

GO terms from the 497 *Arabidopsis thaliana* proteins present within the 289 shared clusters were analyzed by agriGO to identify enriched GO terms. Chi-square test was performed with a p-value cutoff of 0.01.

predicted to be unique to each species lack any associated GO term (Table 8). Proteins predicted to be unique to the plastids of *Arabidopsis* appear to be the best characterized with 98% containing some form of GO information, followed by those of *Prunus persica* with 81.9%. *Solanum lycopersicum* displays the least amount of GO information with only 56.5% of these uniquely plastid-targeted proteins having associated GO terms.

Analysis of proteins conserved between all plastid-targeted proteomes

Two separate analyses were performed to identify proteins which were predicted to be targeted to the plastids of all seven angiosperms studied. First analysis workflow utilized a semi-global approach where UCLUST [33] was used to cluster the proteins of all predicted plastid proteomes at 50% identity. The second approach utilized a global approach where the plastid proteome of *Arabidopsis thaliana* was compared with every other species' predicted plastid proteome at 40% identity and 40% coverage. Those proteins which have a matched protein from all other six species were then determined to be conserved across the plastid proteomes.

The first analysis using UCLUST at 50% identity identified 289 clusters of proteins which had at least one member from all seven species. These 289 clusters contain 497 unique sequences from *Arabidopsis thaliana*, 773 from *Malus × domestica*, 384 from *Vitis vinifera*, 392 from *Fragaria vesca*, 545 from *Populus trichocarpa*, 439 from *Prunus persica*, and 478 from *Solanum lycopersicum*. Blast2GO analysis reveals that these proteins are involved in a large number of biological processes (Figure 4) with the most populous being cellular component organization (GO:0016043, 109 proteins), carbohydrate metabolic process (GO:0005975, 98 proteins), and response to stress (GO:0006950, 95 proteins). The electron transport chain-related molecular functions for this data set reveals that the majority of proteins fit into six main categories; organic cyclic compound binding (GO:0097159 109 proteins), small molecule binding (GO:0036094 78 proteins), transferase activity (GO:0016740 68 proteins), protein binding (GO:0005515 68 proteins), hydrolase activity (GO:0016787 50 proteins), and nucleic acid binding (GO:0003676 40 proteins) (Figure 5). GO term enrichment was performed by selecting a single *Arabidopsis thaliana* protein sequence and utilizing agriGO to compare with those GO terms from the entire *Arabidopsis thaliana* predicted plastid-targeted proteome. This dataset contains 56 GO terms which were enriched with a p-value cut-off of 0.01 (Table 9). The lowest p-values are associated with the GO terms photosynthesis (GO:0015979), carbohydrate metabolic process (GO:0005975), macromolecule modification (GO:0043412), protein modification process (GO:0006464) and thylakoid (GO:0009579) respectively.

Analysis with USEARCH 40/40 identified the presence of 737 unique protein sequences from Arabidopsis thaliana with a matching protein in the predicted plastid-targeted proteomes of *Solanum lycopersicum*, *Prunus persica*, *Vitis vinifera*, *Malus × domestica*, *Fragaria vesca*, and *Populus trichocarpa*. As in the UCLUST50 analysis, the top three biological processes in this dataset as determined by Blast2GO are cellular component organization (GO:0016043, 254 proteins), response to stress (GO:0006950, 251 proteins), and carbohydrate metabolic process (GO:0005975, 210 proteins) (Figure 4). Again, the most populous molecular function GO terms mirror those in the UCLUST 50% with the majority falling into the six categories of: organic cyclic compound binding (GO:0097159 298 proteins), small molecule binding (GO:0036094 215 proteins), transferase activity (GO:0016740 164 proteins), protein binding (GO:0005515 183 proteins), hydrolase activity (GO:0016787 115 proteins), and nucleic acid binding (GO:0003676 107 proteins) (Figure 5). GO term enrichment using agriGO identified 59 GO terms to be enriched with a p-value cut-off of 0.01 (Table 10). The lowest p-values are associated with the GO terms photosynthesis (GO:0015979), thylakoid (GO:0009579), macromolecule modification (GO:0043412), protein modification process (GO:0006464), and generation of precursor metabolites and energy respectively (GO:0006091).

Table 10. GO terms enriched in *Arabidopsis thaliana* members of the 737 proteins with a match in all 6 species as determined by USEARCH4040.

GO Term	Ontology	Description	Number in USEARCH 4040	Number in At Cp-targeted Proteome	p-value	FDR
GO:0015979	P	photosynthesis	127	362	8.90E-22	3.50E-19
GO:0043412	P	macromolecule modification	190	718	7.70E-18	1.00E-15
GO:0006464	P	protein modification process	190	718	7.70E-18	1.00E-15
GO:0006091	P	generation of precursor metabolites and energy	146	515	4.70E-16	3.70E-14
GO:0009987	P	cellular process	634	3796	4.50E-16	3.70E-14
GO:0008152	P	metabolic process	598	3513	1.80E-15	1.20E-13
GO:0044238	P	primary metabolic process	551	3152	4.30E-15	2.40E-13
GO:0044267	P	cellular protein metabolic process	218	925	7.60E-15	3.80E-13
GO:0005975	P	carbohydrate metabolic process	210	911	3.00E-13	1.30E-11
GO:0044237	P	cellular metabolic process	471	2637	1.50E-12	5.90E-11
GO:0009628	P	response to abiotic stimulus	212	940	1.90E-12	6.90E-11
GO:0006519	P	cellular amino acid and derivative metabolic process	191	824	3.60E-12	1.10E-10
GO:0044281	P	small molecule metabolic process	191	824	3.60E-12	1.10E-10
GO:0050896	P	response to stimulus	316	1657	6.00E-10	1.70E-08
GO:0019538	P	protein metabolic process	249	1238	1.40E-09	3.70E-08
GO:0016043	P	cellular component organization	254	1274	2.00E-09	4.60E-08
GO:0065007	P	biological regulation	203	959	2.00E-09	4.60E-08
GO:0006950	P	response to stress	251	1274	8.20E-09	1.80E-07
GO:0044260	P	cellular macromolecule metabolic process	288	1530	2.70E-08	5.70E-07
GO:0016265	P	death	61	203	6.30E-08	1.20E-06
GO:0008219	P	cell death	61	203	6.30E-08	1.20E-06
GO:0019748	P	secondary metabolic process	137	624	3.00E-07	5.40E-06
GO:0009058	P	biosynthetic process	407	2394	5.50E-07	9.50E-06
GO:0048856	P	anatomical structure development	155	749	1.40E-06	2.40E-05
GO:0065008	P	regulation of biological quality	84	344	2.00E-06	3.10E-05
GO:0006629	P	lipid metabolic process	146	703	2.70E-06	4.10E-05
GO:0043170	P	macromolecule metabolic process	318	1842	1.40E-05	2.00E-04
GO:0050789	P	regulation of biological process	134	674	6.70E-05	9.50E-04
GO:0009607	P	response to biotic stimulus	124	618	9.40E-05	1.30E-03
GO:0009653	P	anatomical structure morphogenesis	117	584	1.70E-04	2.30E-03
GO:0022414	P	reproductive process	86	404	2.50E-04	3.10E-03
GO:0007165	P	signal transduction	107	534	3.60E-04	4.30E-03
GO:0050794	P	regulation of cellular process	107	534	3.60E-04	4.30E-03
GO:0023046	P	signaling process	108	545	5.00E-04	5.60E-03
GO:0023060	P	signal transmission	108	545	5.00E-04	5.60E-03
GO:0006810	P	transport	181	1010	6.20E-04	6.50E-03
GO:0051234	P	establishment of localization	181	1010	6.20E-04	6.50E-03
GO:0051179	P	localization	181	1010	6.20E-04	6.50E-03
GO:0090066	P	regulation of anatomical structure size	46	191	8.60E-04	8.20E-03
GO:0016049	P	cell growth	46	191	8.60E-04	8.20E-03
GO:0008361	P	regulation of cell size	46	191	8.60E-04	8.20E-03
GO:0032535	P	regulation of cellular component size	46	191	8.60E-04	8.20E-03
GO:0023052	P	signaling	121	636	9.70E-04	9.00E-03
GO:0042592	P	homeostatic process	40	161	1.10E-03	9.90E-03
GO:0019725	P	cellular homeostasis	40	161	1.10E-03	9.90E-03
GO:0016301	F	kinase activity	95	286	1.10E-14	6.10E-13

Table 10. Cont.

GO Term	Ontology	Description	Number in USEARCH 4040	Number in At Cp-targeted Proteome	p-value	FDR
GO:0016772	F	transferase activity, transferring phosphorus-containing groups	95	286	1.10E-14	6.10E-13
GO:0000166	F	nucleotide binding	215	934	1.50E-13	5.60E-12
GO:0003824	F	catalytic activity	406	2207	1.10E-11	3.10E-10
GO:0005488	F	binding	497	2878	4.20E-11	9.30E-10
GO:0016740	F	transferase activity	164	760	3.90E-08	7.20E-07
GO:0016818	F	hydrolase activity, acting on acid anhydrides, in phosphorus-containing anhydrides	17	31	2.70E-06	2.70E-05
GO:0016817	F	hydrolase activity, acting on acid anhydrides	17	31	2.70E-06	2.70E-05
GO:0003774	F	motor activity	17	31	2.70E-06	2.70E-05
GO:0016462	F	pyrophosphatase activity	17	31	2.70E-06	2.70E-05
GO:0017111	F	nucleoside-triphosphatase activity	17	31	2.70E-06	2.70E-05
GO:0009579	C	thylakoid	154	524	1.90E-18	3.70E-16
GO:0016020	C	membrane	321	1712	2.70E-09	2.60E-07
GO:0009536	C	plastid	460	2749	1.50E-07	1.00E-05

GO terms from one *Arabidopsis thaliana* proteins from each of the 737 shared protein sequences were analyzed by agriGO to identify enriched GO terms. Chi-square test was performed with a p-value cutoff of 0.01.

Comparing the USEARCH40/40 dataset along with that of the UCLUST50 dataset reveals that the two methods agree upon 439 shared plastid-targeted sequences, with 58 present only in UCLUST50 and 298 in USEARCH40/40. Both datasets have similar enriched GO terms with 49 enriched compared to the entire *Arabidopsis thaliana* plastid-targeted proteomes. However, USEARCH40/40 enriched GO terms contain an additional 10 biological process not represented in the UCLUST50 analysis; anatomical structure development (GO:0009653), regulation of biological quality (GO:0065008), anatomical structure morphogenesis (GO:0009653), reproductive process (GO:0022414), cell growth (GO:0016049), regulation of anatomical structure size (GO:0090066), regulation of cell size (GO:0008361), regulation of cellular component size (GO:0032535), homeostatic process (GO:0042592), and cellular homeostasis (GO:0019725).

The UCLUST50 enriched GO terms include five biological processes [catabolic process (GO:0009056), response to external stimulus (GO:0009719), nitrogen compound metabolic process (GO:0006807), nucleobase, nucleoside, nucleotide and nucleic acid metabolic process (GO:0006139), and response to endogenous stimulus (GO:0009719)] and two cellular components [cytoplasmic part (GO:0044444) and cytoplasm (GO:0005737)] not represented in the USEARCH40/40 analysis. A complete list of the 478 *Arabidopsis thaliana* loci from the UCLUST50 dataset and the 737 protein sequences from the USEARCH40/40 dataset along with their respective GO terms are provided in File S5.

Proteins conserved in this study were further compared with those from GreenCut2. GreenCut2 represents a collection of 597 nuclear-encoded proteins determined to be conserved across 20 photosynthetic eukaryotes, but absent in non-photosynthetic organisms [56]. A total of 677 unique loci from *Arabidopsis thaliana* (33 redundant of the original set of 710) were compared with those identified as shared between the seven species examined in this study both with USEARCH 40/40, as well as UCLUST 50% using Venny [57]. This comparison identified 70 proteins present in all three datasets, and a substantial set unique to each dataset. Sequence information from this comparison is provided in File S6.

In order to look at the overlap of homologous proteins between each species, the Jaccard similarity coefficient was determined for the USEARCH40/40 (Table 11). This comparison displays that the plastid-targeted proteomes of apple, strawberry, Arabidopsis and poplar are most similar to that of Peach, while the predicted plastid-targeted proteomes of peach, tomato and grape are most similar to *Arabidopsis*. An additional modified Jaccard similarity coefficient matrix was generated with the UCLUST50 analysis. In this matrix the predicted plastid-targeted of all species are most similar to that of peach, while that of peach is most similar to apple.

Discussion

This study reveals several interesting aspects about the constitution of predicted plastid-targeted proteomes for the species analyzed. A large portion of a plant's nuclear genome is dedicated to plastid-targeted proteins, many of which lack identity to plastid-targeted proteins in other species. Some plastid-targeted proteins have identity with potentially alternatively-targeted proteins in other systems. Of the predicted plastid-targeted proteins in *Arabidopsis*, 737 have significant identity to predicted plastid-targeted proteins in each of six investigated species, suggesting an evolutionarily core conserved set of plastid-targeted proteins. The caveat is that TargetP accuracy is not well defined, and has been shown to differ between experiments [58]. van Wijk and Baginsky determined that TargetP has a 35% false positive rate, suggesting that the predicted datasets established here will be greatly reduced upon future confirmatory experiments. Experimental approaches to characterize the proteome of plastids or some of their

Table 11. Jaccard's similarity coefficient matrix for seven species based on predicted plastid-targeted proteomes.

USEARCH 40/40							
	Strawberry	**Arabidopsis**	**Apple**	**Peach**	**Tomato**	**Poplar**	**Grape**
Strawberry	1.000	0.409	0.365	0.449	0.369	0.322	0.335
Arabidopsis	0.409	1.000	0.391	0.466	0.447	0.378	0.397
Apple	0.365	0.391	1.000	0.423	0.342	0.310	0.303
Peach	0.449	0.466	0.423	1.000	0.427	0.389	0.393
Tomato	0.369	0.447	0.342	0.427	1.000	0.352	0.365
Poplar	0.322	0.378	0.310	0.389	0.352	1.000	0.330
Grape	0.335	0.397	0.303	0.393	0.365	0.330	1.000
UCLUST50							
	Strawberry	**Arabidopsis**	**Apple**	**Peach**	**Tomato**	**Poplar**	**Grape**
Strawberry	1.000	0.194	0.199	0.267	0.171	0.158	0.173
Arabidopsis	0.194	1.000	0.192	0.253	0.222	0.193	0.210
Apple	0.199	0.192	1.000	0.268	0.162	0.156	0.164
Peach	0.267	0.253	0.268	1.000	0.229	0.224	0.240
Tomato	0.171	0.222	0.162	0.229	1.000	0.176	0.198
Poplar	0.158	0.193	0.156	0.224	0.176	1.000	0.184
Grape	0.173	0.210	0.164	0.240	0.198	0.184	1.000

Similarity between each species' plastid-targeted proteomes was represented by calculating the Jaccard coefficient from the UCLUST50 comparison and a USEARCH 4040 analysis.

constituents have relied largely upon mass spectrometry techniques [59,60,61,62]. A study in 2004 focused on the isolation and classification of the constituents of the *Arabidopsis thaliana* chloroplast proteome resulting in the identification of 604 nuclear-encoded proteins [60]. However, TargetP (at that time) was only able to correctly predict the plastid localization of 62.3% of these proteins, with 6.1% predicted to target the mitochondria, 8.1% secreted, and 23.5% predicted to have "any other location". When excluding the envelope proteins, TargetP chloroplast localization accuracy increased to 67.2%. An additional study which identified 241 stromal proteins from *Arabidopsis thaliana* chloroplasts identified through MALDITOF MS and nano-LC-ESI-MS/MS had a much higher predictability to be chloroplast targeted by TargetP with 88% accuracy [61]. Yet another study of 916 nuclear-encoded *Arabidopsis* plastid-targeted proteins revealed that 86% were correctly predicted using TargetP [63]. Such studies indicate that localization prediction methods need to be improved. Some reasons for this could include the complexities of the experimental system, sequence data, dual targeting, splice variants, presence of lesser characterized transport systems, or simply a lack of understanding of mechanisms of localization. As genomes become better characterized and targeting prediction improves, our ability to better understand the commonalities and diversities of plastid compositions and functions will also likely improve.

While it may be expected that plastid-targeted proteins would be highly conserved in this 7-species analysis, a previous study demonstrated a striking lack of similarity between the plastid proteomes of *Arabidopsis* and *Oryza sativa* [20]. In this study, a predicted *Arabidopsis* chloroplast proteome containing 2,100 proteins shared only 900 with the 4,800 plastid proteins in *Oryza sativa*. These 900 proteins were largely involved in transcription, energy, and metabolism. It would not be surprising to see this shared set of proteins shrink substantially as the number of species compared increases. A large focus has been put on the identification of proteins found only in photosynthetic organisms termed the GreenCut [64]. The first draft of this protein set contained 349 proteins conserved in photosynthetic eukaryotes, and absent in non-photosynthetic organisms. This was later updated generating GreenCut2 with 597 conserved proteins [56]. The comparison of the conserved plastid-targeted protein identified in this study by UCLUST50 and USEARCH40/40 methods reveals that there is only a minor overlap with GreenCut2 (Figure 6). Additionally, clusters do not consist of members from either a single species, or all seven. Instead, there are a significant number of clusters in our analysis with members from various combinations of species. These clusters could be helpful in the identification of proteins involved in plastidial pathways or traits conserved within a set of species. The comparison of more closely related angiosperms may in fact yield more similar plastid proteomes, while including more distantly related angiosperms likely reduces this similarity. It is worth noting that these differences could potentially occur due to the loss or gain of chloroplast transit peptides, rearrangement of protein domains or gene duplication, to name a few plausible mechanisms.

Differences in the outcome of comparative analysis projects can be attributed to the utilization of alignment or clustering methods. Previous studies in microbial comparative proteomic and genomic studies have utilized an alignments of 50% identity and 50% coverage to predictively resolve paralogs from orthologs [65,66] and 90% nucleic acid identity in algae (Bayer et al., 2012). A previous study which compared plastid proteomes between Arabidopsis and rice utilized BLASTP with e-value cut-off of $10xe^{-10}$ [20] while that of GreenCut utilized a BLASTP mutual

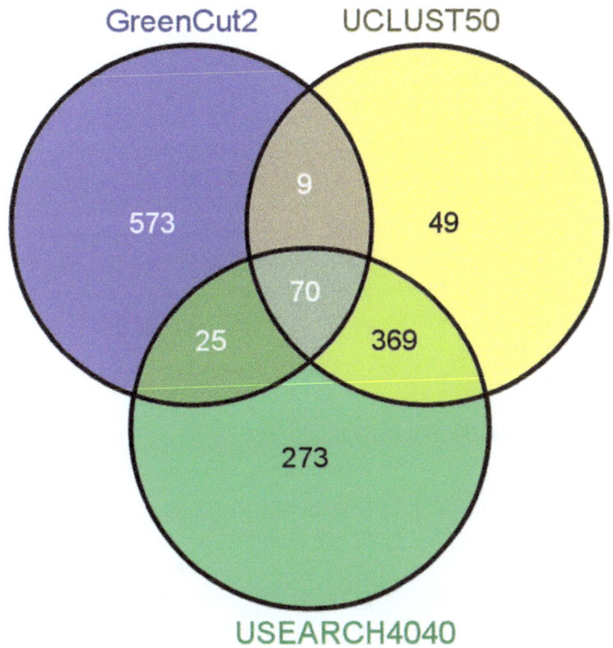

Figure 6. Comparison of conserved plastid-targeted protein datasets with GreenCut2. *Arabidopsis thaliana* loci associated with conserved proteins within the plastid-targeted proteomes identified in this study were compared with those of GreenCut2. A total of 70 sequences were found to be conserved between the three datasets.

best hit between *Chalydomonas* and Arabidopsis and human to identify paralogs, co-orthologs, and orthologs of *Chlamydomonas* [64]. GreenCut2 further expanded upon these parameters again utilizing mutual best hit analysis with BLASTP cut-off of $10xe^{-10}$ to identify orthologs and included sequences with over 50% amino acid identity as in-paralogs [56]. For this study, a predicted plastid-targeted protein was considered "unique" to a species if a global USEARCH alignment at 40% identity over 40% of the query sequence matched no sequences from the other 6 species. These parameters were chosen instead of 50% identity and 50% coverage as more matches were identified with a higher confidence with paralogous proteins removed. UCLUST, based on of global alignments, was then utilized to identify clusters with 50% amino acid identity, presumably including both orthologs and in-paralogs [33]. Of course the functionality of a protein cannot be ascertained through sequence identity alone. This study may not identify genes that have undergone rearrangement yet retain similar gene product functions, or examples of convergent evolution. The assumption in this study is that if such arrangements occur and functions are retained, or divergent plants adapt to create the same function in separate gene sequences, that an appropriate sequence from one of the remaining six other species would retain or have similar sequence identity for a 40/40 alignment to occur. However, the datasets that we present could be further investigated in the future through the use of a BLASTP-based comparison, and with other genomes as they become available.

The unique plastid-targeted proteins within each of the investigated species possess varied levels of functional information. Despite the lack of spatial and temporal transcriptome and proteome expression context, this information has a large referential value for future work. Blast2GO analysis and GO term enrichment analysis provide a glimpse as to what these proteins are likely participating in within the plastids. However, only a few GO terms were found to be enriched, most of which differed between species. This suggests that there are likely not specific classes of proteins which fall into this category of "unique" for the plastid proteomes of each species. As expected, due to substantial research performed on it, a large proportion of the proteins unique to the Arabidopsis plastid proteome (98%) possess associated GO terms (Table 8). However, tomato, which is used in the scientific community to understand and characterize the chloroplast to chromoplast transition, has the lowest percentage of unique proteins without GO terms, at 56.5%. Our analysis reinforces the lack of understanding about plastid biology especially in non-model systems, and the urgent need for further functional characterization of novel biological processes that these organelles harbor.

Plastids play an integral part in plant development, photosynthesis, and several other known biochemical processes. However in fruits their role has remained uncharacterized. Several important biochemical processes for synthesis and storage of pigments, nutraceutical and medically important compounds as well as aromatic compounds are resident in fruit plastids. These components are important both for consumer appeal as well as nutritional value. In addition, various physiological disorders, such as sunscald in apple, are associated with the inability of fruits to adequately quench excessive energy from sunlight. Through furthering our understanding of the plastid function in non-model plant systems and organs such as fruit, novel mechanisms for enhancing photosynthetic efficiency and crop productivity could be discovered.

The results presented in this work indicate that the current state of knowledge regarding plastid biology, mostly derived from model systems, is not comprehensive enough. In each plant species evaluated in this work, plastids are predicted to host a plethora of biological and metabolic processes necessitating subsequent wet-lab validation in non-model systems. New plant genomes are expected to enable the identification of potentially new plastid-targeted proteins that will aid in studying novel roles of plastids in plant development, metabolism and adaptation.

Conclusions

While previous studies have advocated the integration of multiple protein localization prediction techniques [20] it appears that no significant difference exists between a custom analysis utilizing multiple approaches and TargetP analysis for identifying plastid-targeted proteins in apple. Such results suggest that initial data mining with only TargetP may, in fact, be sufficient, depending upon the application. This is with the caveat that TargetP has an approximate 35% false positive rate in detecting plastid-targeted proteins [58]. As the understanding of protein localization improves, and complete genome sequences from larger number of plants become available, such predictive techniques will likely become a more reliable method to generate draft plastid proteomes.

The TargetP-based analysis indicates that a large subset of a plant's nuclear-encoded proteome is predicted to be localized to the plastid. However, the proportion of transcripts encoding plastid-targeted proteins varies, compromising 10–20% of the transcriptome depending upon the species investigated. Of the nuclear-encoded plastid proteome, there appears to be a significant subset that is species-specific. Many of these proteins have homology to proteins not predicted to be plastid-targeted in other systems, indicating that it may be common for proteins to gain or lose targeting peptides during evolution. If this is indeed the case, it would be interesting to investigate the evolutionary and mechanistic context of gain or loss of target peptides across species.

Through using two comparative methods, a USEARCH-based approach as well as a semi-global UCLUST-based approach, we displayed that very few plastid-targeted proteins are conserved between the predicted nuclear-encoded plastid proteomes. This value varied based upon the comparative technique and user parameters, but our predictions identify 497 and 737 Arabidopsis proteins which contain predicted plastid-targeted homologs in all other examined angiosperms. GO term enrichment analysis suggests that specific functions are significantly conserved in plastids, namely photosynthesis, many metabolic processes, transport, and cell death. Knowledge about these conserved proteins can be utilized in future studies to better understand and potentially predict those proteins which are plastid-targeted in other non-model systems and additionally identify novel plastid-targeted proteins.

The expression of genes encoding plastid-proteins appears to be very diverse within the fruit developmental continuum with 64 significant expression patterns detected (those containing five or more genes). However, most of the genes investigated can be clustered into nine expression patterns. These expression patterns can be overlapped with important milestones within the development of fruit to find plastid proteins which may be responsible for novel fruiting milestones or processes. While expression data is available for ~13,000 genes, subsequent developmental transcriptomics, metabolomics, and proteomics investigations are expected to provide a comprehensive understanding of the roles of plastids in apple fruit development.

Supporting Information

File S1 Expression data from Janssen et al. for *Malus* × *domestica* genes predicted to encode plastid-targeted proteins. Expression data for each cluster of co-expressed genes with sequence header and expression data in the form of Log2 of expression relative to the lowest value.

File S2 GO term information for *Malus* × *domestica* plastid-targeted protein clusters.

File S3 Header information for predicted plastid-targeted proteins from the investigated seven species. Excel file containing identifiers of predicted plastid-targeted protein sequences generated through with TargetP analysis.

File S4 Plastid-targeted proteins unique to each examined species with GO term information. Excel file containing the header information and associated GO terms.

File S5 Plastid-targeted proteins shared between all species. Excel file containing the header information for the UCLUST50 analysis separated by species. Information is provided for the 737 *Arabidopsis thaliana* sequences shared in the USEARCH comparative analysis.

File S6 Blast2GO annotation file containing header and GO term information for 70 protein sequences present in GreenCut2, UCLUST50, and USEARCH4040 analyses.

Acknowledgments

The authors would like to thank Ms. Vandhana Krishnan for her guidance in performing clustering analysis.

Author Contributions

Conceived and designed the experiments: SS AD. Performed the experiments: SS AH RR. Analyzed the data: SS AH RR PJ AD. Contributed reagents/materials/analysis tools: PJ AD. Wrote the paper: SS PJ AD.

References

1. Kutschera U, Niklas K (2005) Endosymbiosis, cell evolution, and speciation. Theory Biosci 124: 1–24.
2. Mettal U, Boland W, Beyer P, Kleinig H (1988) Biosynthesis of monoterpene hydrocarbons by isolated chromoplasts from daffodil flowers. Eur J Biochem 170: 613–616.
3. Kiss JZ, Hertel R, Sack FD (1989) Amyloplasts are necessary for full gravitropic sensitivity in roots of *Arabidopsis thaliana*. Planta 177: 198–206.
4. Fischer K, Weber A (2002) Transport of carbon in non-green plastids. Trends Plant Sci 7: 345–351.
5. Martin W, Herrmann RG (1998) Gene transfer from organelles to the nucleus: how much, what happens, and Why? Plant Physiol 118: 9–17.
6. Rochaix JD (1997) Chloroplast reverse genetics: new insights into the function of plastid genes. Trends Plant Sci 2: 419–425.
7. Martin W, Rujan T, Richly E, Hansen A, Cornelsen S, et al. (2002) Evolutionary analysis of Arabidopsis, cyanobacterial, and chloroplast genomes reveals plastid phylogeny and thousands of cyanobacterial genes in the nucleus. Proc Natl Acad Sci U S A 99: 12246–12251.
8. Jarvis P (2008) Targeting of nucleus-encoded proteins to chloroplasts in plants. New Phytol 179: 257–285.
9. Keegstra K, Olsen LJ, Theg SM (1989) Chloroplastic precursors and their transport across the envelope membranes. Annu Rev Plant Physiol Plant Mol Biol 40: 471–501.
10. Vonheijne G, Steppuhn J, Herrmann RG (1989) Domain-structure of mitochondrial and chloroplast targeting peptides. Eur J Biochem 180: 535–545.
11. Zhang XP, Glaser E (2002) Interaction of plant mitochondrial and chloroplast signal peptides with the Hsp70 molecular chaperone. Trends Plant Sci 7: 14–21.
12. Chen MH, Huang LF, Li HM, Chen YR, Yu SM (2004) Signal peptide-dependent targeting of a rice a-amylase and cargo proteins to plastids and extracellular compartments of plant cells. Plant Physiol 135: 1367–1377.
13. Miras S, Salvi D, Ferro M, Grunwald D, Garin J, et al. (2002) Non-canonical transit peptide for import into the chloroplast. J Biol Chem 277: 47770–47778.
14. Schemenewitz A, Pollmann S, Reinbothe C, Reinbothe S (2007) A substrate-independent, 14: 3: 3 protein-mediated plastid import pathway of NADPH: protochlorophyllide oxidoreductase A. Proc Natl Acad Sci U S A 104: 8538–8543.
15. Schein AI, Kissinger JC, Ungar LH (2001) Chloroplast transit peptide prediction: a peek inside the black box. Nucleic Acids Res 29.
16. Bannai H, Tamada Y, Maruyama O, Nakai K, Miyano S (2002) Extensive feature detection of N-terminal protein sorting signals. Bioinformatics 18: 298–305.
17. Emanuelsson O, Brunak S, von Heijne G, Nielsen H (2007) Locating proteins in the cell using TargetP, SignalP and related tools. Nat Protoc 2: 953–971.
18. Emanuelsson O, Nielsen H, Brunak S, von Heijne G (2000) Predicting subcellular localization of proteins based on their N-terminal amino acid sequence. J Mol Biol 300: 1005–1016.
19. Small I, Peeters N, Legeai F, Lurin C (2004) Predotar: A tool for rapidly screening proteomes for N-terminal targeting sequences. Proteomics 4: 1581–1590.
20. Richly E, Leister D (2004) An improved prediction of chloroplast proteins reveals diversities and commonalities in the chloroplast proteomes of Arabidopsis and rice. Gene 329: 11–16.
21. Verde I, Abbott A, Scalabrin S, Jung S, Shu S, et al. (2013) The high-quality draft genome of peach (*Prunus persica*) identifies unique patterns of genetic diversity, domestication and genome evolution. Nat Genet 45: 487–494.
22. Jaillon O, Aury JM, Noel B, Policriti A, Clepet C, et al. (2007) The grapevine genome sequence suggests ancestral hexaploidization in major angiosperm phyla. Nature 449: 463–U465.
23. Sato S, Tabata S, Hirakawa H, Asamizu E, Shirasawa K, et al. (2012) The tomato genome sequence provides insights into fleshy fruit evolution. Nature 485: 635–641.
24. Shulaev V, Sargent DJ, Crowhurst RN, Mockler TC, Folkerts O, et al. (2011) The genome of woodland strawberry (*Fragaria vesca*). Nat Genet 43: 109–U151.
25. Tuskan GA, DiFazio S, Jansson S, Bohlmann J, Grigoriev I, et al. (2006) The genome of black cottonwood, *Populus trichocarpa* (Torr. & Gray). Science 313: 1596–1604.
26. Velasco R, Zharkikh A, Affourtit J, Dhingra A, Cestaro A, et al. (2010) The genome of the domesticated apple (*Malus* × *domestica* Borkh.). Nat Genet 42: 833-+.
27. Kreuz K, Kleinig H (1984) Synthesis of prenyl lipids in cells of spinach leaf – compartmentation of enzymes for formation of isopentenyl diphosphate. Eur J Biochem 141: 531–535.
28. Camara B, Hugueney P, Bouvier F, Kuntz M, Moneger R (1995) Biochemistry and molecular biology of chromoplast development. Int Rev Cytol 163: 175–247.
29. Boyer J, Liu R (2004) Apple phytochemicals and their health benefits. Nutr J 3.
30. Liu YS, Roof S, Ye ZB, Barry C, van Tuinen A, et al. (2004) Manipulation of light signal transduction as a means of modifying fruit nutritional quality in tomato. Proc Natl Acad Sci U S A 101: 9897–9902.
31. Robinson N, Hewitt J, Bennett A (1988) Sink metabolism in tomato fruit. Plant Physiol 87.
32. Phan CT (1973) Chloroplasts of the peel and the internal tissues of apple-fruits. Experientia 29: 1555–1557.
33. Edgar RC (2010) Search and clustering orders of magnitude faster than BLAST. Bioinformatics 26: 2460–2461.
34. Hunter S, Apweiler R, Attwood TK, Bairoch A, Bateman A, et al. (2009) InterPro: the integrative protein signature database. Nucleic Acids Res 37: D211–D215.
35. Krogh A, Larsson B, von Heijne G, Sonnhammer ELL (2001) Predicting transmembrane protein topology with a hidden Markov model: Application to complete genomes. J Mol Biol 305: 567–580.
36. Ostlund G, Schmitt T, Forslund K, Kostler T, Messina DN, et al. (2010) InParanoid 7: new algorithms and tools for eukaryotic orthology analysis. Nucleic Acids Res 38: D196–D203.
37. Youens-Clark K, Buckler E, Casstevens T, Chen C, DeClerck G, et al. (2011) Gramene database in 2010: updates and extensions. Nucleic Acids Res 39: D1085–D1094.
38. Flicek P, Amode MR, Barrell D, Beal K, Brent S, et al. (2011) Ensembl 2011. Nucleic Acids Res 39: D800–D806.
39. Kinsella RJ, Kahari A, Haider S, Zamora J, Proctor G, et al. (2011) Ensembl BioMarts: a hub for data retrieval across taxonomic space. Database.
40. Lu Y, Savage LJ, Larson MD, Wilkerson CG, Last RL (2011) Chloroplast 2010: a database for large-scale phenotypic screening of Arabidopsis mutants. Plant Physiol 155: 1589–1600.
41. Lamesch P, Berardini TZ, Li DH, Swarbreck D, Wilks C, et al. (2012) The Arabidopsis Information Resource (TAIR): improved gene annotation and new tools. Nucleic Acids Res 40: D1202–D1210.
42. Conesa A, Gotz S, Garcia-Gomez JM, Terol J, Talon M, et al. (2005) Blast2GO: a universal tool for annotation, visualization and analysis in functional genomics research. Bioinformatics 21: 3674–3676.
43. Gotz S, Garcia-Gomez JM, Terol J, Williams TD, Nagaraj SH, et al. (2008) High-throughput functional annotation and data mining with the Blast2GO suite. Nucleic Acids Res 36: 3420–3435.
44. Quevillon E, Silventoinen V, Pillai S, Harte N, Mulder N, et al. (2005) InterProScan: protein domains identifier. Nucleic Acids Res 33: W116–W120.
45. Myhre S, Tveit H, Mollestad T, Laegreid A (2006) Additional Gene Ontology structure for improved biological reasoning. Bioinformatics 22: 2020–2027.
46. Kanehisa M (2002) The KEGG database. *In silico* simulation of biological processes 247: 91–103.
47. Kanehisa M, Goto S, Sato Y, Furumichi M, Tanabe M (2012) KEGG for integration and interpretation of large-scale molecular data sets. Nucleic Acids Res 40: D109–D114.
48. Du Z, Zhou X, Ling Y, Zhang Z, Su Z (2010) agriGO: a GO analysis toolkit for the agricultural community. Nucleic Acids Res 38: W64–70.
49. Janssen BJ, Thodey K, Schaffer RJ, Alba R, Balakrishnan L, et al. (2008) Global gene expression analysis of apple fruit development from the floral bud to ripe fruit. BMC Plant Biol 8.
50. Saeed AI, Hagabati NK, Braisted JC, Liang W, Sharov V, et al. (2006) TM4 microarray software suite. DNA Microarrays, Part B: Databases and Statistics 411: 134-+.
51. Saeed AI, Sharov V, White J, Li J, Liang W, et al. (2003) TM4: A free, open-source system for microarray data management and analysis. Biotechniques 34: 374-+.
52. Ben-Dor A, Shamir R, Yakhini Z (1999) Clustering gene expression patterns. J Comput Biol 6: 281–297.
53. Kaul S, Koo HL, Jenkins J, Rizzo M, Rooney T, et al. (2000) Analysis of the genome sequence of the flowering plant *Arabidopsis thaliana*. Nature 408: 796–815.
54. Clijsters H (1969) On the photosynthetic activity of developing apple fruits. Qualitas Plantarum et Materiae Vegetabiles 19: 129–140.
55. Meinke D, Muralla R, Sweeney C, Dickerman A (2008) Identifying essential genes in *Arabidopsis thaliana*. Trends Plant Sci 13: 483–491.
56. Karpowicz SJ, Prochnik SE, Grossman AR, Merchant SS (2011) The GreenCut2 resource, a phylogenomically derived inventory of proteins specific to the plant lineage. J Biol Chem 286: 21427–21439.

57. Oliveros JC (2007) VENNY. An interactive tool for comparing lists with Venn Diagrams.
58. van Wijk KJ, Baginsky S (2011) Plastid proteomics in higher plants: current state and future goals. Plant Physiol 155: 1578–1588.
59. Barsan C, Sanchez-Bel P, Rombaldi C, Egea I, Rossignol M, et al. (2010) Characteristics of the tomato chromoplast revealed by proteomic analysis. J Exp Bot 61: 2413–2431.
60. Kleffmann T, Russenberger D, von Zychlinski A, Christopher W, Sjolander K, et al. (2004) The *Arabidopsis thaliana* chloroplast proteome reveals pathway abundance and novel protein functions. Curr Biol 14: 354–362.
61. Peltier JB, Cai Y, Sun Q, Zabrouskov V, Giacomelli L, et al. (2006) The oligomeric stromal proteome of *Arabidopsis thaliana* chloroplasts. Mol Cell Proteomics 5: 114–133.
62. Wang YQ, Yang Y, Fei Z, Yuan H, Fish T, et al. (2013) Proteomic analysis of chromoplasts from six crop species reveals insights into chromoplast function and development. J Exp Bot.
63. Zybailov B, Rutschow H, Friso G, Rudella A, Emanuelsson O, et al. (2008) Sorting signals, N-terminal modifications and abundance of the chloroplast proteome. PLoS One 3.
64. Merchant SS, Prochnik SE, Vallon O, Harris EH, Karpowicz SJ, et al. (2007) The Chlamydomonas genome reveals the evolution of key animal and plant functions. Science 318: 245–251.
65. Friis C, Wassenaar TM, Javed MA, Snipen L, Lagesen K, et al. (2010) Genomic Characterization of *Campylobacter jejuni* Strain M1. PLoS One 5.
66. Lukjancenko O, Ussery DW, Wassenaar TM (2012) Comparative Genomics of Bifidobacterium, Lactobacillus and Related Probiotic Genera. Microb Ecol 63: 651–673.

Understanding the Relationship between Cotton Fiber Properties and Non-Cellulosic Cell Wall Polysaccharides

Dhivyaa Rajasundaram[1,2], Jean-Luc Runavot[3], Xiaoyuan Guo[4], William G. T. Willats[4], Frank Meulewaeter[3], Joachim Selbig[1,2]*

1 Institute of Biochemistry and Biology, University of Potsdam, Potsdam-Golm, 14476, Germany, **2** Max-Planck Institute of Molecular Plant Physiology, Potsdam-Golm, 14476, Germany, **3** Bayer CropScience NV-Innovation Center, Technologiepark 38, 9052 Gent, Belgium, **4** Department of Plant and Environmental Sciences, Faculty of Sciences, University of Copenhagen, Thorvaldsensvej, 40 1.1871, Fredriksberg C, Denmark

Abstract

A detailed knowledge of cell wall heterogeneity and complexity is crucial for understanding plant growth and development. One key challenge is to establish links between polysaccharide-rich cell walls and their phenotypic characteristics. It is of particular interest for some plant material, like cotton fibers, which are of both biological and industrial importance. To this end, we attempted to study cotton fiber characteristics together with glycan arrays using regression based approaches. Taking advantage of the comprehensive microarray polymer profiling technique (CoMPP), 32 cotton lines from different cotton species were studied. The glycan array was generated by sequential extraction of cell wall polysaccharides from mature cotton fibers and screening samples against eleven extensively characterized cell wall probes. Also, phenotypic characteristics of cotton fibers such as length, strength, elongation and micronaire were measured. The relationship between the two datasets was established in an integrative manner using linear regression methods. In the conducted analysis, we demonstrated the usefulness of regression based approaches in establishing a relationship between glycan measurements and phenotypic traits. In addition, the analysis also identified specific polysaccharides which may play a major role during fiber development for the final fiber characteristics. Three different regression methods identified a negative correlation between micronaire and the xyloglucan and homogalacturonan probes. Moreover, homogalacturonan and callose were shown to be significant predictors for fiber length. The role of these polysaccharides was already pointed out in previous cell wall elongation studies. Additional relationships were predicted for fiber strength and elongation which will need further experimental validation.

Editor: David D. Fang, USDA-ARS-SRRC, United States of America

Funding: This work was supported by the European Union Seventh Framework Programme (FP7 2007–2013) under the WallTraC project (grant agreement No. 263916). This paper reflects the authors' views only. The European Community is not liable for any use that may be made of the information contained herein. The funders had no role in study design, data collection and analysis, decision to publish, or preparation of the manuscript.

Competing Interests: Co-authors Dr. Jean-Luc Runavot and Dr. Frank Meulewaeter are employed by Bayer CropScience NV, Innovation Center. This does not alter the authors' adherence to all the PLOS ONE policies on sharing data and materials.

* Email: jselbig@uni-potsdam.de

Introduction

Cell walls, the key determinant of overall plant growth and development are primarily composed of polysaccharides, namely cellulose, hemicellulose, and pectins, lignin, and structural proteins [1,2]. Cell wall biology has been an area of prominent research over many years with the use of novel technologies to probe these higher order structures in the native state. Since the early 1970's, comparative biochemical analyses revealed that all plant cell walls share several common features. However, they exhibit diversity with respect to their chemical composition [3–5]. Indeed, cell walls are structurally complex as they are constantly remodeled and re-constructed during plant growth and development. Also, walls are modulated according to functional requirements, thereby limiting our knowledge on cell wall design [6–8]. Biochemical analyses complemented by genetic analyses have identified genes and gene products associated with cell wall synthesis. However, an understanding of how these genes are expressed across cells of different tissues and their impact on cell wall design and maintenance is still lacking [9–11]. Furthermore, the glycan-rich cell walls influence the nutritional and processing properties of plant based products such as pulp for paper manufacture, textile fibers, timber products, pharmaceuticals, and materials for fuel and composite manufacture [12–15]. Therefore, understanding the plant cell walls is not only fundamental to plant sciences but also of industrial relevance.

Microarrays are widely used in plant research for the high throughput analysis of nucleotides, proteins and increasingly, carbohydrate [16–18]. Carbohydrate microarrays also referred to as glycan arrays enable hundreds of glycans to be analyzed in parallel. Glycans on the arrays can include oligosaccharides, polysaccharides, glycoproteins and glycolipids [19–21]. Glycan arrays have several biological and medical applications which include glycoproteomic methods to identify new glycoproteins and glycans [22,23], characterization of glycan probes [24], profiling carbohydrate-lectin interactions [25,26], glycosaminoglycans-growth factor and cytokine interactions [27,28], pathogen-induced antibody interaction [29,30], cancer-antibody induced interaction [31,32], carbohydrate-virus interactions [33], quantitative carbohydrate-protein interactions [34], and drug discovery [35,36].

Comprehensive microarray polymer profiling (CoMPP), a microarray based glycan screening method is mostly used for high throughput characterization of plant cell walls. In this technique, generation of microarrays by sequential extraction of cell wall polysaccharides and screening samples against a large number of well-defined cell wall probes such as antibodies, carbohydrate binding proteins and modules is done. This methodology was first described in *Arabidopsis thaliana* and *Physcomitrella patens* [37]. In the study of Singh et al, application of CoMPP to study cotton fibers showed that towards the end of elongation, there was a loss in certain cell wall polymer epitopes [38]. Despite the availability of glycan arrays from several experiments, computational analysis has mostly been restricted to collection of glycobiology information in databases, motif analysis of glycans, and oligosaccharide structure determination [39–41].

In our study, we used the glycan array technology to study cotton fibers, one of the most important raw materials for the textile industry. There are four different domesticated species producing cotton fibers namely *Gossypium hirsutum* ('Upland cotton'), *Gossypium barbadense* ('Pima'or 'Egyptian'cotton), *Gossypium arboreum* ('Tree cotton'), and *Gossypium herbaceum* [42]. The development of cotton fibers occurs in four major stages: initiation, elongation, secondary wall synthesis and maturation. Although much work has already been done on the cotton fiber transcriptome, the key question in cotton fiber research is to link the cell wall profile of different cotton types to the cotton fiber properties and to a better understanding of fiber development [43–46]. Here, we aim to study the relation between fiber properties and non-cellulosic polysaccharide composition using univariate and multivariate regression based approaches on a diverse set of cotton fibers. To this end, we analyzed two datasets for the same cotton fibers: a glycan array profile and the physical fiber properties as determined by HVI and AFIS. We elucidated the usefulness of regression based approaches to determine the functional relationship between the two datasets and we also selected a subset of variables which have a good prediction of the phenotypic traits.

Materials and Methods

Plant material and evaluation of phenotypic traits

In this study, we used 32 different cotton lines of which three are from *Gossypium arboreum*, three from *Gossypium barbadense*, two from *Gossypium herbaceum* and 24 from *Gossypium hirsutum*. The cotton lines used in this study are listed in Table S1, including the plant introduction number (PI number) from the USDA National Plant Germplasm System (http://www.ars-grin.gov/npgs/). Seeds were sown in soil compost and plants were grown at constant conditions in a greenhouse at 26–28°C during a 16 h photoperiod. Mature cotton fibers were collected by harvesting all fully open bolls from several plants. The impact of boll position and plant-to-plant variation was minimized by mixing the fiber from all harvested bolls. Two types of analyses were performed on these fibers, the first being the glycan array measurements (Table S2) and the second being fiber characteristics/phenotype measurements (Table S1). For each line, High Volume Instrument (HVI) and Advanced Fiber Information System (AFIS) measurements were performed on 40 g of mature cotton fiber by CIRAD (France) according to the standard methods ASTM D3818-92 and D5867-05. These measurements were done on 6 and 5 replicates for HVI and AFIS, respectively, except for micronaire where only 2 replicates were performed. Five fiber characteristics which include length from HVI and AFIS, strength, elongation and micronaire were selected for further analysis due to their importance for textile processing. Length HVI refers to the average fiber length of the longer 50% of fibers in a given sample. Length AFIS (W) L deduces length parameters from individual fiber measurements. Strength of the cotton fiber refers to the force required to break a bundle of fibers 1 tex in size (1 tex equals the weight in grams of 1000 meters of fibers). Elongation of the cotton fibers is the measurement of the elasticity of cotton fibers with a higher number indicating more elasticity. Micronaire is obtained by measuring the resistance of the fibers to airflow and depends on the fiber fineness and degree of maturation.

Comprehensive Microarray Polymer Profiling (CoMPP) of mature cotton fiber cell wall

CoMPP analysis was performed on mature cotton fibers as previously described by [38] with minor modifications. Mature cotton fiber samples were extracted sequentially in 50 mM cyclohexanediamine tetraacetic acid (CDTA) and 4 M Sodium hydroxide (NaOH) with 1% (v/v) sodium tetrahydridoborate ($NaBH_4$). These two solvents were used to extract pectins and non-cellulosic polysaccharides, respectively. For each line, 300 ul of solvent was added to 10 mg of sample and incubated with shaking for 2 h. After centrifugation, supernatant from each extraction was printed in four replicates and four dilutions (1:2, 1:6, 1:18 and 1:54 [v/v] dilutions). Cadoxen extraction was omitted because it is mainly used to extract cellulose which we do not aim to analyse in our study. The array was probed with eleven monoclonal antibodies (mAbs) recognizing different carbohydrate epitopes as listed out in Table 1. A heat map was generated to display the relative intensity of each signal to the maximum signal observed within each antibody detection (Table S2). CoMPP is a semi-quantitative technique and should not be taken to obtain absolute amounts. Practically speaking, we set the maximum value in the whole data sheet as 100 and the other values are divided by this maximum value and multiplied by 100 to obtain numbers comprised between 0 and 100. When the quantification is done, the arrays are manually checked to make sure that there are clear dots on it and not only background or noise. The negative control is an array incubated with 5% milk in PBS and probed with secondary antibody and then developed as the others.

Pre-processing of the data

For the statistical analysis, we use R version 3.1.0 on a 64 bit linux platform [56]. The numerical values from both datasets were of different physical quantities and on different scales of magnitude. Moreover, there is no external knowledge that variables with higher numeric variation should be considered more important. Standardization of the raw data was done by computing z- scores of the raw data. Z- scores were calculated for each data point by subtracting the mean and dividing by the standard deviation of all data points.

Linear methods to delineate the relationship between the two datasets

Multiple regression models the relationship between a single scalar response variable and a set of explanatory (or independent) variables. Here, we used multiple regression analysis to model which of the cell wall probes were associated to the fiber

Table 1. List of probes used in the glycan array.

Probes used in the analysis	Specificity of the probes	Reference	Source
BS-400-2	(1,3)-β-D-glucan (callose)	[47]	Purchased from Biosupplies (Australia)
JIM5	Partially methyl-esterified homogalacturonan (HG)	[48]	Paul Knox lab
LM19	Un-esterified homogalacturonan (HG)	[49]	Paul Knox lab
JIM13	Arabinogalactan (AGP)	[50]	Paul Knox lab
JIM20	Extensin glycoproteins	[51]	Paul Knox lab
LM11	Xylan	[52]	Paul Knox lab
LM15	XXXG xyloglucans (XG) epitope	[53]	Paul Knox lab
LM24	XXLG and XLLG xyloglucan (XG) epitopes	[24]	Paul Knox lab
LM25	XXLG and XLLG xyloglucan (XG) epitopes	[24]	Paul Knox lab
BS-400-4	Mannan	[54]	Purchased from Biosupplies (Australia)
LM21	Mannan	[55]	Paul Knox lab

characteristics. This allowed us to determine the overall fit (variance explained) of the model and the relative contribution of each of the cell wall probes to the total variance explained. The results from the analysis were reported in the coefficients and ANOVA tables. Summary of the fitted model object gave an account of the residuals, the estimates of the intercept, the slope (with the results of a t-test), the residual standard error, the R^2 statistic and the results of an F-test. The terms used in the output of regression analysis are defined as follows: residual standard error is the standard deviation of the data about the regression line. The squared multiple correlation coefficient (R^2) is the proportion of variability in the response that is fitted in the model and the F value is a test statistic to decide whether the model as a whole has statistically significant predictive capability. p values give the statistically significant predictive capability in the presence of other variables [57,58]. Based on this, five models were selected to determine which of the cell wall polysaccharides play an important role in determining that particular fiber characteristic.

In addition to the multiple regression analysis, relationships between multiple dependent and independent variables were investigated simultaneously using canonical correlation analysis (CCA). The two sets of data were represented by matrices X (dimension $n \times p$) and Y (dimension $n \times q$) and columns in X and Y denote the variables p (glycan measurements) and q (fiber characteristics) respectively. Classification of variables as dependent or independent is of little importance for the statistical estimation of the canonical functions as canonical correlation finds linear combinations of sets of multiple dependent and independent variables which are maximally correlated [59,60].

The first step in CCA was to derive one or more canonical function between the glycan and phenotypic measurements. Each function consisted of a pair of variates, one representing the cell wall probes and the other representing the fiber characteristics. The maximum number of canonical variates (functions) that could be extracted from the sets of variables equals the number of variables in the smallest data set, independent or dependent. As a result, the first pair of canonical variates was derived so as to have the highest intercorrelation possible between the glycan array and the fiber measurements. Technically, the second pair of canonical variates exhibits the maximum relationship between the two sets of variables (variates) not accounted for by the first pair of variates and successive pairs of canonical variates were based on residual variance. Therefore each of the pairs of variates is orthogonal and independent of all other variates derived from the same set of data. The strength of the relationship between the pairs of variates obtained from both datasets was determined by the canonical correlation. An estimate of shared variance between the canonical variates was provided by the squared canonical correlations, also called canonical roots or eigenvalues. The statistical significance of each canonical function was assessed using multivariate tests of significance namely Wilk's lambda, Hotelling's trace, Pillai's trace and Roy's greatest characteristic criterion (Roy's gcr). The statistically significant canonical functions were then interpreted using canonical loadings, cross-loadings and redundancy index [61–65]. We used the "mixOmics" package [66] in R to perform the canonical correlation analysis.

Sparse partial least square regression to predict the cell wall probes associated to fiber characteristics

Partial least squares (PLS), a well-known regression technique dealing with collinear matrices, clearly has an edge over other regression techniques [67]. Unlike CCA, the PLS latent variables are linear combinations of the variables based on the maximization of covariance but do not allow feature selection. There are many variants of PLS of which we focused on a sparse partial least squares approach (sPLS) which includes a built-in feature to select variables while integrating the data. We used the "mixOmics" package [66] in the regression mode. Specifically, we use a two block data setup, X be the nxp matrix and Y be the nxq matrix where n denotes the samples, variables p and q denote the glycan measurements and fiber characteristics respectively. Sparse PLS, based on lasso regression penalizes the loading vectors using singular value decomposition and has an additional advantage to perform better even when the covariates are highly correlated. We used sPLS in the regression mode and the aim was to model the relationship between the variables and also predict one group of variables from the other [68–71].

Results

Standardization of the raw data

In this study, we attempted to assess the relationship between the cell wall polysaccharides and the physical fiber properties of mature cotton fibers, the data of which are provided as Table S1 and S2. The glycan array values used for the regression analysis were the sums from the CDTA and the NaOH extractions as performing the analysis using the individual values gave the same correlations. For the fiber characteristics dataset, the values were in different units and scales such as mm (for length), g/tex (for strength), and percentage (for elongation). To make the fiber characteristics dataset compliant to the glycan array, the raw data were jointly standardized using z scores prior to the analysis.

Modelling the fiber properties using linear regression models

We investigated the linear relationship between the fiber properties and their corresponding array values by a series of regression analyses. Multiple regression models were built considering one fiber characteristic at a time as the dependent variable and multiple probes as the independent variables. Five such models were predicted for the phenotypic traits and the overall model prediction result (Table 2) shows that the model for length HVI, length AFIS and micronaire are statistically significant. The significant predictor variables of length HVI are BS-400-2, LM19 and the ones for length AFIS include BS-400-2, JIM5, JIM20, LM15, and LM19. LM15, LM19, LM24 and LM25 are the significant predictor variables for the model predicting cotton fiber micronaire and the overall model has a p value of 4.906e-06. The models for strength and elongation do not show any statistical significance.

Assessing the relationship between multiple probes and all of the fiber characteristics simultaneously using canonical correlation analysis

The multiple regression analysis can predict the value of a single (metric) dependent variable from a linear function of a set of independent variables. However, to explore the relationship of sets of multiple predictor variables (probe measurements) to sets of multiple response variables (phenotypic traits) CCA was used. As CCA uses information from all the variables in both the predictor and response sets, it serves as a more efficient approach than methods routinely used, such as multiple linear regression.

For the CCA analysis, the glycan array measurements (probed by 11 antibodies) are designated as the set of independent variables. The fiber characteristics namely length AFIS, length HVI, strength, elongation and micronaire were specified as the set of dependent variables (Figure 1). However, it is of little importance to classify the variables as independent or dependent as the technique aims to maximize the correlation between the two sets of variables. In Figure 1, the terms r_{x1} to r_{x11} represent the canonical loadings which reflect the variance that the eleven variables from the glycan array shares with the independent canonical variate U_1. Similarly the terms r_{y1} to r_{y5} represent the canonical loadings which reflect the variance that the five phenotypic variables share with the dependent canonical variate V_1. The canonical correlation between the independent and dependent canonical variates is measured by the canonical functions which are represented by R^2_{c1} to R^2_{c5}. The statistical problem involved identifying any latent relationships (relationships between composites of variables rather than the individual variables themselves) between the glycan and the fiber measurements.

The canonical correlation which is based on the linear relationship of the glycan array data and fiber characteristics was computed to derive five canonical functions. Each of these functions consists of a pair of variates, one for the glycan array data and the other for the fiber characteristics. Since the study includes 11 independent variables and 5 dependent variables, the maximum number of canonical functions which could be derived is five (Table 3).

In addition to tests of each canonical function separately, multivariate tests of these five functions simultaneously were also performed. The test statistics employed include Wilks' lambda, Pillai's criterion, Hotelling's trace, and Roy's gcr. Table 4 details the p-values from the multivariate test statistics, which all indicate that only the first canonical function, taken collectively, is statistically significant at 1% level.

From the results of these tests, we proceeded to interpret other aspects of the analysis based on the first canonical function. A redundancy index was calculated for the independent and dependent variates of the first function in Table 5. The redundancy index is calculated as the average loading squared times the canonical R^2. As can be seen, the redundancy index for the dependent (0.191) and independent variates (0.200) is quite low. The low values result from the relatively low shared variance in the dependent variates (0.214) and independent variates (0.225), not the canonical R^2. With such a small percentage, this is an example of a statistically significant canonical function that does not have practical significance because it does not explain a large proportion of the dependent variables' variance.

The interpretations involve examining the canonical functions to determine the relative importance of each of the original variables in deriving the canonical relationships (Table 6). The three methods for interpretation are (1) canonical weights (standardized coefficients), (2) canonical loadings (structure correlations), and (3) canonical cross-loadings.

Table 6 contains the standardized canonical weights for each canonical variate for both dependent and independent variables. As mentioned earlier, the magnitude of the weights represent their relative contribution to the variate. Based on the size of the weights, the order of contribution of independent variables to the

Table 2. Summary statistics of the five possible multiple regression models.

Fiber characteristics	Residual standard error	Multiple R-squared	Adjusted R-squared	F-statistic	p-value	Significant predictors
Length HVI	0.696	0.706	0.545	4.372 on 11 and 20 DF	0.002	BS-400-2, LM19
Length AFIS	0.632	0.720	0.566	4.677 on 11 and 20 DF	0.001	BS-400-2, JIM5, JIM20, LM15, LM19
Strength	0.940	0.378	0.036	1.107 on 11 and 20 DF	0.404	JIM20
Elongation	0.825	0.376	0.033	1.098 on 11 and 20 DF	0.410	–
Micronaire	0.469	0.851	0.769	10.4 on 11 and 20 DF	4.906e-06	LM15, LM19, LM24, LM25

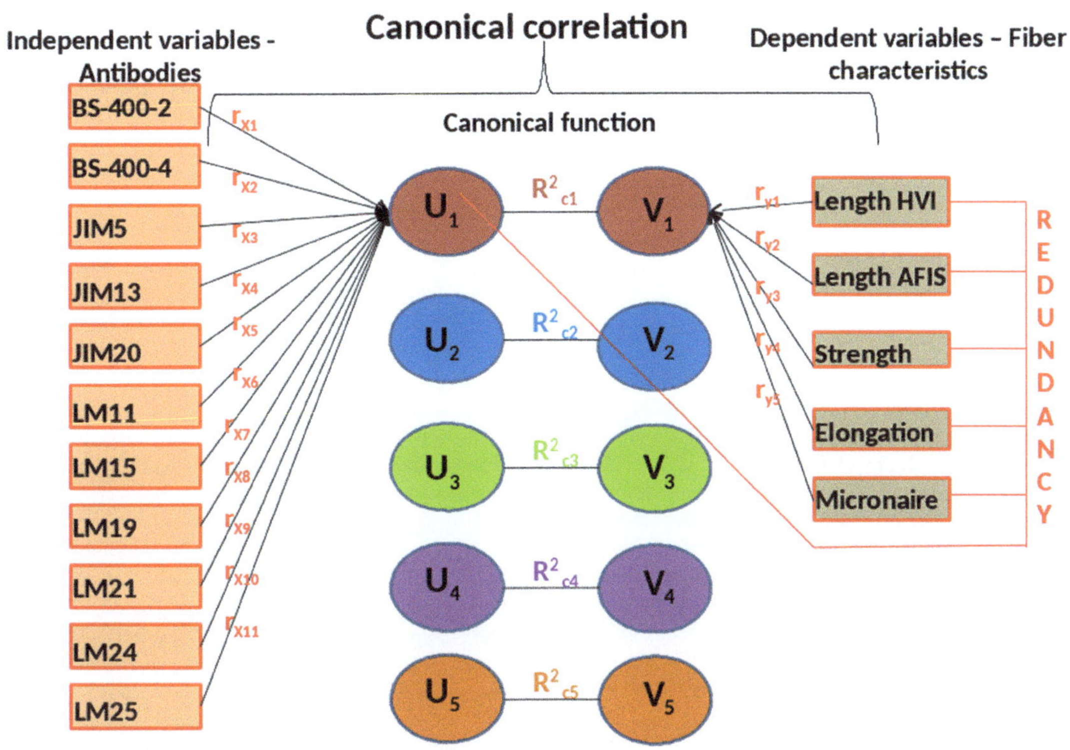

Figure 1. Canonical correlation analysis maximizes the correlation between the linear combination of the cell wall polysaccharides in the glycan array and the fiber properties. In this figure, given a linear combination of *X* variables: $U_1 = f_{1\times1} + f_{2\times2} + \ldots + f_p X_p$ and a linear combination of *Y* variables: $V_1 = g_1Y_1 + g_2Y_2 + \ldots + g_qY_q$, the first canonical correlation is the maximum correlation coefficient between U_1 and V_1, for all U_1 and V_1.

first variate is LM19, LM25, JIM5, LM15, BS-400-4, LM21, LM24, JIM13, and JIM20 and the dependent variable order on the first variate is micronaire followed by length AFIS, length HVI, strength and elongation. Because canonical weights are typically unstable, particularly in instances of multicollinearity, owing to their calculation solely to optimize the canonical correlation, the canonical loading and cross-loadings are considered more appropriate.

Table 6 also contains the canonical loadings for the dependent and independent variates for the first canonical functions. In the first dependent variates, all the five variables had different values of loadings resulting in low shared variance (0.214). This indicates a low degree of inter-correlation among the five dependent variables. Observing the independent variates, there is a different pattern and loading values ranged from 0.06 to 0.77. The variables with the highest loadings on the independent variate are LM25, LM19, LM15, and JIM5. We also observed some loadings with negative values which include those of BS-400-4, JIM20, and LM11.

In case of the cross loadings, micronaire has a value of −0.890 and interestingly has a negative loading. Length AFIS to some extent has a loading value of 0.387 while those of the other variables is low. By squaring these terms, we find the percentage of the variance for each of the variables explained by function 1. The results show that 79.21 percent of the variance in micronaire, 14.97 percent of the variance in length AFIS is explained by function 1 whereas strength, elongation and length HVI have very low values. Similarly for the independent variables' cross loadings, variables LM25, LM19, LM15, JIM5 have high correlations of 0.73, 0.67, 0.61, and 0.61 respectively. From this information, approximately 51.8% of the variance in LM25, 45.1% of the variance in LM19, 36.3% of the variance in LM15, and 35.7% of the variance in JIM5 is explained by the dependent canonical variates.

Table 3. Canonical Correlation analysis relating probe signals and fiber characteristics with the measure of overall model fit.

Canonical function	Canonical correlation	Canonical R^2	F statistics	p-value
1	0.945	0.883	2.85	1.57338e-05
2	0.868	0.753	1.88	1.072479e-02
3	0.803	0.645	1.30	2.035076e-01
4	0.523	0.273	0.59	8.715784e-01
5	0.342	0.116	0.34	9.040993e-01

Table 4. Multivariate tests of significance for the canonical functions.

Canonical function	Wilks' Lambda, using F-approximation (Rao's F):	Hotelling-Lawley Trace, using F-approximation:	Pillai-Bartlett Trace using F-approximation	Roy's largest root using F-approximation
1	1.57338e-05	2.666759e-07	0.00	8.732348e-12
2	1.072479e-02	1.221924e-03	0.042	
3	2.035076e-01	5.475660e-02	0.285	
4	8.715784e-01	8.365373e-01	0.801	
5	9.040993e-01	8.855773e-01	0.848	

The final step of interpretation is examining the signs of the cross-loadings. Examining the signs of the independent variables' cross loadings, those with high correlations have a positive direct relationship whereas BS-400-4, JIM20 and LM11 have an inverse relationship. The four highest cross-loadings of the first independent variate correspond to the variables with the highest canonical loadings as well. Observing the cross loadings of the dependent variables, we see that micronaire has the highest canonical loading and an inverse relationship. Also, elongation is observed to have an inverse relation but since it is of very low value, it was not taken into account.

sPLS approach to predict specific cell wall polysaccharides involved in fiber properties

sPLS was computed in the regression mode and the input for the analysis included the 11 cell wall probes along with the five fiber characteristics The number of dimensions H to be retained was estimated with the Qh^2 criterion, for which a value below the threshold 0.0975 indicates a significant contribution for the prediction purpose. The Qh^2 values calculated for each dimension of the sPLS showed that 2 dimensions were enough to capture the whole information. From Figure 2, we can interpret the results from the sPLS via the correlation circle plot where the predictor variables are in red and the response variables are represented in blue. A correlation circle plot is a graphic tool to represent variables of two different data-types and examine the relationships between the variables and variates. In this plot, variables namely cell wall probes and fiber measurements can be represented as vectors. The relationship between these two data-types is approximated by the inner product between the associated vectors which is defined as the product of the two vector lengths and their cosine angle. For better interpretation, two circles of radii 0.5 and 1 are represented to visualize the variables. The longer the distance to the origin, the stronger is the relationship between the variables.

Using the interpretation which is detailed, we find that BS-400-4, LM21, and JIM13 share a positive relationship with elongation characteristic of cotton fibers. We were also able to attribute the strength of the cotton fibers to JIM20, LM11 and LM24. Interestingly, LM19, JIM5, LM15, and LM25 were projected diametrically opposite to that of the micronaire in the correlation circle, thereby indicating a strongly negative relationship. Length HVI and length AFIS share a negative relation to BS-400-2. To estimate the significance of the predicted relationships, the root mean squared error prediction (RMSEP) values were computed for each response variable (fiber properties) and ranked according to the absolute value of their loadings in v_2. The lower the RMSEP value, the better the prediction of the model is. In this case, the model for micronaire was the best one (RMSEP of 0.71), followed by that of length AFIS (1.13), strength (1.14), elongation (1.15), and length HVI (1.21).

Figure 3 displays the graphical representation of the cotton lines in dimension 1 and 2. This plot shows that some of the lines are clustered together, with Acala SJ1, Germains Acala (GC 352 and GC 362), TAM-90C-19 S, and FM966 forming one cluster, Acala red okra, okra leaf, multiple marker, Tidewater, and TTU 202-1107B forming a second cluster and PIMAS7, Lankart 57, IV4F-91057, GA161, Ting tao tzu ching chung mien, Brymer brown, Malla guza, Selection of SHIH, China 10, Texas rust brown, Tex 1000 and 30834 (A1660) forming a third cluster. Strikingly, some of these clusters contain lines from different *Gossypium* species and lines from one species often belong to multiple clusters. The

Table 5. Redundancy analysis of dependent and independent variates for the first canonical function.

Standardized variance of the dependent variables explained by				
Their own Canonical variates (shared variance)			The opposite canonical variates (Redundancy)	
Percentage	Cumulative percentage	Canonical R^2	Percentage	Cumulative percentage
0.214	0.214	0.883	0.191	0.191
Standardized variance of the independent variables explained by				
Their own Canonical variates (shared variance)			**The opposite canonical variates (Redundancy)**	
Percentage	**Cumulative percentage**	**Canonical R^2**	**Percentage**	**Cumulative percentage**
0.225	0.225	0.883	0.200	0.200

Table 6. Canonical weights, loadings, and cross-loadings for the first canonical function.

	Canonical weights	Canonical loadings	Canonical cross-loadings
Dependent variables			
Length HVI	−0.636	0.127	0.120
Strength	−0.226	0.040	0.038
Elongation	−0.033	0.056	0.053
Micronaire	−0.843	−0.941	−0.890
Length AFIS	0.810	0.409	0.387
Independent variables			
BS-400-2	0.184	0.362	0.342
BS-400-4	−0.487	−0.119	−0.113
JIM5	−0.823	0. 0.632	0.598
JIM13	−0.290	0.288	0.272
JIM20	−0.204	−0.376	−0.355
LM11	−0.114	−0.360	−0.340
LM15	−0.719	0.638	0.603
LM19	1.243	0.712	0.672
LM21	0.356	0,275	0.260
LM24	−0.324	0.066	0.062
LM25	1.082	0.767	0.724

variation in fiber characteristics and composition is thus clearly not species-specific. However, one should be careful in interpreting the results from the individual lines as the study was designed to discover correlations between fiber properties and composition and not to study properties of individual lines.

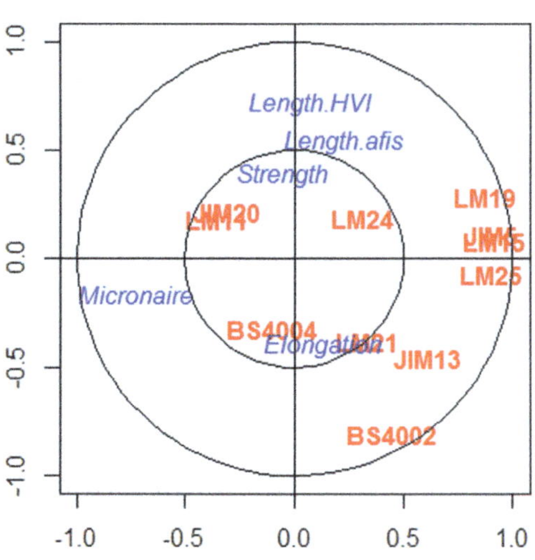

Figure 2. Graphical representation of the variables selected by sPLS on the first two dimensions predicts specific cell wall polysaccharides linked to the fiber properties. The coordinates of each variable are obtained by computing the correlation between the latent variable vectors and the original dataset. The selected variables are then projected onto correlation circles where highly correlated variables cluster together. These graphics help to identify association between the two datasets. The correlation between two variables is positive if the angle is sharp $\cos(\alpha)>0$, negative if the angle is obtuse $\cos(\theta)<0$, and null if the vectors are perpendicular $\cos(\beta)\sim 0$.

Discussion

Understanding the genetics and physiology of cotton fibers is of importance to the textile industry. There have been numerous studies, both profiling and sequencing based experiments to study cotton fiber development at the transcriptional level. The high degree of transcriptional complexity in the development of cotton fibers has been the focus of these studies [43,72–75]. We used the CoMPP technique in our analysis to study directly the glycan composition of cotton lines from different species. The work presented here demonstrates the potential of glycan microarrays in combination with multivariate statistical approaches for understanding the cell wall composition responsible for the fiber characteristics. Specifically, the use of regression based approaches in our study helps to predict models for each of the fiber trait under study.

We studied the association between glycan array measurements and their relation with fiber characteristics using linear approaches like multiple regression, CCA and sPLS. From the results of multiple regression (Table 2), we were able to predict three models for length HVI, length AFIS and micronaire of cotton fibers but not for strength and elongation characteristics. Moreover, to extend our understanding of the data to situations involving more than one fiber characteristic at a time, CCA was used as it simultaneously models effects of multiple independent variables on multiple dependent variables. As CCA uses information from all the variables in both the exposure and outcome variable sets and maximizes the estimation of the relationship between the two sets, CCA may offer a more efficient approach for assessing the relationship of the cell wall probes with fiber characteristics than methods routinely used such as multiple linear regression. CCA starts with simultaneous consideration of both glycan array measurements and the phenotype measures, limiting the inefficiencies that may accompany conventional multiple testing, and thus, reducing type-1 error. The resulting procedure gives a global view of association between indicators of both datasets. Thus,

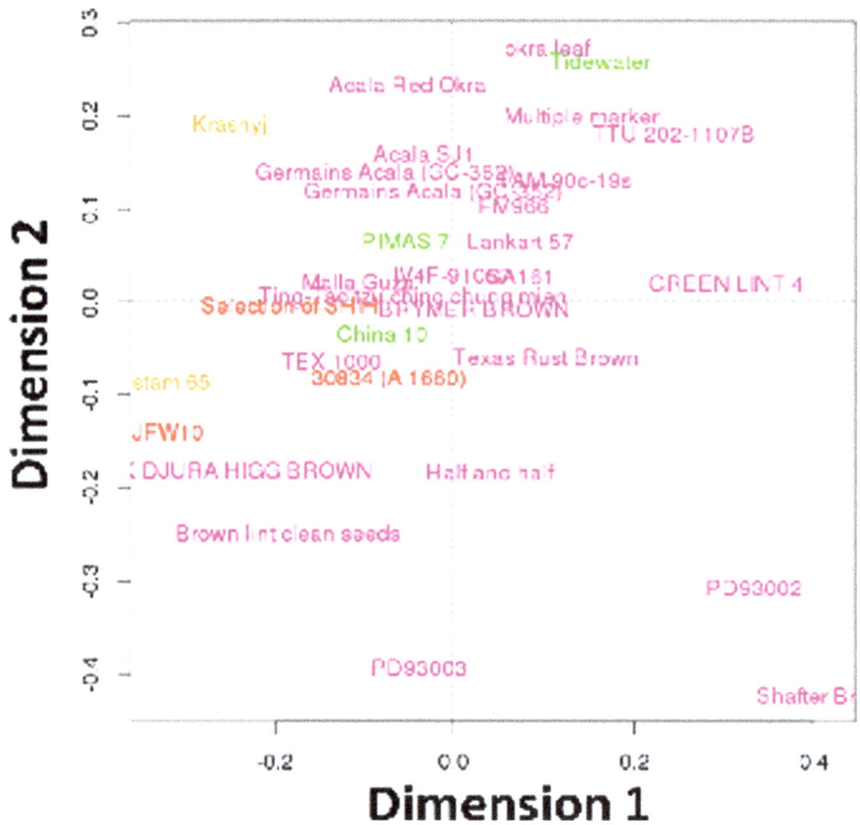

Figure 3. Graphical representation of the cotton lines on the first two sPLS dimensions shows the trend in clustering of specific cotton lines across different species. Four different species of cotton are shown in different colors. *Gossypium hirsutum* is colored in magenta, *Gossypium barbadense* in green, *Gossypium herbaceum* in orange and *Gossypium arboreum* in red.

CCA could be used as a comprehensive approach to extract information from data simultaneously. Another major advantage of using the CCA to multiple regression analysis is to deal with the issue of multicollinearity. In multiple regression, the interpretation is usually based on the significance of weights, which is highly influenced by multicollinearity. If two variables have a high correlation one of them will be completely eliminated even if both have a high correlation to the outcome. In our analysis this is illustrated by JIM5 and LM19 (both detecting homogalacturonan), with both showing a high correlation with micronaire in CCA but only LM19 being identified as a predictor of micronaire in the linear regression model. From the results of the CCA, we obtained an overall picture of associations between the glycan and phenotype measurements, with information about the relative contribution of the variables to that particular canonical variate through canonical loadings. The canonical analysis revealed that the canonical correlation was statistically significant at 1%. However, canonical correlation based methods are statistically difficult to assess as they do not fit into a regression framework. In this context penalized CCA adapted with elastic net (CCA-EN) could be used but the elastic net is similar to a lasso soft-thresholding penalization and the algorithm uses partial least squares and not canonical correlation computations [63]. From [63] it is evident that sPLS made a good compromise between all of these approaches and includes variable selection. Additionally, we used the sPLS approach to be able to predict specific cell wall polysaccharides linked with fiber characteristics. Moreover, sPLS maximizes the covariance between the latent variables whereas the canonical correlation based methods maximize the correlation.

There were both unique and common findings from the three types of regression analysis. The major and most significant finding in common to all these analyses is that micronaire is negatively correlated with the xyloglucan (XG) and homogalacturonan (HG) probes. One possible explanation for this observation is that cotton fiber with a high micronaire usually has a very thick secondary cell wall resulting in very high levels of cellulose and lower levels of the non-cellulosic components. However, we do not find a negative correlation of micronaire with other non-cellulosic compounds suggesting that increased cellulose levels of high micronaire fibers affect the XG and HG epitopes in a different way than the other non-cellulosic epitopes. For instance, it could specifically decrease extractability of the XG and HG epitopes. As micronaire measures a combination of fiber fineness and maturity, we wanted to understand whether the observed correlation is with maturity or fineness or a combination of both. We tested this using linear regression models once again and built models for fineness and maturity of the fibers. We observed that the regression models for fineness had an adjusted R^2 value of 0.803 with JIM5, LM19, and LM25 as significant predictors at a 1% threshold. The regression model for maturity was also significant at the 1% threshold but with no particular significant predictors thereby suggesting that the observed correlation is attributed to fiber fineness. This indicates that this correlation is linked to the thickness rather than the shape of the fiber, which is consistent with a link to the cellulose levels.

Since only the first canonical function of the CCA analysis is statistically significant and this function explains only for micronaire a large fraction of the variance, the results of the CCA analysis are not informative with respect to the other fiber properties. For these fiber properties, the correlation between fiber length and callose is the only one that was detected in both the linear regression and the sPLS analysis. Callose has been described to play a role in cotton fiber elongation. Indeed, it was reported that plasmodesmatal closure was positively correlated with the rapid fiber elongation and that callose was involved in the gating of these plasmodesmata [76]. However, this observation involves transient callose detection, only after 5 dpa and already significantly reduced at 20 dpa, what makes it unlikely to be detected in mature fibers. Other callose deposition was reported by [77]. This callose is supposed to be deposited in the secondary cell wall and remains in the fiber. From the results of the multiple regression models (Table 2), a positive correlation between several of the homogalacturonan probes and length property of the fibers is apparent. The link between pectins and the elongation of cell walls is already observed in several plant systems [78] and studies in flax stems, pea stems and maize coleoptiles revealed a negative correlation between pectin levels and cell elongation. In cotton fibers and trichomes, there exists a positive correlation between pectic sheath and elongation [79] and recent studies by [80] have established that pectic polysaccharides and xyloglucan containing uronic acids were the major polysaccharides extracted during elongation. Hence, our results are in agreement with various studies which state that pectin biosynthesis promotes fiber elongation [81] and that the degree of esterification is a key factor in controlling the elongation [38,82]. The correlation between length and HG was not detected in the sPLS analysis most likely because the stronger (negative) correlation of HG with micronaire.

Furthermore, relationships between fiber strength or elongation and specific carbohydrate epitopes could be deduced from the results of the sPLS analysis (Figure 2). For instance, fiber strength was associated both with the xylan (LM11) and the extensin (JIM20) epitope. A role of xylan in fiber strength would be consistent with the function of heteroxylan in other cell types which is commonly related to the strengthening of cell walls as revealed by defects in cellulose deposition in xylan mutants [83]. A role of extensin in fiber strength is less expected and would need experimental validation. In the linear regression analysis, extensin was identified as a significant predictor for length AFIS but not for length HVI. A role for extensin in determining cotton fiber length would be more consistent with its role in other plant cell types [84]. Finally, AGP glycan (JIM13) and mannan (BS-400-4 and LM21) epitopes were found to predict cotton fiber elongation from the sPLS model. Interestingly, studies have indicated that AGPs are important players during fiber development. Immunofluorescence assays by JIM 13 showed distinct patterns in developing fiber cells indicating that polysaccharide chains of AGPs are involved in initiation and elongation stages of cotton fibers [85–87]. However, it is not clear how these AGPs would affect the elongation property of the mature fiber. These unexpected correlations present thus interesting hypotheses for further structure-function relationship studies of the cotton fiber.

Overall, CoMPP assays of cell wall polysaccharides from cotton fibers suggest that it will be a powerful tool in detecting and quantifying the differences between large sets of cotton lines thus gathering lot of information which is necessary for a proper statistical approach. With the use of predictive statistical approaches to integrate different kinds of datasets, this analysis has thus discovered some correlations that are in line with already known biological functions and others for which the biological relevance still has to be tested. Also, it confirmed the relevance of this type of analysis to enable a detailed understanding of the data from CoMPP assays of cell wall polysaccharides. However, the use of mature cotton fibers in this analysis only allows detecting relevant correlations for components that are still present at maturity. In addition, many changes in polysaccharide composition occur between the fiber elongation stage and maturity. One would thus expect to identify only a fraction of the relationships between polysaccharide composition and fiber properties by analysis of mature fibers, especially for fiber properties such as length that are determined in the early stages of development. Hence it would be interesting to perform a similar kind of analysis using the polysaccharide composition of developing fibers to see whether additional relationships with fiber properties can be determined. The panel of cotton lines used in this study was selected to have maximal diversity in fiber properties and composition. Applying this type of analysis to commercially important cotton lines would allow to understand whether differences in polysaccharide composition affect properties of commercial cotton in the same way as observed in this study and to get insight into the developmental polysaccharides that are essential to obtain high quality cotton fibers. With the sequencing of the *G. hirsutum* genome, cotton fiber research is an exciting field and the work presented here will provide a base for future studies, with potential to translate this study on the developing fibers.

Supporting Information

Table S1 Fiber characteristics/phenotype measurements for the 32 cotton lines used in the study. The plant introduction number (PI number) from the USDA national plant germplasm is also included for each cotton line.

Table S2 Comprehensive microarray polymer profiling (CoMPP) heat map of CDTA and NaOH extractions of mature cotton fibers from 32 cotton lines. References for probe specificity are listed in Table 1.

Acknowledgments

We would like to thank Prof. JP Paul Knox, University of Leeds for the cell wall epitopes used in the glycan array experiment.

Author Contributions

Conceived and designed the experiments: DR JLR XG WW FM JS. Performed the experiments: JLR XG. Analyzed the data: DR. Contributed reagents/materials/analysis tools: DR JLR XG. Contributed to the writing of the manuscript: DR JLR. Read and approved the manuscript: DR JLR XG WW FM JS.

References

1. Heredia A, Jiménez A, Guillén R (1995) Composition of plant cell walls. Z Für Lebensm-Unters -Forsch 200: 24–31.
2. Keegstra K (2010) Plant Cell Walls. Plant Physiol 154: 483–486. doi:10.1104/pp.110.161240.
3. Somerville C, Bauer S, Brininstool G, Facette M, Hamann T, et al. (2004) Toward a Systems Approach to Understanding Plant Cell Walls. Science 306: 2206–2211. doi:10.1126/science.1102765.

4. Minorsky PV (2002) The wall becomes surmountable. Plant Physiol 128: 345–353. doi:10.1104/pp.900022.
5. Carpita NC, Gibeaut DM (1993) Structural models of primary cell walls in flowering plants: consistency of molecular structure with the physical properties of the walls during growth. Plant J Cell Mol Biol 3: 1–30.
6. Roberts K (2001) How the Cell Wall Acquired a Cellular Context. Plant Physiol 125: 127–130. doi:10.1104/pp.125.1.127.
7. McCann M, Rose J (2010) Blueprints for Building Plant Cell Walls. Plant Physiol 153: 365–365. doi:10.1104/pp.110.900324.
8. Pilling E, Höfte H (2003) Feedback from the wall. Curr Opin Plant Biol 6: 611–616.
9. Somerville C (2006) Cellulose synthesis in higher plants. Annu Rev Cell Dev Biol 22: 53–78. doi:10.1146/annurev.cellbio.22.022206.160206.
10. Mutwil M, Debolt S, Persson S (2008) Cellulose synthesis: a complex complex. Curr Opin Plant Biol 11: 252–257. doi:10.1016/j.pbi.2008.03.007.
11. Ellis M, Egelund J, Schultz CJ, Bacic A (2010) Arabinogalactan-Proteins: Key Regulators at the Cell Surface? Plant Physiol 153: 403–419. doi:10.1104/pp.110.156000.
12. Chapple C, Carpita N (1998) Plant cell walls as targets for biotechnology. Curr Opin Plant Biol 1: 179–185. doi:10.1016/S1369-5266(98)80022-8.
13. Thakur BR, Singh RK, Handa AK (1997) Chemistry and uses of pectin-a review. Crit Rev Food Sci Nutr 37: 47–73. doi:10.1080/10408399709527767.
14. Sticklen MB (2008) Plant genetic engineering for biofuel production: towards affordable cellulosic ethanol. Nat Rev Genet 9: 433–443. doi:10.1038/nrg2336.
15. Morris G, Kök S, Harding S, Adams G (2010) Polysaccharide drug delivery systems based on pectin and chitosan. Biotechnol Genet Eng Rev 27: 257–284.
16. Schena M (1996) Genome analysis with gene expression microarrays. BioEssays News Rev Mol Cell Dev Biol 18: 427–431. doi:10.1002/bies.950180513.
17. Ekins R, Chu FW (1999) Microarrays: their origins and applications. Trends Biotechnol 17: 217–218.
18. Wang D (2003) Carbohydrate microarrays. Proteomics 3: 2167–2175. doi:10.1002/pmic.200300601.
19. Park S, Lee M-R, Shin I (2008) Carbohydrate microarrays as powerful tools in studies of carbohydrate-mediated biological processes. Chem Commun Camb Engl: 4389–4399. doi:10.1039/b806699j.
20. Shin I, Park S, Lee M (2005) Carbohydrate microarrays: an advanced technology for functional studies of glycans. Chem Weinh Bergstr Ger 11: 2894–2901. doi:10.1002/chem.200401030.
21. Wang R, Liu S, Shah D, Wang D (2005) A practical protocol for carbohydrate microarrays. In: Zanders ED, editor. Methods Mol Biol Clifton NJ 310: 241–252.
22. Hsu T-L, Hanson SR, Kishikawa K, Wang S-K, Sawa M, et al. (2007) Alkynyl sugar analogs for the labeling and visualization of glycoconjugates in cells. Proc Natl Acad Sci U S A 104: 2614–2619. doi:10.1073/pnas.0611307104.
23. Hanson SR, Hsu T-L, Weerapana E, Kishikawa K, Simon GM, et al. (2007) Tailored glycoproteomics and glycan site mapping using saccharide-selective bioorthogonal probes. J Am Chem Soc 129: 7266–7267. doi:10.1021/ja0724083.
24. Pedersen HL, Fangel JU, McCleary B, Ruzanski C, Rydahl MG, et al. (2012) Versatile high-resolution oligosaccharide microarrays for plant glycobiology and cell wall research. J Biol Chem: jbc.M112.396598. doi:10.1074/jbc.M112.396598.
25. Uchiyama N, Kuno A, Koseki-Kuno S, Ebe Y, Horio K, et al. (2006) Development of a lectin microarray based on an evanescent-field fluorescence principle. Methods Enzymol 415: 341–351. doi:10.1016/S0076-6879(06)15021-1.
26. Gupta G, Surolia A, Sampathkumar S-G (2010) Lectin microarrays for glycomic analysis. Omics J Integr Biol 14: 419–436. doi:10.1089/omi.2009.0150.
27. Gama CI, Tully SE, Sotogaku N, Clark PM, Rawat M, et al. (2006) Sulfation patterns of glycosaminoglycans encode molecular recognition and activity. Nat Chem Biol 2: 467–473. doi:10.1038/nchembio810.
28. De Paz JL, Noti C, Seeberger PH (2006) Microarrays of Synthetic Heparin Oligosaccharides. J Am Chem Soc 128: 2766–2767. doi:10.1021/ja057584v.
29. Ratner DM, Seeberger PH (2007) Carbohydrate microarrays as tools in HIV glycobiology. Curr Pharm Des 13: 173–183.
30. Wang L-X, Ni J, Singh S, Li H (2004) Binding of high-mannose-type oligosaccharides and synthetic oligomannose clusters to human antibody 2G12: implications for HIV-1 vaccine design. Chem Biol 11: 127–134. doi:10.1016/j.chembiol.2003.12.020.
31. Huang C-Y, Thayer DA, Chang AY, Best MD, Hoffmann J, et al. (2006) Carbohydrate microarray for profiling the antibodies interacting with Globo H tumor antigen. Proc Natl Acad Sci U S A 103: 15–20. doi:10.1073/pnas.0509693102.
32. Lawrie CH, Marafioti T, Hatton CSR, Dirnhofer S, Roncador G, et al. (2006) Cancer-associated carbohydrate identification in Hodgkin's lymphoma by carbohydrate array profiling. Int J Cancer J Int Cancer 118: 3161–3166. doi:10.1002/ijc.21762.
33. Blixt O, Head S, Mondala T, Scanlan C, Huflejt ME, et al. (2004) Printed covalent glycan array for ligand profiling of diverse glycan binding proteins. Proc Natl Acad Sci U S A 101: 17033–17038. doi:10.1073/pnas.0407902101.
34. Liang P-H, Wang S-K, Wong C-H (2007) Quantitative analysis of carbohydrate-protein interactions using glycan microarrays: determination of surface and solution dissociation constants. J Am Chem Soc 129: 11177–11184. doi:10.1021/ja072931h.
35. Bryan MC, Wong C-H (2004) Aminoglycoside array for the high-throughput analysis of small molecule-RNA interactions. Tetrahedron Lett 45: 3639–3642. doi:10.1016/j.tetlet.2004.03.035.
36. Disney MD, Barrett OJ (2007) An aminoglycoside microarray platform for directly monitoring and studying antibiotic resistance. Biochemistry (Mosc) 46: 11223–11230. doi:10.1021/bi701071h.
37. Moller I, Marcus SE, Haeger A, Verhertbruggen Y, Verhoef R, et al. (2008) High-throughput screening of monoclonal antibodies against plant cell wall glycans by hierarchical clustering of their carbohydrate microarray binding profiles. Glycoconj J 25: 37–48. doi:10.1007/s10719-007-9059-7.
38. Singh B, Avci U, Eichler Inwood SE, Grimson MJ, Landgraf J, et al. (2009) A specialized outer layer of the primary cell wall joins elongating cotton fibers into tissue-like bundles. Plant Physiol 150: 684–699. doi:10.1104/pp.109.135459.
39. Aoki-Kinoshita KF, Kanehisa M (2006) Bioinformatics approaches in glycomics and drug discovery. Curr Opin Mol Ther 8: 514–520.
40. Von der Lieth C-W, Bohne-Lang A, Lohmann KK, Frank M (2004) Bioinformatics for glycomics: status, methods, requirements and perspectives. Brief Bioinform 5: 164–178.
41. Marchal I, Golfier G, Dugas O, Majed M (2003) Bioinformatics in glycobiology. Biochimie 85: 75–81.
42. Wendel JF, Brubaker C, Alvarez I, Cronn R, Stewart JM (2009) Evolution and Natural History of the Cotton Genus. In: Paterson AH, editor. Genetics and Genomics of Cotton. Plant Genetics and Genomics: Crops and Models. Springer US. 3–22.
43. Wang QQ, Liu F, Chen XS, Ma XJ, Zeng HQ, et al. (2010) Transcriptome profiling of early developing cotton fiber by deep-sequencing reveals significantly differential expression of genes in a fuzzless/lintless mutant. Genomics 96: 369–376. doi:10.1016/j.ygeno.2010.08.009.
44. Al-Ghazi Y, Bourot S, Arioli T, Dennis ES, Llewellyn DJ (2009) Transcript Profiling During Fiber Development Identifies Pathways in Secondary Metabolism and Cell Wall Structure That May Contribute to Cotton Fiber Quality. Plant Cell Physiol 50: 1364–1381. doi:10.1093/pcp/pcp084.
45. Gou J-Y, Wang L-J, Chen S-P, Hu W-L, Chen X-Y (2007) Gene expression and metabolite profiles of cotton fiber during cell elongation and secondary cell wall synthesis. Cell Res 17: 422–434. doi:10.1038/sj.cr.7310150.
46. Avci U, Pattathil S, Singh B, Brown VL, Hahn MG, et al. (2013) Cotton fiber cell walls of Gossypium hirsutum and Gossypium barbadense have differences related to loosely-bound xyloglucan. PloS One 8: e56315. doi:10.1371/journal.pone.0056315.
47. Meikle PJ, Bonig I, Hoogenraad NJ, Clarke AE, Stone BA (1991) The location of (1→3)-β-glucans in the walls of pollen tubes of Nicotiana alata using a (1→3)-β-glucan-specific monoclonal antibody. Planta 185: 1–8. doi:10.1007/BF00194507.
48. Willats WG, Limberg G, Buchholt HC, van Alebeek GJ, Benen J, et al. (2000) Analysis of pectic epitopes recognised by hybridoma and phage display monoclonal antibodies using defined oligosaccharides, polysaccharides, and enzymatic degradation. Carbohydr Res 327: 309–320.
49. Verhertbruggen Y, Marcus SE, Haeger A, Ordaz-Ortiz JJ, Knox JP (2009) An extended set of monoclonal antibodies to pectic homogalacturonan. Carbohydr Res 344: 1858–1862. doi:10.1016/j.carres.2008.11.010.
50. Yates EA, Valdor JF, Haslam SM, Morris HR, Dell A, et al. (1996) Characterization of carbohydrate structural features recognized by anti-arabinogalactan-protein monoclonal antibodies. Glycobiology 6: 131–139.
51. Smallwood M, Beven A, Donovan N, Neill SJ, Peart J, et al. (1994) Localization of cell wall proteins in relation to the developmental anatomy of the carrot root apex. Plant J 5: 237–246. doi:10.1046/j.1365-313X.1994.05020237.x.
52. McCartney L, Marcus SE, Knox JP (2005) Monoclonal antibodies to plant cell wall xylans and arabinoxylans. J Histochem Cytochem Off J Histochem Soc 53: 543–546. doi:10.1369/jhc.4B6578.2005.
53. Marcus SE, Verhertbruggen Y, Hervé C, Ordaz-Ortiz JJ, Farkas V, et al. (2008) Pectic homogalacturonan masks abundant sets of xyloglucan epitopes in plant cell walls. BMC Plant Biol 8: 60. doi:10.1186/1471-2229-8-60.
54. Pettolino FA, Hoogenraad NJ, Ferguson C, Bacic A, Johnson E, et al. (2001) A (1–>4)-beta-mannan-specific monoclonal antibody and its use in the immunocytochemical location of galactomannans. Planta 214: 235–242.
55. Marcus SE, Blake AW, Benians TAS, Lee KJD, Poyser C, et al. (2010) Restricted access of proteins to mannan polysaccharides in intact plant cell walls. Plant J Cell Mol Biol 64: 191–203. doi:10.1111/j.1365-313X.2010.04319.x.
56. R Core Team 2013 (2013) A language and environment for statistical computing. R Foundation for Statistical Computing, Vienna, Austria. URL http://www.R-project.org/.
57. Tabachnick BG, Fidell LS (2012) Using Multivariate Statistics. 6 edition. Boston: Pearson. 1024 p.
58. Schneider A, Hommel G, Blettner M (2010) Linear Regression Analysis. Dtsch Ärztebl Int 107: 776–782. doi:10.3238/arztebl.2010.0776.
59. Hair JF (2010) Multivariate data analysis. Upper Saddle River, NJ: Prentice Hall.
60. Lutz JG, Eckert TL (1994) The Relationship between Canonical Correlation Analysis and Multivariate Multiple Regression. Educ Psychol Meas 54: 666–675. doi:10.1177/0013164494054003009.
61. Tenenhaus A, Philippe C, Guillemot V, Cao K-AL, Grill J, et al. (2014) Variable selection for generalized canonical correlation analysis. Biostatistics. doi:10.1093/biostatistics/kxu001.

62. Witten DM, Tibshirani RJ (2009) Extensions of Sparse Canonical Correlation Analysis with Applications to Genomic Data. Stat Appl Genet Mol Biol 8.
63. Le Cao K-A, Martin PG, Robert-Granie C, Besse P (2009) Sparse canonical methods for biological data integration: application to a cross-platform study. BMC Bioinformatics 10: 34. doi:10.1186/1471-2105-10-34.
64. Rencher AC (2003) Canonical Correlation. Methods of Multivariate Analysis. John Wiley & Sons, Inc. 361–379.
65. Thompson B (1984) Canonical correlation analysis uses and interpretation. Beverly Hills, Calif.: Sage Publications.
66. Dejean S, Gonzalez I, Lê Cao K-A with contributions from Monget P, Coquery J, Yao F, Liquet B and Rohart F (2013) mixOmics: Omics Data Integration Project. R package version 5.0-1. Available: http://CRAN.R-project.org/package=mixOmics.
67. Boulesteix A-L, Strimmer K (2007) Partial least squares: a versatile tool for the analysis of high-dimensional genomic data. Brief Bioinform 8: 32–44. doi:10.1093/bib/bbl016.
68. Lê Cao K-A, Rossouw D, Robert-Granié C, Besse P (2008) A sparse PLS for variable selection when integrating omics data. Stat Appl Genet Mol Biol 7: Article 35. doi:10.2202/1544-6115.1390.
69. GonzáLez I, DéJean S, Martin PGP, GonçAlves O, Besse P, et al. (2009) Highlighting relationships between heterogeneous biological data through graphical displays based on regularized canonical correlation analysis. J Biol Syst 17: 173–199. doi:10.1142/S0218339009002831.
70. González I, Cao K-AL, Davis MJ, Déjean S (2012) Visualising associations between paired "omics" data sets. BioData Min 5: 19. doi:10.1186/1756-0381-5-19.
71. Chun H, Keles S (2010) Sparse partial least squares regression for simultaneous dimension reduction and variable selection. J R Stat Soc Ser B Stat Methodol 72: 3–25. doi:10.1111/j.1467-9868.2009.00723.x.
72. Bowman MJ, Park W, Bauer PJ, Udall JA, Page JT, et al. (2013) RNA-Seq Transcriptome Profiling of Upland Cotton (Gossypium hirsutum L.) Root Tissue under Water-Deficit Stress. PloS One 8: e82634. doi:10.1371/journal.-pone.0082634.
73. Lacape J-M, Claverie M, Vidal RO, Carazzolle MF, Guimarães Pereira GA, et al. (2012) Deep Sequencing Reveals Differences in the Transcriptional Landscapes of Fibers from Two Cultivated Species of Cotton. PLoS ONE 7: e48855. doi:10.1371/journal.pone.0048855.
74. Rambani A, Page JT, Udall JA (2014) Polyploidy and the petal transcriptome of Gossypium. BMC Plant Biol 14: 1–14. doi:10.1186/1471-2229-14-3.
75. Gilbert MK, Turley RB, Kim HJ, Li P, Thyssen G, et al. (2013) Transcript profiling by microarray and marker analysis of the short cotton (Gossypium hirsutum L.) fiber mutant Ligon lintless-1 (Li 1). BMC Genomics 14: 403. doi:10.1186/1471-2164-14-403.
76. Ruan Y-L, Xu S-M, White R, Furbank RT (2004) Genotypic and Developmental Evidence for the Role of Plasmodesmatal Regulation in Cotton Fiber Elongation Mediated by Callose Turnover. Plant Physiol 136: 4104–4113. doi:10.1104/pp.104.051540.
77. Salnikov VV, Grimson MJ, Seagull RW, Haigler CH (2003) Localization of sucrose synthase and callose in freeze-substituted secondary-wall-stage cotton fibers. Protoplasma 221: 175–184. doi:10.1007/s00709-002-0079-7.
78. Goldberg R, Morvan C, Jauneau A, Jarvis MC (1996) Methyl-esterification, de-esterification and gelation of pectins in the primary cell wall. In: J Visser and A.G.J Voragen, editor. Progress in Biotechnology. Pectins and Pectinases Proceedings of an International Symposium. Elsevier, Vol. Volume 14. 151–172.
79. Vaughn KC, Turley RB (1999) The primary walls of cotton fibers contain an ensheathing pectin layer. Protoplasma 209: 226–237. doi:10.1007/BF01453451.
80. Tokumoto H, Wakabayashi K, Kamisaka S, Hoson T (2002) Changes in the sugar composition and molecular mass distribution of matrix polysaccharides during cotton fiber development. Plant Cell Physiol 43: 411–418.
81. Haigler CH, Betancur L, Stiff MR, Tuttle JR (2012) Cotton fiber: a powerful single-cell model for cell wall and cellulose research. Front Plant Sci 3. Available: http://www.ncbi.nlm.nih.gov/pmc/articles/PMC3356883/. Accessed 24 Jun 2014.
82. Wang H, Guo Y, Lv F, Zhu H, Wu S, et al. (2010) The essential role of GhPEL gene, encoding a pectate lyase, in cell wall loosening by depolymerization of the de-esterified pectin during fiber elongation in cotton. Plant Mol Biol 72: 397–406. doi:10.1007/s11103-009-9578-7.
83. Hao Z, Mohnen D (2014) A review of xylan and lignin biosynthesis: foundation for studying Arabidopsis irregular xylem mutants with pleiotropic phenotypes. Crit Rev Biochem Mol Biol 49: 212–241. doi:10.3109/10409238.2014.889651.
84. Sadava D, Chrispeels MJ (1973) Hydroxyproline-rich cell wall protein (extensin): Role in the cessation of elongation in excised pea epicotyls. Dev Biol 30: 49–55. doi:10.1016/0012-1606(73)90047-X.
85. Huang G-Q, Gong S-Y, Xu W-L, Li W, Li P, et al. (2013) A fasciclin-like arabinogalactan protein, GhFLA1, is involved in fiber initiation and elongation of cotton. Plant Physiol 161: 1278–1290. doi:10.1104/pp.112.203760.
86. Qin L-X, Rao Y, Li L, Huang J-F, Xu W-L, et al. (2013) Cotton GalT1 Encoding a Putative Glycosyltransferase Is Involved in Regulation of Cell Wall Pectin Biosynthesis during Plant Development. PLoS ONE 8: e59115. doi:10.1371/journal.pone.0059115.
87. Bowling AJ, Vaughn KC, Turley RB (2011) Polysaccharide and glycoprotein distribution in the epidermis of cotton ovules during early fiber initiation and growth. Protoplasma 248: 579–590. doi:10.1007/s00709-010-0212-y.

16

Phylogeny Reconstruction and Hybrid Analysis of *Populus* (Salicaceae) Based on Nucleotide Sequences of Multiple Single-Copy Nuclear Genes and Plastid Fragments

Zhaoshan Wang[1]✉, Shuhui Du[1]✉, Selvadurai Dayanandan[2], Dongsheng Wang[1], Yanfei Zeng[1], Jianguo Zhang[1]*

1 State Key Laboratory of Tree Genetics and Breeding, Key Laboratory of Silviculture of the State Forestry Administration, Research Institute of Forestry, Chinese Academy of Forestry, Beijing, People's Republic of China, **2** Forest and Evolutionary Genomics Laboratory, and the Centre for Structural and Functional Genomics, Biology Department, Concordia University, Montreal, Quebec, Canada

Abstract

Populus (Salicaceae) is one of the most economically and ecologically important genera of forest trees. The complex reticulate evolution and lack of highly variable orthologous single-copy DNA markers have posed difficulties in resolving the phylogeny of this genus. Based on a large data set of nuclear and plastid DNA sequences, we reconstructed robust phylogeny of *Populus* using parsimony, maximum likelihood and Bayesian inference methods. The resulting phylogenetic trees showed better resolution at both inter- and intra-sectional level than previous studies. The results revealed that (1) the plastid-based phylogenetic tree resulted in two main clades, suggesting an early divergence of the maternal progenitors of *Populus*; (2) three advanced sections (*Populus*, *Aigeiros* and *Tacamahaca*) are of hybrid origin; (3) species of the section *Tacamahaca* could be divided into two major groups based on plastid and nuclear DNA data, suggesting a polyphyletic nature of the section; and (4) many species proved to be of hybrid origin based on the incongruence between plastid and nuclear DNA trees. Reticulate evolution may have played a significant role in the evolution history of *Populus* by facilitating rapid adaptive radiations into different environments.

Editor: Damon P. Little, The New York Botanical Garden, United States of America

Funding: Financial support for this research was provided by Key Project of Research Institute of Forestry, Chinese Academy of Forestry (ZD200911), the Special Research Program for Public-welfare Forestry of China (201004035), corporation project on science and postgraduate education in Beijing and collaborative innovation plan of Jiangsu higher education. The funders had no role in study design, data collection and analysis, decision to publish, or preparation of the manuscript.

Competing Interests: The authors have declared that no competing interests exist.

* Email: zhangjg@caf.ac.cn

✉ These authors contributed equally to this work.

Introduction

The genus *Populus*, distributed throughout the northern hemisphere from subtropical to boreal forests [1] and one of the most economically and ecologically important genera of forest trees [2], is well known for its rapid growth, profuse vegetative propagation, environmental stress tolerance and the numerous uses of its wood [3]. Furthermore, the genus has become an excellent research model due to its small genome size and the completion of the genome sequence of *P. trichocarpa* [4]. A clear understanding of the evolutionary relationships of *Populus* species will provide an important foundation for biological studies and genetic breeding programs.

The relationships among sections as well as relationships within each section remain controversial and/or poorly resolved because of the extensive interspecific hybridization and high degree of morphological variation among species [3,5]. The combination of these two features results in a major disagreement in the number of species and their delimitation [5,6]. Eckenwalder [5] recognized 29 species of *Populus* grouped into six sections (*Abaso*, *Aigeiros*, *Leucoides*, *Populus*, *Tacamahaca*, *Turanga*) based on morphological similarity and crossability. However, more than 60 species (plus a number of hybrids, varieties and forms) are described in the Flora of China [6]. The morphology-based phylogenetic tree of *Populus* demonstrated that section *Abaso* and *Turanga* were sister groups to the other four sections followed by section *Leucoides*, but the relationship among the other three sections remain unresolved [5]. Phylogenetic analysis based on 5.8S RNA and ITS sequences suggested that section *Populus* was sister to *Leucoides*, *Tacamahaca* and *Aigeiros* [7]. On the other hand, a phylogenetic tree based on plastid RFLP data showed an opposite trend with section *Populus* as an advanced clade occupying the terminal position [8]. Cervera [9] proposed that *P. lasiocarpa* and *P. violascens* of section *Leucoides* should be classified into section *Tacamahaca*. Furthermore, the systematic placement of species in section *Aigeiros* is inconsistent [8,10]. Cervera [9] proposed that *P. nigra* should be classified into a new section or as a subsection of section *Tacamahaca*. Most interspecific relationships within each section, in particular species in sections *Populus*, *Tacamahaca* and *Aigeiros*, are poorly resolved.

In addition to the lack of highly variable orthologous single-copy DNA markers, the complex reticulate evolution in *Populus* poses difficulty in resolving the phylogeny of *Populus* [8,10,11]. Species of *Populus* show extensive hybridization within sections as well as between closely-related sections. The species of section *Aigeiros*, *Tacamahaca* and *Leucoides* can intercross freely [12,13]. The evidence for hybridization in *Populus* has been documented based on molecular markers, for example, in the phylogenetic study based on maternally inherited plastid and biparentally inherited nuclear DNA sequences, *P. nigra* showed different affinity to sections *Populus* and *Tacamahaca*, which suggested a possible hybrid origin for *P. nigra* [8,10]. *P. tomentosa* has also been suspected to be a hybrid for a long time, but its exact parents remain unknown. The inconsistent systematic position of section *Populus* mentioned above may also be an indication of ancient hybridization in *Populus*.

The phylogenetic topology derived from maternally inherited plastid [14] data sets represents the maternal genealogy while nuclear DNA phylogeny mirrors biparental evolutionary history. Comparative analysis of DNA sequences from the nuclear and the plastid offers an effective way to parse out reticulate evolutionary events [8,15]. The completion of the whole genome of *P. trichocarpa* [4] provides a means to find highly variable single-copy nuclear DNA sequences to assess the evolutionary relationships of *Populus*. Here, we utilized 24 single-copy nuclear DNA sequences and 12 plastid fragments to reconstruct the phylogeny of *Populus* with an emphasis on the reticulate evolution in the genus.

Materials and Methods

Ethics Statement

No special permits were required for this study and this study did not involve endangered or protected species.

Species sampling, DNA extraction, PCR amplification and sequencing

The sampling was based on the classification proposed by Eckenwalder [5] and the Flora of China [6]. During the fieldwork from 2005 to 2013, we sampled twenty-six species representing 5 sections of *Populus* and 4 species of *Salix* as outgroups for the phylogeny reconstruction. The fresh leaves were dried and stored in silica gel. Information about the sampled species is given in (File S1 in Supporting Information S1).

Using a high throughput comparative genomic approach, Duarte et al. [16] identified 959 single-copy genes that were shared among *P. trichocarpa*, *Arabidopsis thaliana*, *Vitis vinifera* and *Oryza sativa* from which fifteen sequences were selected and characterized in our previous paper (Du et al., in press). The remaining nine pairs of primers were developed following the methods as described in Du et al. (in press). Twelve pairs of primers were selected from previous studies [17–21] and used for plastid amplification. All the primers used for amplifying and sequencing are listed in (File S2 in Supporting Information S1).

Total genomic DNA was isolated from silica-gel-dried leaves using a modified method [22]. Polymerase chain reaction (PCR) was performed in a total volume of 30 μL containing 5–50 ng of genomic DNA, 3 μL 10×PCR Buffer, 2.0 mM $MgCl_2$, 0.8 μM of each dNTP, 2.4 μM of each primer and 0.15 U Ex Taq DNA polymerase (TaKaRa, Shiga, Japan). Amplification was carried out in a temperature gradient 96 U thermocycler (Applied Biosystems, Forster City, CA, USA), using following thermal cycling profiles: 4 min at 95°C followed by 30 cycles of 30 s at 94°C, 30 s at 52°C to 60°C (depending on the optimal annealing temperature of specific primers), 90 s at 72°C and a final extension at 72°C for 10 min. After purifying using a DNA Purification kit (Amersham Pharmacia Biotech, Piscataway, USA), the PCR product was sequenced using an ABI 3730 DNA analyzer (Applied Biosystems, Foster City, California, USA) with the same primers used for amplification. For the samples where direct sequencing failed, the purified PCR products were cloned into pGEM -T easy Vector System II (Promega, Madison, WI, USA). For each sample, 6–12 positive clones were randomly picked and sequenced in both directions using standard T7 and SP6 primers.

Data analysis

The assembled contigs of each individual were aligned using CLUSTAL X [23] and refined manually in BioEdit [24]. For all the loci, regions with more than 5 mononucleotide or microsatellite repeats were excluded because of the uncertainty of homology which could be exacerbated by potential inaccuracies of enzymatic process during PCR and sequencing [25,26]. All indels were excluded in the following phylogenetic analyses after coded as binary characters according to the simple indel coding method [27] using FastGap 1.2 [28].Phylogenetic relationships among species were reconstructed using parsimony, maximum likelihood (ML) and Bayesian methods. The parsimony analysis was conducted in PAUP* 4.0b10* [29], with all characters equally weighted and treated as unordered. Heuristic search was performed with MULPARS option, tree-bisection-reconnection (TBR) branch swapping, RANDOM stepwise addition with 1000 replicates and the number of trees held in RAM was set to be 100000. Bootstrap analysis was conducted to assess topological robustness with 1000 replicates using simple taxon addition [30]. An appropriate nucleotide substitution model for each sequence was determined using jModeltest [31]. The models were chosen according to the Akaike information criterion (AIC) and used for subsequent ML and Bayesian analysis. ML analysis was conducted in PAUP* 4.0b10* [29] with random taxon addition of 1000 replicates, TBR branch swapping, MULPARS option, 100000 trees held in RAM and 100 replication of bootstrap analysis. Bayesian inference was performed with MrBayes 3.2.1 [32]. Two independent runs of Metropolis-coupled MCMC were conducted simultaneously, with each run being one cold chain and three incrementally heated chains and all started randomly in the parameters space. All other parameters were set to default. 1,000,000 generations were run and trees were sampled once every 100 generations. The program Tracer v1.5 [33] was utilized to check for stationary. The first 25% of sampled trees were discarded as burn-in and the posterior probabilities were calculated from the remaining trees. All the phylogenetic trees were viewed in the program FigTree v 1.3.1 [34]. The homogeneity across nuclear DNA loci was tested using the Shimodaria-Hasegawa test [35] in CONSEL [36].

Results

Sequence Characteristics

We successfully obtained all 24 nuclear DNA sequences and 12 plastid fragments from all 30 species except for the locus DSH22 in *S. raddeana* and DSH11 in *P. grandidentata*, which failed to amplify and were treated as missing data in the subsequent phylogenetic analyses. After removing regions with mononucleotide repeats and microsatellite sequences, the aligned length of the nuclear DNA ranged from 222 bp to 1106 bp with a total length of 15732 bp, in which exon sequences consisted of 10184 bp (64.7%). As shown in (File S3 in Supporting Information S1), the number of variable sites ranged from 38 (locus DSH6) to 160 (locus DSH10) and that of informative sites ranged from 27 (locus DSH6) to 111 (DSH22). The aligned length of plastid fragments varied between 532 bp and 2620 bp with a total length of 14197 bp, the exon sequences of which only occupied 4592 bp (32.3%). The most appropriate models fitted each locus decided by jModeltest [31] was presented in (File S3 in Supporting Information S1).

Phylogenetic analysis of the plastid fragments

Phylogenetic relationships among species remained poorly resolved based on individual plastid sequences (File S5 in Supporting Information S1). Because all plastid gene sequences are effectively inherited as one locus, they were concatenated into a single contiguous sequence for the phylogenetic analysis. The best fitting evolutionary model for the combined plastid data set was TVM+G in ML and Bayesian analyses. 50% majority-rule consensus parsimony and ML trees were generated from 8 most parsimonious trees and 6 most likely trees based on the combined plastid data set. The phylogenetic trees generated by parsimony, ML and Bayesian methods were similar to each other, with only a few differences in bootstrap support (BS) or posterior probability (PP) values in some clades. For instance, the sister relationship between *P. tremula* and *P. grandidentata*, and between *P. tomentosa* and *P. tremula/P. grandidentata* in the parsimony and ML trees were not supported by Bayesian methods (Figure 1).

As shown in Figure 1, all *Populus* species formed a fully supported monophyletic group comprising two major clades. In the first clade, all species of section *Populus* and *P. nigra* of section *Aigeiros* formed a single group with *P. tremuloides* sister to other species. Within this group, *P. davidiana*, *P. hopeiensis* and *P. tomentosa* grouped together sister to *P. tremula* and *P. grandidentata*, meanwhile, *P. alba*, *P. nigra* and *P. adenopoda* showed close phylogenetic relationships to each other. Unexpectedly, *P. cathayana*, *P. simonii* and *P. przewalskii* of section *Tacamahaca*, *P. lasiocarpa* of section *Leucoides* and *P. afghanica* of section *Aigeiros* formed the other highly supported group. In the second clade, *P. pruinosa* and *P. euphratica* of section *Turanga* clustered together as sister taxa to other species. Three American poplars, *P. balsamifera*, *P. trichocarpa* and *P. deltoides* were more closely related to each other than to other Asiatic *Tacamahaca* species which further divided into two subclades with *P. szechuanica* sister to a group of species comprising *P. koreana*, *P. suaveolens*, *P. laurifolia*, *P. ussuriensis*, *P. pamirica* and *P. talassica*.

Phylogenetic analysis of the nuclear DNA

Phylogenetic relationships based on individual nuclear DNA loci were not fully resolved (File S6 in Supporting Information S1). The Shimodaria-Hasegawa test showed that there was no significant incongruence among most of the individual nuclear DNA loci ($P>0.05$), but slight incongruence was detected in some cases (File S4 in Supporting Information S1). After excluding these incongruent loci from the phylogeny reconstruction, topology of the phylogenetic tree remained the same with only some differences in BS or PP value. We, therefore, combined the 24 individual nuclear DNA sequences to a single data set to reconstruct the phylogeny of *Populus* and to make direct comparison with the cpDNA phylogeny.

The best fitting evolutionary model for the combined nuclear DNA data set was GTR+R in ML and Bayesian analyses. The topologies of parsimony and ML trees are from 50% majority-rule consensus trees, which were generated from 4 most parsimonious trees and 6 most likely trees, respectively, based on combined nuclear DNA data set. Nuclear DNA phylogenetic trees generated from parsimony, ML and Bayesian methods showed similar topologies to each other with the only differences within section *Populus* of Bayesian tree (Figure 2). In the phylogenetic tree generated using nuclear data set (Figure 2), monophyly of *Populus* was strongly supported. Section *Turanga* was sister to other sections, followed by section *Populus* with high support value, in which a native North America poplar, *P. grandidentata* was sister to other species. The remaining species of section *Populus* subdivided into two groups: *P. tomentosa*, *P. alba*, *P. hopeiensis* and *P. adenopoda* clustered together with slightly lower resolution in the first group and three tremble aspens, *P. davidiana*, *P. tremuloides* and *P. tremula* showed close relationship to each other with high statistical support in the second. Section *Leucoides* (*P. lasiocarpa*) followed section *Populus* and was sister to sections *Tacamahaca* and *Aigeiros*. All the species of these two sections subdivided into two subclades. In the first subclade, *P. cathayana*, *P. simonii* and *P. przewalskii* clustered together sister to section *Aigeiros*, within which all the three species clustered as a single clade. The second subclade consisted of two sister groups: the two North America species (*P. balsamifera/P. trichocarpa*) and the other 7 species.

Discussion

Phylogenetic relationships of *Populus*

With a large combined data set of 24 nuclear DNA sequences and 12 plastid fragments, we reconstructed maternal and biparental phylogenies of *Populus* (Figure 1 and 2). All species of *Populus* clustered together as one clade separated from outgroup species, which supported the results of previous studies [5,10,37,38]. Species of each section grouped together in both nuclear and plastid DNA phylogenies with certain exceptions. Species of section *Aigeiros* clustered within different clades in plastid tree. *P. simonii*, *P. cathayana* and *P. przewalskii* of section *Tacamahaca* grouped with *P. lasiocarpa* separating from other balsam poplars in the plastid phylogeny.

There has been an argument about the oldest lineage of *Populus* for a long time. Phylogenetic analysis based on AFLP and ITS sequence proposed that section *Populus* was the oldest lineage in *Populus* [9]. However, fossil records and morphological phylogenetic analysis showed that section *Turanga* was sister to other sections [5]. *Populus wilmottae*, one of the most ancient fossil species of *Populus* [39], bears three-valved capsules, a feature considered primitive based on its predominance in the Violales, including many Flacourtiaceae, which is shared by section *Turanga* [5,39]. Furthermore, the modern *Turanga* lineage is strikingly heteroblastic, with willow-like juvenile leaves strongly differentiating it from other species of *Populus* — thought to have developed early in the evolutionary history of this genus [5]. This suggested that section *Turanga* is an ancestral lineage as observed in phylogenetic trees with high BS or PP support value.

Intersectional relationships of *Populus* have been poorly resolved because of inadequate samples and/or insufficient resolution of molecular markers used in previous studies. Eckenwalder [5] reconstructed the phylogeny of *Populus* using 76 morphological characters. Section *Abaso* and *Turanga* were sister to other sections followed by section *Leucoides* and the relationship among the three advanced sections remained unresolved [5]. In an ITS sequence-based phylogeny of *Populus*, section *Populus* was monophyletic and species of sections *Tacamahaca* and *Aigeiros* mixed together in one clade with low statistical support [40].The relationship between the two sections did not resolve in *trnT-trnF* phylogeny [10]. In the phylogenetic tree reconstructed based on AFLP data, Cervera et al. [9] found

Figure 1. Phylogeny of *Populus* obtained from the combined 12 plastid fragments using parsimony method, (A). Numbers next to nodes sequentially indicated ML/parsimony/BI support values. The branches without numbers indicate 100% statistical support. (B) The topology difference derived from Bayesian analysis while (C) is from ML analysis. A, *Aigeiros*; L, *Leucoides*; P, *Populus*; Tac, *Tacamahaca*; Tur, *Turanga*.

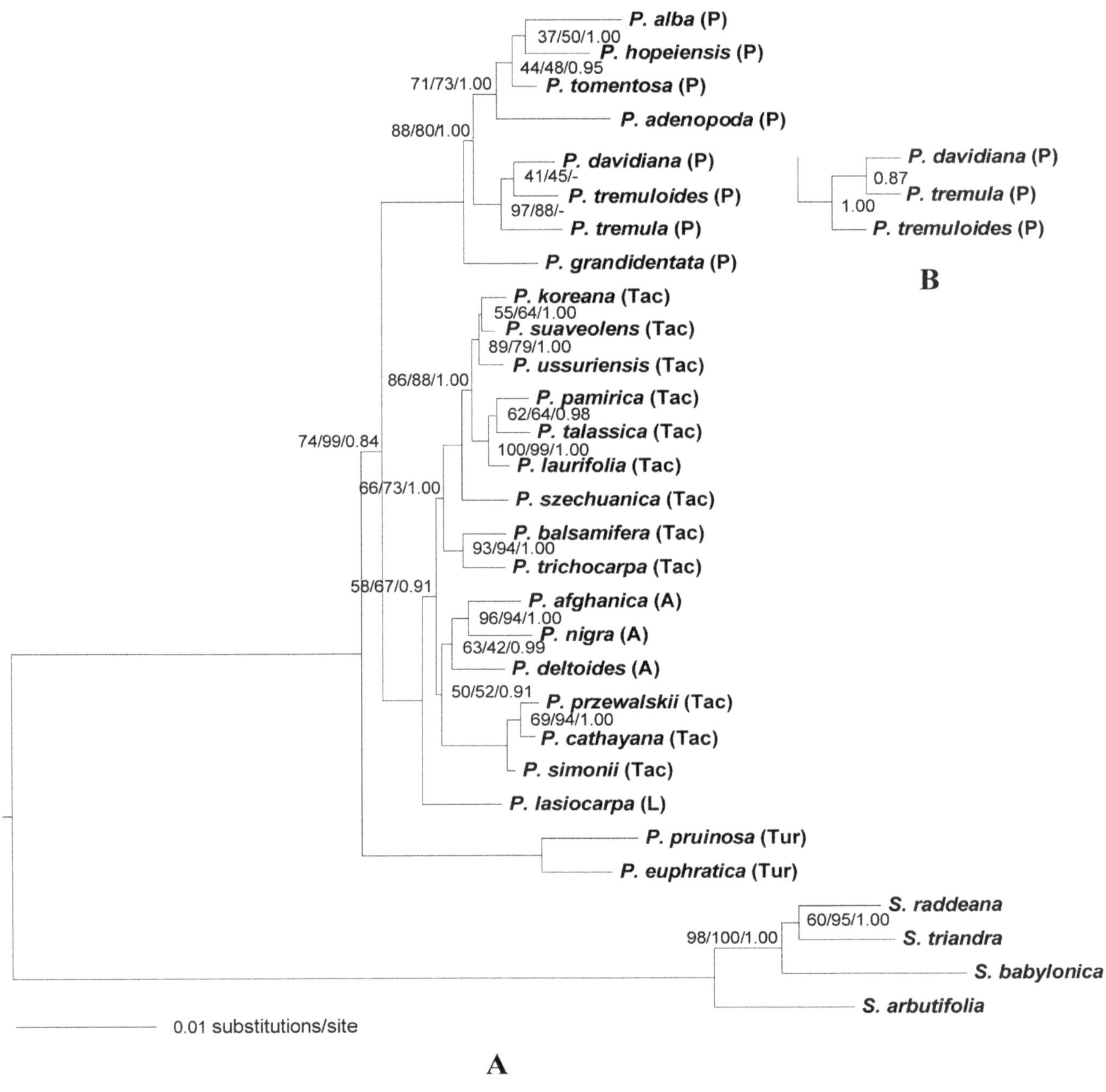

Figure 2. Phylogeny of Populus obtained from the combined 24 single-copy nuclear DNA sequences using ML method, (A). Numbers next to nodes sequentially indicate ML/parsimony/BI support values. The branches without numbers indicate 100% statistical support. (B) The topology difference derived from Bayesian analysis. A, *Aigeiros*; L, *Leucoides*; P, *Populus*; Tac, *Tacamahaca*; Tur, *Turanga*.

that section *Leucoides* showed close relationships with sections *Tacamahaca* and *Aigeiros*. Moreover he proposed that *P. ciliata*, *P. lasiocarpa* and *P. violascens* of section *Leucoides* should be classified into section *Tacamahaca* [9]. ISSR-based phylogeny of *Populus* also revealed close intersectional relationship between section *Tacamahaca* and *Aigeiros* [37].

Based on the nuclear DNA phylogeny, section *Populus* and *Leucoides* were successive to *Turanga*, section *Tacamahaca* and *Aigeiros* occupied the terminal position. This phylogenetic order more or less followed previously published patterns [5]. The close affinity between section *Tacamahaca* and *Aigeiros* observed in the present study is in agreement with previous studies [5,10,32,34]. For the first time, we found that species of *Populus* divided into two major clades in the plastid phylogeny, which reflected an early divergence of the maternal progenitors of *Populus*.

Intrasectional phylogenetic relationships

In previous studies, interspecific relationships within sections of *Populus* were rarely addressed or poorly resolved [9,10]. Based on our data, intrasectional relationships among species were relatively better resolved. Because only one species of section *Leucoides* and two species of section *Turanga* were utilized in analysis, interspecific relationships of these two sections cannot be assessed.

Furthermore, interspecific relationships of section *Aigeiros* is discussed below with an emphasis on hybrid origin.

Section Populus. The results from both nuclear and plastid DNA analyses suggest that section Populus is monophyletic. In the nuclear DNA tree, P. grandidentata was sister to other species which subdivided into two clades. The North American trembling aspen P. tremuloides showed close genetic affinity to Eurasian aspen P. tremula and P. davidiana with high BS or PP value. These three species are similar to each other with respect to morphological characters [6,41]. Eckenwalder [5] even proposed that P. tremuloides, P. tremula and P. davidiana should be merged into a single species. The rationalization of this hypothesis and the accurate time of origin and differentiation among the three species require further analysis with a larger sample size. The remaining four species group in a single clade in keeping with their morphological similarity. Unlike the nuclear DNA tree, another North American aspen, P. tremuloides, separated from other species and formed a single clade in the plastid tree, which indicated that P. tremuloides was of an ancestral maternal origin. As in the nuclear DNA phylogeny, P. adenopoda and P. alba show closest affinity to each other whereas they are widely distributed in southern China and central Eurasia, respectively. It is of great interest to pay attention to allopatric speciation resulted from geographic isolation between these two species in future.

Section Tacamahaca. The relationships of species within section Tacamahaca are known to be the most complicated. In present study, P. cathayana, P. przewalskii and P. simonii clustered with section Leucoides in the plastid-based tree but sister to section Aigeiros in nuclear DNA tree as a single highly supported group. The classification of P. cathayana, P. przewalskii and P. simonii either in section Tacamahaca or in other sections requires further investigation based on comprehensive sample with morphological and molecular methods. The other balsam poplars formed two sister clades. Two North American balsam poplars P. trichocarpa and P. balsamifera formed a strongly supported clade sister to the Asian ones in the plastid and nuclear DNA phylogeny considering their morphological, ecogeographic similarity and recent divergence [42]. A similar situation was shown in Asiatic species of section Tacamahaca; however, their distribution is allopatric in China. We speculate that these species derived from vicariance and allopatric divergence from a once widely-distributed ancestral species of section Tacamahaca. The polyphyly of section Tacamahaca was clearly verified based on our plastid and nuclear DNA phylogeny [10], which was in agreement with morphology-based phylogenetic analysis [5].

Reticulate evolution in *Populus*

The most ancient undisputed fossil record of *Populus* dated to late Paleocene (about 58 Ma) is considered to be related to section *Turanga* because of the morphological similarity [39,43]. Through combined fossil record and phylogenetic analyses we ascertain that section *Turanga* is more primitive in *Populus*. Species of the section *Leucoides* which inhabit permanent swamps first appeared in North America in late Eocene [5]. Combining fossil records and the phylogenetic tree based on 76 morphological characters, Eckenwalder [5] speculated that the temperate habitats of *Populus* were invaded by an ancestral member of section *Leucoides*, and following this, there was a rapid radiation (effectively simultaneous) driven by ancient hybridization events into the distinct habitats along with appearance of other advanced sections. The newly appeared sections can adapt to more extreme environments, for example, section *Populus* can tolerate aridity and coldness. The subsequent evolution within each section were partly influenced by hybridization [1].

Section Populus. Based on our plastid and nuclear DNA phylogeny, the ancestor of section Populus originated from the hybridization of two ancestral sections (section Turanga and Leucoides) with section Leucoides as the maternal parent. In the plastid tree, P. tremuloides was sister to other species in section Populus. In the nuclear DNA tree, P. tremuloides clustered with the Asiatic trembling aspen. According to leaf morphological variation and the fossil record, P. tremuloides was hypothesized to be a species of complex origin derived from multiple ancestral hybridization [5,44]. It is inferred that hybridization occurred between ancestors of P. davidiana and P. tremula dispersed from Eurasia to North America as the paternal parent and P. tremuloides.

P. tomentosa, one of the cultivated species widely distributed in China, has attracted attention from taxonomists and geneticists [45]. Based on morphological traits and molecular evidence, *P. tomentosa* is considered as a complex hybrid species involving more than two species. Both RAPD and AFLP analysis suggests that *P. tomentosa* has closest affinity to *P. adenopoda* and was possibly a natural hybrid of *P. alba* and *P. adenopoda* [46]. The nuclear DNA phylogeny also revealed the close relationship between *P. tomentosa* and *P. adenopoda*. However, *P. adenopoda* and *P. tomentosa* clustered in two different clades within section *Populus* in the plastid phylogeny. In the hybridization event giving rise to *P. tomentosa*, the ancestor of *P. davidiana* and *P. hopeiensis* served as the maternal parent and *P. adenopoda* as the paternal role. Based on morphological similarity to *P. hopeiensis* and the closer genetic affinity, we infer that *P. tomentosa* may have been domesticated from *P. hopeiensis*, which shows sympatric distribution in China.

Section Tacamahaca. Twelve species of section Tacamahaca were used in this study. P. cathayana, P. przewalskii and P. simonii clustered with section Leucoides based on plastid phylogeny and sister to section Aigeiros in nuclear DNA tree, which indicated that their maternal lineage derived from section Leucoides followed by high gene flow with section Aigeiros. The remained nine species of section Tacamahaca showed a close affinity to section Turanga in the plastid tree and sister to section Leucoides in nuclear DNA tree, which suggested these species also derived from a hybridization event in which section Turanga and section Leucoides played the maternal and paternal roles, respectively.

Section Aigeiros. Three species of section Aigeiros were used in this study. They clustered together in the nuclear DNA tree, however, in the plastid tree P. nigra clustered with P. alba, P. deltoids grouped with P. balsamifera and P. trichocarpa while P. afghanica showed close affinity to P. lasiocarpa. This implied that all of the three species of section Aigeiros were hybrid origin with different maternal parent.

Despite large number of morphological, molecular and phylogenetic studies on the hybrid origin of *P. nigra*, the maternal parent of this species remains uncertain. Phylogenetic analysis based on plastid data clustered *P. nigra* within section *Populus* and the plastid of *P. nigra* show close affinity to either *P. alba* [8] or common ancestors of *P. tremula* and *P. davidiana* [10]. Our results confirm that the plastid of *P. nigra* is inherited from *P. alba* or a common ancestor shared with *P. alba*. A close relationship between *P. deltoides* and section *Tacamahaca* was observed previously [9,10,37,47]. In the plastid tree, *P. deltoides* along with two native North American balsam poplars of section *Tacamahaca* form a monophyletic group with high BS or PP values. The similarity of floral morphology [48], overlapping distribution [49] and hybridization [50–52] of these three species in North America suggest that they may have diverged from a common maternal ancestor following a hybridization event giving rise to *P. deltoides*. A similar situation was also seen in *P. afghanica*, which shares a common maternal ancestor with *P. lasiocarpa*.

The taxonomic position of species in section *Aigeiros* is contentious. Rajora & Dancik [47] proposed a new section *Nigrae*, consisting only of *P. nigra*. Cervera *et al.* [9] pointed out that *P. deltoides* should be separated from the consectional *P. nigra*. Moreover, Eckenwalder [53] suggested that sections *Tacamahaca* and *Aigeiros* should be merged into a single section because of their close evolutionary relationships, which was supported by phylogenetic analysis based on ITS and plastid *trnT-trnF* sequences of *Populus* [10]. Nevertheless, considering the highly supported relationship in our nuclear DNA phylogeny and morphological similarity among the species of section *Aigeiros*, it is reasonable to retain this section which is a good model for further research about hybrid speciation in *Populus*.

In all, species of section *Populus* and *Tacamahaca* played a part in the origin of species in section *Aigeiros*, which suggests that section *Aigeiros* may have originated later than the former two sections. This is consistent with the paleontological results, and fossils of the section *Aigeiros* occur later in the sediments [5].

Conclusions

Based on nucleotide sequences of 24 single-copy nuclear genes and 12 plastid fragments, two robust phylogenies of *Populus* (Salicaceae) were reconstructed. The genus *Populus* was monophyletic in both phylogenetic trees and section *Turanga* was an ancestral lineage within the genus *Populus*. Comparative analyses of these two phylogenetic trees revealed reticulate evolutionary patterns in this genus. Three advanced sections (*Populus*, *Aigeiros* and *Tacamahaca*) were of hybrid origin. A detailed study involving more species (especially section *Abaso*) and genes are needed to further infer the origin, dispersal and hybridization in *Populus*.

Supporting Information

Supporting Information S1 Combined supporting information file. File S1. The information of species (names, altitude, longitude, latitude, collector, collection number and place of voucher deposition). File S2. The information of primers. File S3. The information of nuclear and plastid DNA sequences. File S4. Results of the Shimodaira-Hasegawa test that are significant incongruence ($P<0.05$). File S5. Phylogenetic trees reconstructed based on plastid fragments using parsimony, most likelihood and Bayesian inference methods. Numbers next to nodes indicated bootstrap support value or posterior probabilities. File S6. Phylogenetic trees reconstructed based on nuclear DNA fragments using parsimony, most likelihood and Bayesian inference methods. Numbers next to nodes indicated bootstrap support value or posterior probabilities.

Acknowledgments

We thank Dr. Ge Song, Dr. Zhang Fumin, Dr. Zou Xinhui and Dr. Tang Liang in State Key Laboratory of Systematic and Evolutionary Botany, Institute of Botany, Chinese Academy of Sciences, Beijing, China, for their suggestion on data analysis. We also thank Li Yunxiao and Li Shanshan in Zhang's group for data analysis and experimental work. We appreciate two anonymous reviewers who gave valuable advice on this paper. **Data archiving:** Sequence data used in this study have been submitted to GenBank: accession numbers KF940053-KF941130.

Author Contributions

Conceived and designed the experiments: JZ ZW. Performed the experiments: ZW S. Du. Analyzed the data: ZW S. Du DW. Contributed reagents/materials/analysis tools: ZW S. Du YZ. Wrote the paper: ZW S. Du S. Dayanandan.

References

1. DiFazio SP, Slavov GT, Joshi CP (2011) *Populus*: a premier pioneer system for plant genomics. In: Joshi CP, editor. Genetics, genomics and breeding of poplar. Lebanon: Science Publishers, Inc. pp: 1–28.
2. Stettler RF Zsuffa L, Wu R (1996) The role of hybridization in the genetic manipulation of Populus. In: Stettler RF, Bradshaw HD, Heilman PE, Hinckler TM editors. Biology of *Populus* and its implications for management and conservation. Ottawa: Canadian Government Publishing. pp. 87–112.
3. Cronk Q (2005) Plant eco-devo: the potential of poplar as a model organism. New Phytologist 166: 39–48.
4. Tuskan GA, Difazio S, Jansson S, Bohlmann J, Grigoriev I, et al. (2006) The genome of black cottonwood, *Populus trichocarpa* (Torr. & Gray). Science 313: 1596–1604.
5. Eckenwalder JE (1996) Systematics and evolution of *Populus*. In: Stettler RF, Bradshaw HD, Heilman PE, Hinckler TM editors.Biology of *Populus* and its implications for management and conservation. Ottawa: Canadian Government Publishing. pp. 7: 32.
6. Wu ZY (1999) Flora of China. Volume 20, Fascicule 2. Science Press. pp. 7–78.
7. Leskinen E, Alström-Rapaport C (1999) Molecular phylogeny of Salicaceae and closely related Flacourtiaceae: Evidence from 5.8 S, ITS 1 and ITS 2 of the rDNA. Plant Systematics and Evolution 215: 209–227.
8. Smith RL, Sytsma KJ (1990) Evolution of *Populus nigra* (sect. *Aigeiros*): introgressive hybridization and the chloroplast contribution of *Populus alba* (sect. *Populus*). American journal of botany: 1176–1187.
9. Cervera M, Storme V, Soto A, Ivens B, Van Montagu M, et al. (2005) Intraspecific and interspecific genetic and phylogenetic relationships in the genus *Populus* based on AFLP markers. Theoretical and Applied Genetics 111: 1440–1456.
10. Hamzeh M, Dayanandan S (2004) Phylogeny of *Populus* (Salicaceae) based on nucleotide sequences of chloroplast *trnT-trnF* region and nuclear rDNA. American journal of botany 91: 1398–1408.
11. Slavov GT, Zhelev P (2010) Salient biological features, systematics, and genetic variation of *Populus*. In: Jansson S, Bhalerao R, Groover A, editors. Genetics and Genomics of *Populus*. Berlin: Springer. pp. 15–38.
12. Zsuffa L (1975) A summary review of interspecific breeding in the genus *Populus* L. In: Proceedings of the 14th annual meeting of the Canadian Tree Improvement Association, part 2, 107–123. Canadian Forest Service, Ottawa, Ontario, Canada.Willing R and Pryor L
13. Willing RR, Pryor LD (1976) Interspecific hybridisation in poplar. Theoretical and Applied Genetics 47: 141–151.
14. Mejnartowicz M (1991) Inheritance of chloroplast DNA in *Populus*. Theoretical and Applied Genetics 82: 477–480.
15. Page RDM (2000) Extracting species trees from complex gene trees: reconciled trees and vertebrate phylogeny. Molecular Phylogenetics and Evolution 14: 89–106.
16. Duarte JM, Wall PK, Edger PP, Landherr LL, Ma H, et al. (2010) Identification of shared single copy nuclear genes in *Arabidopsis*, *Populus*, *Vitis* and *Oryza* and their phylogenetic utility across various taxonomic levels. BMC Evolutionary Biology 10: 61.
17. Bobowski B, Hole D, Wolf P, Bryant L (1999) Identification of roots of woody species using polymerase chain reaction (PCR) and restriction fragment length polymorphism (RFLP) analysis. Molecular Ecology 8: 485–491.
18. Demesure B, Sodzi N, Petit R (1995) A set of universal primers for amplification of polymorphic noncoding regions of mitochondrial and chloroplast DNA in plants. Molecular Ecology 4: 129–134.
19. Huang SS, Hwang SY, Lin TP (2002) Spatial pattern of chloroplast DNA variation of Cyclobalanopsis glauca in Taiwan and East Asia. Molecular Ecology 11: 2349–2358.
20. Shaw J, Lickey EB, Beck JT, Farmer SB, Liu W, et al. (2005) The tortoise and the hare II: relative utility of 21 noncoding chloroplast DNA sequences for phylogenetic analysis. American journal of botany 92: 142–166.
21. Shaw J, Lickey EB, Schilling EE, Small RL (2007) Comparison of whole chloroplast genome sequences to choose noncoding regions for phylogenetic studies in angiosperms: the tortoise and the hare III. American journal of botany 94: 275–288.
22. Doyle JJ (1987) A rapid DNA isolation procedure for small quantities of fresh leaf tissue. Phytochem Bull 19: 11–15.
23. Thompson JD, Gibson TJ, Plewniak F, Jeanmougin F, Higgins DG (1997) The CLUSTAL X windows interface: flexible strategies for multiple sequence alignment aided by quality analysis tools. Nucleic acids research 25: 4876–4882.

24. Hall TA (1999) BioEdit: a user-friendly biological sequence alignment editor and analysis program for Windows 95/98/NT. Nucleic acids symposium series 41: 95–98.
25. Kelchner SA (2000) The evolution of non-coding chloroplast DNA and its application in plant systematics. Annals of the Missouri Botanical Garden 87: 482–498.
26. Zhu Q, Ge S (2005) Phylogenetic relationships among A-genome species of the genus *Oryza* revealed by intron sequences of four nuclear genes. New Phytologist 167: 249–265.
27. Simmons MP, Ochoterena H (2000) Gaps as characters in sequence-based phylogenetic analyses. Systematic Biology 49: 369–381.
28. Borchsenius F (2009) FastGap 1.2. Department of Biological Sciences, University of Aarhus, Aarhus, Denmark.
29. Swofford DL (2003) PAUP*: phylogenetic analysis using parsimony, version 4.0 b10. for Macintosh. Sinauer, Sunderland, Massachusetts, USA.
30. Felsenstein J (1985) Confidence limits on phylogenies: an approach using the bootstrap. Evolution: 783–791.
31. Posada D (2008) jModelTest: phylogenetic model averaging. Molecular Biology and Evolution 25: 1253–1256.
32. Ronquist F, Huelsenbeck JP (2003) MrBayes 3: Bayesian phylogenetic inference under mixed models. Bioinformatics 19: 1572–1574.
33. Drummond AJ, Rambaut A (2007) BEAST: Bayesian evolutionary analysis by sampling trees. BMC Evolutionary Biology 7: 214.
34. Rambaut A (2008) FigTree v1.1.1. Published by the author.
35. Shimodaira H, Hasegawa M (1999) Multiple comparisons of log-likelihoods with applications to phylogenetic inference. Molecular Biology and Evolution 16: 1114–1116.
36. Shimodaira H, Hasegawa M (2001) CONSEL: for assessing the confidence of phylogenetic tree selection. Bioinformatics 17: 1246–1247.
37. Hamzeh M, Périnet P, Dayanandan S (2006) Genetic Relationships among species of *Populus* (Salicaceae) based on nuclear genomic data. The Journal of the Torrey Botanical Society 133: 519–527.
38. Chase MW, Zmarzty S, Lledó MD, Wurdack KJ, Swensen SM, et al. (2002) When in doubt, put it in Flacourtiaceae: a molecular phylogenetic analysis based on plastid rbcL DNA sequences. Kew Bulletin 57: 141–181.
39. Manchester SR, Dilcher DL, Tidwell WD (1986) Interconnected reproductive and vegetative remains of *Populus* (Salicaceae) from the middle Eocene Green River Formation, northeastern Utah. American journal of botany: 156–160.
40. Quanliang S, Qiang Z, Minren H, Mingxiu W (2001) Phologenetic relationship of *Populus* sections by ITS sequnece analysis. Acta Botanaca Sinica 43: 323–325.
41. Morin NR (2000) Flora of North America. Flora of North America Association. Volume 7: 5–22.
42. Levsen ND, Tiffin P, Olson MS (2012) Pleistocene speciation in the genus *Populus* (Salicaceae). Systematic Biology 61: 401–412.
43. Manchester SR, Judd WS, Handley B (2006) Foliage and fruits of early poplars (Salicaceae: *Populus*) from the Eocene of Utah, Colorado, and Wyoming. International Journal of Plant Sciences 167: 897–908.
44. Barnes BV (1967) Indications of possible mid-Cenozoic hybridization in the aspens of the Columbia Plateau. Rhodora 69.
45. Zhang D, Zhang Z, Yang K, Li B (2004) Genetic mapping in (*Populus tomentosa*×*Populus bolleana*) and *P. tomentosa* Carr. using AFLP markers. Theoretical and Applied Genetics 108: 657–662.
46. Kuan-yu L, Min-ren H, Ming-xiu W (1997) Study on Origin of *Populus tomentosa* Carr. Acta Phytotaxonomica Sinica 35: 1.
47. Rajora O, Dancik B (1995) Chloroplast DNA variation in *Populus*. II. Interspecific restriction fragment polymorphisms and genetic relationships among *Populus deltoides*, *P. nigra*, *P. maximowiczii*, and *P. x canadensis*. Theoretical and Applied Genetics 90: 324–330.
48. Eckenwalder JE (1984) Natural intersectional hybridization between North American species of *Populus* (Salicaceae) in sections *Aigeiros* and *Tacamahaca*. II. Taxonomy. Canadian Journal of Botany 62: 325–335.
49. Little EL Jr, Viereck LA (1971) Conifers and important hardwoods. In: United States. Forest Service, Little EL Jr., Viereck LA, editors. Atlas of United States trees. Washington: Miscellaneous Publications United States Department of Agriculture. Volume 1.
50. Talbot P, Thompson SL, Schroeder W, Isabel N (2011) An efficient single nucleotide polymorphism assay to diagnose the genomic identity of poplar species and hybrids on the Canadian prairies. Canadian Journal of Forest Research 41: 1102–1111.
51. Hamzeh M, Sawchyn C, Perinet P, Dayanandan S (2007) Asymmetrical natural hybridization between *Populus deltoides* and *P. balsamifera* (Salicaceae) This note is one of a selection of papers published in the Special Issue on Poplar Research in Canada. Botany 85: 1227–1232.
52. Thompson SL, Lamothe M, Meirmans PG, Perinet P, Isabel N (2010) Repeated unidirectional introgression towards *Populus balsamifera* in contact zones of exotic and native poplars. Molecular Ecology 19: 132–145.
53. Eckenwalder JE (1977) Systematics of *Populus* L.(Salicaceae) in southwestern North America with special reference to sect. *Aigeiros* Duby. Ph.D. dissertation, University of California, Berkeley, California, USA.

Permissions

All chapters in this book were first published in PLOS ONE, by The Public Library of Science; hereby published with permission under the Creative Commons Attribution License or equivalent. Every chapter published in this book has been scrutinized by our experts. Their significance has been extensively debated. The topics covered herein carry significant findings which will fuel the growth of the discipline. They may even be implemented as practical applications or may be referred to as a beginning point for another development.

The contributors of this book come from diverse backgrounds, making this book a truly international effort. This book will bring forth new frontiers with its revolutionizing research information and detailed analysis of the nascent developments around the world.

We would like to thank all the contributing authors for lending their expertise to make the book truly unique. They have played a crucial role in the development of this book. Without their invaluable contributions this book wouldn't have been possible. They have made vital efforts to compile up to date information on the varied aspects of this subject to make this book a valuable addition to the collection of many professionals and students.

This book was conceptualized with the vision of imparting up-to-date information and advanced data in this field. To ensure the same, a matchless editorial board was set up. Every individual on the board went through rigorous rounds of assessment to prove their worth. After which they invested a large part of their time researching and compiling the most relevant data for our readers.

The editorial board has been involved in producing this book since its inception. They have spent rigorous hours researching and exploring the diverse topics which have resulted in the successful publishing of this book. They have passed on their knowledge of decades through this book. To expedite this challenging task, the publisher supported the team at every step. A small team of assistant editors was also appointed to further simplify the editing procedure and attain best results for the readers.

Apart from the editorial board, the designing team has also invested a significant amount of their time in understanding the subject and creating the most relevant covers. They scrutinized every image to scout for the most suitable representation of the subject and create an appropriate cover for the book.

The publishing team has been an ardent support to the editorial, designing and production team. Their endless efforts to recruit the best for this project, has resulted in the accomplishment of this book. They are a veteran in the field of academics and their pool of knowledge is as vast as their experience in printing. Their expertise and guidance has proved useful at every step. Their uncompromising quality standards have made this book an exceptional effort. Their encouragement from time to time has been an inspiration for everyone.

The publisher and the editorial board hope that this book will prove to be a valuable piece of knowledge for researchers, students, practitioners and scholars across the globe.

List of Contributors

Purbasha Sarkar, Elena Bosneaga and Manfred Auer
Energy Biosciences Institute, University of California, Berkeley, California, United States of America
Life Sciences Division, Lawrence Berkeley National Laboratory, Berkeley, California, United States of America

Edgar G. Yap Jr., Wen-Ting Tsai, Erica Neuhaus, Michael Joo, Roseann Csencsits and Kenneth H. Downing
Life Sciences Division, Lawrence Berkeley National Laboratory, Berkeley, California, United States of America

Jyotirmoy Das, Angelo Cabal, Dolonchampa Maji and Shailabh Kumar
Energy Biosciences Institute, University of California, Berkeley, California, United States of America

Sergey Yakovlev
Materials Sciences Division, Lawrence Berkeley National Laboratory, Berkeley, California, United States of America

Zeyun Yu
Department of Computer Science, University of Wisconsin, Milwaukee, Wisconsin, United States of America

Chandrajit Bajaj
Department of Computer Sciences & The Institute of Computational Engineering and Sciences, University of Texas, Austin, Texas, United States of America

Rocío Calderón, Carlos Lucena, JoséL. Trapero-Casas, Pablo J. Zarco-Tejada and Juan A. Navas-Cortés
Instituto de Agricultura Sostenible (IAS), Consejo Superior de Investigaciones Científicas (CSIC), Apartado 4084, Campus de Excelencia Internacional Agroalimentario, Córdoba, Spain

Vincent Billard, Alain Ourry, Anne Maillard and Philippe Etienne
Normandie Université, Caen, France
UNICAEN, UMR 950 Ecophysiologie Végétale, Agronomie et nutritions N, C, S, Caen, France
INRA, UMR 950 Ecophysiologie Végétale, Agronomie et nutritions N, C, S, Caen, France

Maria Garnica and José-Maria Garcia-Mina
Timac Agro Spain, Poligono de Arazuri-Orcoyen, Orcoyen, Spain

Laurent Coquet and Thierry Jouenne
Plateforme de protéomique PISSARO, UMR6270 CNRS Facultédes Sciences de Rouen, Mont-Saint-Aignan, France

Florence Cruz and Jean-Claude Yvin
Centre de Recherche International en Agroscience, CRIAS-TAI, Groupe Roullier, Dinard, France

Juan Viruel
Departamento de Agricultura y Economía Agraria, Escuela Politécnica Superior de Huesca, Universidad de Zaragoza, Huesca, Spain

Pilar Catalán
Departamento de Agricultura y Economía Agraria, Escuela Politécnica Superior de Huesca, Universidad de Zaragoza, Huesca, Spain
Department of Botany, Institute of Biology, Tomsk State University, Tomsk, Russia

José Gabriel Segarra-Moragues
Departamento de Ecología, Centro de Investigaciones sobre Desertificación (CIDE), Consejo Superior de Investigaciones Científicas (CSIC), Moncada, Valencia, Spain

Huize Chen and Rong Han
Higher Education Key Laboratory of Plant Molecular and Environmental Stress Response (Shanxi Normal University) in Shanxi Province, Linfen, China
School of Life Science, Shanxi Normal University, Linfen, China

Yan Gong
Higher Education Key Laboratory of Plant Molecular and Environmental Stress Response (Shanxi Normal University) in Shanxi Province, Linfen, China
School of Chemistry and Material Science, Shanxi Normal University, Linfen, China

Humberto J. Debat and Daniel A. Ducasse
Instituto de Patología Vegetal, Centro de Investigaciones Agropecuarias, Instituto Nacional de Tecnología Agropecuaria (IPAVE-CIAP-INTA), Córdoba, Argentina

Mauro Grabiele, Patricia M. Aguilera and Dardo A. Marti
Instituto de Biología Subtropical, Universidad Nacional de Misiones (IBS-UNaM-CONICET), Posadas, Misiones, Argentina
Instituto de Biotecnología de Misiones, Facultad de Ciencias Exactas Químicas y
Naturales, Universidad Nacional de Misiones (INBIOMIS-FCEQyN-UNaM), Misiones, Argentina

Rosana E. Bubillo
Estación Experimental Cerro Azul, Instituto Nacional de Tecnología Agropecuaria (EEA Cerro Azul-INTA), Misiones, Argentina

Mónica B. Otegui and Pedro D. Zapata
Instituto de Biotecnología de Misiones, Facultad de Ciencias Exactas Químicas y
Naturales, Universidad Nacional de Misiones (INBIOMIS-FCEQyN-UNaM), Misiones, Argentina

Temel Oguz
SOCIB, Balearic Islands Coastal Ocean Observing and Forecasting System, Palma de Mallorca, Spain
Institute of Marine Sciences, Middle East Technical University, Erdemli, Mersin, Turkey

Diego Macias
Institute for Environment and Sustainability, Joint Research Center, European Commission, Ispra, Italy

Jesus Garcia-Lafuente
Departamento de Física Aplicada II, Universidad de Málaga, Málaga, Spain

Ananda Pascual
IMEDEA (CSIC-UIB), Esporles, Palma de Mallorca, Spain

Joaquin Tintore
SOCIB, Balearic Islands Coastal Ocean Observing and Forecasting System, Palma de Mallorca, Spain
IMEDEA (CSIC-UIB), Esporles, Palma de Mallorca, Spain

Vokkaliga Thammegowda Harshavardhan, Christiane Seiler and Markus Kuhlmann
Research Group Abiotic Stress Genomics, Interdisciplinary Center for Crop Plant Research (IZN), Halle (Saale), Germany, and Leibniz Institute of Plant Genetics and Crop Plant Research (IPK) Gatersleben, Stadt Seeland, OT Gatersleben, Germany

Le Van Son
Research Group Gene Regulation, Leibniz Institute of Plant Genetics and Crop Plant Research (IPK) Gatersleben, Stadt Seeland, OT Gatersleben, Germany
National Key Laboratory of Gene Technology, Institute of Biotechnology Vietnam, Academy of Science and Technology, Hanoi, Vietnam

Astrid Junker, Kathleen Weigelt-Fischer and Thomas Altmann
Research Group Heterosis, Leibniz Institute of Plant Genetics and Crop Plant Research (IPK) Gatersleben, Stadt Seeland, OT Gatersleben, Germany

Christian Klukas
Research Group Image Analysis, Leibniz Institute of Plant Genetics and Crop Plant Research (IPK) Gatersleben, Stadt Seeland, OT Gatersleben, Germany

Nese Sreenivasulu
Research Group Abiotic Stress Genomics, Interdisciplinary Center for Crop Plant Research (IZN), Halle (Saale), Germany, and Leibniz Institute of Plant Genetics and Crop Plant Research (IPK) Gatersleben, Stadt Seeland, OT Gatersleben, Germany
Grain Quality and Nutrition Center, International Rice Research Institute (IRRI), Metro Manila, Philippines

Helmut Bäumlein
Research Group Gene Regulation, Leibniz Institute of Plant Genetics and Crop Plant Research (IPK) Gatersleben, Stadt Seeland, OT Gatersleben, Germany

Anne Bouchut and Amy E. DeRocher
Seattle Biomedical Research Institute, Seattle, WA, United States of America

Jennifer A. Geiger and Marilyn Parsons
Seattle Biomedical Research Institute, Seattle, WA, United States of America
Dept. of Global Health, University of Washington, Seattle, WA, United States of America

Nadja Korotkova and Thomas Borsch
Institut für Biologie/Botanik, Systematische Botanik und Pflanzengeographie, Freie Universität Berlin, Berlin, Germany
Dahlem Centre of Plant Sciences (DCPS), Berlin, Germany
Botanischer Garten und Botanisches Museum Berlin-Dahlem, Berlin, Germany

Lars Nauheimer
Institut für Biologie/Botanik, Systematische Botanik und Pflanzengeographie, Freie Universität Berlin, Berlin, Germany **David Benavente**
Dahlem Centre of Plant Sciences (DCPS), Berlin, Germany

Hasmik Ter-Voskanyan
Botanischer Garten und Botanisches Museum Berlin-Dahlem, Berlin, Germany

Martin Allgaier
The Berlin Center for Genomics in Biodiversity Research (BeGenDiv), Berlin, Germany

Tingting Liu, Chaofeng Shen, Yi Wang, Canke Huang and Jiyan Shi
Institute of Environmental Science and Technology, College of Environmental and Resource Sciences, Zhejiang University Hangzhou, P.R. China

Lei Zhang, Huijuan Ma, Tingting Chen, Jun Pen, Shuxun Yu and Xinhua Zhao
State Key Laboratory of Cotton Biology, Institute of Cotton Research of CAAS, Anyang, P. R. Chinass

Jūratė Lesutienė
Marine Science and Technology Center, Klaipėda University, Klaipėda, Lithuania
Department of Biology, Karlstad University, Karlstad, Sweden

Elena Gorokhova
Department of Applied Environmental Sciences, Stockholm University, Stockholm, Sweden

Daiva Stankevičienė
Institute of Ecology, Nature Research Centre, Vilnius, Lithuania

Eva Bergman and Larry Greenberg
Department of Biology, Karlstad University, Karlstad, Sweden

Scott Schaeffer and Amit Dhingra
Department of Horticulture, Washington State University, Pullman, WA, United States of America
Molecular Plant Science Graduate Program, Washington State University, Pullman, WA, United States of America

Artemus Harper
Department of Horticulture, Washington State University, Pullman, WA, United States of America

Rajani Raja and Pankaj Jaiswal
2082 Cordley Hall, Department of Botany and Plant Pathology, Oregon State University, Corvallis, OR, United States of America

Index

A

Angiosperms, 2, 125-126, 133-134, 136, 138-140, 191, 195-196, 216

Antioxidant Assays, 63

Antioxidant Enzyme, 60, 67, 154-156, 163-164, 166-167

Arabidopsis, 1-3, 5-8, 10-16, 34, 38, 40-41, 43-45, 71-72, 74-76, 82, 86-87, 96-97, 100-103, 105, 107, 109-115, 117, 119, 121-124, 144-146, 149, 153, 168-169, 171-180, 182-198, 200, 209, 211, 216

Arabidopsis Thaliana, 3, 15, 38, 40-41, 43-45, 71, 74, 76, 87, 96-97, 100, 110-115, 123-124, 145-146, 169, 177-180, 182-193, 195-198, 200, 211

Arabinogalactan (agp), 201

Arbequina, 17-19, 21-26, 29-30

B

Brassica Napus, 33-41, 43-45, 101, 112, 145, 152

C

Cadmium Telluride Quantum Dots (cdte-qds), 60-61, 63, 65, 67, 69, 71

Canonical Function, 201-205, 207

Carbohydrate Partitioning, 168, 172, 176

Carbohydrate Quantification, 170

Cell Wall Polysaccharide, 141-142, 144, 149, 151, 178

Chlorogenic Acid, 79, 88

Chlorophyll Content, 17-19, 21, 23, 25, 27, 29-31, 34, 36-37, 39, 61-62, 65, 71, 102, 105-106, 109-110, 155, 159, 164, 177

Chloroplastic Proteins, 33

Comparative Analysis, 3, 15, 37, 125, 139, 179, 181, 183, 185, 187-189, 191, 193, 195-197, 211

Comprehensive Microarray Polymer Profiling (compp), 200

Confocal Fluorescence, 63

Copper Remobilization, 33, 35, 37, 39, 41, 43, 45

Copper Stress, 45, 141, 143, 150, 152

Copper-deficiency, 33, 35, 37, 39, 41, 43, 45

Cotton Fiber Properties, 199, 201, 203, 205, 207, 209

Crude Mitochondria Isolation, 116

Cryo-immobilized Plant, 1, 3, 5, 7, 9, 11, 13-15

Cryphonectria Parasitica, 29

Cyanidioschyzon Merolae, 89-91, 93, 95, 97-99

Cytosolic, 96, 99, 142, 148, 162, 167-169, 171-178

D

De Novo Transcriptome Assembly, 73-74, 87

Dioecious Crop Tree, 73

Dioscorea Humilis, 46-57

Drought Tolerance, 100

E

Electron Tomography, 1-3, 5-16

Electrophoresis, 35-36, 44, 64, 67, 82, 86, 116, 124, 126

Elsholtzia Splendens, 141-143, 145, 147, 149-151, 153

F

Fluorescence Microscopy, 91-94

G

Genome Comparisons, 125, 127, 136

Gossypium Hirsutum, 154-155, 157, 159, 161, 163, 165-167, 200, 206, 208-209

Guaiacol Peroxidase (gpox), 63

H

Haplotype Polymorphism, 47

Heat-shock Inducible Gene, 89, 91, 93-95, 97-99

High-pressure Freezing (hpf), 2-3

Higher Plant Genomes, 179, 181, 183, 185, 187, 189, 191, 193, 195, 197

Hybrid Analysis, 210-211, 213, 215, 217

I

Ilex Paraguariensis, 73, 75-88

L

Last Glacial Maximum (lgm), 46-47, 56

Latitudinal Environmental Niches, 46-47, 49, 51, 53, 55, 57, 59

Lignocellulosic Biofuel, 1

M

Macromolecular Architecture, 1

Malondialdehyde (mda), 61-62, 155, 162-163

Mature Plant Cell Walls, 1

Methane Metbolisma, 148

Micronutrient, 33, 44, 141

Microsatellite Analysis, 47

Mitochondrial Biogenesis, 117-118, 123

Mitochondrial Nad2, 114-115, 117, 119, 121, 123-124

Mitochondrial Transcription Termination Factors (mterfs), 114

Molybdenum Accumulation, 33, 35, 37, 39, 41, 43, 45

Morphological And Physiological Responses, 154-155, 157, 159, 161, 163, 165, 167

N

Nitrogen Metabolism, 148

Non-cellulosic Cell Wall, 199, 201, 203, 205, 207, 209

Nuclear Genes, 210, 216-217

Nucleotide, 48, 74, 77-78, 83-85, 90, 92, 98, 127, 136, 139-140, 169, 177-178, 185, 190, 193, 210-211, 213, 215-217

O

Olive Cultivars, 17-19, 21, 23, 25-27, 29-32

Organelle Genome Assembly, 83

P

Pheophytin Content, 105-106, 111

Phosphoglucomutase (pgm), 168, 178

Phylogenetic Signal, 134, 137-139

Phylogeny Reconstruction, 210-213, 215, 217

Phylogeographical Patterns, 47

Phylogeography, 46-47, 49, 51, 53, 55, 57-59

Plastid Genomic Regions, 125, 127, 129, 131, 133-135, 137-139

Plastidial Phosphoglucomutase Activities, 174

Polysaccharide, 2, 141-145, 147, 149, 151-153, 158, 178, 199-200, 207-209

Populus (salicaceae), 136, 210-211, 213, 215-217

Proteomic Modifications, 37

Pyrus (rosaceae), 125

R

Reactive Oxygen Species (ros), 60, 115, 141

Redundancy Analysis, 204

Riverine Barriers, 46-47, 49, 51, 53, 55, 57, 59

Room Temperature (rt), 1, 3, 169

Root Development, 45, 168

S

Self-pressurized Rapid Freezing (sprf), 3-4, 12, 16

Semiconductor Nanoparticles (nps), 60

Soil Temperature, 17-19, 21-27, 29-31

Soluble Protein Concentrations, 62, 64

Sphingolipid Metabolism, 148

Subtropical Rainforest, 73

T

Tetramethylrhodamine Methyl Ester Perchlorate (tmrm), 115

Transcriptome Analysis, 87, 105

Trypan Blue Staining, 171, 175

V

Verticillium Dahliae Pathotypes, 17,
19, 21, 23, 25, 27, 29-31

W

Wheat Seedlings, 60-65, 67, 71

Y

Yerba Mate, 73-88

Yerba Mate Caffeine, 79